Hazard Assessment of Chemicals
Current Developments

Volume 3

Advisory Board

Contributors to This Volume

James R. Beall
David R. Bevan
John J. Black
S. J. de Mora
James A. Frazier
Roy M. Harrison
Judith M. Hushon
Han K. Kang
Hans Konietzko
Mary Rose Kornreich
Joseph R. Lakowicz
A. J. Owens
David A. Savitz
J. M. Steed
Andrew G. Ulsamer
G. F. Westlake
A. Yokozeki

Hazard Assessment of Chemicals

Current Developments

VOLUME 3

Edited by

JITENDRA SAXENA
Criteria and Standards Division
Office of Drinking Water
Environmental Protection Agency
Washington, D.C.

1984

ACADEMIC PRESS, INC.
(Harcourt Brace Jovanovich, Publishers)

Orlando San Diego San Francisco New York London
Toronto Montreal Sydney Tokyo São Paulo

ACADEMIC PRESS, INC.
Orlando, Florida 32887

United Kingdom Edition published by
ACADEMIC PRESS, INC. (LONDON) LTD.
24/28 Oval Road, London NW1 7DX

ISBN 0-12-312403-4
ISSN 0730-5427

PRINTED IN THE UNITED STATES OF AMERICA

84 85 86 87 9 8 7 6 5 4 3 2 1

Contents

Contributors ix

Preface xi

Contents of Previous Volumes xv

Physicochemical Speciation of Inorganic Compounds in Environmental Media

S. J. de MORA and ROY M. HARRISON

I. Introduction 1
II. Airborne Particles 3
III. Metal Vapors and Alkyls in Air 14
IV. Street Dusts 16
V. Sediments and Soils 18
VI. Natural Waters 24
VII. Conclusions 52
References 54

Scoring Systems for Hazard Assessment

JUDITH M. HUSHON and MARY ROSE KORNREICH

I. Introduction 63
II. Survey of Scoring Systems 64
III. Factors to Consider in Designing or Selecting a Scoring System 64
IV. Steps Involved in Scoring 80
V. Criteria for Scoring 91
VI. Comparison of Scoring System Capabilities 97
VII. Applications of Scoring Systems 106
References 107

The Role of Medical Records in Evaluating Hazardous Chemical Exposures

DAVID A. SAVITZ

I. Introduction 111
II. Potential Uses of Medical Record Systems 113
III. Characteristics of Medical Records as a Data Source 115
IV. Attributes of Medical Records Determining Their Effectiveness in Evaluating Hazardous Chemical Exposures 119
V. Methodological Considerations in Implementing a Record-Based Hazard Evaluation System 124
VI. Case Studies 128
VII. Recommendations 136
References 138

Mediation of Toxicological Properties of Chemicals by Particulate Matter

DAVID R. BEVAN and JOSEPH R. LAKOWICZ

I. Introduction 142
II. Tobacco Smoke 143
III. Polynuclear Aromatic Hydrocarbons 147
IV. N-Nitroso Compounds 164
V. Pesticides 167
VI. Gases and Vapors 169
VII. Metal Compounds 172
VIII. General Summary 173
IX. Prospectus 173
References 174

Aquatic Animal Neoplasia as an Indicator for Carcinogenic Hazards to Man

JOHN J. BLACK

I. Introduction 181
II. Environmental Pollution 183

III. Environmental Carcinogenesis 184
IV. Diagnostic Considerations 189
V. Assessment of Exposure to Chemical Agents 195
VI. Epizootiologic Studies of Spontaneous Neoplasms in Aquatic Animals 196
VII. Linkage of Environmental Pollution to Aquatic Animal Neoplasia 218
VIII. Conclusion 225
References 226

Behavioral Effects of Industrial Chemicals on Aquatic Animals

G. F. WESTLAKE

I. Introduction 233
II. Variability in Behavior of Aquatic Animals 235
III. Preference or Avoidance of Industrial Chemicals by Aquatic Animals 237
IV. Stress Behavior for Biological Monitoring 241
V. Effects of Industrial Chemicals on Other Types of Animal Behavior 243
VI. Conclusions 244
References 244

Stratospheric Ozone Modification by Man's Influence

A. J. OWENS, A. YOKOZEKI, and J. M. STEED

I. Introduction 252
II. Chemistry and Physics of Ozone 255
III. Modeling the Ambient Atmosphere 269
IV. Model Calculations to Characterize Potential Changes 289
V. A Comprehensive Approach—Multiple Perturbations 306
VI. Ozone Measurements 320
VII. Prospects for Future Research 328
References 330

Overview of Health Effects of Formaldehyde

ANDREW G. ULSAMER, JAMES R. BEALL, HAN K. KANG, and JAMES A. FRAZIER

I. Introduction 338
II. Regulatory Activities 340
III. Chemical Properties 341
IV. Sources and Exposure 343
V. Metabolism 348
VI. General Toxicology 353
VII. Hypersensitization 359
VIII. Teratogenic and Reproductive Effects 361
IX. Genetic Effects 367
X. Carcinogenicity 369
XI. Epidemiology 375
XII. Summary and Conclusions 389
References 392

Chlorinated Ethanes: Sources, Distribution, Environmental Impact, and Health Effects

HANS KONIETZKO

I. General Remarks 402
II. Monochloroethane (CH_3CH_2Cl) 408
III. 1,1-Dichloroethane (CH_3CHCl_2) 411
IV. 1,2-Dichloroethane (CH_2ClCH_2Cl) 414
V. 1,1,1-Trichloroethane (CH_3CCl_3) 420
VI. 1,1,2-Trichloroethane ($CHCl_2CH_2Cl$) 426
VII. 1,1,1,2-Tetrachloroethane (CH_2ClCCl_3) 429
VIII. 1,1,2,2-Tetrachloroethane ($CHCl_2CHCl_2$) 431
IX. Pentachloroethane ($CHCl_2CCl_3$) 435
X. Hexachloroethane (CCl_3CCl_3) 438
XI. General Conclusions 441
References 442

Chemical Substance Index 449

Subject Index 457

Contributors

Numbers in parentheses indicate the pages on which the authors' contributions begin.

James R. Beall (337), United States Department of Energy, Germantown, Maryland 20545

David R. Bevan (141), Department of Biochemistry and Nutrition, Virginia Polytechnic Institute and State University, Blacksburg, Virginia 24061

John J. Black (181), Department of Experimental Biology, Roswell Park Memorial Institute, Buffalo, New York 14263

S. J. de Mora (1), Department of Environmental Sciences, University of Lancaster, Lancaster, LA1 4YQ, England

James A. Frazier (337), National Research Council, Washington, D.C. 20418

Roy M. Harrison (1), Department of Environmental Sciences, University of Lancaster, Lancaster, LA1 4YQ, England

Judith M. Hushon (63), Bolt Beranek and Newman, Inc., Arlington, Virginia 22209

Han K. Kang (337), United States Veterans Administration, Washington, D.C. 20420

Hans Konietzko (401), Institute of Occupational and Social Medicine, University of Mainz, Mainz, Federal Republic of Germany

Mary Rose Kornreich (63), 1326 Buttermilk Lane, Reston, Virginia 22090

Joseph R. Lakowicz (141), Department of Biological Chemistry, University of Maryland School of Medicine, Baltimore, Maryland 21201

A. J. Owens (251), Engineering Department, E. I. du Pont de Nemours and Co., Inc., Wilmington, Delaware 19898

David A. Savitz (111), Department of Preventive Medicine and Biometrics, University of Colorado School of Medicine, Denver, Colorado 80262

J. M. Steed[1] (251), Petrochemicals Department, E. I. du Pont de Nemours and Co., Inc., Wilmington, Delaware 19898

Andrew G. Ulsamer (337), United States Consumer Product Safety Commission, Bethesda, Maryland 20207

G. F. Westlake (233), Environment Canada, 45 Alderney Drive, Dartmouth, Nova Scotia, Canada B2Y 2N6

A. Yokozeki (251), Petrochemicals Department, E. I. du Pont de Nemours and Co., Inc., Wilmington, Delaware 19898

[1]Present address: Petrochemicals Department, Corpus Christi Plant, E. I. du Pont de Nemours and Co., Inc., Ingleside, Texas 78362.

Preface

Assessment of the environmental and health hazards arising from chemicals requires a multidisciplinary approach. One needs to consider chemical economics, production, usage, environmental release, monitoring data, environmental behavior, and health and environmental effects. Predictions can often be made concerning environmental and health hazards based upon the structure–activity relationship and the physicochemical characteristics.

A vast amount of new information about new pollutants, new effects, and new measures to deal with the problem of the increasing presence of chemicals in the environment is accumulating continuously. This serial publication provides a single forum for comprehensive and authoritative articles about new and significant developments in the area of chemical hazard assessment. A unique feature of *Hazard Assessment of Chemicals—Current Developments* is that each volume, in addition to subject reviews, contains case studies of chemicals/chemical classes or chemical spills. Case studies provide a comprehensive review and evaluation of all pertinent information on topical (or spilled) chemicals to the reader and offer examples of the application of available methods and approaches to environmental and health hazard assessment of chemicals.

Volume 3 of the series presents a collection of seven subject reviews and two case histories. The scope of this volume is as broad as chemical hazard assessment itself. The topics range from the assessment of toxicological hazards to man from chemicals to the chemistry of the stratospheric ozone modification by atmospheric pollution. All, however, relate to the interactions and effects that take place when chemicals enter the environment, and the methods for their measurement.

To cope with toxic chemical problems with limited resources, researchers and decision makers have always searched for methods to rank chemicals for testing and/or regulation according to their relative capacity for inflicting adversity upon man and his environment. A hazard priority ranking system (or scoring system) improves efficiency by focusing attention on the most important concerns. Hushon and Kornreich have provided an exhaustive review of the existing scoring systems, along with a critical review of their capabilities. Also included in this article is a list of available sources for obtaining monitoring, exposure, and health and environmental effects data for use in the scoring systems.

A wealth of human health data collected on patients and recorded in medical records provides information on health status which potentially can be linked to environmental exposure. Savitz's article provides recommendations for the optimal use of medical records for the purpose of making inferences and utilizes case studies as examples to illustrate these recommendations. One of the more difficult problems in toxicology today is to assess the effect of exposure to multiple agents. The effect of multiple agents may be synergistic, antagonistic, or simply the sum of their individual effects. Bevan and Lakowicz in their article focus on one of these interactions, namely the effect of particulates on the toxic manifestations of chemicals with emphasis on the interaction between particulates and tobacco smoke which contains numerous hazardous compounds. Black, in his article, has summarized the current status of the application of aquatic organisms as indicator systems for early detection of waterborne human carcinogens. He discusses several reported instances of neoplasia in wild, free-living populations of aquatic organisms and their significance to the carcinogenic hazard to humans. Westlake contributed a very interesting and timely article on the significance and measurement of subtle behavioral changes in aquatic animals in response to chemical exposure. The olfactory capabilities of most aquatic animals are believed to be extraordinary and, thus, behavioral responses may be particularly suited for examining the effect of low and environmentally significant concentrations of chemicals.

In the area of environmental chemistry are articles by de Mora and Harrison on techniques for metal speciation, and by Owens, Yokozeki, and Steed on the current understanding of the potential impact of fluorocarbons and other contaminants with long atmospheric lifetimes on stratospheric ozone. In all environmental media, the physicochemical form in which the chemical exists is an important determinant of both the toxicity and environmental mobility of the chemical. Therefore, the techniques for investigation of speciation in air, sediment, soil, street dust, and natural waters are extremely valuable and a complete article is devoted to these techniques. The presence of ozone in the stratosphere is acknowledged to be virtually essential to the existence of life on land. While the release of nitrogen oxide from high-flying supersonic transports was one of the first human activities identified as potentially influencing stratospheric ozone, currently we know of numerous potential perturbants. The article by Owens *et al.* brings us up-to-date on this subject. Presented in this article are a discussion of the underlying physical and chemical processes and atmospheric models which are valuable for understanding these interactions and the extrapolation of the model calculations into the future.

The two chemical case histories in this volume concern formaldehyde and chlorinated ethanes. Both of these chemicals are focal points of the research and regulatory communities and the public. The United States Environmental Protection Agency has proposed drinking water standards for many members of the family of chlorinated ethanes. Konietzko's article provides information on chemistry, production, usage, and environmental and biological data on the chlorinated ethanes, which are of commercial and/or environmental significance. The published literature on formaldehyde is too voluminous to be covered comprehensively in one review. Ulsamer *et al.* in their article concentrate on the health effects of formaldehyde with a discussion of relevant chemistry, sources of exposure, and the current regulatory status of the compound in the United States and abroad.

We are confident that with these subject reviews and case histories, Volume 3 will prove to be a valuable addition to the series.

Contents of Previous Volumes

Volume 1

Assessment of Toxic Substances Information Sources
ANTHONY LEE

Preconcentration of Trace Metals from Aquatic Environmental Samples
A. CHOW and H. D. GESSER

The Reproductive Toxicology of Aquatic Contaminants
WESLEY J. BIRGE, JEFFREY A. BLACK, and
BARBARA A. RAMEY

Partition Coefficient and Water Solubility in Environmental Chemistry
CARY T. CHIOU

Chemical Carcinogens: *In Vitro* Metabolism and Activation
EDMOND J. LAVOIE and STEPHEN S. HECHT

Modeling of Toxic Spills into Waterways
BARRY A. BENEDICT

Environmental and Laboratory Rates of Volatilization of Toxic Chemicals from Water
DONALD MACKAY

Estimation of Exposure to Hazardous Chemicals
JUDITH M. HUSHON and ROBERT J. CLERMAN

Structure–Activity in Hazard Assessment
PAUL N. CRAIG and KURT ENSLEIN

Azaarenes: Sources, Distribution, Environmental Impact, and Health Effects
JOSEPH SANTODONATO and PHILIP H. HOWARD

Chemical Substance Index

Subject Index

Volume 2

Use of Models for Assessing Relative Volatility, Mobility, and Persistence of Pesticides and Other Trace Organics in Soil Systems
WILLIAM A. JURY, WILLIAM F. SPENCER, and WALTER J. FARMER

Microcosms for Assessment of Chemical Effects on the Properties of Aquatic Ecosystems
JEFFREY M. GIDDINGS

Disposition of Chemical Contaminants in Maternal–Embryonic/Fetal Systems
M. R. JUCHAU

Epidemiologic Approaches to Chemical Hazard Assessment
JOHN R. WILKINS III and NANCY A. REICHES

The National Toxicology Program's Research and Testing Activities
L. G. HART, J. E. HUFF, J. A. MOORE, and D. P. RALL

Accidental Release of Vinyl Chloride: The Train Derailment near MacGregor, Manitoba
J. CHARLTON, A. CHOW, and H. D. GESSER

Anatomy of a TCDD Spill: The Seveso Accident
G. REGGIANI

Chemical Substance Index

Subject Index

Hazard Assessment of Chemicals

Current Developments

Volume 3

Physicochemical Speciation of Inorganic Compounds in Environmental Media

S. J. de Mora and Roy M. Harrison

Department of Environmental Sciences
University of Lancaster
Lancaster, England

I. Introduction 1
II. Airborne Particles 3
  A. X-Ray Powder Diffraction (XRD) 5
  B. Single-Particle Techniques 8
  C. Atmospheric Sulfates 12
  D. Miscellaneous Techniques for Particulate Air Pollutants 13
III. Metal Vapors and Alkyls in Air 14
  A. Mercury 14
  B. Lead 15
IV. Street Dusts 16
  A. X-Ray Powder Diffraction (XRD) 16
  B. Sequential Extractions 17
  C. Scanning Electron Microscopy (SEM) 18
V. Sediments and Soils 18
VI. Natural Waters 24
  A. Introduction 24
  B. Physical Techniques for Generic Speciation 25
  C. Chemical Techniques for Generic Speciation 32
  D. Species-Specific Techniques 43
  E. Comprehensive Speciation Schemes 48
VII. Conclusions 52
  References 54

I. INTRODUCTION

The study of physicochemical speciation is a recent development in environmental chemistry. Physicochemical speciation is a term used both as a noun to

HAZARD ASSESSMENT OF CHEMICALS:
Current Developments, Vol. 3

ISBN 0-12-312403-4

describe the precise physicochemical forms in which chemicals exist in an environmental sample and as a verb to describe the processes of determining those forms.

Both physical and chemical factors must be considered in describing speciation. Thus the physical nature of a substance, e.g., dissolved, colloidal, or particulate, is as significant as the chemical bonding, e.g., metal sulfide, ion pair, free ion. There are many possible physicochemical forms of inorganic substances in water, varying from, at the smallest and simplest end of the spectrum, solvated free ions and ion pairs, to, at the other end of the size spectrum, large particulate precipitates. In air, also, both particle size and chemical combination are pertinent factors in speciation studies.

Physicochemical speciation is an extremely important determinant of both the toxicity and environmental mobility of a substance. In a review dealing with heavy metals, Astruc *et al.* (*14*) list three important aspects relating to aqueous systems determined by speciation.

1. Speciation identifies pathways and sinks in natural water bodies and the risks of metal resolubilization from sediment sinks. Thus, metal associated with suspended sediment of large grain size may rapidly enter the bottom sediment where its availability for remobilization is determined by its solubility under existing redox conditions and the strength of its association with the sediment grains. Metal present in water as free metal ions or ion pairs is potentially far more mobile and may be transported over great distances.
2. Speciation determines aspects of toxicity toward aquatic organisms. Various workers (*8, 56*) have shown that complexed forms of metals can be far less toxic to aquatic organisms than free metal ions. Conversely, the presence of complexing agents may act to mobilize essential trace elements such as iron from otherwise unavailable colloidal forms, hence benefiting dependent organisms (*158*).
3. Speciation determines the efficiency of water or sewage treatment plants. Particulate forms of inorganic substances may be removed from water by filtration, as in water treatment, or by sedimentation, as in sewage treatment. Removal of colloidal or dissolved forms may be more difficult. Some forms of a trace metal may be removed in secondary sewage treatment by adsorption onto the biological floc; others are not effectively removed (*45, 247*).

The physicochemical form of a trace element in a soil has an important influence upon availability for plant uptake. The presence of highly available forms, as is the case for metals in a soil of low pH, may result in toxic effects upon the plant or in an unacceptable degree of incorporation of metal into the plant tissues from the viewpoint of subsequent human consumption.

The chemical nature of an ingested substance influences its solubility in the gastrointestinal system and thus the efficiency of absorption into the body. Experiments with labeled lead compounds show highly efficient gastrointestinal absorption of the soluble compound lead chloride, whereas the relatively insoluble lead sulfide is absorbed only with low efficiency (*7*). Similar considerations apply to the inhalation of aerosols. In this case, however, the particle size determines the site of deposition of the aerosol particles. Very fine particles (< 2 μm) tend to deposit most effectively in the alveolar region of the lungs, from which direct absorption into the bloodstream occurs with very high efficiency. The chemical form of the deposited material influences the rate of absorption (*57*). Larger particles (2–10 μm) deposit in a significant proportion in the tracheobronchial part of the respiratory system from which mucociliary escalation carries them to the stomach in which, again, the chemical form plays an important role in influencing absorption. In a study of heavy metals in the atmosphere of a primary zinc–lead smelter, O'Neill *et al.* (*206*) predicted that gastrointestinal absorption may dominate the overall absorption of metals by workers exposed at most sites within the smelter. This allows the possible use of protective diets, which aim to alter the physicochemical form of the metal thus rendering it less available for gastrointestinal uptake.

Thus, in all environmental media the speciation of inorganic substances is an important determinant of both mobility and toxicity. The techniques available for the investigation of metal speciation will be reviewed in this article.

II. AIRBORNE PARTICLES

Before a discussion of speciation methodology, some consideration of the physical characteristics of atmospheric aerosols is of benefit.

Much man-made pollutant material is emitted in the "nuclei" mode (size range). This covers particle diameters < ~ 0.1 μm, which are subject to rapid coagulation in the atmosphere. Such particles arise from condensation of vapors (e.g., vehicle-emitted lead) and gas-to-particle conversion processes (e.g., homogeneous gas-phase oxidation of sulfur dioxide to sulfuric acid aerosol). Coagulation of particles in the nuclei range causes formation of particles of diameter 0.1–2 μm, known as the "accumulation" mode. Particles of this size are of low Brownian diffusivity and are hence subject to only very slow further coagulation or deposition to surfaces. They are also little affected by deposition processes such as gravitational settling and impaction which are important for larger particles. Thus an aerosol in the accumulation mode is very stable and has a long atmospheric lifetime, probably of the order of weeks. Larger aerosol particles (> 2 μm) are generally formed by mechanical disintegration and erosion processes and, because of their greater size, are subject to more rapid deposition from the air. Most, but not all, particulate material in the "coarse" mode is of natural origin (e.g., soil-derived minerals).

If, as in most cases, the accumulation mode is that of prime interest to the air chemist, it is instructive to estimate the numerical particle loading of the air. A typical accumulation mode particle is of diameter rather less than 1 μm. Taking 1 μm as diameter, the mass of such a particle of density 2 g/cm^3 is $\sim 10^{-2}$ g. Thus each microgram of monodisperse aerosol would contain 10^6 particles, and 1 m^3 of air at a realistic particle loading of 100 $\mu g/m^3$ would contain 10^8 particles per m^3 for such a monodisperse aerosol. In fact, a polydisperse atmospheric aerosol will contain a substantially greater numerical loading of particles since the smaller particles, although contributing little to aerosol mass, are numerically very abundant.

Techniques for speciation of atmospheric particles come into two prime categories: macroscopic and microscopic. Techniques such as X-ray powder diffraction (XRD)[1] are macroscopic and examine the bulk properties of an aerosol sample of milligram proportions. Electron spectroscopy (ESCA) also is a macroscopic technique, looking simultaneously at the properties of large numbers of particles, but in this case penetrating only a thin surface layer. The most important microscopic technique is scanning electron microscopy with X-ray energy spectroscopy (SEM/XES). This is used to examine the morphology and chemical composition of individual particles. The method is rather time consuming and hence by using the individual particle analysis mode it is possible to obtain semiquantitative information on only, at most, tens of particles within an aerosol sample in a reasonable length of time. Thus, unless the sample is highly uniform, the danger exists that the information generated is not representative of the macroscopic composition of the sample, as only a tiny proportion of the total particles in an air sample can possibly be examined. The use of raster modes to scan a larger area of sample and establish interparticle distributions of selected elements can give valuable supplementary information of chemical composition

[1]Abbreviations: AAS, atomic absorption spectrophotometry; ASV, anodic stripping voltametry; dc, direct current; DMA, dimethyl arsonate; DPP, differential pulse polarography; ECD, electron capture detector; EDAA, electrodeposition atomic absorption spectrophotometry; EDTA, ethylenediaminetetraacetic acid; E_p, peak potential; ESCA, electron spectroscopy for chemical analysis; FID, flame ionization detector; GC, gas chromatography; GFC, gel filtration chromatography; HMDE, hanging mercury drop electrode; HPLC, high-performance liquid chromatography; IDA, iminodiacetic acid; IEC, ion-exchange chromatography; i_p, peak current; ISE, ion-selective electrode; LAMMA, laser microprobe mass spectrometry; MMA, monomethyl arsonate; MWCO, molecular weight cutoff; NTA, nitrilotriacetic acid; PTFE, polytetrafluoroethylene; SCE, saturated calomel electrode; SEM, scanning electron microscopy; STEM, scanning transmission electron microscopy; TEM, transmission electron microscopy; TFE, thin-film electrode; TML, tetramethyllead; XES, X-ray energy spectroscopy; XRD, X-ray diffraction.

on a larger scale, but does not entirely overcome the problem of collection of truly representative data.

The most frequently used techniques of analysis of inorganic components of atmospheric aerosols yield no information whatever of chemical speciation. Hence, methods of metal analysis such as atomic absorption, X-ray fluorescence, and neutron activation can provide only the concentration of total metal in sample and cannot directly provide information on speciation at even the crudest level, as, for example, discrimination of Cr(III) and Cr(VI). Still less can they identify molecular forms of metals. Consequently, a range of alternative techniques has been developed that have provided at least a partial insight into the problem.

A. X-Ray Powder Diffraction (XRD)

When a collimated monochromatic X-ray beam is focused upon a crystalline material, diffracted X rays may be detected leaving the sample at angles that can be related to the interplanar spacings within the crystalline sample by use of the Bragg equation. Hence, after X-ray diffraction examination of a known sample, the interplanar spacings (known as *d*-spacings) are estimated from the diffraction angles, and the magnitudes and relative intensities of the *d*-spacings of the sample are compared with known spectra using a search manual. Complex mixtures may be characterized by this technique as long as there are few overlaps of diffraction lines.

The major drawback of XRD methods in the examination of environmental samples is the requirement for crystalline material. Although many substances exist in the atmosphere as highly crystalline particles, others are present in poorly crystalline or amorphous forms and give little or no response in XRD analysis. The technique is also of limited sensitivity, and is hence responsive only to rather major crystalline components of a sample.

Early use of XRD methods was severely hampered by the presence in air samples of natural mineral materials such as α quartz and calcite, which have strong diffraction patterns that tend to obscure the response of the man-made components of the aerosol. This problem may be overcome by size fractionation of the aerosol to remove coarse ($> 2\ \mu m$) particles (*31*) or by fractionation into several discrete size ranges by impaction, with subsequent XRD analysis of each fraction (*30*). An alternative approach is to use density fractionation techniques (*115*).

Air samples for XRD work may be most conveniently collected by filtration or impaction. In either case, a collection substrate is necessary and some prior consideration of the optimum material is worthwhile. It is often possible to obtain diffraction patterns from samples while unseparated from the filter substrate. In this case, the use of substrate that itself gives little diffraction is important, and O'Connor and Jaklevic (*201*) considered the properties of cellulose ester, polycarbonate, and PTFE for this purpose. All three substrates were found to give a substantial background, with several discrete peaks in the case of PTFE but a more diffuse hump for the other materials. None of the patterns was so intense as to rule out use as a substrate, although subtractive techniques were necessary when identifying unknown samples. In our work, glass fiber and polystyrene substrates also have proved useful. Silver membranes are used for collection of asbestos in industrial hygiene applications for subsequent XRD analysis. These give a very clean background, but are rather expensive for routine use. Particles may be stripped ultrasonically from a filter substrate prior to XRD analysis (*30, 31*). This has the advantage of allowing concentration onto a smaller area of filter, but is not fully efficient and also tends to disintegrate the filter causing incorporation of filter material into the separated sample.

As noted earlier, analysis of atmospheric samples involves examination of extremely fine particulate material. It is well known that broadening of diffraction lines is associated with diminishing crystallite sizes, and this could provide a limitation to atmospheric analysis as broadened lines showed markedly diminished intensity. This effect was systematically investigated by O'Connor and Jaklevic (*202*) for the common atmospheric component ammonium sulfate. These workers concluded that line broadening did not cause troublesome smearing of the diffraction lines unless the particles were formed from component single crystal domains of size substantially below 0.1 μm. This work indicates that for a typical ambient aerosol sample line broadening is unlikely to be a problem.

It is theoretically possible to quantify the concentrations of individual components of an air filter sample by calibration of line intensities. Davis and Cho (*73*) and Davis (*71*) have described a technique applicable to high-volume filter samples that uses the addition of a thin layer of "reference" component prior to XRD examination. The resulting line intensities must then be corrected for the masking effect of the reference component, the absorption of the matrix, and the transparency of both reference and sample components. This method requires a rather detailed knowledge of the aerosol composition and is not readily applied to typical ambient air samples. The question of quantification is of continuing importance and requires further attention.

Published data on airborne particles are largely restricted to qualitative identifications. Some of the more intense areas of study are reviewed in Sections II,A,1–3.

1. Lead Compounds

These have been of particular interest due to the widespread use of leaded gasoline additives. Biggins and Harrison (*29–31*) have determined lead com-

TABLE I

Lead Compounds Identified in Ambient Air Using X-Ray Diffraction

Compound	References
Automotive lead	
$PbSO_4 \cdot (NH_4)_2SO_4$	*31, 203*
$PbSO_4$	*31*
$PbBrCl \cdot (NH_4)_2BrCl$	*31*
α-$2PbBrCl \cdot NH_4Cl$	*31*
$PbBrCl \cdot 2NH_4Cl$	*31*
$PbBrCl$	*31*
Industrial (smelter) lead	
PbS	*112, 184*
$PbSO_4$	*92, 112, 184*
$PbO \cdot PbSO_4$	*92, 112, 184*

pounds in air samples from a range of roadside sites. The compounds identified are listed in Table I. Three compounds, $PbBrCl$, $PbBrCl \cdot 2NH_4Cl$, and α-$2PbBrCl \cdot NH_4Cl$ are known components of vehicle exhaust particulates. The other compounds, $PbSO_4 \cdot (NH_4)_2SO_4$ (the most abundant compound), $PbSO_4$, and $PbBrCl \cdot (NH_4)_2BrCl$ are formed in the atmosphere by reaction of automotive $PbBrCl$ with airborne sulfates after coagulation of the aerosol (*30*). Identification of these compounds was achieved using both a Guinier camera and a diffractometer. The compound $PbSO_4 \cdot (NH_4)_2SO_4$ has also been reported by O'Connor and Jaklevic (*203*) in air particulate samples collected with a dichotomous sampler.

Foster and Lott (*112*) and Lott and Foster (*184*) have identified lead compounds emitted from a lead smelter and collected by high-volume air filtration. The structural assignments are shown in Table I together with those reported by Eatough *et al.* (*92*) in the flue dusts of lead smelters.

A larger range of compounds, including compounds of lead, zinc, and cadmium, has been identified within the atmosphere of a major primary zinc–lead smelter (*137*). Specific compounds were found to relate closely to the source of airborne metal at any particular work station. The phases identified appear in Table II.

2. *Sulfate Compounds*

Many sulfates apparently exist in air in a crystalline form readily amenable to XRD analysis. Table III shows compounds reported by O'Connor and Jaklevic (*203*) and by Biggins and Harrison (*32*), who proposed mechanisms of formation and a scheme of classification for atmospheric sulfates.

TABLE II

Metal Phases Identified in the Atmosphere of a Primary Zinc–Lead Smelter Using X-Ray Diffraction[a]

Zinc	Lead	Cadmium
β-ZnS	PbS	CdO
ZnO	PbO (litharge)	Cd^0 [b]
	PbO (massicot)	$Cd(OH)_2$ [b]
	$PbSO_4$	
	Pb^0	
	$PbO \cdot PbSO_4$	

[a] From Ref. *137*.
[b] Detected in floor dusts only.

3. Natural Minerals

As mentioned earlier, many natural minerals give strong diffraction patterns and, as a consequence, may be identified in ambient air samples. Some examples from the literature are listed in Table IV.

B. Single-Particle Techniques

These are divided into two main groupings: electron beam methods and laser methods. Electron beam methods include the electron microprobe, the scanning electron microscope (SEM), and transmission and scanning transmission electron

TABLE III

Sulfate Compounds Identified in Ambient Air Using X-Ray Diffraction

Compound	References
$Fe_2(SO_4)_3 \cdot 3(NH_4)_2SO_4$	*32*
$2CaSO_4 \cdot (NH_4)_2SO_4$	*32*
$PbSO_4 \cdot (NH_4)_2SO_4$	*32*
$PbSO_4$	*32*
$CaSO_4 \cdot 2H_2O$	*32, 72*
Na_2SO_4	*32*
$(NH_4)_2SO_4$	*32*
$(NH_4)_2SO_4 \cdot NH_4HSO_4$	*203*
$ZnSO_4 \cdot (NH_4)_2SO_4 \cdot 6H_2O$	*203*
$MgSO_4 \cdot 7H_2O$	*115*

TABLE IV

Some Natural Minerals Identified in Ambient Air by X-Ray Diffraction

Mineral	References
NaCl (halite)	*30*
Na_2SO_4	*30*
α-SiO_2 (α quartz)	*30*
$CaCO_3$ (calcite)	*30*
$CaSO_4 \cdot 2H_2O$ (gypsum)	*30, 72*
Biotite	*72*
Muscovite	*72*
Kaolinite	*72*
Plagioclase	*72*
$MgCO_3 \cdot CaCO_3$ (dolomite)	*72*
Fe_2O_3 (hematite)	*72, 115*
$FeO \cdot Fe_2O_3$ (magnetite)	*72*
$CaSO_4 \cdot \frac{1}{2}H_2O$ (hemihydrate gypsum)	*115*
Chlorite	*115*
$MgSO_4 \cdot 7H_2O$ (Epsom salt)	*115*

microscopes (TEM and STEM). X Rays, stimulated by the electron beam, are used to provide elemental analysis of the material in the beam. The electron microprobe is optimized to provide analytical data, whereas the other microscope methods give useful morphological information in addition to semiquantitative elemental analyses.

Butler and co-workers (*52*) have used the SEM with X-ray analysis to investigate atmospheric particles collected near heavy traffic in the urban environment. They report that particles in the 0.5- to 1.0-μm range can be focused and analyzed with a minimum detection limit of 0.9% by weight for transition elements. The results were reported in terms of the identification of particles individually rich in the elements titanium, vanadium, zinc, and lead. Although this work can provide interesting insights into the sources and character of atmospheric particles, it suffers from two major detractions. First, the number of particles examined is very small, and, as discussed earlier, the results are thus of little statistical significance. Second, only elements of atomic number $\geqslant 11$ are detected, and then only semiquantitatively. Thus, no exact assignments of chemical composition and structure are possible.

Some benefits are gained from the use of transmission electron microscope methods. Smaller particles down to < 0.1 μm may be examined using a STEM (*36*), and selective area electron diffraction patterns may be obtained allowing firm structural assignments for crystalline materials (*4, 36, 271*). Good results have been obtained with this technique by Bloch and co-workers (*36*), and a summary of their findings appears in Table V.

TABLE V

Particles Detected in Aerosol Samples from Two Sites[a]

Identification	Elements detected and (relative concentration)	Crystal structure	Site	Mean projected diameter (μm)
$(NH_4)_2SO_4$	S (1.00)	$(NH_4)_2SO_4$	UIA, E3	1.1
Na_2SO_4	Na (0.17), S (1.00)	Na_2SO_4	UIA, E3	1.7
K_2SO_4	K (1.00), S (0.33)	K_2SO_4	UIA	1.9
$CaSO_4$	Ca (1.00), S (0.83)	$CaSO_4 \cdot 2H_2O$	UIA, E3	1.4
NaCl	Na (0.54), Cl (1.00)	NaCl	UIA, E3	4.6
Fly ash	Al (0.67), Si (1.00), S (0.26), K (0.26), Ca (0.07), Fe (0.72)	Amorphous	UIA, E3	2.5
Fe oxide spheres	Fe (1.00)	Amorphous	UIA	1.2
Goethite	Fe (1.00)	Goethite	E3	2.0
TiO_2	Ti (1.00)	Amorphous	UIA, E3	0.8
Condensation aggregates	Al (0.07), Si (0.16), S (0.81), K (0.12), Ca (0.25), Fe (0.33), Pb (1.00)	Amorphous	UIA, E3	5.1
S-Containing particles	S (1.00)	—	E3	—
$PbBr_xCl_y$	Pb (1.00), Br (0.27), Cl (0.10)	Crystalline	E3	1.7
Quartz	Si (1.00)	Quartz	UIA, E3	1.6
Calcite	Ca (1.00)	Calcite	UIA, E3	2.2

[a] From Ref. *36*.

The electron microprobe has been used by Ter Haar and Bayard to examine lead compounds in vehicle exhaust and in ambient air (*254*). A range of very specific structural assignments was made, which included such compounds as $(PbO)_2PbBrCl$, Pb(OH)Br, and $PbCl_2$. The relative proportions of different compounds were also reported. This work was criticized subsequently by Heidel and Desborough (*144*) who cast doubt upon structural assignments that were in some cases dependent upon highly precise elemental analyses. It is interesting to note that the compounds reported and the atmospheric reactions postulated by Ter Haar and Bayard (*254*) are very different from those reported by Biggins and Harrison (*31*) who carried out structural assignments by XRD.

A recent development in the field of single-particle analysis uses the laser microprobe mass spectrometer (LAMMA). A high-intensity laser pulse with a spatial resolution of ~1 μm is focused upon an individual particle collected on a filter or impactor and is used to volatilize and ionize the particle (*82*). The ion fragments are mass analyzed in a time-of-flight mass spectrometer, which may be used to generate both positive and negative ion spectra. Rather involatile inorganic substances may be analyzed by this technique, giving characteristic ion spectra.

Adams *et al.* (*4*) report the application of LAMMA to tracing the source of antimony smelter emissions and to source tracing of unpolluted aerosol. It is evident that given substances generate characteristic spectra, but that the ion spectra generated by environmental particles are often more complex indicating the presence of more than one chemical component. Adams *et al.* (*4*) conclude

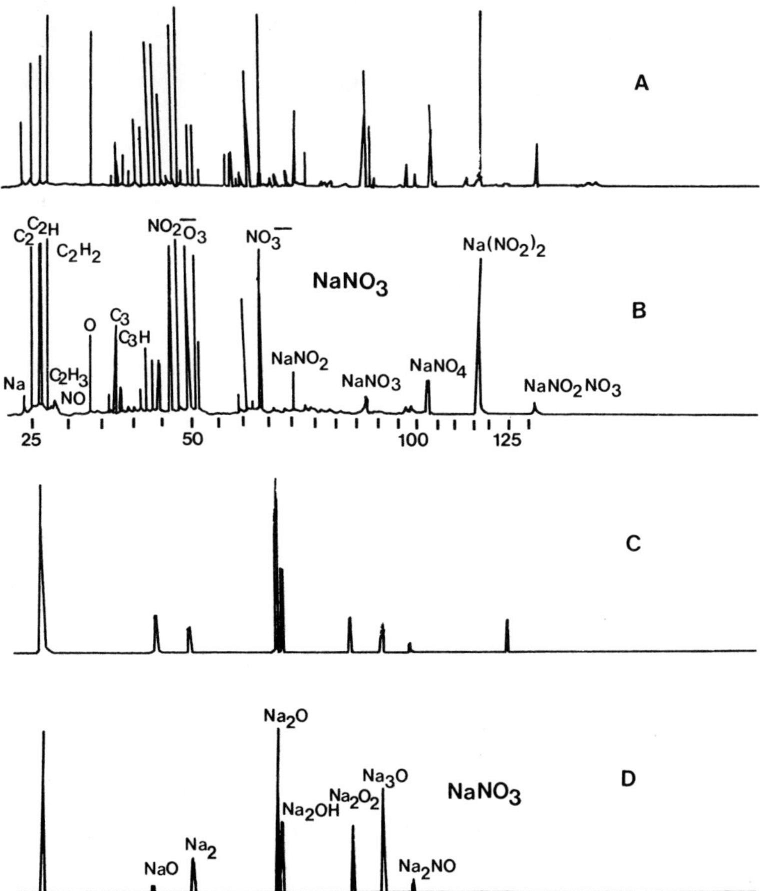

Fig. 1. Laser microprobe mass spectra of (A) atmospheric particle sampled at Chapiquina, Chile (negative ion spectrum); (B) $NaNO_3$ standard (negative ion); (C) atmospheric particle (positive ion spectrum); and (D) $NaNO_3$ standard (positive ion). From Ref. *4*.

that although quantitative interpretation of the spectra is at present impossible and the qualitative identification of lines is complicated by mass spectral interferences, the method is fast and provides an insight into the chemical nature of atmospheric aerosol particles. Their results justify this conclusion, but suggest that only rather simple chemical substances can be identified with any certainty. Figure 1 shows a good example in which a sodium nitrate-rich atmospheric particle collected in a remote area of Chile is compared with a standard sample of this compound.

C. Atmospheric Sulfates

1. Solvent Extraction

In the section dealing with XRD (Section II,A) it has been shown that the polluted atmosphere contains a considerable range of discrete sulfate compounds. Without a doubt, the most important sulfates in most polluted air samples, in terms of both abundance and environmental significance, are H_2SO_4 and its atmospheric neutralization products NH_4HSO_4 and $(NH_4)_2SO_4$. Whereas $(NH_4)_2SO_4$ may be characterized by XRD, it cannot be readily quantified, and the other compounds exist as solution droplets (except for NH_4HSO_4 at very low humidities) that are not amenable to this technique.

Quantification of these sulfate species has been achieved by solvent extraction. Leahy *et al.* (*174*) demonstrated that sulfuric acid could be selectively extracted using benzaldehyde. Subsequent extraction with isopropanol yields NH_4HSO_4, and a final water extraction gives other water-soluble sulfates, assumed to be primarily $(NH_4)_2SO_4$. Criticism of this technique was made by Appel *et al.* (*13*) who reported that, in the presence of other atmospheric particles, benzaldehyde extraction of H_2SO_4 is $\leq 30\%$ efficient. In application of benzaldehyde extraction to smelter flue dust, Eatough *et al.* (*93*) report that bivalent metal bisulfate salts and some sulfite salts could be extracted, and if benzoic acid is present as an impurity in the benzaldehyde bivalent metal sulfates may also be extracted.

2. Other Techniques for Sulfuric Acid

In most ambient air samples, the only strong acid present in particulate form is H_2SO_4, as HNO_3 and HCl exist primarily in the gaseous state. Thus microtitration of an aqueous extract of atmospheric particles using the Gran plot method may be used to quantify the concentration of H^+ and hence the H_2SO_4 concentration (*43, 44*). Chemical interactions within the aqueous extract and the presence of weak acids are the major limitations upon this technique.

A glass fiber filter impregnated with 2-perimidinylammonium bromide will react selectively with H_2SO_4 to form the sulfate. Subsequent pyrolysis releases stoichiometric amounts of SO_2, which are quantified (*187, 256*). In a more recent method, the 2-perimidinylammonium sulfate is quantified colorimetrically by reaction with HNO_3 to form 2-amino-4,6,9-trinitroperimidine (*70*).

Sulfuric acid will react also with secondary amines, and may be reacted with diethylamine vapor on a filter surface subsequent to collection. The absorbed diethylamine is determined colorimetrically (*155*). Alternatively, if [^{14}C]trimethylamine is used, the absorbed amine may be assayed by β-counting, hence giving a measure of the sulfuric acid collected (*91*).

A qualitative discrimination of $(NH_4)_2SO_4$ from H_2SO_4 was reported by Ahlberg *et al.* (*5*) who measured the particle size distribution of sulfur in air both at low relative humidity (10%) and after humidification to 95%. Such an increase in humidity is associated with pronounced deliquescence of $(NH_4)_2SO_4$, which shifts to larger particle sizes.

In a sophisticated instrumental design, Tanner *et al.* (*88, 253*) used a flame photometric analyzer as an ambient aerosol sulfur monitor after removal of gaseous SO_2 in a diffusion denuder. By heating the air stream, the Na_2CO_3-coated denuder is made to selectively remove H_2SO_4, but not other sulfates. Thus, by cyclic addition of ammonia to the gas stream prior to the denuder, a measure of H_2SO_4 concentrations may be achieved.

D. Miscellaneous Techniques for Particulate Air Pollutants

Keyser *et al.* (*164*) have described several techniques capable of analysis of the surface of atmospheric particles or of giving a depth profile of element concentration within an individual particle. Fly ash and large auto exhaust particles were examined by secondary ion mass spectrometry, Auger electron spectroscopy, and electron microprobe X-ray emission spectroscopy. Although these techniques do not identify specific valence states of elements, the examination of depth profiles for several elements permits inferences regarding the possible chemical associations between elements. These techniques are highly sophisticated, but generate information of rather limited application to speciation.

Electron spectroscopy also is a surface analytical technique, giving penetration of about 2.5 nm. It can be used to examine a bulk sample, but provides information upon only the first few surface atom layers. When the results of ESCA analysis of atmospheric particles are compared with those of a bulk analytical method (e.g., atomic absorption), surface enrichments of some trace elements may be inferred (*16*). The energy of photoelectrons is dependent upon the valence state of the atom that will then arise. This has provided some limited speciation information, most notably for sulfur compounds in aerosols. Thus, for example, Eatough *et al.* (*92*) found evidence of sulfate, sulfite, elemental sulfur, and sulfide [i.e., S(+6), S(+4), S(0), and S(−2)] in smelter flue dusts.

Using X-ray absorption spectroscopy, Jaklevic *et al.* (*160*) were able to discriminate ZnO, $ZnSO_4$, and $(NH_4)_2Zn(SO_4)_2$ in atmospheric aerosols by scanning the K X-ray absorption edge. This method may be of wider application (the authors report examination only of Zn, Fe, and Cu compounds), but will be of very limited use due to the requirement of a synchrotron radiation source.

One technique with the capability of identifying molecular species of both crystalline and noncrystalline character is laser-excited Raman spectroscopy (*95*, *221*). Identification of calcium sulfate (anhydrite) and elemental carbon has proved possible, but this technique is in need of further development before routine use is possible.

III. METAL VAPORS AND ALKYLS IN AIR

A. Mercury

Mercury may exist in air in a variety of physicochemical forms. These include particulate inorganic salts, vapor-phase inorganic salts, vapor-phase monoalkyl and dialkyl derivatives [e.g., CH_3HgCl and $(CH_3)_2Hg$], and elemental mercury vapor.

Most techniques for the analysis of mercury in air make no attempt at separation of these forms, and hence provide data of very limited value. The method of Trujillo and Campbell (*261*) does, however, provide species selective data. A prefilter is used to collect particulate mercury, and this is followed by tubes containing Carbosieve B (collects alkyl- and dialkylmercury compounds) and silvered Chromosorb P (collects elemental mercury vapor). Thus, three fractions are trapped, but it is unclear where volatile inorganic forms of mercury (e.g., $HgCl_2$ vapor) will be collected. Analysis of trapped mercury is by thermal desorption and atomic absorption.

Selectivity is also attained by an alternative sequential sampler proposed by Henriques *et al.* (*145*). A Millipore prefilter collects particulate mercury and is followed in sequence by a gold filter coated with a gold–silicon alloy (collects elemental Hg vapor), a potassium permanganate scrubber (collects monomethylmercury and other readily oxidized compounds), and finally a pure gold filter (collects dimethylmercury and other gold-soluble compounds).

The most selective sampler is that described by Braman and Johnson (*39*). Prefiltration is by a glass wool filter (collects particulate Hg compounds), which is followed in sequence by adsorption tubes containing siliconized Chromosorb W treated with hydrogen chloride (collects vapor-phase mercuric chloride, etc.), Chromosorb W treated with caustic soda (monoalkylmercury compounds), silvered glass beads (elemental Hg), and gold-coated glass beads (dialkylmercury). Collected mercury is thermally desorbed using helium carrier gas, and in a detection system containing a direct current (dc) discharge, emission at 257.3 nm is measured. This gave a detection limit of about 0.01 ng.

B. Lead

The major proportion of lead in air is in an inorganic form, and considerable attention has been given to the speciation of inorganic forms of lead in other sections of this article. A small proportion of lead in urban air, perhaps 1–6%, exists as vapor-phase alkyllead compounds (*135*), and a great deal of work has been carried out to speciate this form of lead. It has generally been assumed that alkyllead in air exists as tetraalkyllead, although recent work has suggested that a vapor-phase trialkyllead component also may be important (*80*).

Total vapor-phase alkyllead in air may be analyzed by collection in iodine monochloride after prefiltration of particulate lead. In iodine monochloride solution, alkylleads are converted to dialkyllead which may be selectively extracted by dithizone in carbon tetrachloride after addition of EDTA to mask inorganic lead (*34*). After back extraction of the dialkyllead into a dilute HNO_3/H_2O_2 solution, analysis is by atomic absorption. This technique may be used routinely for 24- and 48-h sampling of ambient air (*34*).

The tetraalkyllead compounds used as gasoline additives that may be present in polluted air are tetramethyllead, $(CH_3)_4Pb$; trimethylethyllead, $(CH_3)_3C_2H_5Pb$; dimethyldiethyllead, $(CH_3)_2(C_2H_5)_2Pb$; triethylmethyllead, $(C_2H_5)_3CH_3Pb$; and

TABLE VI

Speciation of the Atmospheric Tetraalkyllead Concentrations at the Different Measuring Sites[a]

Site	Percentage of total tetraalkyllead contents[b]				
	Me_4Pb	Me_3EtPb	Me_2Et_2Pb	$MeEt_3Pb$	Et_4Pb
Gasoline station A	18 ± 10	12 ± 9	5 ± 4	2 ± 2	66 ± 21
Gasoline station B	46 ± 11	31 ± 5	12 ± 5	7 ± 6	5 ± 3
Car-repair workshop	53 ± 5	27 ± 4	11 ± 2	1.8 ± 0.5	7 ± 4
Underground tunnel	69 ± 10	13 ± 4	5 ± 3	2 ± 2	12 ± 10
Highway crossing	35 ± 14	25 ± 3	13 ± 3	4 ± 3	24 ± 12
Central-city street	61 ± 7	13 ± 2	5 ± 1	1.2 ± 0.4	19 ± 6
Residential area[c]	76 ± 16	15 ± 10	6 ± 2	nd[d]	nd
Rural area	85 ± 12	20 ± 6	nd	nd	nd

[a] From Ref. *79*.
[b] Average of all measurements.
[c] Two values for methyltriethyllead and tetraethyllead above the detection limit not included.
[d] nd, Not detected.

tetraethyllead, $(C_2H_5)_4Pb$. These may be trapped from the air by using cooled gas chromatographic column-packing materials (*136*) or glass beads (*77*). After subsequent thermal desorption, the individual tetraalkyllead compounds may be separated by gas chromatography (GC) and detected by flameless atomic absorption (*77*), microwave plasma (*212*), or mass spectrometry (MS) (*168*). When the latter technique is used, it is important to establish the existence of normal isotope ratios for lead-containing ion fragments, as interferences may be occurring otherwise (*210*). Some typical analytical data (*79*) obtained in Antwerp, Belgium are shown in Table VI.

It will be appreciated from Section VI on water analysis that several metals and metalloids may exist as volatile alkyl species in the environment. Little attention has so far been given to their analysis in air, but this topic is likely to be a focus for future research.

IV. STREET DUSTS

The dusts that accumulate in gutters and on road surfaces and pavements are known as street dusts. There has been a considerable interest in such dusts arising from their possible importance as a source of toxic metal ingestion for children. The major component of street dusts appears from microscopic examination to be soil. This is enriched by vehicle corrosion and abrasion products such as rust and tire rubber and by pollutants deposited from the atmosphere.

The major interest in street dusts has centered upon their lead content. This lies typically within the range 1000–5000 mg Pb/kg for motor-vehicle influenced dusts but may extend up to nearly 50% in some dusts sampled around industrial sources, most notably secondary lead smelters.

A. X-Ray Powder Diffraction (XRD)

Despite the considerable concentrations of lead in street dusts, the identification of crystalline components by XRD is by no means easy. Some enrichment of lead, normally by density separation techniques, is necessary.

Biggins and Harrison (*33*) modified the lead enrichment methods developed by Olson and Skogerboe (*205*) for soil analysis. Samples of street dust were subjected first to a magnetic separation, after which the nonmagnetic fraction was placed in diiodomethane and a dense fraction ($\rho > 3.32$ g/cm^3) separated. Lead compounds were identified in the dense nonmagnetic fraction only. The range of compounds identified in all fractions is listed in Table VII. Of the lead compounds, $PbSO_4 \cdot (NH_4)_2SO_4$ is abundant in air. It is deposited into the dusts and converted to $PbSO_4$, the most commonly observed compound, by rainwater leaching (*33*). Elemental lead, Pb^0, was observed only in car parks and is thought

TABLE VII

Chemical Phases Identified in Street Dusts by X-Ray Diffraction[a]

$PbSO_4$	α-Fe_2O_3
Pb^0	Fe_3O_4
$PbSO_4 \cdot (NH_4)_2SO_4$	α-Al_2O_3
Pb_3O_4	$AlPO_4$
$PbO \cdot PbSO_4$	α-SiO_2
$2PbCO_3 \cdot Pb(OH)_2$	$CaCO_3$
Feldspars	
Albite	
Anorthite	
Sanidine	
Micas/clay minerals	
Chlorite	
Muscovite	
Biotite	

[a] According to Refs. *33* and *134*.

to be associated with emissions from cold, choked vehicles. Pb_3O_4, $PbO \cdot PbSO_4$, and $2PbCO_3 \cdot Pb(OH)_2$ all are used in leaded paints.

B. Sequential Extractions

In their work with XRD examination of street dusts, Biggins and Harrison (*33*) concluded that only a small proportion of lead in street dusts is present in a crystalline form amenable to XRD identification. It appears that weathering processes fairly rapidly convert crystalline lead compounds deposited from the atmosphere into noncrystalline forms more typical of a soil.

TABLE VIII

Sequential Extraction of Metals from Street Dusts[a]

	Pb		Cd		Cu		Zn	
Fraction	Mean (%)	Range	Mean (%)	Range	Mean (%)	Range	Mean (%)	Range
Exchangeable	1.5	1–3	20	15–32	7	1.5–11	2	1–3
Carbonate	43	27–62	38	22–52	18	5–30	44	29–68
Fe–Mn Oxide	38	24–56	28	22–38	4.5	2–10	43	23–54
Organic	7.5	5.5–9.5	8	2.5–19	58	47–68	7.5	4.5–14
Residual	10	7–16	6	4–8	12	2.5–29	4.5	1–11

[a] From Ref. *134*.

Thus Harrison *et al.* (*134*) used a sequential extraction scheme (Section V) to gain information on chemical associations of lead, cadmium, zinc, and copper in samples of street dust. The results are summarized in Table VIII. The chemical significance of the fractions used and of the results is discussed in greater detail in Section V.

C. Scanning Electron Microscopy (SEM)

The use of an SEM with energy dispersive X-ray analyzer for examination of street dusts has been reported by Linton *et al.* (*181*). The instrument was used in two modes: a *raster* mode in which the electron beam is scanned over a field of particles and the X-ray emission characteristics of a single element monitored, and a *spot* mode in which the electron beam is held stationary upon a given particle while the X-ray spectrum is recorded.

Tracer elements were sought as an aid to source identification: Br and Cl for automotive lead, and Al, Ca, Si, and Ti for nonautomotive sources. By examination of particle morphology and interelement concentration ratios it was possible to identify tentatively lead of both automotive and paint origin. Nonetheless, no specific compounds were identifiable. It has been our experience with this technique that many particles in street dusts are aggregates and that many apparently single particles contain a wide spectrum of elements and hence even tentative identification of individual compounds is not possible.

V. SEDIMENTS AND SOILS

Aquatic sediments have been the focus of a considerable research effort as they are a major sink for waterborne trace metals, and since for many metals sediments may contain a historical record of deposition. As with other environmental media, speciation studies are important as an indicator of potential toxicity and mobility of inorganic pollutants.

Salomons and Förstner (*225*) have elaborated five mechanisms for metal accumulation on sedimentary particles: (1) adsorptive bonding on fine-grained substances, (2) precipitation of discrete metal substances, (3) coprecipitation of metals with hydrous iron and manganese oxides and carbonates, (4) associations with organic components, and (5) incorporation in crystalline material (i.e., mineral lattices). Any speciation technique should aim to separate metals in these various chemical associations. Table IX summarizes common methods used to extract selectively given phases from sediments. No technique is fully specific to one phase, and this has led to the development of sequential extraction schemes in which the sediment is extracted with progressively more potent reagents, each intended to remove one further phase from the sample.

TABLE IX

Summary of Common Methods for the Extraction of Metals Associated with Different Chemical Phases in Sediments[a]

Adsorption and cation exchange
  Extraction with $BaCl_2$, $MgCl_2$, NH_4OAc, NaCl
Manganese and iron phases: reducible, easily and moderately reducible phases
  Extractions (in approximate order of release of iron): acidified NH_2OH, ammonium oxalate, NH_2OH/HOAc, dithionite/citrate
Carbonate phases
  Extractions with CO_2 treatment, acidic cation exchange, NaOAc/HOAc (pH 5)
Organic phases: humic and fulvic acids, solid organic material
  Extractions with H_2O_2, H_2O_2/NH_4OAc, H_2O_2/HNO_3, organic solvents, 0.5 *N* NaOH, 0.1 *N* $NaOH/H_2SO_4$, NaOCl–dithionite/citrate, $HNO_3/HClO_4$
Residual
  $HClO_4$/HF

[a] Based on Ref. *225*.

A classification of metals associated with sediments and soils has been listed by Harrison *et al.* (*134*) and is given in Table X. Any sequential extraction scheme should aim to use chemical reagents such as those in Table IX to remove as selectively as possible metals in the associations listed. One of the best researched schemes is that of Tessier *et al.* (*255*) listed in Table XI. It distinguishes five phases, or fractions, identical to those in Table X except that the "soluble" and "exchangeable" classes are removed together into one extractant. The chemical basis for use of the reagents in Table XI is indicated in the last column of Table X.

Some illustrative data from the sequential analysis of river sediments are given in Fig. 2. Four fractions were analyzed: ammonium acetate (exchangeable),

TABLE X

Classification of Metals Associated with Dusts, Sediments, and Soils[a]

Classification	Form of association[b]	Extraction technique
Soluble	Metal ppt, pore water	Release to pore water or river water
Exchangeable	Specifically adsorbed, ion exchangeable	Exchange with excess cations
Carbonate phase	ppt or co-ppt	Release by mild acid
Fe–Mn oxide phase	Specifically adsorbed, co-ppt	Reduction
Organic phase	Complexed, adsorbed	Oxidation
Residual	In mineral lattices	Digestion with strong acids

[a] From Ref. *134*.
[b] ppt, Precipitate.

TABLE XI

Sequential Extraction Procedure[a]

Nominal fraction extracted	Procedure[b]
Exchangeable	1 *M* $MgCl_2$ (8 ml), pH 7, 1 h, ca. 20°C, continuous agitation
Carbonate	1 *M* NaOAc (8 ml), pH 5.0, 5 h, ca. 20°C, continuous agitation
Fe–Mn oxides	0.04 *M* $NH_2OH \cdot HCl$ in 25% acetic acid (20 ml), 6 h, 96°C, occasional agitation
Organic	0.02 *M* HNO_3 (3 ml) + 30% H_2O_2 (5 ml), pH 2.0, 2 h, 85°C, occasional agitation; further 30% H_2O_2 (3 ml), pH 2.0, 3 h, 85°C, occasional agitation; then 3.2 *M* NH_4OAc in 20% HNO_3 (5 ml), 0.5 h, ca. 20°C, continuous agitation
Residual/total	[2 × 70% HNO_3 (5 ml) to dryness, total only]; 40% HF (10 ml)/72% $HClO_4$ (2 ml) to near dryness, 40% HF (10 ml)/$HClO_4$ (1 ml) to near dryness, 72% $HClO_4$ (1 ml) to white fumes, taken up in HCl (2 ml), diluted to 25 ml

[a] From Ref. *134*.
[b] One-gram sample.

hydroxylamine hydrochloride (carbonates and Fe–Mn oxides), hydrogen peroxide/hydrochloric acid (organic), and resistant (mineral lattice). The trace metals (Zn, Cu, Ni, Pb, and Cd) reside primarily in the resistant fraction in unpolluted rivers (e.g., Orinoco and Great Ruaha Rivers), consistent with their derivation primarily from natural minerals. Pollutant metal, however, is less strongly bound, being associated with the surface of sediment grains, and is removed by the milder extractants (e.g., Meuse and Somme Rivers).

As indicated above, there are many reagents available for sequential extractions and indeed many full schemes are reported. Even though two schemes may aim to separate the same fractions from a sample, they will never generate the same result, and it is important to recognize that the findings from sequential extractions are very useful in a qualitative or semiquantitative sense, but may not be regarded as fully quantitative. The chemical associations of trace metals in environmental sediments are far too diverse to allow simple separation into a few discrete fractions. These fractions do, however, serve to provide valuable indicators of speciation and of likely environmental mobility and bioavailability for a trace metal. It should also be borne in mind that Rendell *et al.* (*213*) have demonstrated readsorption of released metal during sequential extractions, and this factor requires further that results be regarded with due caution.

As reported earlier, Harrison *et al.* (*134*) have applied the sequential extraction scheme of Tessier *et al.* (*255*) to the speciation of Pb, Cd, Zn, and Cu in street dusts. In the same paper, the analysis of roadside soils by this scheme is also

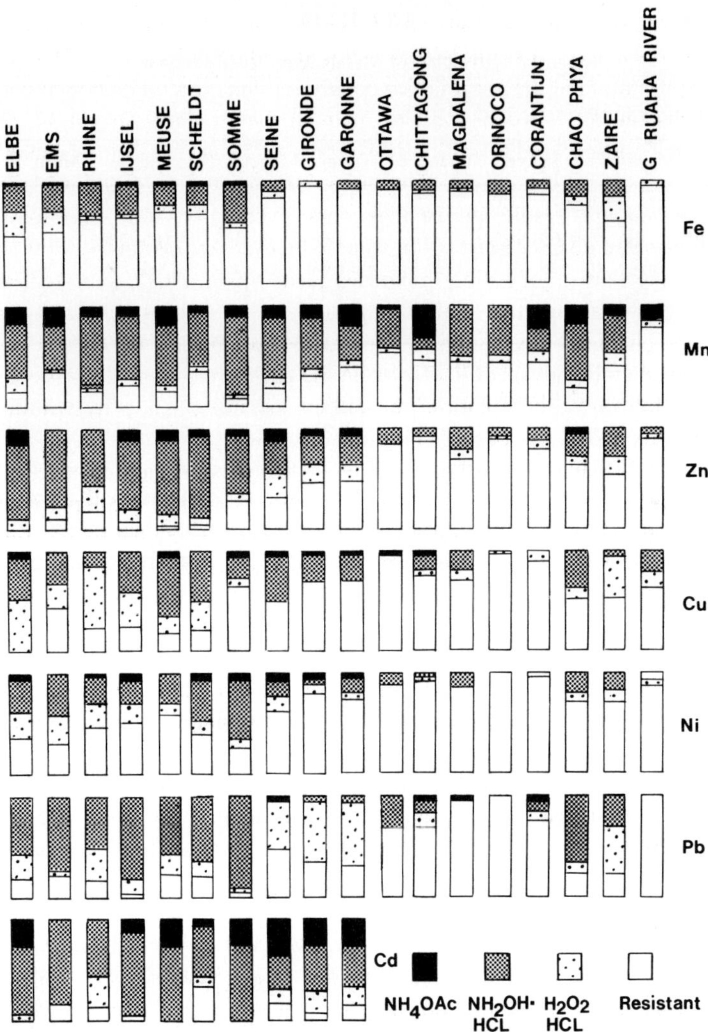

Fig. 2. The speciation of trace metals in 18 different river sediments, arranged according to their approximate geographic position from north to south. Most tropical rivers contained low Cd levels, and no reliable data were obtained. For the Rio Magdalena and Orinoco River, insufficient material was available for determination of the "exchangeable" (NH_4OAc) fraction, and this is contained in the hydroxylamine extract. Reproduced with permission from Salomons and Förstner (*225*).

TABLE XII

Sequential Extraction of Metals from Roadside Soils[a,b]

	Pb		Cd		Cu		Zn	
Fraction	Mean (%)	Range	Mean (%)	Range	Mean (%)	Range	Mean (%)	Range
Exchangeable	1	0–1.5	26	6–54	2	1–3	3	0–6
Carbonate	26	8.5–56	24	13–28	1.5	0.5–2.5	31	8.5–76
Fe–Mn Oxide	44	29–60	25	22–30	12	4–18	34	5–62
Organic	12	2.5–20	8	3–25	48	36–62	8.5	2–16
Residual	17	11–23	18	6.5–42	37	33–46	23	6–56

[a] From Ref. *134*.
[b] Total of four samples in pH range 6.9–8.4.

reported. These results are summarized in Table XII. The patterns of speciation are generally very similar to those for street dusts with the association of metals with the "carbonate" phase slightly higher in street dusts than in soils. This is consistent with XRD identification of calcite ($CaCO_3$) as a major component of most of the analyzed street dusts, although it was only a minor component of soils (*134*); powdered limestone is used as a component of road surfacing materials. The results for soils show some interesting features. Cadmium is the only trace metal present substantially in the exchangeable fraction, which is consistent with the relatively facile uptake of this metal by growing plants. Conversely, the other metals are more strongly bound, consistent with their known immobility in soils. Copper is associated largely with the organic fraction, a common finding for this metal in waters, sediments, and soils generally.

The use of XRD for speciation of lead in street dusts has been mentioned earlier. Olson and Skogerboe (*205*) developed XRD techniques for soils: lead compounds were initially preconcentrated by density and magnetic separation. The results of XRD analyses are shown in Table XIII. These results are substantially consistent with the later findings of Biggins and Harrison (*33*) in street dust analysis. However, Biggins and Harrison (*33*) concluded that the major lead compound, $PbSO_4$, could account for at most only a few percent of lead in the samples. This question was later put to the test when Harrison *et al.* (*134*) spiked a soil sample with finely ground $PbSO_4$ and subjected it to a sequential extraction scheme. The results (Table XIV) showed that the lead appeared primarily in the first three extracts, being especially abundant in the first, consistent with the moderate solubility of $PbSO_4$ in aqueous media. The pattern of extraction was in marked contrast to that in roadside soils (Table XII). It was concluded that $PbSO_4$ was not of importance in the soil and street dust samples examined, and that under British environmental conditions it is rapidly weathered to non-

TABLE XIII

Lead Compounds Identified in Soils and Street Dusts by X-Ray Diffraction[a]

Sample identity	Soil fraction	Compounds found	Concentration estimates[b]
Fort Collins 1	Magnetic	$PbSO_4$	Major
and Fort Collins 2	Nonmagnetic	$PbSO_4$	Major
		$PbO \cdot PbSO_4$	Minor
		PbO_2	Trace
		PbO[c]	Trace
Denver 1	Magnetic	$PbSO_4$	Major
	Nonmagnetic	$PbSO_4$	Major
Chicago 10	Magnetic	$PbSO_4$	Major
	Nonmagnetic	Pb^0	Major
		$PbSO_4$	Minor
Chicago 20, 30, and 40	Magnetic	$PbSO_4$	Major
	Nonmagnetic	$PbSO_4$	Major
Missouri 1	Magnetic	None[d]	—
	Nonmagnetic	PbS	Major
		$PbSO_4$	Minor

[a] From Ref. *205*.

[b] Major indicates the principal portion of lead present in the soil fraction indicated and therefore the principal portion of the soil sample; minor refers to approximately 1–10% of the Pb in the respective fractions; trace quantities are less than approximately 1% of the total in each fraction.

[c] Assignment is based on the presence of only the most intense *d*-spacing and is therefore questionable.

[d] Complex *d*-spacing pattern obtained with all intensities low; positive assignment of any one compound or group of compounds questionable.

TABLE XIV

Recovery of $PbSO_4$ Spike from a Soil Sample[a,b]

Fraction	Recovery (%)
Exchangeable	42.5
Carbonate	18.5
Fe–Mn oxide	20.4
Organic	0.1
Residual	4.7

[a] From Ref. *134*.

[b] Mean of two samples in which a soil of 84 μg Pb/g was spiked with 3730 μg Pb/g and 3270 μg Pb/g as $PbSO_4$.

crystalline forms of chemical association with the soil. Thus, sequential extractions are more appropriate to the speciation of metals in this type of sample than is XRD.

VI. NATURAL WATERS

A. Introduction

The recent interest in examining the physicochemical speciation of trace metals in natural waters arises in part from the increasing awareness that only a fraction of the total metal present may invoke biochemical and geochemical responses (*111, 158, 177, 191, 192, 244*), and from the development of sufficiently sensitive analytical techniques. Excellent reviews are available discussing the speciation of several elements on an individual basis (*102, 109*). This treatment will outline the range of techniques amenable to speciation studies.

The methods considered can provide information regarding either specific species or the partitioning of an element into readily identifiable fractions, termed here generic speciation. Generic speciation may be investigated using both physical and chemical procedures. Size fractionation techniques discussed include centrifugation, filtration, ultrafiltration, dialysis, and gel filtration chromatography (GFC). The environmental application of such methods has been extensively reviewed by de Mora and Harrison (*81*), clearly illustrating the importance in examining the size distribution of metal species in natural waters. Chemical investigations of generic speciation initially concentrated on establishing the fraction of metal associated with organic material. More recently, a number of procedures have been developed to identify a "technique-labile" metal fraction which must be considered individually.

Such metal fractions are *operationally defined* because the concentration determined is dependent upon the technique utilized [i.e., Chelex labile, anodic stripping voltametry (ASV) labile]. Furthermore, procedural variations within a particular technique determine different metal fractions. This is exemplified in using chelating resins in either column or batch mode and ASV measurements as a function of solution pH. Species-specific analyses generally rely upon either ion-selective electrodes or chromatographic separations.

Redox speciation has not been considered separately; however, the discussion of techniques outlined above incorporates the applicability of differentiating oxidation states where appropriate. Bioassay and isotopic fractionation techniques are beyond the scope of this article.

Both sampling and storage methods for speciation studies have been examined in considerable detail (*22, 28, 170, 189*). Teflon or polythene samplers and storage containers are recommended and can be decontaminated by soaking in

10% nitric acid for at least 48 h. Acidification, freeze-drying, and freezing can induce irreversible changes in trace metal species; however, samples may be stored at 4°C for as long as 3 weeks with no apparent deleterious effects (*23, 101, 114*).

Care must be exercised in subsampling from a large water sampler to ensure that each aliquot is representative of the sample as a whole. Differential settling of particles can cause the erroneous determination of suspended sediment loads, and hence, particulate metal levels (*90, 183*). The incomplete recovery of suspended material from 30-liter Niskin water samplers has been observed (*54, 121*). A similar effect was observed by de Mora and Hay (unpublished results) using a 1-liter NIO water sampler whereby successive 250-ml aliquots of an estuarine sample showed a tendency to exhibit lower concentrations of suspended matter.

B. Physical Techniques for Generic Speciation

1. Centrifugation

Centrifugation may be used to differentiate size fractions of suspended particulate material in natural waters. As a first approximation, the sedimentation rate is a function of the rotation rate and time of centrifugation. All particles with diameter ⩾ 190 nm and specific gravity about 2.5 were separated from river and lake water by centrifugation for 30 min at 3000 rpm (*28*): conditions that would not remove the humic material (*25*). Similarly, the molecular weight of cadmium and lead organic complexes deposited from freshwater samples by centrifugation at 40,000 rpm for 5 h was estimated to be less than 50,000 (*48*).

Settling rates also are influenced by the specific density of the particulate material, and, hence, particles are not fractionated strictly by size. Duinker *et al.* (*90*) observed that particulate metal concentrations for elements associated with dense particles (Al, Fe, K, Mn, and Ti) were higher when determined by centrifugation as opposed to filtration. Such an influence clearly depends upon the contribution of small dense particles and illustrates that the two techniques fractionate different species.

2. Filtration

Chemical constituents in natural waters are operationally defined as "dissolved" (filterable, soluble) or "particulate" (nonfilterable) depending upon their ability to pass through a filter with a nominal pore size in the range of 0.4 to 0.5 μm (*123*). A single filtration step is often the only size characterization carried out in examining the speciation of trace metals in natural waters (*19, 101, 142, 175*). However, the material passing through a filter includes polymers and colloids together with species in true solution (*249*). Several workers have shown

TABLE XV

Size Distribution of Trace Metal Species in Natural Waters

Size range	Metal species	Examples	Phase state
<1 nm	Free metal ions	Mn^{2+}, Cd^{2+}	Soluble
1–10 nm	Inorganic ion pairs, inorganic complexes, low-molecular-weight organic complexes	$NiCl^{+}$, $HgCl_4^{2-}$, Zn-fulvates	Soluble
10–100 nm	High-molecular-weight organic complexes	Pb-humates	Colloidal
100–1000 nm	Metal species adsorbed onto inorganic colloids, metals associated with detritus	Co–MnO_2, Pb–$Fe(OH)_3$	Particulate
>1000 nm	Metals adsorbed into living cells, metals adsorbed onto or incorporated into mineral solids and precipitates	Cu–clays, $PbCO_3$ (s)	Particulate

that the concentrations of "dissolved" aluminum, iron, manganese, and titanium decrease with filtration through decreasing nominal pore sizes (*156, 163, 265*). Because the different size fractions may contain trace metals in particular associations (Table XV), the size distribution of metals can give insight into the relative importance of these different species. The physicochemical speciation scheme of Laxen and Harrison (*171*) utilizes this approach. Samples are filtered in parallel through five filters with pore size ranging from 12 to 0.015 μm. The technique is applicable to drinking water (*133*), river and lake waters, and various industrial effluents (*170–172*).

The applicability of a particular filter for trace metal speciation studies depends upon the size selectivity and the susceptibility toward contamination and adsorptive effects. These characteristics are determined by the composition of the filter and will be considered individually here.

In order to investigate the size distribution of metallic species, a filter must efficiently separate two size fractions rather than merely remove particulate material from suspension. To achieve this, the filter pores should be relatively uniform in size and remain relatively constant throughout the filtration. Furthermore, the stated nominal pore size must approximate the "effective" pore size. This has been defined as either the diameter of spherical particles removed with 90% efficiency from 1 liter of a 1 mg/liter suspension (*69*) or the size at which 50% by number of the suspended particles is retained (i.e., the median retention diameter) (*229*).

Filtration characteristics are best evaluated by utilizing a retention curve as

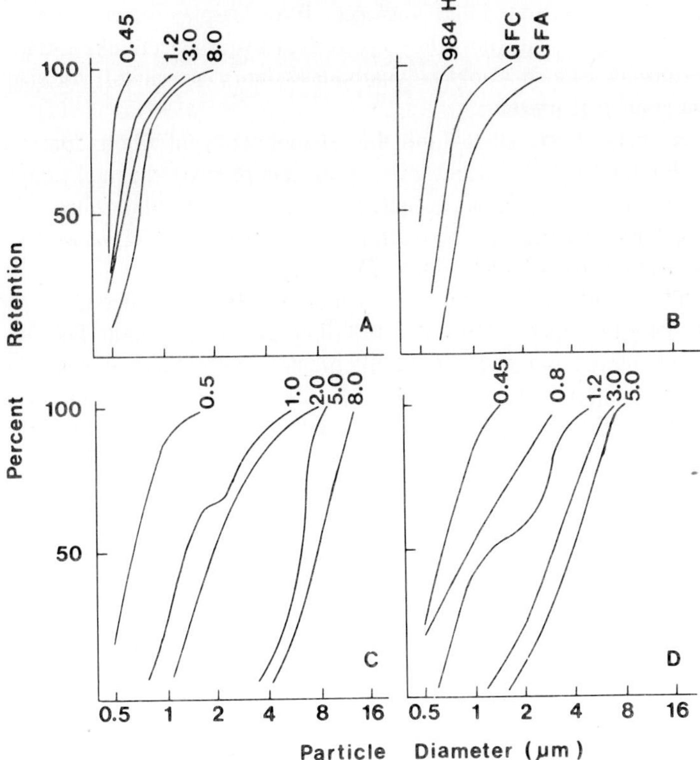

Fig. 3. Retention curves for (A) cellulose filters (Millipore), (B) glass fiber filters (Whatman GFC and GFA, Reeve Angel 984 H), (C) polycarbonate filters (Nuclepore), (D) silver filters (Flotronics). The numbers above the curves are the pore sizes in micrometers given by the manufacturer. Reproduced with permission from Sheldon (*229*).

depicted in Fig. 3. The median retention diameter is immediately apparent. The slope of the curve reflects the degree of uniformity of the pores such that a steep slope is indicative of a highly selective filter.

Although cellulose acetate, glass fiber, and silver filters may be used for total particle retention, they are not applicable for size fractionation (*69, 229, 230*). In general, the effective pore size may be considerably smaller than that stated, and decreases further with filter loading. As is evident in Fig. 3, Millipore filters exhibited the best selectivity characteristics, but filters with stated pore sizes from 0.45 to 8.0 μm exhibited comparably effective pore sizes. Nuclepore polycarbonate filters exhibited median retention diameters similar to the stated size in the range 0.5–8.0 μm. The effective pore size also remained relatively constant with filter loading until the pores were blocked. However, such filters are easily overloaded and must be used with care. Laxen and Chandler (*169*) have recently demonstrated that Nuclepore filters display better size discrimination at smaller pore sizes (0.4–0.015 μm) than Sartorius cellulose nitrate filters. Such results indicate that only polycarbonate filters are suitable for size differentiation studies at present.

Care must be exercised to limit the contamination of filters due to the extremely low trace metal concentrations prevalent in environmental samples. Filters can be effectively decontaminated by rinsing with dilute nitric acid (*22*). This procedure removes zinc from cellulose acetate filters (*272*) as well as copper and lead from polycarbonate filters (*258, 266*).

Adsorption during filtration poses a greater, but less understood, problem. Several metal ions may adsorb onto the filter and/or filter unit. The adsorptive loss of both copper and lead is generally greater onto glass than onto polycarbonate filtration equipment (*119, 120, 154, 258*). However, Florence (*102*) suggests that a Millipore all-glass apparatus may be acceptable for use with natural samples since refiltration of river and seawater up to four times through the same filter unit caused no significant decrease in the concentrations of cadmium, copper, lead, or zinc. He suggests that the observed serious adsorptive losses of several metals into all-glass equipment occurs from synthetic solutions (*198, 219, 224*) but not from natural waters (*189, 198*). However, significant adsorptive losses onto glass filter units have been observed for copper and lead from seawater (*50, 190*) and cadmium from tap water (*117*). As discussed by de Mora and Harrison (*81*), such discrepancies arise in part from difficulties in the interpretation of adsorption experiments.

Experiments investigating adsorptive losses of trace metals generally involve the filtration of either a synthetic solution or a previously filtered natural water. In the first case, inorganic constituents can be properly modeled and it has been demonstrated that Group I and II metal cations effectively reduce the adsorption of heavy metals (*120, 198, 224*). However, the role of organic material and organometallic complexes in adsorption processes has not been fully evaluated. Cranston and Buckley (*69*) have shown that Nuclepore filters strongly adsorb humic acids. Thus, the filtration of model solutions may not be representative of environmental samples. Similarly, the refiltration of a natural water sample is subject to ambiguous interpretation. The size selectivity of filters is not 100% efficient and the effective pore size may decrease with filter clogging. Hence, material that passed through the primary filter could be trapped during the second filtration. Geometric retention may have a comparable effect such that rod-shaped particles could be retained during the refiltration. Finally, the initial filtration process could perturb solution equilibria and promote the precipitation and/or flocculation of subsequently nonfilterable material. Thus, variations in metal concentration before and after the secondary filtration may reflect real size fractionation rather than adsorption phenomena.

Several procedures have been proposed to overcome the problem of adsorption. Filters and equipment may be either conditioned with copious volumes of sample (*242*) or preconditioned with 0.1 *M* $Ca(NO_3)_2$ (*171*). Alternatively, a double filtration method has been used to recover lead adsorbed onto glass equipment (*219*).

Adsorption onto the filter itself is determined by the composition which in turn defines the applicability of the filter to a particular task. Although cellulose acetate filters adsorb inorganic mercury from seawater, glass fiber filters suffer only slight filtration losses ($< 7\%$) and can be initially decontaminated by ignition at 500°C (*214*). Alternatively, only Nuclepore polycarbonate filters are acceptable for other heavy metal analyses as they adsorb negligible amounts of such metals (*119, 125, 258*). Also, the recovery of cadmium, copper, iron, manganese, and lead from prefiltered tap water and synthetic river water is independent of pore size for Nuclepore filters in the range of 0.015–12.0 μm according to Laxen and Harrison (*171*). Teflon filters are excellent for trace metal studies (*22, 214*) but are not yet available in a sufficient range of pore sizes for use in comprehensive size differentiation studies.

Water filtration may be performed with either positive or negative applied pressure. As noted by Batley and Gardner (*22*), large pressure differentials must be avoided in order to prevent cell rupture which could otherwise cause dissolved metal enrichment and perturb speciation equilibria. Positive pressure filtration is generally preferred as this maintains the integrity of anoxic samples with respect to redox conditions (*125*), eliminates pH fluctuations associated with CO_2 evolution during vacuum filtration (*175*), and reduces the risk of airborne contamination (*242*). Stirred filtration units should reduce filter clogging thereby allowing faster filtration rates and better size discrimination; however, Laxen and Chandler (*169*) suggest that no benefit is derived from using stirred as opposed to unstirred filter cells.

3. Ultrafiltration

Ultrafiltration, pertaining here to filtration through pore sizes less than 15 nm, has been used to discriminate further the size continuum of organic material and trace metal species [see Hoffmann *et al.* (*150*) and references therein]. Problems encountered are similar to those experienced during filtration, that is, size selectivity, contamination, and adsorption. These difficulties have not been well resolved with respect to ultrafiltration.

Ultrafilters are available with nominal pore sizes in the range 1.2–14 nm but are usually designated by a molecular weight cutoff (MWCO). The MWCO is operationally defined as the weight of a globular solute that is retained with an efficiency of 90%. This designation cannot be rigorously applied because particle discrimination is actually achieved by size fractionation and can be influenced by geometric configuration. Globular proteins can be separated at approximately accurate MWCO values; however, linear polyethylene glycols can exhibit complete diffusivity through membranes for molecular weights in excess of the stated limits (*35*). Alternatively, material with molecular weights less than the nominal MWCO could be retained due to adsorption effects, aggregation of the solute, geometric retention, or filter clogging (*167*). The poor size discrimination of ultrafilters has also been observed for naturally occurring organic material. Fulvic acid from lake water was fractionated in the molecular weight range 10,000–300,000 but its true molecular weight is normally considered to be only 1600 (*6*). Kwak *et al.* (*167*) found not only that fulvic acid from soil extracts was retained by the Amicon XM-50 ultrafilter (MWCO ~ 10,000) but also that the fulvates exhibited better retention characteristics than the higher molecular weight humates.

As is the case with filtration, the adsorptive loss of trace metal species during ultrafiltration is extremely difficult to evaluate. Organometallic complexes are susceptible to the problems of poor size differentiation outlined above. Furthermore, inorganic species may adsorb onto the ultrafilter and/or the retained organic material. Alberts *et al.* (*6*) have demonstrated using electron spin resonance techniques that hydrated manganous ions associated with retained organic material were physically adsorbed rather than chemically complexed. The Amicon UM-2 ultrafilter (MWCO ~ 1000) adsorbs copper, lead, and manganese from synthetic solutions of bicarbonate and EDTA/bicarbonate (*127*). Similarly, adsorptive losses of aluminum, chromium, iron, and zinc from river water may be substantial using PM-10 (MWCO ~ 10,000) ultrafilters (*27*). Laxen and Harrison (*171*) observed that the tap water concentrations of cadmium, copper, lead, and zinc were invariably higher in XM-300 (nominal pore size 14 nm) ultrafiltrate than in the 15-nm Nuclepore filtrate.

Metal contamination of ultrafilters is not recognized as a major difficulty, although this may not be the case for sodium, calcium, and cobalt (*27*). The leaching of low-molecular-weight hydrocarbons may be prevented by storage in 0.1% sodium azide and flushing with distilled water prior to use (*241*).

Buffle *et al.* (*46*) extensively examined ultrafiltration procedures in order to optimize size fractionation of organic material and in particular to isolate fulvic and humic acids. Diafiltration (the washing technique) gave more reproducible results than the concentration technique because the concentration of organic material in the cell remains relatively constant. Molecular aggregation occurs during the concentration procedure if the organic content is high. They recommended a cascade system (sequential ultrafiltration) to reduce filter clogging and the adsorptive effect of high-molecular-weight compounds when samples were filtered directly through membranes with small pore sizes. Hoffmann *et al.* (*150*) adapted these procedures to investigate organometallic complexes in river water. They utilized the concentration technique in conjunction with sequential ultrafiltration and avoided large concentration gradients by reducing the volume by only 50% at each step.

4. *Dialysis*

Dialysis techniques have been used in attempts to separate colloidal material from species in true solution. However, some relatively high-molecular-weight material may diffuse through the membrane because pore sizes vary from 1 to 5 nm. The size discrimination depends upon the dialysis membrane used. VisKing tubing (nominal pore size 4.8 nm) has an MWCO equivalent to PM-10 ultrafilters, that is, approximately 10,000 (*27*). Alternatively, Spectra/Por 6 dialysis bags (MWCO ~ 1000) are impermeable to soil-derived fulvic acid (*259*).

Trace metal contamination has been cited as a problem in dialysis experiments (*107*); however, membranes can be decontaminated by soaking in mineral acids (*27, 128, 142, 259*). Adsorption of trace metals from synthetic solutions has not been observed (*142, 259*). Guy and Bourque (*126*) add aluminum to the diffusate solution to prevent cation adsorption onto the internal surface of dialysis membranes.

Beneš and Steinnes (*27*) used equilibration times of 1–4 weeks for the *in situ* dialysis of river water, but usually equilibrium is achieved within 24 h (*128, 259*). Hart and Davies (*140, 142*) recycled the diffusate solution through an ion-exchange resin and removed trace metals from synthetic solutions with 100% efficiency in only 5 h. Guy and Bourque (*126*) used a Donnan dialysis procedure to achieve separation of several trace metals from synthetic solutions with preconcentration factors as high as 100 within 2 h. They utilized Nafion tubing, a cation-exchange membrane impermeable to anionic complexes and humic colloids, and concentrated the metals in an internal solution of sodium tartrate or sodium nitrate. The recovery efficiency decreases as the ionic strength of the sample increases but this effect can be counteracted by increasing the ionic strength of the diffusate solution.

5. *Gel Filtration Chromatography*

A complete size distribution of trace metal species can be obtained with GFC, known also as gel chromatography, gel filtration, gel permeation chromatography, exclusion chromatography, and molecular-sieve chromatography. The principles of the technique will be briefly outlined here but have been discussed at length in a number of reviews and texts (*9, 85, 100, 165*).

Porous polymeric beads are swollen with a solvent and packed vertically in a chromatographic column. Upon applying a sample to the column, size fractionation is achieved due to the flow of the solute being retarded in relation to the bulk flow of eluent. This process is size related and depends upon the diffusion rate of a molecule into and out of the interstitial cavities of the beads. Large particles completely excluded from the beads are eluted first followed by molecules in a decreasing order of size. Different beads exhibit varying exclusion limits and hence different size ranges may be investigated.

GFC columns are calibrated with model compounds, usually globular proteins. Steric effects will tend to overestimate the molecular weight. In contrast to ultrafiltration techniques, adsorption processes will cause an underestimation of the molecular weight. The main advantage of GFC over ultrafiltration is the potential to determine a continuous size distribution of trace metal species rather than concentrations within discrete size ranges. However, high dilution factors and blanks limit the application of GFC to natural waters exhibiting relatively high metal concentrations.

GFC experiments for speciation studies must be conducted with careful control of the eluent with respect to pH, ionic strength, composition, and temperature in order to limit the dissociation of metal complexes and adsorption effects (*3*). Distilled water has been used as an eluent (*243, 251*) but more commonly used are dilute solutions of Group I and II metal cations (*53, 245*). Sterritt and Lester (*245*) observed that attempts to decontaminate Sephadex with nitric acid greatly reduced the recovery efficiency toward copper and hence they preconditioned columns with 100 μg/liter metal solutions and copious volumes of eluent.

C. Chemical Techniques for Generic Speciation

1. Organically Associated Trace Metals

The earliest attempts to investigate the chemical speciation of trace metals in natural waters focused on determining the amount associated with organic material. Slowey *et al.* (*240*) were able to extract up to 60% of the total copper in seawater with chloroform. However, Florence and Batley (*109*) note that the interpretation of such observations proves difficult since charged copper complexes and copper associated with organic colloids may not be completely recovered while some copper associated with inorganic colloids may be partially extracted. In contrast to the above observations, Rosen and Williams (*222*) were unable to extract copper complexes from seawater with ether, chloroform, or ethyl acetate.

A more acceptable procedure involves metal analysis before and after the oxidative destruction of the organic matter. Such methods rely upon analytical techniques which essentially determine inorganic metallic forms (i.e., chelation–solvent extraction, Chelex recovery, ASV). Several oxidation procedures have been examined. Wet chemical techniques include digestion with perchloric acid (*7*) and peroxydisulfuric acid (*239*). Williams (*268*) recommended the photooxidation of organic matter by ultraviolet radiation. This technique generally yields lower blank values than acid digestion and has been employed extensively in speciation studies of trace metals in seawater (*20, 177*). In examining some fresh water samples, a brown precipitate has been observed following UV irradiation and ASV-labile metal concentrations have decreased (*101, 103, 171*).

Blutstein and Smith (*37*) observed a similar deposit with estuarine samples and assumed it to be polymerized organic material. More likely, the oxidative destruction of organic envelopes on colloidal material leads to the destabilization of the colloids and consequent precipitation of iron(III) hydroxide. Adsorptive effects and coprecipitation would significantly disrupt the speciation of other trace elements. Furthermore, UV irradiation unavoidably causes a temperature rise which in turn leads to a pH rise due to the loss of dissolved CO_2. Attempts to control pH with both acetate and phosphate buffers failed to prevent a decrease in the ASV-labile metal concentration. Similarly, acidification to the pH range 5.5–6 with 0.1 *M* $HClO_4$ prior to UV irradiation did not necessarily prevent the deposition of iron(III) hydroxide (*171*). Hence, photooxidation techniques may be applicable only with waters exhibiting low iron concentrations.

Batley and Farrar (*18*) compared UV and high-energy γ-irradiation. Similar releases of bound cadmium, copper, lead, and zinc were achieved from seawater and storm runoff samples.

Ozonolysis has been suggested as a method to destroy organic sequestering agents in natural waters. Clem and Hodgson (*64*) used ASV following ozone treatment to analyze cadmium and lead in sewage effluents and seawater from San Francisco Bay. L'Hopitault *et al.* (*151*) similarly observed that ozone treatment enhanced ASV-labile concentrations of cadmium, lead, and zinc in spiked lake water. In contrast, Batley and Farrar (*18*) and Laxen and Harrison (*171*) found a reduction in the concentration of ASV-labile cadmium, copper, lead, and zinc in natural and tap waters following ozonolysis. Ozone treatment is used to remove trace metal contaminants from industrial effluents (*228*). This procedure solubilizes metals associated with organic material and apparently promotes the oxidative precipitation of several metals including iron, manganese, and lead. These precipitates scavenge other elements in solution. Thus, the behavioral discrepancies outlined above undoubtedly parallel those observed in photooxidizing organic material by UV irradiation. This suggests that ozonolysis may be an acceptable oxidative procedure for speciation studies, but only in circumstances in which the precipitation of iron and manganese oxyhydroxides is unlikely to occur.

Adsorbent resins such as Amberlite XAD have long been used to investigate organic material in seawater (*218, 248*) and potable waters (*49*). More recently, Sugimura *et al.* (*252*) examined the speciation of several trace metals in seawater using XAD-2 resin. They applied seawater at natural pH to retain neutral and basic organometallic species and at pH 3 to isolate the acidic organic complexes. Mackay (*186*) suggests that XAD-1 resin is unsuitable for quantitative measurements of trace metal speciation because the retention of organometallic complexes is not reproducible and significant amounts of inorganic species may be adsorbed. Florence (*103*) used a similar resin, Bio-Rad SM2, to determine lipid-soluble metal fractions. Adsorption of free metal ions from the sample was prevented by the addition of a citrate buffer of pH 5.7.

Gel permeation chromatography also has been used to characterize metal–organic complexes. Such techniques have been discussed in Section VI,B,5 as they also crudely estimate molecular weights.

2. Ion-Exchange Chromatography

The removal of ions from aqueous solution may be accomplished by contact with an ion-exchange resin. Trace metals of environmental interest may be retained on a resin either by cation-exchange or more commonly due to complexation with a chelating functional group incorporated in the polymeric macroporous resin. A great selection of functional groups has been investigated with respect to many different metals. Although not all these resins have been utilized in speciation studies per se, several exhibit potential value in supplementing present schemes involving ion-exchange techniques. Chelex 100 with an iminodiacetate (IDA) functional group has been the most widely used chelating resin due to its ability to extract several metal ions simultaneously and its commercial availability in a relatively pure, inexpensive form. Recent emphasis has been placed on chelators utilizing sulfur rather than oxygen and nitrogen binding sites. Applications of IDA and other resin types are listed in Tables XVI and XVII, respectively.

Initial interest in chelating resins arose due to the ability to preconcentrate several metals from dilute solution. Since such investigations generally involved total metal analyses as opposed to speciation studies, procedures were developed to ensure maximum metal extractions. Riley and Taylor (*215*) demonstrated that the sample pH and flow rate through the H form of Chelex 100 affected the

TABLE XVI

Investigations of Trace Metals in Natural Waters Using Iminodiacetic Acid Chelating Resins

Functional form	Metals studied	Reference
H^+	Ag, Al, As, Be, Bi, Cd, Ce, Co, Cr, Cs, Cu, Hg, In, Mn, Mo, Ni, Pb, Sc, Se, Th, Tl, Sn, W, U, V, Y, Zn	*215*
	Cd, Cu, Fe, Mn, Mo, Ni, V, Zn	*218*
	Cd, Cu, Pb, Zn	*104, 105*
NH_4^+	Mo, V, Zn	*216*
	Cd, Cu, Pb, Zn	*104*
Na^+	Cd, Cu, Pb, Zn	*104*
	Cd, Cu, Fe, Pb, Zn	*140*
Ca^{2+}	Cd, Co, Cu, Ni, Pb, Zn	*2, 97*
	Cd, Cu, Pb, Zn	*1*

TABLE XVII

The Use of Chelating Resins (Other Than Iminodiacetic Acid Type) to Analyze Trace Metals

Functional group	Metals investigated	Reference
Chitosan	Cd, Cu, Pb, Zn	*197*
	Cu	*195*
	Mo	*194*
	V	*196*
Cysteine	Ag, Au, Hg, Pt	*182*
Diamine	As, Cr, Mn, Mo, Se, V, W	*179*
	Ag, Cd, Co, Cr, Cu, Fe, Hg, Mn, Pb, Zn	*178*
Dithiocarbamate	Ag, Cd, Cu, Hg, Pb, Sb	*86*
	Ag, Co, Cu, Fe, Hg, Mn, Ni, Pb, Zn	*178*
Oxine	Al, Cd, Co, Cu, Fe, Mn, Ni, Pb, Zn	*236*
	Al, Co, Cu, Fe, Ni, Mo, Nb, Ti, V, W, Zn, Zr	*264*
	Cd, Co, Cu, Fe, Mn, Ni, Pb, Zn	*250*
	Cd, Cu, Pb, Zn	*103*
Polyamine–polyurea	Co, Cu, Ni, Zn	*87*
	Co, Cr, Cu, Fe, Ni, Zn	*180*
Poly(maleic anhydride)	Pb	*81a, 222a*
Salicylic acid	Al, Ca, Cd, Co, Cr, Cu, Fe, La, Mg, Mn, Ni, Pb, Y, Zn	*237*
Silyl xanthate	Cu, Ni, Zn	*178*
Thiols	Ag, As, Bi, Cu, Hg, Sb, Pt	*238*
	Ag, Al, Bi, Cd, Hg, Pb, Sb, Sn, U	*209*
	Ag, Ca, Cd, Co, Cu, Fe, Hg, Mn, Na, Ni, Pb, U	*460*
	Cd, Cu, Pb, Zn	*103, 110*
	Zn	*459*

recovery of several elements from spiked seawater. The ionic form of the resin could also significantly influence the retention efficiency. Initial zinc leakage from seawater observed when using the H form of Chelex 100 was eliminated by using an NH_4 form (*216*).

Such effects prompted Florence and Batley (*104, 105*) to examine the influence of resin form and pH on metal retention. When using H-Chelex with poorly buffered water samples, the initial uptake of divalent cations liberates sufficient H^+ to reduce the pH to a level at which the extraction of trace metals is no longer quantitative. This pH effect may be prevented by using the Chelex initially in the Na form (*20, 140*), the NH_4 form (*101, 216*), or the Ca form (*2, 97, 171*). Ca-Chelex exhibits a further advantage over NH_4 and Na forms in that steady flow rates may be maintained when using the column method because this resin does not shrink following the uptake of divalent cations.

In most instances, the application of ion-exchange chromatography (IEC) to speciation studies relies upon the presence of trace metal species inert to the exchange reaction. Trace metal components in solution are thereby differentiated as labile (i.e., Chelex-labile) or nonlabile. Although pH adjustment may be necessary to achieve maximum recovery for total metal analyses, in speciation studies the environmental samples are generally examined in an unmodified state in an attempt to limit perturbations to the established equilibria.

Muzzarelli and Rocchetti (*195*) demonstrated the existence of inert metal constituents in natural waters. Both Dowex A-1 (also an IDA resin) and chitosan chelating resins extracted copper from seawater with an efficiency of 100% only following the oxidative destruction of the organic matter. Although this implies that the nonlabile material is organic in nature, several mechanisms may be responsible. Incomplete retention may be due in part to the molecular exclusion of trace metals associated with colloidal material larger than the pore size of the resin (*105*). Colloidal iron(III) hydroxide and large organic molecules were quantitatively rejected by H-Chelex with an estimated pore size of 1.5 nm (*101*). Smaller metal species may not be retained if they exist as complexes with slow dissociation rates relative to the solution/resin contact time (*97, 98*). The implications of such kinetic influences will be considered below. Alternatively, the retention of trace metals associated with organometallic complexes having stability constants greater than that of the functional group of the chelating resin will not be quantitative, but dependent upon mass action effects. Since functional groups exhibiting very high affinities toward transition metals are chosen in order to attain maximal metal recovery from solution, some authors suggest that naturally occurring organometallic complexes are unlikely to have stability constants higher than the synthetic chelators (*101*). However, McKnight and Morel (*191, 192*) suggest that siderophores may sequester a significant proportion of copper and iron in natural waters. Also, since siderophores may be biosynthesized by phytoplankton under conditions of limited trace metal availability, the relative importance of such chelators to the overall speciation may be further enhanced.

Considerable procedural variations exist in the application of chelating resins to trace metal studies. Speciation schemes utilize both the column mode (*20*) and batch mode (*140–142, 170–173*) to distinguish labile and nonlabile metal fractions. Early workers advocated the use of columns so that large volumes of sample could be applied, thereby giving large preconcentration factors. Subsequent analysis of the retained metals can follow their elution with mineral acid or ammonia solution (*194–196, 215–217*). Alternatively, metals can be determined directly on the resin by X-ray fluorescence (*178, 180*), X-ray photoelectron spectroscopy (*148*), or graphite furnace atomic absorption spectrometry (*235*).

Hart and Davies (*140*) developed a batch technique to overcome possible effects of slow kinetics of cation chelation on the resin, thereby optimizing the recovery of Chelex-labile metals. Comparable ion-exchangeable concentrations of Cd, Cu, Pb, and Zn were exhibited by natural river water samples following Chelex equilibration times of 16 and 168 h. Figura and McDuffie (*98, 99*) examined the effect of shorter time scales. They compared column and batch techniques for Ca-Chelex, identifying moderately labile and slowly labile metal fractions, respectively. They suggested that the variations in metal uptake between the two procedures result from differences in the dissociation rates of the metal complexes. The speciation scheme of trace metals in aqueous solution proposed by these authors therefore includes the concept of a metal-binding spectrum.

Ion-exchange chromatography is seldom utilized in speciation schemes simply to provide labile and nonlabile fractions. Some assessment of the organometallic fraction may be made by comparing Chelex lability before and after UV irradiation (*20, 105, 107*). However, in some instances UV irradiation may destabilize colloids. The ensuing precipitation of iron(III) hydroxide may scavenge other trace metals thereby reducing the Chelex-labile and ASV-labile fractions (*37, 101, 103, 171*).

The greatest potential for IEC in speciation studies arises from Figura and McDuffie's (*98, 99*) suggestion of examining the metal-binding spectrum of trace metals. They defined four metal fractions (inert, slowly labile, moderately labile, and very labile) by using both column and batch IEC techniques in conjunction with ASV determinations. Hence, metal concentrations are determined in categories with relatively well-defined dissociation constants, a significant improvement over "operationally defined" metal fractions. Such data allow better assessment of bioavailable metal levels and provide some much-needed information on metal–organics for thermodynamic speciation models.

As an alternative technique, metal-binding data may be acquired by the parallel use of chelating resins with different stability constants. Several resin types are available (Table XVII) but have seldom been used in parallel. Muzzarelli and Rocchetti (*195*) compared Dowex A-1 (an IDA resin) with chitosan. Copper recovery from seawater was 20 and 50%, respectively, thereby suggesting different metal affinities. Anion- and cation-exchange resins have been used to differentiate charged species of cadmium and lead in river water (*175*). Pankow *et al.* (*208*) utilized cation- and anion-exchange resins to determine Cr(III) and Cr(VI) species, respectively, in river water. Florence and Batley (*103, 110*) compared Chelex 100 with a thiol type chelating resin and suggested that the latter may provide biologically more meaningful data. The greater affinity toward copper than toward cadmium, lead, and zinc by the thiol resin may in part mimic the higher biological toxicity of this element.

Finally, de Mora and Harrison (*81a*) have examined the extraction of lead from tap water as a function of size using both Ca-Chelex and poly(maleic anhydride) chelating resins. The respective recovery efficiencies were 73 and 80% for unfiltered samples rising to 81 and 100% for 0.08-μm-filtered aliquots.

3. *Polarography and Voltametry*

Both polarography and ASV have been increasingly utilized in physicochemical speciation studies. The greater sensitivity of ASV favors its use over polarography. The theory and instrumentation of these techniques have been reviewed elsewhere (*48, 76, 188, 198*); environmental applications and procedural variations will be discussed here with reference to trace metals in natural waters. Only those metals that form an amalgam with mercury are amenable to such techniques. Hence, most studies are restricted to cadmium, copper, lead, and zinc (*101, 118, 172, 200*), although other elements such as tin (*185*) and cobalt (*68*) have been investigated.

Several problems are encountered in the ASV analysis of trace metals in natural waters. Overlapping peaks may cause difficulties in the determination of lead in the presence of tin or thallium and of cadmium with thallium present (*19, 185*). The formation of intermetallic compounds in the mercury may diminish peak stripping currents (i_p). Both Cu–Zn and Cu–Ni interactions have been observed (*106, 118*). Such interferences can be prevented in copper analyses by the judicious choice of preelectrolysis (plating, deposition) potential. Zinc determinations can be performed following the addition of gallium to the sample, which promotes Ga–Cu intermetallics in preference to Cu–Zn interactions (*67*). Organic fouling of the electrode can depress the peak current, shift the peak potential (E_p) to more positive values, and broaden the peak (*41*). The adsorption of organic surfactants may also produce tensametric waves during operation in the differential pulse mode which may interfere with metal stripping peaks (*21, 159*). The influence of both intermetallic formation and adsorption of organics is more severe with the thin-film electrode (TFE) than with the hanging mercury drop electrode (HMDE). However, the TFE does display greater sensitivity and better resolution (*19*).

Numerous methodologies of ASV have been employed to study trace metal speciation. The technique can be used to determine total and electroactive metal concentrations. Complexation can be investigated using either metal or ligand titrations. Alternatively, pseudopolarograms may be obtained by altering the deposition potential or pH.

In the first instance, the low detection limit makes the techniques particularly suitable for total metal analyses. Such measurements are usually performed at a low pH following UV irradiation (*20*) or acid digestion (*20, 58, 171*) in order to ensure that all of the element of interest exists as electroactive species. Similarly, total metal concentrations may be determined in various fractions separated by chemical techniques, such as solvent extraction (*103*) or ion-exchange chromatography (*105*), and by physical techniques, such as ultrafiltration (*171*) or dialysis (*140*). Futhermore, these electroanalytical techniques measure only one oxidation state and are amenable to the investigation of redox speciation for some elements. The concentration of Mn^{2+} and Fe^{2+} in hypolimnetic waters and oxic estuaries has been determined by differential pulse polarography (DPP) by Davison (*74*) and Knox and Turner (*165*). Henry *et al.* (*146*) determined As(III) and total As by DPP before and after SO_2 reduction.

In a fashion analagous to ion-exchange chromatography, ASV is more commonly utilized in speciation studies to distinguish labile (ASV labile, "free," or electroactive) and bound (electroinactive) metal fractions (*20, 23, 101, 107, 170–173*). It must be stressed that this distinction is *operationally defined* and dependent upon specified instrumental and solution parameters. The ASV-labile component consists of free metal ions and some metal complexes (both organic and inorganic) which may themselves be directly reduced at the electrode surface. A minor contribution might be expected from metal pseudocolloids within the diffusion layer, that is, metal species desorbed from colloidal material (*103*). Also included would be kinetically labile species, the metal ions derived from the dissociation of inorganic and organic complexes within the diffusion layer. As discussed by Davison (*75*), this contribution is dependent upon neither the pre-electrolysis time nor the thermodynamic stability of the complex but rather upon the dissociation rate of the complex relative to the time scale of the measurement. Figura and McDuffie (*98, 99*) utilize ASV to identify "very labile" metal species in their scheme based upon the metal-binding spectrum.

The ASV-labile metal concentration is usually determined by standard addition rather than comparison with a standard calibration curve because the peak current is sensitive to variations in the ionic strength, pH, and salt matrix composition. Furthermore, speciation may vary greatly even between samples of comparable inorganic composition due to organic complexation. Metal titration curves of environmental samples generally assume a bilinear character, as illustrated in Fig. 4, due to the presence of naturally occurring chelators. The initial shallow slope results from the partial sequestration of the metal spike and is used to determine the stability constants of metal complexes. Once the complexing capacity of the sample is exceeded, all of the added metal remains electroactive and therefore the slope increases. This steep slope may be used to quantify the initial electroactive metal concentration. Though total metal concentrations may be determined using mixed metal titrants, ASV-labile fractions are best measured with single metal titrants due to competition for complexation sites.

This calibration technique presupposes that the added metal spike achieves equilibrium with the metal species present in the sample. Due to the kinetics of complexation, equilibration times may vary from sample to sample. During the consecutive analysis at 3-min intervals of copper in seawater at pH 8.1, Duinker and Kramer (*89*) observed a gradual decline in the peak current which only stabilized 20 min after the metal addition. Allowing insufficient time for equilibration may cause nonlinear calibration curves (*106*).

The determination of ASV-labile concentrations by metal titration assumes also that the added metal will contribute to a single peak, a further manifestation of equilibration. While some qualitative information concerning complexation may be gleaned, the concurrent growth of multiple peaks renders quantification impossible. As exemplified in Fig. 5, copper complexes may be particularly susceptible to such complexities (*138*). Multiplets may arise due to organic

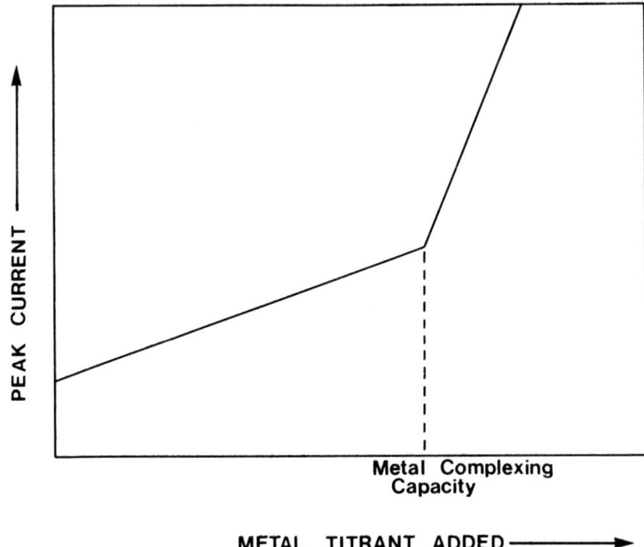

Fig. 4. An idealized metal titration curve obtained by anodic stripping voltametry. The breakpoint in the curve is a measure of the metal complexing capacity of the solution and the ASV-labile metal concentration is determined using the final steep slope.

fouling of the electrode, the presence of different ASV-labile complexes, or the oxidation of both Cu^{+} and Cu^{2+} from the mercury drop.

As emphasized previously, the ASV-labile metal concentration is an operationally defined entity. Substantial variations in technique, particularly with regard to solution parameters, can result in the measurement of different fractions which are similarly but unharmoniously defined. Several workers determined ASV-labile metal levels in a solution buffered with acetate at about pH 4.8 (*20, 23, 26, 58*). Florence and Batley (*108*) justify this approach suggesting that the acetate acts as both a pH buffer and supporting electrolyte. However, natural equilibria may be perturbed. Skogerboe *et al.* (*234*) point out that the acetate ion may effectively compete with other ligands and complexes may dissociate at the relatively low pH, compared with about 8 for natural waters. Similarly, the pH dependence of trace metal adsorption onto particles in natural waters is well understood (*153, 226, 269*). The decrease in pH could promote significant desorption thereby increasing "free" metal levels. Such deleterious effects may be prevented by analyzing solutions at natural pH levels (*37, 89, 171, 207*). The pH in poorly buffered natural waters may be controlled by sample deaeration with a suitably adjusted N_2/CO_2 gas mixture (*171*). Nürnberg and Raspor (*199*) control CO_2 levels in the gas mixture by bubbling Ar/CO_2 through a suspension of magnesium hydroxide carbonate in a borate buffer of pH 8. Finally, the use of

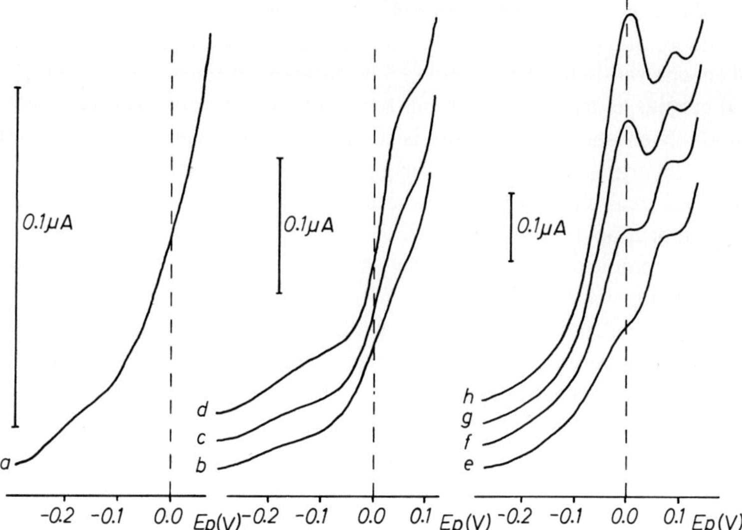

Fig. 5. Anodic stripping voltametric analysis of copper in <1-μm filtrate of a highway drainage sample. Standard additions (ppb) are as follows: *a*, 0; *b*, 25; *c*, 50; *d*, 100; *e*, 200; *f*, 300; *g*, 400; and *h*, 500. Reproduced with permission from Harrison and Wilson (*138*).

thin film mercury electrodes formed *in situ* (*20*) also has been criticized because the added Hg^{2+} competes for free ligands in solution thereby affecting metal speciation (*234*). The HMDE can be used as an alternative. Such variations in methodologies clearly illustrate that the ASV-labile concentration is an operationally defined entity.

The breakpoint in the metal titration curve, as indicated previously in Fig. 4, is a measure of the complexing capacity of the solution. Accordingly, ASV has often been used to evaluate metal binding capacities in natural waters. Such applications have been reviewed recently by Hart (*139*). The initial shallow slope yields information regarding stability constants (*207, 233*). This procedure assumes that the metal complex formed does not contribute to the peak stripping current. However, in some cases a kinetic contribution has been observed for which a correction must be applied in order to evaluate the stability constant (*231, 232*). Tuschall and Brezonik (*262*) report instances for which cadmium and copper organic complexes are reducible and therefore stability constants could not be determined.

Alternatively, complexation may be investigated by measuring the decrease in the ASV-labile metal concentration as a function of added chelator concentration. Ernst *et al.* (*94*) investigated copper and lead carbonates and lead humates by this technique. Similarly, complexation of nitrilotriacetic acid (NTA) with lead, cadmium, and zinc has been studied at environmental metal concentrations by Nürnberg and co-workers (*199, 200*). In this approach, the initial free ionic metal concentration must be calculated. Reasonable thermodynamic models exist to hypothesize inorganic speciation; however, the composition, concentration, and stability constants of naturally occurring organic chelators are poorly understood. Nürnberg and Raspor (*199*) suggest that organic chelates in Adriatic Sea samples are negligible and will not affect the calculated free ionic metal concentration. But several authors (*46, 109, 171, 259, 260*) have observed metal–organic complexation in natural waters indicating that this simplification is not universally valid.

Finally, considerable insight into trace metal speciation may be gained from "pseudopolarograms." Pseudopolarograms may be obtained by measuring ASV peak stripping currents (i_p) as a function of either the deposition potential or pH. In the first instance, the i_p for trace metals may increase with a progressively more negative deposition potential due to the reduction of an organometallic complex. This has been used to indentify impurities in commercially available NTA (*98*) and to analyze lead alkyls in environmental samples (*197a*). In obtaining pH pseudopolarograms, because the i_p is sensitive to pH changes, the electroactive metal concentration must be determined by standard addition at each pH. The ASV-labile metal concentration increases with decreasing pH (*89, 94, 114*). This may be due to the dissociation of metal complexes and desorption (or dissolution at low pH) from colloidal and particulate material. Fukai and Huynh-ngoc (*114*) have noted that comparable "labile" zinc concentrations were determined using Chelex at natural seawater pH and ASV measurements at pH 4. Such observations reinforce the conclusion that electroactive concentrations determined by ASV in acetate buffer at pH < 5 do not necessarily represent "labile" metal levels in waters at natural pH.

4. *Electrodeposition*

Electroactive metal concentrations may be determined using electrodeposition onto a graphite tube with subsequent analysis by atomic absorption spectrometry (EDAA). Preconcentration of the analyte together with the elimination of matrix interferences significantly improve the sensitivity of AAS. In a fashion similar to ASV, the speciation of some metals may be examined by differential electrodeposition under adjusted solution conditions or at different potentials.

Batley (*17*) measured electroactive cadmium and lead in seawater by AAS following an electrodeposition procedure. Metals could be electrodeposited *in situ* at natural pH and in the presence of oxygen. Frick and Tallman (*113*) described a flow cell for the EDAA analysis of mercury in river water. Methyl mercury was not deposited under an applied potential of -1.00 or -1.40 V. Inorganic mercury could be determined before and after sample digestion thereby yielding organomercury concentrations by difference.

Chromium redox speciation has been investigated using selective electrodeposition at different potentials by Batley and Matousek (*24*). Both Cr(III) and Cr(VI) are preconcentrated at a potential of −1.8 V vs SCE from solution at pH 4.7. However, at −0.3 V and pH 4.7, Cr(VI) is reduced to Cr(III) and accumulates by adsorption.

D. Species-Specific Techniques

1. *Ion-Selective Electrode*

The apparent simplicity in measuring "free" metal ion activity makes ion-selective electrodes (ISE) particularly attractive for speciation studies. But in practice, several difficulties limit their applicability for natural waters. Ion-selective electrodes are relatively insensitive; Cd and Pb ISE exhibit a non-Nernstian response below approximately 10^{-5} *M* (*117, 223*). The Cd ISE is subject to Cu contamination (*117*) and the Pb ISE is susceptible to air oxidation (*223*). Several electrodes may be fouled with organic material present in natural waters (*223*). Finally, the electrodes may exhibit poor selectivity (*102*). The Cu ISE can respond to inorganic copper complexes and organic sequestering agents thereby limiting its application as a species-specific technique.

For these reasons, ion-selective electrodes have seen only limited application in environmental studies. The copper ISE has been applied to both river water (*246*) and seawater (*161*). Complexation of Cd, Cu, Hg, and Pb in some natural waters has been investigated (*116, 122, 211*). The stability constants of Cd, Cu, and Pb humates and fulvates have been evaluated but usually utilizing metal concentrations significantly higher than those encountered in the environment (*47, 117, 128*).

2. *Gas/Liquid Chromatography*

A chromatographic technique coupled to a sensitive metal detector system provides a powerful and versatile tool to examine trace metal speciation in natural waters. Unlike the methods previously described which provide information regarding the partitioning or generic speciation of metals, chromatographic separations have been used to determine the concentrations of specific metal complexes. Such procedures are dependent upon either the existence of naturally occurring volatile metallic species (i.e., Hg, Pb) or the volatilization of elements by hydride generation (i.e., As, Bi, Ge, Pb, Sb, Se, and Sn). Lee (*176*) has recently described a comparable technique to analyze subnanogram quantities of nickel in seawater following reduction and volatilization as a carbonyl complex. Detection limits in the parts per billion (10^{-9}) range can be attained and thus several metal species may be determined at environmental levels. Furthermore, conditions may be adjusted to examine different oxidation states for some elements.

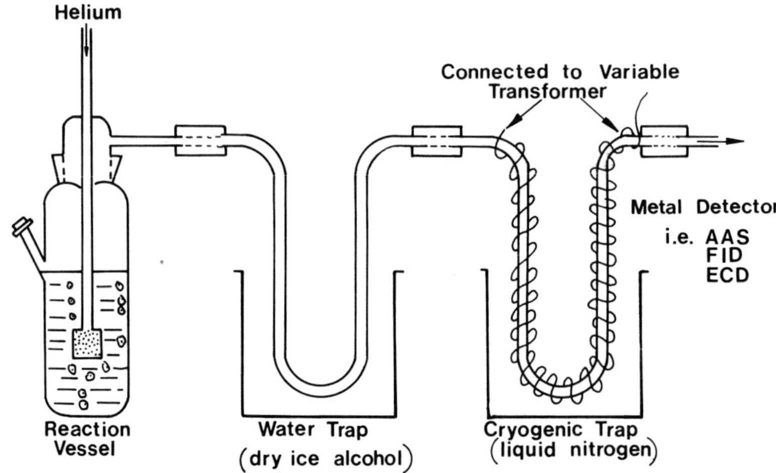

Fig. 6. Chromatographic equipment used for the separation of naturally occurring volatile metallic species or metal hydrides generated in the reaction vessel.

The analytical procedure involves three main processes: (1) production of volatile metallic species, (2) trapping and subsequent differential release of such species, and (3) metal detection. These three steps will be considered individually for the range of elements amenable to such techniques. A schematic outline of the equipment is illustrated in Fig. 6.

A proportion of arsenic, mercury, lead, and selenium in natural waters may exist as volatile species, usually fully alkylated complexes. In this case, the generation of volatile species is superfluous and the alkylated complexes may be removed from solution by scrubbing with helium (*10, 61, 62*). For other elements, reduction with sodium borohydride generates volatile hydrides and alkyl hydrides which are subsequently removed from the reaction vessel with helium. This procedure has been adopted for speciation studies of tin (*40, 149*), antimony (*11*), selenium (*220*), and arsenic (*10, 51*) as well as for the total determination of bismuth (*130*) and germanium (*12, 129, 130*). Tellurium also should be amenable to this technique.

Information regarding elements exhibiting more than one oxidation state can be obtained by utilizing selective reduction steps. Both As(III) and Sb(III) species are reduced with $NaBH_4$ at near neutral pH while As(V) and Sb(V) complexes are reduced at about pH 1 (*10, 11*). Similarly, whereas Se(IV) may be determined by hydride generation, Se(VI) will not be reduced by $NaBH_4$ and is determined by difference following both Se(IV) and total Se analyses (*220*).

The volatile species are often trapped cryogenically following the removal of water vapor. The U-tube traps are cooled in either liquid nitrogen or a dry

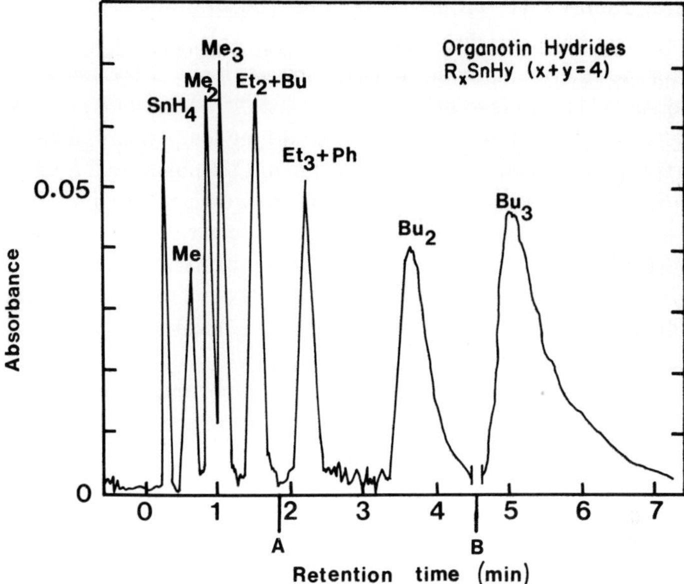

Fig. 7. Chromatogram of hydrides generated from a known mixture of Sn(IV) and nine organotin halides. Me, Methyl; Et, ethyl; Bu, *n*-butyl; and Ph, phenyl. Note that Et_2SnH_2 and $BuSnH_3$ have the same retention time as do Et_3SnH and $PhSnH_3$. Approximate concentrations of the reactants that produced this chromatogram are Sn(IV), 6 ng; $MeSnCl_3$, 14 ng; Me_2SnCl_2, 14 ng; Me_3SnCl, 18 ng; Et_2SnCl_2, 33 ng; Et_3SnBr or $PhSnCl_3$, 40 ng; Bu_2SnCl_2, 110 ng; and Bu_3SnCl, 470 ng. Chromatogram terminated at A with water trap immersed in dry ice/2-propanol. Tri-*n*-butyltin hydride is released from the hydride trap after di-*n*-butyltin dihydride if trap immersed in 80°C water bath (B). Reproduced with permission from Hodge *et al.* (*149*).

ice–methanol mixture and generally packed with either Chromosorb W or silanized chromatographic quality glass wool. Subsequently, the trap is warmed and the separation of the metallic species is achieved due to differences in their boiling points. Retention times are recorded subsequent to time $t = 0$, the elution of the metal hydride, as illustrated in Fig. 7. In some instances a conventional gas chromatographic separation may be utilized (*78, 193, 227*).

The greatest variability in these organometallic analyses arises from the variation and optimization of the metal detection system. Atomic absorption spectrometry provides an extremely sensitive detector and the utilization of such systems has been reviewed by Fernandez (*96*) and Van Loon (*263*). Coker (*66*) and Chau *et al.* (*63*) used GC–flame AAS to determine tetraalkyllead concentrations in gasoline and air, respectively. Using a hydrogen/air mixture, a quartz cuvette burner has been developed to determine arsenic (*10, 132*), tin (*59, 149, 257*), and antimony (*11, 132*).

Several flameless AAS techniques have been investigated in order to attain greater sensitivity. Chau *et al.* (*61, 62*) developed a silica furnace heated to approximately 1000°C for Pb and Se analyses. A graphite furnace AAS has been utilized for the determination of Mn (*65*), Cr and Pb (*78, 227*), Sb (*11, 132*), Ge (*12*), and As (*51*). Haring *et al.* (*132*) proposed the addition of a nickel solution to arsenic samples to allow a relatively high ashing temperature, thereby removing interfering substances. Roden and Tallman (*220*) also used a GC–graphite furnace AAS interface to determine Se in groundwater. Prior to analysis the solution was passed through an XAD-8 anion-exchange column at pH 1.6–1.8 to remove organic material, presumed to be humic in nature, that suppressed the release of selenium hydride. Hatch and Ott (*143*) described a quartz cold vapor trap for the flameless AAS analysis of mercury. Gonzalez and Ross (*124*) interfaced a GC with a similar detector to determine alkylated mercury levels in fish tissue. Oda and Ingle (*204*) also used a cold vapor trap to examine mercury speciation following the selective reduction of inorganic and organic mercury complexes by $SnCl_2$ and $NaBH_4$, respectively.

Several detector systems other than AAS have been examined. Braman and Tompkins (*40*) analyzed tin species in a hydrogen/air flame by atomic emission spectroscopy. Hahn *et al.* (*130*) analyzed hydride derivatives of arsenic, bismuth, germanium, antimony, selenium, and tin using an inductively coupled argon plasma polychromator. Both flame ionization detectors (FID) and electron capture detectors (ECD) were investigated by Andreae (*10*) for the determination of arsenic complexes, the latter proving to be far more sensitive. Measures and Burton (*193*) also utilized a GC–ECD system for the analysis of selenium in seawater.

In contrast to the general procedure outlined above, some GC metal determinations incorporate an initial extraction step. Tetraalkyllead complexes in water, sediment, and fish samples were extracted into hexane prior to analysis by GC–AAS (*60*). Chau *et al.* (*59*) extracted tin(IV) and methylated tin(IV) species into a benzene/tropolone mixture. These compounds are then butylated via a Grignard reaction and the tetraalkylated derivatives analyzed by GC–AAS. Measures and Burton (*193*) determined dissolved selenium(IV) in seawater by reaction with 4-nitro-*o*-phenylenediamine forming 5-nitropiazselenol which was extracted into toluene and analyzed by GC–ECD. Total selenium was similarly determined following a photooxidation procedure, and thereby giving the concentration of selenium(VI) by difference.

3. *High-Performance Liquid Chromatography (HPLC)*

Recent advances in chromatography–AAS coupled systems have taken advantage of the enhanced resolution and speed available with high-pressure (performance) liquid chromatography (HPLC). Such applications have been reviewed by Fernandez (*96*) and cataloged by Horlick (*152*).

Considering first HPLC–flame AAS techniques, the column eluent is normally pumped directly into the flame, thereby allowing flow rates to be controlled by the HPLC program. Metal detection limits in flame AAS are solvent dependent and eluent mixtures may be varied to achieve maximum sensitivity while maintaining necessary peak resolution. Jones and Manahan (*162*) analyzed synthetic mixtures of several organochromium complexes and observed detection limits in the order of 40 ng per compound. Botre *et al.* (*38*) separated tetramethyllead (TML) and tetraethyllead obtaining a detection limit of 0.39 μg of TML per 1 μl injection without attempting to optimize the system. Detection limits for alkyltin complexes in a nitrous oxide/acetylene flame varied from 11 to 19 μg according to Thorburn-Burns *et al.* (*257*).

Greater sensitivity is possible with flameless AAS but generally noncontinuous chromatograms are obtained. The HPLC–AAS linkage may involve subsampling the column eluent (*42, 270*) or stopping the eluent flow for each analysis (*55*). Arsenic speciation in soil extracts, drinking waters, and synthetic solutions has been investigated and detection limits as low as 10 ppb As have been obtained (*42, 157, 270*). Thornburn-Burns *et al.* (*257*) also examined an HPLC/on line hydride generator/quartz tube pyrolysis system; however, the sensitivity was not as good as with the gas/liquid chromatography hydride techniques mentioned previously.

4. *Ion-Exchange Chromatography*

As discussed in Section VI,C,2, ion-exchange chromatography can provide considerable information about the generic speciation of trace elements in natural waters, including the identification of metal fractions with varying labilities. However, in some instances specific organometallic complexes may be isolated for subsequent analysis.

Baltisberger and Knudsen (*15*) determined nanogram quantities of mercury(I), mercury(II), and methylmercury(II), relying upon the differential elution of retained material from a cation-exchange resin. Samples were first reacted with isothiocyanatopentaaquochromium(III) to form $CH_3HgSCNCr^{3+}$, $CrNSCHg_2^{4+}$, and $(CrNCS)_2Hg^{6+}$. The derived polynuclear species were retained on Bio-Rad AG 50W-X8 resin. The methylmercury(II) and mercury(I) cationic species were eluted sequentially with 1.0 *M* $HClO_4$ while a mixture of 1.2 *M* HCl–1.8 *M* H_2SO_4 removed the mercury(II) complex.

Similarly, Henry and Thorpe (*147*) used differential elution procedures with both cation- and anion-exchange resins to examine arsenic speciation. An anion-exchange resin (Dowex 50W-X8 in the H form) removed inorganic As(III) and As(V) species together with monomethyl arsonate (MMA) and dimethyl arsonate (DMA). Whereas 0.02 *M* acetic acid removed inorganic forms and MMA, DMA was recovered for subsequent analysis by elution with 1.0 *M* NaOH. A cation-exchange resin (Bio-Rad AG 1-X8 in the acetate form) retained only MMA and As(V) species, which could subsequently be extracted sequentially with 0.1 *M* sodium acetate at pH 4.7. The arsenicals were determined by differential pulse polarography as As(III) following perchloric acid digestion and reduction with aqueous sulfur dioxide. Complementary data for As(III) and total As, and therefore As(V) by difference, could also be obtained by DPP before and after SO_2 reduction (*146*).

E. Comprehensive Speciation Schemes

Several species-specific techniques and methods for the investigation of physical and chemical generic speciation have been considered above. Only in rare instances is a single procedure utilized to investigate trace metals in natural waters. More commonly, a number of techniques are combined in order to elucidate the distribution of the total concentration of an element into the various component fractions, namely the physicochemical speciation. Several combinations of methods are possible and the resulting schemes will obviously differentiate particular suites of metal forms.

A number of comprehensive physicochemical speciation schemes will be discussed here together with examples of their application to environmental samples. As stressed previously, metal fractions may be operationally defined and this must be considered in comparing different schemes. As a general criticism of the schemes presently available, the individual ion activity for free metal ions cannot be evaluated. This is especially important since this may comprise the bioavailable, and hence the toxic, metal concentration. As a consequence, most schemes have been designed to isolate a trace metal fraction that closely approximates the free metal ion component.

1. Florence and Batley

The first comprehensive chemical speciation scheme was proposed by Batley and Florence (*20*). Dissolved Cd, Cu, and Pb (i.e., filtered through 0.45-μm membrane filter) were subdivided into seven categories according to their ASV and Chelex labilities. ASV-labile (in acetate buffer at pH 4.8) and total (ASV-labile following 0.16 *M* HNO_3 digestion) metal concentrations were determined in subsamples that had been (1) untreated, (2) UV irradiated, (3) passed through a Chelex resin, and (4) passed through a Chelex resin following UV irradiation.

The scheme has been applied to relatively unpolluted surface seawater collected near Sydney, Australia (*20*). Most of the Cd, Cu, and Pb was associated with ASV-labile organics but a significant proportion of the Cu and Pb was associated with the fraction that was neither ASV nor Chelex labile even after UV irradiation. This metal fraction was assumed to consist of inorganic complexes and metals adsorbed onto inorganics.

Estuarine waters also have been examined (*23*). A significant proportion of the lead (45–70%) and copper (40–60%) was associated with the colloidal matter. However, a smaller percentage of the cadmium (15–75%) was so bound. Florence (*101*) has investigated fresh waters. Lead and copper exhibited significant organic complexation whereas cadmium and zinc existed largely as ASV-labile (possibly ionic) forms. This speciation scheme cannot be universally applied to fresh waters. As mentioned in Section VI,C,1, UV irradiation may promote the oxidative precipitation of iron and manganese. The concurrent scavenging of other trace metals obscures speciation information.

As discussed in Section VI,C,3, ASV measurements in acetate buffer at pH 4.8 with TFE formed *in situ* have been criticized (*138, 234*). Differences in methodology arise due in part to varying applications of the data. ASV analysis at pH 4.8 cannot be extrapolated to natural pH in order to provide speciation information. However, Florence (*103*) promotes the use of ASV to determine the toxic fraction of a metal and notes that the ideal ASV pH must therefore be established by comparison with bioassay techniques. Further to the interest in bioavailable metal concentrations, recent innovations to the basic scheme outlined above include the use of oxine and thiol chelating resins and the determination of lipid-soluble metal fractions (*103, 110*).

2. *Hart and Davies*

A similar speciation scheme has been developed by Hart and Davies (*142*). Particulate cadmium, copper, iron, lead, and zinc are determined following pressure filtration through Nuclepore 0.4-μm membrane filters. The filterable (i.e., dissolved) metal fraction is subdivided into three categories. The dialyzable metal is determined following a combined dialysis/ion–exchange procedure (*140*). Ion-exchangeable and bound (i.e., difference between total dissolved and ion-exchangeable) metal levels are determined following a batch Chelex procedure (*141*). One major advantage of this speciation scheme is that its applicability is not limited to those metals that can be determined by ASV.

This scheme has been applied to three Australian lakes (*142*). The cadmium, copper, lead, and zinc in each case were predominantly in filterable forms and, in turn, largely ion exchangeable. This metal fraction is composed of free metal ions and simple inorganic and organic complexes including fulvates. Alternatively, the iron existed mainly in bound forms, probably iron oxyhydroxides, but possibly including metal organics with very high stability constants such as iron siderophores.

3. *Figura and McDuffie*

As discussed in Section VI,C,2, Figura and McDuffie (*97–99*) have devised a physicochemical speciation scheme based on the binding spectrum of metals. The insoluble (i.e., particulate) trace metal concentrations are determined by the difference between total metal analyses before and after filtration through Nuclepore 0.4-μm polycarbonate filters. Four metal fractions in the filtered sample are characterized: (1) "very labile"—the metal fraction determined by ASV in a solution buffered at pH 6.3 with acetate; (2) "moderately labile"—the sample is passed through a column of Ca-Chelex after which the retained metals are recovered by elution with nitric acid and a correction is applied to subtract the contribution of the ASV-labile metal fraction which would also have been retained; (3) "slowly labile"—the metal fraction is determined by a batch Ca-Chelex extraction (3 days) on the effluent from the column extraction; and (4) "inert"—the metal fraction not recovered during the batch Chelex procedure.

Figura and McDuffie (*99*) have examined the speciation of cadmium, copper, lead, and zinc in secondary sewage effluents, river waters, and an estuarine sample. The cadmium and zinc existed predominantly in the very and moderately labile fractions. Copper was found mainly in the moderately and slowly labile fractions and the lead was slowly labile with a significant inert contribution in some cases.

This scheme is particularly attractive because a range of dissociation rate constants can be estimated for species in each category. Modifications to the scheme could include the concurrent use of several types of chelating resins and possibly eliminating the need for an ASV-labile fraction which limits the applicability of the scheme to only a few trace metals.

4. Laxen and Harrison

In contrast to those methods outlined above, Laxen and Harrison (*171*) have designed a physicochemical speciation scheme in which the size distribution of metals is examined in conjunction with ASV and Ca-Chelex labilities. Total metal analyses are performed on untreated water samples following pressure filtration in parallel through five Nuclepore polycarbonate filters of the following nominal pore sizes: 12, 1, 0.4, 0.08, and 0.015 μm. Six size fractions are thereby defined of which two (1- and 0.08-μm filtrates) are examined in greater detail. Chelex-labile metal concentrations are determined following a 48-h batch extraction (*141*). ASV-labile metal concentrations are determined before and after UV irradiation to define a strongly associated organic contribution. As with other schemes, ASV measurements are limited to Cd, Cu, Pb, and Zn but Mn and Fe in the size-fractionated samples also are determined.

This scheme has been applied to river waters (*171*) and used to investigate the impact on metal speciation in the receiving river arising from the discharge of effluent from a sewage treatment works (*172*) and a lead-acid battery manufacturer (*173*). The sewage effluent increases concentrations of Cd, Cu, and Pb but does not affect Fe and Mn levels. The contributed Cd is predominantly in the $<$0.015-μm fraction while the lead is in the $>$12-μm fraction. In contrast,

TABLE XVIII

The Size Distributions of Lead in Various Tap Waters

	Bentham, first draw		Bentham, well flushed		Yarmouth, first draw	
Size fraction (μm)	ppb	Percentage of lead in sample	ppb	Percentage of lead in sample	ppb	Percentage of lead in sample
>12	42	20	1.3	25	8.0	26
1–12	7	3	0.5	10	10.0	32
0.4–1	2	1	0	0	3.0	10
0.08–0.4	20	10	0	0	0.3	1
0.015–0.08	24	12	2.4	47	2.9	9
<0.015	111	54	0.9	18	6.8	22
Total	206		5.1		31.0	

copper concentrations increase in both the <0.08 and the >12-μm size fractions. The battery effluent raises lead levels from 33 to 43 μg/liter, but again virtually all the added lead occurs in the particulate fraction > 12 μm.

This scheme has also been used to examine the physicochemical speciation of lead in tap waters (*81b, 133*). The size distribution of lead in potable waters may be particularly important in examining quality control of plumbosolvent waters and identifying the mechanism responsible for lead contamination. Elevated lead levels in the small size ranges (i.e., Table XVIII, first-draw Bentham tap water) may indicate some dissolution of lead piping or, more likely in this case, leaching of lead from soldered joints. Well-flushed samples will exhibit significantly lower lead levels. The lead associated with the larger size fractions may result from the removal of pipe scale (i.e., basic lead carbonate) in response to changes in pressure or pH. Alternatively, lead may be associated with colloidal particles due to coprecipitation or adsorption processes. de Mora and Harrison (*81b*) have observed that the size spectrum of lead correlates well with that of iron and aluminum for Scottish tap waters from Stirling and Glasgow, respectively.

This scheme also suffers from the difficulties generally associated with the UV irradiation of fresh waters. The oxidative precipitation of iron and concurrent trace metal scavenging preclude the determination of an organically associated metal fraction. However, direct evidence for lead organic complexes in 15-nm-filtered tap water has been obtained using gel filtration chromatography (*81b*). Two distinct peaks were resolved using Sephadex G-50 but the deleterious effects of adsorption were not eliminated. Further developments of this physicochemical speciation scheme include the use of poly(maleic anhydride) chelating resins to complement Chelex-labile measurements (*81a*).

VII. CONCLUSIONS

The investigation of physicochemical speciation of inorganic components in environmental media encompasses consideration of both the physical nature of the component (i.e., dissolved, colloidal, particulate) and chemical information as regards ion pair formation and complexation. The physicochemical speciation determines the environmental mobility and toxicity of a substance. Geochemical cycling and transfer rates will be influenced by the size spectrum of a constituent. For a given total amount of an element, the chemical form determines its bioavailability and hence its toxicity. Both physical and chemical characteristics can affect the efficiency of water and sewage treatment plants.

The techniques applicable to speciation studies have been reviewed. Two basic categories exist. First, some techniques may render species-specific information. Such is the case with X-ray diffraction whereby specific crystalline components may be identified and chromatography–atomic absorption spectrophotometry systems by which the concentration of individual alkylmetals may be determined. Second, some methods may provide information concerning the partitioning of an element into a readily identifiable and experimentally reproducible fraction in which the element may exist in a variety of unknown forms. This we have termed generic speciation and both physical and chemical procedures may be utilized. Physical techniques such as filtration and centrifugation will examine the size distribution of a substance whereas chemical methods such as the differential extraction of sediments or the Chelex extraction of trace metals identify operationally defined fractions.

Techniques for speciation of airborne particles concentrate on either macroscopic or microscopic properties. Macroscopic techniques examine the bulk properties of an aerosol sample. X-Ray powder diffraction has been utilized for the qualitative identification of lead and sulfate compounds as well as natural minerals. Microscopic techniques such as scanning electron microscopy with X-ray energy spectroscopy and laser microprobe mass spectrometry are used to examine the morphology and chemical composition of individual particles. The time-consuming nature of the technique restricts the number of particles in any sample that can be analyzed with the subsequent danger that the information obtained may not be representative of the sample as a whole. Particulate sulfates have been determined by solvent extraction and microtitration methods.

The generic speciation of trace metals in sediments, soils, and street dusts has often been examined by means of a sequential extraction scheme. The five phases into which trace metal constituents are generally differentiated include exchangeable, carbonate, Fe/Mn oxides, organic, and residual. Several reagents have been utilized to distinguish each of these phases. However, it must be appreciated that although considerable attention is paid to ensure the selectivity of a particular extraction technique, there is some degree of overlap among the fractions. This is true particularly when comparing results obtained using different fractionation schemes. Elemental lead and specific lead oxides and sulfates have been identified in street dusts and soils using X-ray powder diffraction.

Size fractionation techniques used in generic speciation studies of substances in natural waters have been reviewed. Centrifugation, filtration, ultrafiltration, dialysis, and gel filtration chromatography have been considered with respect to contamination, adsorption effects, and size selectivity. Filtration is the most commonly used size differentiation method, but generally only dissolved and particulate fractions are distinguished. The size distribution of components may be determined by parallel filtration through Nuclepore polycarbonate filters of various nominal pore sizes. Size fractionation during centrifugation is impeded due to density effects on particle settling rates. Dialysis may isolate species in true solution whereas gel filtration chromatography allows the continuous size resolution of colloidal material.

Chemical techniques for generic speciation distinguish nonspecific components that exhibit similar chemical behavior under a clearly designated procedure. Such operationally defined fractions include organically associated trace metals isolated by either solvent extraction or adsorbent resins such as Amberlite XAD. Ion-exchange chromatography can identify labile metal concentrations. Several chelating resins are available with different stability constants and their use in column and batch modes can provide kinetic information regarding the dissociation rate of metal complexes. Electrochemical methods can determine electroactive metal concentrations. Anodic stripping voltametry is widely employed due to its high sensitivity toward Cd, Cu, Pb, and Zn.

Though ion-selective electrodes have had limited application in environmental studies, several species-specific analytical schemes have made use of a chromatographic separation coupled to a detector, most often atomic absorption spectrophotometry. For instance, GC–AAS systems can resolve and quantify individual alkyl derivatives of As, Hg, Pb, and Se in atmospheric and aqueous samples. Elements susceptible to volatilization by hydride generation (As, Bi, Ge, Sb, Se, Sn) also may be speciated by such techniques. Incorporation of selective reduction steps may also render information regarding the oxidation states of the element of concern. Enhanced resolution and speed of analysis may be possible for some elements using high-performance liquid chromatography coupled to atomic absorption spectrophotometry.

Techniques for physical and chemical generic speciation investigations and species-specific analyses have been reviewed. The complete appreciation of the physicochemical speciation of a substance in an environmental sample generally requires the combination of several complementary procedures. Some comprehensive speciation schemes employed in the examination of trace metals in natural waters have been discussed in detail. Only with the elucidation of physicochemical speciation of substances will their chemical behavior in environmental media be fully understood.

References

1. Abdullah, M. I., El-rayis, O. A., and Riley, J. P. (1976). *Anal. Chim. Acta* **84,** 363–368.
2. Abdullah, M. I., and Royle, L. G. (1972). *Anal. Chim. Acta* **58,** 283–288.
3. Acher, A., Pistol, Y., and Yaron, B. (1981). *In* "Developments in Arid Ecology and Environmental Quality" (H. Shuval, ed.), pp. 211–220. Balaban ISS, Philadelphia.
4. Adams, F., Bloch, P., Natusch, D. F. S., and Surkyn, P. (1981). *Int. Conf. Environ. Pollut., Proc., Thessaloniki, Greece,* 122–142.
5. Ahlberg, M. S., Leslie, A. C. D., and Winchester, J. W. (1978). *Nucl. Instrum. Methods* **149,** 451–455.
6. Alberts, J. J., Schindler, J. E., Nutter, D. E., and Davis, E. (1976). *Geochim. Cosmochim. Acta* **40,** 369–372.
7. Alexander, J. E., and Corcoran, E. F. (1967). *Limnol. Oceanogr.* **12,** 236–242.
8. Allen, H. E., Hall, R. H., and Brisbin, T. D. (1980). *Environ. Sci. Technol.* **14,** 441–443.
9. Altgelt, K. H. (1968). *In* "Advances in Chromatography" (J. C. Gidding and R. A. Keller, eds.), Vol. 7, pp. 1–46. Dekker, New York.
10. Andreae, M. O. (1977). *Anal. Chem.* **49,** 820–823.
11. Andreae, M. O., Asmodé, J., Foster, P., and Van't dack, L. (1981). *Anal. Chem.* **53,** 1766–1771.
12. Andreae, M. O., and Froelich, P. N. (1981). *Anal. Chem.* **53,** 287–291.
13. Appel, B. R., Wall, S. M., Haik, M., Kothny, E. L., and Tokiwa, Y. (1980). *Atmos. Environ.* **14,** 559–563.
14. Astruc, M., Lecomte, J., and Mericam, P. (1981). *Environ. Technol. Lett.* **2,** 1–8.
15. Baltisberger, R. J., and Knudsen, C. L. (1975). *Anal. Chem.* **47,** 1402–1406.
16. Barbaray, B., Contour, J. P., Mouvier, G., Barde, R., Maffiolo, G., and Millancourt, B. (1979). *Environ. Sci. Technol.* **13,** 1530–1532.
17. Batley, G. E. (1981). *Anal. Chim. Acta* **124,** 121–129.
18. Batley, G. E., and Farrar, Y. J. (1978). *Anal. Chim. Acta* **99,** 283–292.
19. Batley, G. E., and Florence, T. M. (1974). *J. Electroanal. Chem.* **55,** 23–43.
20. Batley, G. E., and Florence, T. M. (1976). *Anal. Lett.* **9,** 379–388.
21. Batley, G. E., and Florence, T. M. (1976). *J. Electroanal. Chem.* **72,** 121–126.
22. Batley, G. E., and Gardner, D. (1977). *Water Res.* **11,** 745–756.
23. Batley, G. E., and Gardner, D. (1978). *Estuarine Coastal Mar. Sci.* **7,** 59–70.
24. Batley, G. E., and Matousek, J. P. (1980). *Anal. Chem.* **52,** 1570–1574.
25. Beneš, P., Gjessing, E. T., and Steinnes, E. (1976). *Water Res.* **10,** 711–716.
26. Beneš, P., Koc, J., and Štulík, K. (1979). *Water Res.* **13,** 967–975.
27. Beneš, P., and Steinnes, E. (1974). *Water Res.* **8,** 947–953.
28. Beneš, P., and Steinnes, E. (1975). *Water Res.* **9,** 741–749.
29. Biggins, P. D. E., and Harrison, R. M. (1978). *Nature (London)* **272,** 531–532.
30. Biggins, P. D. E., and Harrison, R. M. (1979). *Atmos. Environ.* **13,** 1213–1216.
31. Biggins, P. D. E., and Harrison, R. M. (1979). *Environ. Sci. Technol.* **13,** 558–565.
32. Biggins, P. D. E., and Harrison, R. M. (1977). *J. Air Pollut. Control Assoc.* **29,** 838–840.
33. Biggins, P. D. E., and Harrison, R. M. (1980). *Environ. Sci. Technol.* **14,** 336–339.

34. Birch, J., Harrison, R. M., and Laxen, D. P. H. (1980). *Sci. Total Environ.* **14,** 31–42.
35. Blatt, W. F., Hudson, B. G., Robinson, S. M., and Zipilivan, E. M. (1967). *Nature* (*London*) **216,** 511–513.
36. Bloch, P., Adam, F., Van Landuyt, J., and Van Goethem, L. (1979). *Physico-chem. Beh. Atmos. Pollutants, Proc. Eur. Symp.*, 307–321.
37. Blutstein, H., and Smith, J. D. (1978). *Water Res.* **12,** 119–125.
38. Botre, C., Cacace, F., and Cozzani, R. (1976). *Anal. Lett.* **9,** 825–830.
39. Braman, R. S., and Johnson, D. L. (1974). *Environ. Sci. Technol.* **8,** 996–1003.
40. Braman, R. S., and Tompkins, M. A. (1978). *Anal. Chem.* **51,** 12–19.
41. Brezonik, P. L., Brauner, P. A., and Stumm, W. (1976). *Water Res.* **10,** 605–612.
42. Brinckman, F. E., Jewett, K. L., Iverson, W. P., Irgolic, K. J., Ehrhardt, K. C., and Stockton, R. A. (1980). *J. Chromatogr.* **191,** 31–46.
43. Brosset, C. (1978). *Atmos. Environ.* **12,** 25–38.
44. Brosset, C., and Ferm, M. (1978). *Atmos. Environ.* **12,** 909–916.
45. Brown, M. J., and Lester, J. N. (1979). *Water Res.* **13,** 817–837.
46. Buffle, J., Deladoey, P., and Haerdi, W. (1978). *Anal. Chim. Acta* **101,** 339–357.
47. Buffle, J., Greter, F., and Haerdi, W. (1977). *Anal. Chem.* **49,** 216–222.
48. Buffle, J., Greter, F. L., Nembrini, G., Paul, J., and Haerdi, W. (1976). *Z. Anal. Chem.* **282,** 339–350.
49. Burnham, A. K., Calder, G. V., Fritz, J. S., Junk, G. A., Svec, H. J., and Willis, R. (1972). *Anal. Chem.* **44,** 139–142.
50. Burrell, D. C. (1974). "Atomic Spectrometric Analysis of Heavy-metal Pollutants in Water," p. 331. Ann Arbor Sci. Publ., Ann Arbor, Michigan.
51. Burton, J. D., Maher, W. A., Measures, C. I., and Statham, P. J. (1980). *Thalassia Jugosl.* **16,** 155–164.
52. Butler, J. D., MacMurdo, S. D., and Stewart, C. J. (1976). *Int. J. Environ. Stud.* **9,** 93–103.
53. Butterworth, F. E., and Alloway, B. J. (1981). *Heavy Metals Environ., Int. Conf., 3rd, 1981,* 713–716.
54. Calvert, S. E., and McCartney, M. J. (1979). *Limnol. Oceanogr.* **24,** 532–536.
55. Cantillo, A. Y., and Segar, D. A. (1977). *Int. Conf. Heavy Metals Environ.* (*Symp. Proc.*), *1st, 1975,* 183–204.
56. Chakoumakos, C., Russo, R. C. and Thurston, R. V. (1979). *Environ. Sci. Technol.* **13,** 213–219.
57. Chamberlain, A. C., Heard, M. J., Little, P., Newton, D., Wells, A. C., and Wiffin, R. D. (1978). "U.K.A.E.A. Report," AERE-9198. HM Stationery Office, London.
58. Chau, Y. K., and Lum-Shue-Chan, K. (1974). *Water Res.* **8,** 383–388.
59. Chau, Y. K., Wong, P. T. S., and Bengert, G. A. (1982). *Anal. Chem.* **54,** 246–249.
60. Chau, Y. K., Wong, P. T. S., Bengert, G. A., and Kramar, O. (1979). *Anal. Chem.* **51,** 186–188.
61. Chau, Y. K., Wong, P. T. S., and Goulden, P. D. (1975). *Anal. Chem.* **47,** 2279–2281.
62. Chau, Y. K., Wong, P. T. S., and Goulden, P. D. (1977). *Int. Conf. Heavy Metals Environ.* (*Symp. Proc.*), *1st, 1975,* 295–302.
63. Chau, Y. K., Wong, P. T. S., and Saitoh, H. (1976). *J. Chromatogr. Sci.* **14,** 162–164.
64. Clem, R. G., and Hodgson, A. T. (1978). *Anal. Chem.* **50,** 102–110.
65. Coe, M., Cruz, R., and Van Loon, J. C. (1980). *Anal. Chim. Acta* **120,** 171–176.
66. Coker, D. T. (1975). *Anal. Chem.* **47,** 386–389.
67. Copeland, T. R., Osteryoung, R. A., and Skogerboe, R. K. (1974). *Anal. Chem.* **46,** 2093–2097.
68. Ćosović, B., Degobbis, D., Bilinski, H., and Branica, M. (1982). *Geochim. Cosmochim. Acta* **46,** 151–158.

69. Cranston, R. E., and Buckley, D. E. (1972). "Bedford Institute Report," BI-2-72-7, p. 14. Bedford Institute of Oceanography, Bedford, Nova Scotia, Canada.
70. Dasgupta, P. K., Lundquist, G. L., and West, P. W. (1979). *Atmos. Environ.* **13,** 767–774.
71. Davis, B. L. (1978). *Atmos. Environ.* **12,** 2403–2406.
72. Davis, B. L. (1981). *Atmos. Environ.* **15,** 613–618.
73. Davis, B. L., and Cho, N. K. (1977). *Atmos. Environ.* **11,** 73–85.
74. Davison, W. (1977). *Limnol. Oceanogr.* **22,** 746–753.
75. Davison, W. (1978). *J. Electroanal. Chem.* **87,** 395–404.
76. Davison, W., and Whitfield, M. (1977). *J. Electroanal. Chem.* **75,** 763–789.
77. De Jonghe, W., Chakraborti, D., and Adams, F. C. (1980). *Anal. Chem.* **52,** 1974–1977.
78. De Jonghe, W., Chakraborti, D., and Adams, F. C. (1980). *Anal. Chim. Acta* **115,** 89–101.
79. De Jonghe, W., Chakraborti, D., and Adams, F. C. (1981). *Environ. Sci. Technol.* **15,** 1217–1222.
80. De Jonghe, W., Jiang, S., and Adams, F. C. (1981). *Int. Conf. Environ. Pollut. Proc., Thessaloniki, Greece,* 183–189.
81. de Mora, S. J., and Harrison, R. M. (1983). *Water Res.* **17,** 723–733.
81a. de Mora, S. J., and Harrison, R. M. (1983). *Anal. Chim. Acta.* **153,** 307–311.
81b. de Mora, S. J., and Harrison, R. M. (1983). *Heavy Metals Environ. Int. Conf. 4th, 1983,* 1207–1210.
82. Denoyer, E., Van Grieken, R., Adams, F., and Natusch, D. F. S. (1982). *Anal. Chem.* **54,** 26A–32A.
83. Deratani, A., and Sebille, B. (1981). *Anal. Chem.* **53,** 1742–1746.
84. Deratani, A., and Sebille, B. (1981). *Makromol. Chem.* **182,** 1875–1888.
85. Determan, H. (1969). *In* "Advances in Chromatography" (J. C. Gidding and R. A. Keller, eds.), Vol. 8, pp. 1–45. Dekker, New York.
86. Dingman, J. F., Gloss, K. M., Milano, E. A., and Siggia, S. (1974). *Anal. Chem.* **46,** 774–777.
87. Dingman, J. F., Siggia, S., Barton, C., and Hiscock, K. B. (1972). *Anal. Chem.* **44,** 1351–1357.
88. D'Ottavio, T., Garber, R., Tanner, R. L., and Newman, L. (1981). *Atmos. Environ.* **15,** 197–203.
89. Duinker, J. C., and Kramer, C. J. M. (1977). *Mar. Chem.* **5,** 207–228.
90. Duinker, J. C., Nolting, R. F., and Van Der Sloot, H. A. (1979). *Neth. J. Sea Res.* **13,** 282–297.
91. Dzubay, T. G., Snyder, G. K., Reutter, D. J., and Stevens, R. K. (1979). *Atmos. Environ.* **13,** 1209–1213.
92. Eatough, D. J., Eatough, N. L., Hill, M. W., Mangelson, N. F., Ryder, J., Hansen, L. D., Meisenheimer, R. G., and Fischer, J. W. (1979). *Atmos. Environ.* **13,** 489–506.
93. Eatough, D. J., Izatt, S., Ryder, J., and Hansen, L. D. (1978). *Environ. Sci. Technol.* **12,** 1276–1279.
94. Ernst, R., Allen, H. E., and Mancy, K. H. (1975). *Water Res.* **9,** 969–979.
95. Etz, E. S., and Rosasco, G. J. (1976). *NBS Special Publ.* **464,** 343–346.
96. Fernandez, F. J. (1977). *At. Absorpt. Newsl.* **16,** 33–36.
97. Figura, P., and McDuffie, B. (1977). *Anal. Chem.* **49,** 1950–1953.
98. Figura, P., and McDuffie, B. (1979). *Anal. Chem.* **51,** 120–125.
99. Figura, P., and McDuffie, B. (1980). *Anal. Chem.* **52,** 1433–1439.
100. Fischer, L. (1969). "An Introduction to Gel Chromatography." North-Holland Publ., Amsterdam.
101. Florence, T. M. (1977). *Water Res.* **11,** 681–687.
102. Florence, T. M. (1982). *Talanta* **29,** 345–364.

103. Florence, T. M. (1982). *Anal. Chim. Acta.* **141,** 73–94.
104. Florence, T. M., and Batley, G. E. (1975). *Talanta* **22,** 201–204.
105. Florence, T. M., and Batley, G. E. (1976). *Talanta* **23,** 179–186.
106. Florence, T. M., and Batley, G. E. (1977). *J. Electroanal. Chem.* **75,** 791–798.
107. Florence, T. M., and Batley, G. E. (1977). *Talanta* **24,** 151–158.
108. Florence, T. M., and Batley, G. E. (1980). *Anal. Chem.* **52,** 1962–1963.
109. Florence, T. M., and Batley, G. E. (1980). *CRC Crit. Rev. Anal. Chem.* **9,** 219–296.
110. Florence, T. M. and Batley, G. E. (1981). *Heavy Metals Environ., Int. Conf., 3rd, 1981,* 599–602.
111. Förstner, U., and Salomons, W. (1980). *Environ. Technol. Lett.* **1,** 494–505.
112. Foster, R. L., and Lott, P. F. (1980). *Environ. Sci. Technol.* **14,** 1240–1244.
113. Frick, D. A., and Tallman, D. E. (1982). *Anal. Chem.* **54,** 1217–1219.
114. Fukai, R., and Huynh-ngoc, L. (1975). *J. Oceanogr. Soc. Japan* **31,** 179–191.
115. Fukasawa, T., Iwatsuki, M., Kawakubo, S., and Miyazaki, K. (1980). *Anal. Chem.* **52,** 1184–1187.
116. Gamble, D. S., Underdown, A. W., and Langford, C. H. (1980). *Anal. Chem.* **52,** 1901–1908.
117. Gardiner, J. (1974). *Water Res.* **8,** 23–30.
118. Gardiner, J., and Stiff, M. J. (1975). *Water Res.* **9,** 517–523.
119. Gardner, M. J. (1982). "WRC Technical Report," TR-172, p. 40. Water Research Centre, Medmenham, England.
120. Gardner, M. J., and Hunt, D. T. E. (1981). *Analyst* **106,** 471–474.
121. Gardner, W. D. (1977). *Limnol. Oceanogr.* **22,** 764–768.
122. Giesy, J. P., Leversee, G. L., and Williams, D. R. (1977). *Water Res.* **11,** 1013–1020.
123. Goldberg, E. D., Baker, M., and Fox, D. L. (1952). *J. Mar. Res.* **11,** 194–202.
124. Gonzalez, J. G., and Ross, R. T. (1972). *Anal. Lett.* **5,** 683–694.
125. Grasshoff, K. (1976). "Methods of Seawater Analysis," p. 317. Verlag Chemie, Weinheim.
126. Guy, R. D., and Bourque, C. (1981). *Heavy Metals Environ., Int. Conf., 3rd, 1981,* 577–580.
127. Guy, R. D., and Chakrabarti, C. L. (1977). *Int. Conf. Heavy Metals Environ. (Symp. Proc.), 1st, 1975,* 275–294.
128. Guy, R. D., and Chakrabarti, C. L. (1976). *Can. J. Chem.* **54,** 2600–2611.
129. Hahn, M. H., Mulligan, K. J., Jackson, M. E., and Caruso, J. A. (1980). *Anal. Chim. Acta* **118,** 115–122.
130. Hahn, M. H., Wolnik, K. A., Fricke, F. L., and Caruso, J. A. (1982). *Anal. Chem.* **54,** 1048–1052.
131. Hancock, S., and Slater, A. (1975). *Analyst* **100,** 422–429.
132. Haring, B. J. A., van Delft, W., and Bom, C. M. (1982). *Fresenius Z. Anal. Chem.* **310,** 217–223.
133. Harrison, R. M., and Laxen, D. P. H. (1980). *Nature (London)* **286,** 791–793.
134. Harrison, R. M., Laxen, D. P. H., and Wilson, S. J. (1981). *Environ. Sci. Technol.* **15,** 1378–1383.
135. Harrison, R. M., and Perry, R. (1977). *Atmos. Environ.* **11,** 847–852.
136. Harrison, R. M., Perry, R., and Slater, D. H. (1974). *Atmos. Environ.* **8,** 1187–1194.
137. Harrison, R. M., Williams, C. R., and O'Neill, I. K. (1981). *Environ. Sci. Technol.* **15,** 1197–1204.
138. Harrison, R. M., and Wilson, S. J. (1982). *Pergamon Ser. Environ. Sci.* **7,** 301–314.
139. Hart, B. T. (1981). *Environ. Technol. Lett.* **2,** 95–110.
140. Hart, B. T., and Davies, S. H. R. (1977). *Aust. J. Mar. Freshwater Res.* **28,** 105–112.
141. Hart, B. T., and Davies, S. H. R. (1977). *Aust. J. Mar. Freshwater Res.* **28,** 397–402.
142. Hart, B. T., and Davies, S. H. R. (1981). *Aust. J. Mar. Freshwater Res.* **32,** 175–189.

143. Hatch, W. R., and Ott, W. L. (1968). *Anal. Chem.* **40,** 2085–2087.
144. Heidel, R. H., and Desborough, G. A. (1975). *Environ. Pollut.* **8,** 185–191.
145. Henriques, A., Isberg, J., and Kjellgren, D. (1973). *Chem. Scr.* **4,** 139–142.
146. Henry, F. T., Kirch, T. O., and Thorpe, T. M. (1979). *Anal. Chem.* **51,** 215–218.
147. Henry, F. T., and Thorpe, T. M. (1980). *Anal. Chem.* **52,** 80–83.
148. Hercules, S. H., and Hercules, D. M. (1975). *Int. J. Environ. Anal. Chem.* **4,** 155–166.
149. Hodge, V. F., Seidel, S. L., and Goldberg, E. D. (1979). *Anal. Chem.* **51,** 1256–1259.
150. Hoffmann, M. R., Yost, E. C., Eisenreich, S. J., and Maier, W. J. (1981). *Environ. Sci. Technol.* **15,** 655–661.
151. l'Hopitault, J. C., Philippo, A., Pommery, J., Thomas, P., and Erb, F. (1981). *J. Fr. Hydrol.* **12,** 7–27.
152. Horlick, G. (1981). *Anal. Chem.* **54,** 176R–293R.
153. Huang, C. P., Elliot, H. A., and Ashmead, R. M. (1977). *J. Water Pollut. Control Fed.* **49,** 745–756.
154. Hunt, D. T. E. (1979). "WRC Technical Report," TR-104, p. 17. Water Research Centre, Medmenham, England.
155. Huygen, C. (1975). *Atmos. Environ.* **9,** 315–319.
156. Hydes, D. J., and Liss, P. S. (1977). *Estuarine Coastal Mar. Sci.* **5,** 755–769.
157. Iadevaia, R., Aharonson, N., and Woolson, E. (1980). *J. Assoc. Off. Anal. Chem.* **63,** 742–746.
158. Jackson, G. A., and Morgan, J. J. (1978). *Limnol. Oceanogr.* **23,** 268–282.
159. Jacobsen, E., and Lindseth, H. (1976). *Anal. Chim. Acta* **86,** 123–127.
160. Jaklevic, J. M., Kirby, J. A., Ramponi, A. J., and Thompson, A. C. (1980). *Environ. Sci. Technol.* **14,** 437–441.
161. Javinski, R., Trachtenberg, I., and Andrychuk, D. (1974). *Anal. Chem.* **46,** 364–369.
162. Jones, R. R., and Manahan, S. E. (1975). *Anal. Lett.* **8,** 569–574.
163. Kennedy, V. C., and Zellweger, G. W. (1974). *Wat. Resour. Res.* **10,** 785–790.
164. Keyser, T. R., Natusch, D. F. S., Evans, C. A., and Linton, R. W. (1978). *Environ. Sci. Technol.* **12,** 768–773.
165. Knox, S., and Turner, D. R. (1980). *Estuarine Coastal Mar. Sci.* **10,** 317–324.
166. Kremmer, T., and Boross, L. (1979). "Gel Chromatography," p. 299. Wiley, New York.
167. Kwak, J. C. T., Nelson, R. W. P., and Gamble, D. S. (1977). *Geochim. Cosmochim. Acta* **41,** 993–996.
168. Laveskog, A. (1970). *Proc. Int. Clean Air Congr., 2nd, 1970,* 549–557.
169. Laxen, D. P. H., and Chandler, I. M. (1982). *Anal. Chem.* **54,** 1350–1355.
170. Laxen, D. P. H., and Harrison, R. M. (1981). *Anal. Chem.* **53,** 345–350.
171. Laxen, D. P. H., and Harrison, R. M. (1981). *Sci. Total Environ.* **19,** 59–82.
172. Laxen, D. P. H., and Harrison, R. M. (1981). *Water Res.* **15,** 1053–1065.
173. Laxen, D. P. H., and Harrison, R. M. (1983). *Water Res.* **17,** 71–80.
174. Leahy, D., Siegel, R., Klotz, P., and Newman, L. (1975). *Atmos. Environ.* **9,** 219–229.
175. Lecomte, J., Mericam, P., and Astruc, M. (1981). *Heavy Metals Environ., Proc. Int. Conf., 3rd, 1981,* 678–681.
176. Lee, D. S. (1982). *Anal. Chem.* **54,** 1182–1184.
177. Lewis, A. G., Ramnarine, A., and Evans, M. S. (1971). *Mar. Biol.* **11,** 1–4.
178. Leyden, D. E., and Luttrell, G. H. (1975). *Anal. Chem.* **47,** 1612–1617.
179. Leyden, D. E., Luttrell, G. H., Nonidez, W. K., and Werho, D. B. (1976). *Anal. Chem.* **48,** 67–70.
180. Leyden, D. E., Patterson, T. A., and Alberts, J. J. (1975). *Anal. Chem.* **47,** 733–735.
181. Linton, R. W., Natusch, D. F. S., Solomon, R. L., and Evans, C. A. (1980). *Environ. Sci. Technol.* **14,** 159–164.

182. Liu, C. Y., and Sun, P. J. (1981). *Anal. Chim. Acta* **132,** 187–193.
183. Loring, D. H., and Rantala, R. T. T. (1977). Fisheries and Marine Service Tech. Rep. No. 700, pp. 58. Bedford Institute of Oceanography, Bedford, Nova Scotia, Canada.
184. Lott, P. F., and Foster, R. L. (1977). *NBS Spec. Publ.* **464,** 351–366.
185. Macchi, G., and Pettine, M. (1980). *Environ. Sci. Technol.* **14,** 815–818.
186. Mackey, D. J. (1982). *Mar. Chem.* **11,** 169–181.
187. Maddalone, R. F., Thomas, R. L., and West, P. W. (1979). *Environ. Sci. Technol.* **10,** 162–169.
188. Mark, H. B., and Mattson, J. S. (1981). "Water Quality Measurement: The Modern Analytical Techniques," p. 485. Dekker, New York.
189. Mart, L. (1979). *Z. Anal. Chem.* **296,** 350–357.
190. Marvin, K. T., Proctor, R. R., and Neal, R. A. (1970). *Limnol. Oceanogr.* **15,** 320–325.
191. McKnight, D. M., and Morel, F. M. M. (1979). *Limnol. Oceanogr.* **24,** 823–837.
192. McKnight, D. M., and Morel, F. M. M. (1980). *Limnol. Oceanogr.* **25,** 62–71.
193. Measures, C. I., and Burton, J. D. (1980). *Anal. Chim. Acta* **120,** 177–186.
194. Muzzarelli, R. A. A., and Rocchetti, R. (1973). *Anal. Chim. Acta* **64,** 371–379.
195. Muzzarelli, R. A. A., and Rocchetti, R. (1974). *Anal. Chim. Acta* **69,** 35–42.
196. Muzzarelli, R. A. A., and Rocchetti, R. (1974). *Anal. Chim. Acta* **70,** 283–289.
197. Muzzarelli, R. A. A., and Sipos, L. (1971). *Talanta* **18,** 853–858.
197a. Noden, F., and Hodges, D. J. (1979). *Heavy Metals Environ., Proc. Int. Conf. 2nd, 1979,* 408–411.
198. Nürnberg, H. W. (1977). *Electrochim. Acta* **22,** 935–949.
199. Nürnberg, H. W., and Raspor, B. (1981). *Environ. Technol. Lett.* **2,** 457–483.
200. Nürnberg, H. W., Valenta, P., Mart, L., Raspor, P., and Sipos, L. (1976). *Z. Anal. Chem.* **282,** 357–367.
201. O'Connor, B. H., and Jaklevic, J. M. (1980). *X-Ray Spectrom.* **9,** 60–65.
202. O'Connor, B. H., and Jaklevic, J. M. (1981). *Atmos. Environ.* **15,** 19–22.
203. O'Connor, B. H., and Jaklevic, J. M. (1981). *Atmos. Environ.* **15,** 1681–1690.
204. Oda, C. E., and Ingle, J. D. (1981). *Anal. Chem.* **53,** 2305–2309.
205. Olson, K. W., and Skogerboe, R. K. (1975). *Environ. Sci. Technol.* **9,** 227–230.
206. O'Neill, I. K., Harrison, R. M., and Williams, C. R. (1982). *Trans. Inst. Min. Metall.* **91,** C84–C90.
207. O'Shea, T. A., and Mancy, K. M. (1976). *Anal. Chem.* **48,** 1603–1607.
208. Pankow, J. F., Leta, D. P., Lin, J. W., Ohl, S. E., Shum, W. P., and Janauer, G. E. (1977). *Sci. Total Environ.* **7,** 17–26.
209. Phillips, R. J., and Fritz, J. S. (1978). *Anal. Chem.* **50,** 1504–1508.
210. Radzuik, B., Thomassen, Y., Van Loon, J. C., and Chau, Y. K. (1979). *Anal. Chim. Acta* **105,** 255–262.
211. Ramamoorthy, S., and Kushner, D. J. (1975). *J. Fish. Res. Board Can.* **32,** 1755–1766.
212. Reamer, D. C., Zoller, W. H., and O'Haver, T. C. (1978). *Anal. Chem.* **50,** 1449–1453.
213. Rendell, P. S., Batley, G. E., and Cameron, A. J. (1980). *Environ. Sci. Technol.* **14,** 314–318.
214. Riley, J. P. (1975). *In* "Chemical Oceanography" (J. P. Riley and G. Skirrow, eds.), Vol. 3 (2nd Ed.), pp. 193–514. Academic Press, New York.
215. Riley, J. P., and Taylor, D. (1968). *Anal. Chim. Acta* **40,** 479–485.
216. Riley, J. P., and Taylor, D. (1968). *Anal. Chim. Acta* **41,** 175–178.
217. Riley, J. P., and Taylor, D. (1969). *Anal. Chim. Acta* **46,** 307–309.
218. Riley, J. P., and Taylor, D. (1972). *Deep-Sea Res.* **19,** 307–317.
219. Robbe, D., Marchandise, P., Baudet, D., and Magnin, A. (1980). *Environ. Technol. Lett.* **1,** 283–290.

220. Roden, D. R., and Tallman, D. E. (1982). *Anal. Chem.* **54,** 307–309.
221. Rosen, H., and Novakov, T. (1977). *Nature (London)* **266,** 708–710.
222. Rosen, W., and Williams, P. M. (1978). *Geochem. J.* **12,** 21–27.
222a. Rowley, A. G., Law, I. A., and Husband, F. M. (1982). *Anal. Chim. Acta* **143,** 265–268.
223. Saar, R. A., and Weber, J. H. (1980). *Environ. Sci. Technol.* **14,** 877–880.
224. Salim, R., and Cooksey, B. G. (1980). *J. Electroanal. Chem.* **106,** 251–262.
225. Salomons, W., and Förstner, U. (1980). *Environ. Technol. Lett.* **1,** 506–517.
226. Scrudato, R. J., and Estes, E. L. (1975). *Environ. Geol.* **1,** 167–170.
227. Segar, D. A. (1974). *Anal. Lett.* **7,** 89–95.
228. Shambaugh, R. L., and Melnyk, P. B. (1978). *J. Water Pollut. Control Fed.* **50,** 113–121.
229. Sheldon, R. W. (1972). *Limnol. Oceanogr.* **17,** 494–498.
230. Sheldon, R. W., and Sutcliffe, W. H. (1969). *Limnol. Oceanogr.* **14,** 441–444.
231. Shuman, M. S., and Cromer, J. L. (1979). *Environ. Sci. Technol.* **13,** 543–545.
232. Shuman, M. S., and Woodward, G. P. (1973). *Anal. Chem.* **45,** 2032–2035.
233. Shuman, M. S., and Woodward, G. P. (1977). *Environ. Sci. Technol.* **11,** 809–813.
234. Skogerboe, R. K., Wilson, S. A., and Osteryoung, J. G. (1980). *Anal. Chem.* **52,** 1960–1962.
235. Slovák, Z. (1979). *Anal. Chim. Acta* **110,** 301–306.
236. Slovák, Z., Slováková, S., and Smrž, M. (1975). *Anal. Chim. Acta* **75,** 127–138.
237. Slovák, Z., Slováková, S., and Smrž, M. (1976). *Anal. Chim. Acta* **87,** 149–155.
238. Slovák, Z., Smrž, M., Dočekal, B., and Slováková, S. (1979). *Anal. Chim. Acta* **111,** 243–249.
239. Slowey, J. F., and Hood, D. W. (1971). *Geochim. Cosmochim. Acta* **35,** 121–138.
240. Slowey, J. F., Jeffrey, L. M., and Hood, D. W. (1967). *Nature (London)* **214,** 377–378.
241. Smith, R. G. (1976). *Anal. Chem.* **48,** 74–76.
242. Spencer, D. W., and Brewer, P. C. (1969). *Geochim. Cosmochim. Acta* **33,** 325–339.
243. Steinberg, C. (1980). *Water Res.* **14,** 1239–1250.
244. Sterritt, R. M., and Lester, J. N. (1980). *Sci. Total Environ.* **14,** 5–17.
245. Sterritt, R. M., and Lester, J. N. (1982). *Environ. Pollut. Ser. A* **27,** 37–44.
246. Stiff, M. J. (1971). *Water Res.* **5,** 585–599.
247. Stoveland, S., Lester, J. N., and Perry, R. (1979). *Water Res.* **13,** 949–965.
248. Stuermer, D. H., and Harvey, G. R. (1974). *Nature (London)* **250,** 480–481.
249. Stumm, W., and Bilinski, H. (1973). *Adv. Water Pollut. Res., Proc. Int. Conf., 6th, 1972,* 39–49.
250. Sturgeon, R. E., Berman, S. S., Willie, S. N., and Desaulniers, J. A. H. (1981). *Anal. Chem.* **53,** 2337–2340.
251. Sugai, S. F., and Healy, M. L. (1978). *Mar. Chem.* **6,** 291–308.
252. Sugimura, Y., Suzuki, Y., and Miyake, Y. (1978). *J. Oceanogr. Soc. Jpn.* **34,** 93–96.
253. Tanner, R. L., D'Ottavio, T., Garber, R., and Newman, L. (1980). *Atmos. Environ.* **14,** 121–127.
254. Ter Haar, G. L., and Bayard, M. A. (1971). *Nature (London)* **232,** 553.
255. Tessier, A., Campbell, P. G. C., and Bisson, M. (1979). *Anal. Chem.* **51,** 844–851.
256. Thomas, R. L., Dharmarajan, V., Lundquist, G. L., and West, P. W. (1976). *Anal. Chem.* **48,** 639–642.
257. Thorburn-Burns, D., Glocking, F., and Harriott, M. (1981). *Analyst* **106,** 921–930.
258. Truitt, R. E., and Weber, J. H. (1979). *Anal. Chem.* **51,** 2057–2059.
259. Truitt, R. E., and Weber, J. H. (1981). *Anal. Chem.* **53,** 337–342.
260. Truitt, R. E., and Weber, J. H. (1981). *Environ. Sci. Technol.* **15,** 1204–1208.
261. Trujillo, P. E., and Campbell, E. E. (1975). *Anal. Chem.* **47,** 1629–1634.
262. Tuschall, J. R., and Brezonik, P. L. (1981). *Anal. Chem.* **53,** 1986–1989.
263. Van Loon, J. C. (1979). *Anal. Chem.* **51,** 1139A–1150A.

264. Vernon, F., and Nyo, K. M. (1978). *J. Inorg. Nucl. Chem.* **40,** 887–891.
265. Wagemann, R., and Graham, B. (1974). *Water Res.* **8,** 407–412.
266. Wallace, G. T., Fletcher, I. S., and Duce, R. A. (1977). *J. Environ. Sci. Health, Part A* **12,** 493–506.
267. Whitfield, P. H., and Lewis, A. G. (1976). *Estuarine Coastal Mar. Sci.* **4,** 255–266.
268. Williams, P. M. (1969). *Limnol. Oceanogr.* **14,** 156–158.
269. Wold, J., and Pickering, W. F. (1981). *Chem. Geol.* **33,** 91–99.
270. Woolson, E. A., and Aharonson, N. (1980). *J. Assoc. Off. Anal. Chem.* **63,** 523–528.
271. Yakowitz, H., Jacobs, M. H., and Hunneyball, P. D. (1972). *Micron* **3,** 498–505.
272. Zirino, A. R., and Healy, M. L. (1971). *Limnol. Oceanogr.* **16,** 773–778.

Scoring Systems for Hazard Assessment

Judith M. Hushon

Bolt Beranek and Newman, Inc.
Arlington, Virginia

Mary Rose Kornreich

Reston, Virginia

I. Introduction 63
II. Survey of Scoring Systems 64
III. Factors to Consider in Designing or Selecting a Scoring System 64
  A. Needs of the Program 79
  B. Availability of Data 79
  C. Role of Expert Judgment 79
  D. Bias 80
IV. Steps Involved in Scoring 80
  A. Compiling a Candidate List 80
  B. Truncating 82
  C. Grouping 82
  D. Collecting Data 82
  E. Scoring 87
  F. Weighting and Combining Scores 89
V. Criteria for Scoring 91
  A. Criteria for Assessment of Exposure 92
  B. Criteria for Assessment of Biological Effects 95
  C. Criteria for Assessment of Environmental Effects 97
VI. Comparison of Scoring System Capabilities 97
  A. Systems Focused on a Particular Medium 98
  B. Systems Focused on Only One Aspect of Hazard Assessment 100
  C. Comprehensive Hazard Evaluation Systems 104
VII. Applications of Scoring Systems 106
References 107

I. INTRODUCTION

Scoring systems are mechanisms to rank compounds according to their relative capacity for inflicting adversity upon man and his environment. Scoring is usu-

HAZARD ASSESSMENT OF CHEMICALS:
Current Developments, Vol. 3

ISBN 0-12-312403-4

ally viewed as a screening tool to assist at an early stage of the hazard assessment process in setting priorities among chemical substances that should receive more intensive scientific review before being selected for research, testing, or regulation. Scoring to set priorities is a preliminary step meant to improve efficiency by focusing subsequent attention on the most important concerns. The scoring process itself must be highly efficient to avoid using an overly large portion of program resources or delaying the start of the research or regulatory program. A cost-effective balance should be achieved between resources devoted to scoring for priority setting and resources devoted to implementation of the program.

Development and application of a scoring system for hazard assessment involve selection of basic criteria for determining the degree of hazard, defining these criteria in terms of specific parameters that can be used as factors in scoring, identifying the universe of chemicals to be scored, assembling data on which scoring is to be based, and performing the actual scoring. A large number of existing scoring systems have been surveyed to determine how these steps are being performed, what problems arise when various scoring systems are applied, and how these problems can be resolved.

II. SURVEY OF SCORING SYSTEMS

In order to identify scoring systems already developed, the following approach was used. A bibliography of systems for chemical selection compiled by the United States Environmental Protection Agency's (EPA's) Office of Toxics Integration served as the starting point for this survey (*37*). A literature search of the National Technical Information Service (NTIS) data base also was performed since many of the scoring systems have been developed with United States government funds. In addition, data were added on other systems with which the authors were familiar.

The scoring systems in Table I have been reviewed to identify why they were developed, by whom, for whom, the factors scored, the algorithms by which scores are combined, and the universe of the substances to which they have been applied. The scoring systems described here serve as the basis for the conclusions reached in the remainder of this article.

III. FACTORS TO CONSIDER IN DESIGNING OR SELECTING A SCORING SYSTEM

Development of a scoring system requires consideration of a number of key factors, which are discussed in this section.

TABLE I

Summary of Existing Chemical Scoring Systems

Title/purpose of scoring method	Developer	User	Criteria used for scoring	Algorithms used	Substances scored	References
Pesticide Manufacturing Air Prioritization/to characterize airborne exposures to synthetic organic pesticides	Monsanto	EPA/IERL	Population exposed, number of sources, impact potential, standards, emission rate	$I_x = \sum_{j=1}^{K_x} P_j \left[\sum_{i=1}^{n} \frac{(\bar{X}_{ij})^2}{F_i} \frac{X'_{ij}}{S_i} \right]^{1/2}$ where I_x = impact factor, person/km^2; K_x = number of sources emitting materials associated with source x; P_j = population density in region; F_i = environmental hazard material g/m^3 = TLV(8/24)(1/100); S_i = standard (for criteria materials); $\bar{X}_{ij}$ = calculated time average maximum ground-level concentrations; X'_{ij} = ambient concentration	80 major pesticides	*1*
Sequential Testing for Toxicity Classification/to rank the toxicity of a new chemical by a	Eastman Kodak Co.		Amount released/year, number of discharge sites, number of discharges/h, mean rat oral LD_{50}, reversibility of	Each criterion ranked on a scale of 1 to 3 for each chemical; they are then totaled and the total used to indicate the level of testing required	500 chemicals	2

(*continued*)

TABLE I (*Continued*)

Title/purpose of scoring method	Developer	User	Criteria used for scoring	Algorithms used	Substances scored	References
variety of routes and test systems			effects, 5-h inhibition of microorganism growth, LC_{50} fish, octanol/water partition coefficient			
Index of Exposure/to indicate the relative potential for exposure associated with a given use of each chemical	Auerbach Associates	EPA	Route of human exposure, form of human exposure, mode of environmental exposure, number of people exposed, frequency of exposure, extent of contact, duration of exposure, amount of environmental exposure	Each factor scored on a scale of 0 to 3 and weights assigned Weights multiplied by normalized factors and all data added		*3*
Chemical Hazard Ranking System/to rank chemical components of consumer products by probable health impact	IIT Research Institute	CPSC	Toxicity, dose per person, size of population exposed	Score $= \sum_q \left(S \times \sum_i P(C_i) \times H(D/T_i) \right)$ where S = size of exposed population $P(C_i)$ = probability that the chemical belongs to the ith toxic strength class H = selected dose–response function D = mean dose of exposed		*5*

				person from all products containing the target chemical T_i = mean toxic strength of the ith toxic strength class		
System for Evaluation of the Hazards of Bulk Water Transportation of Industrial Chemicals/to identify hazards of chemicals being transported by water	NAS	United States Coast Guard	Fire hazard rating, rating for skin/eyes, rating for vapor inhalation, rating for gases inhalation, rating for repeated inhalation, water pollution hazard to humans, aquatic toxicity, water reactivity, self-reactivity	Each factor scored on a scale of 0 to 4 Combining of scores was discouraged by author		*6*
Barring Model/to develop a system to rank dumpsite chemicals as to whether they represent a hazard	Booz–Allen		Disposal medium, toxic effects, explosion potential, ecological effects, production volume distribution	Factors ranked from 1 to 3 Factors multiplied by weightings and summed to give a total effects rating The total effects rating is then multiplied by the hazard extent to produce a hazard rating		*7*
Select Organic Compounds Hazardous to the Environment/to identify high-exposure compounds for review by NSF panel	SRI	NSF	Release rate, chemical class, production quality, environmental impact, human health hazard	Release rates requested of industry $R = (P + I + E)F_D + P(F_{PL})$ where P = production quantity F_{PL} = fraction of production lost	Panel members identified 337 chemicals for scoring; based on scores, dossiers prepared on 80 with highest exposure and these	*8, 33*

(*continued*)

TABLE I (*Continued*)

Title/purpose of scoring method	Developer	User	Criteria used for scoring	Algorithms used	Substances scored	References
concerning potential to damage human health or the environment				I = quantity imported E = quantity exported F_D = fraction of product to dispersive uses R = release factor	were ranked for effects by experts	
Ranking Algorithm for CEC Water Pollutants/to select a subset of chemicals present in the aquatic environment for further study	SRI	CEC	Production, uses, water half-life degradation, aquatic toxicity, concentration in fish, human toxicity, consumption patterns	The following are modeled: Quantity discharged to water Persistence in water Concentrations in water Concentrations in fish Acute human risk index Overall risk index	1500 chemicals (Biokon List)	*9*
Setting Priorities for Research and Development on Army Chemicals/to select research priorities	SRI	USAMRDC Fort Detrick	Effects, weight for severity of effect, exposure, population at risk, concentration (sources, transport, transformation, transfer)	Separate concentration models for air, surface, and groundwater pollutants All factors used to evaluate environmental exposure Severity, incidence of effects, exposure, and population at risk used for occupational exposure evaluation	35 chemicals	*10*
System for Rapid Ranking of Environmental Pollutants/to choose chemicals on	SRI	EPA/ORD	Production and use, environmental distribution, transformations, human health effects, ecological effects	Expert judgment combined with a more objective screening model Each expert ranks chemicals on a scale of 1 to 10	10 chemicals	*11*

which to prepare scientific and technical reports (STARs)				Experts in a variety of areas Experts scores combined Scientific review panel sets final priority		
Estimating the Hazard of Chemical Substance to Aquatic Life/to determine what impact chemicals will have on aquatic life	ASTM Committee D-19		Physical/chemical properties, rate constants, partition coefficients, aquatic environment description, rate of substance input, toxicity test results	$BE \times SF = EC$ where BE = estimate of concentration causing no biological effects SF = safety factor (1/10 to 1/1000) EC = predicted environmental concentration causing no effects to aquatic life		*12*
Estimation of Toxic Hazard—A Decision Tree Approach/to identify potentially dangerous food constituents for additional testing	Flavor and Extract Manufacturer's Association	Industry	Chemical structure occurrence in body tissues and fluids, natural occurrence, metabolism, toxicity	33 questions to be answered yes/no Divide substances into 3 categories: low, moderate, or severe toxicity	Food additives (method validated by application to 247 substances known to cause cancer)	*13*
TSCA–ITC Scoring System Workshop/ to develop an improved, integrated health and environmental effects scoring system to identify chemicals	Enviro Control, Inc.	EPA/ITC	Mutagenicity screening test results, reproductive and teratogenic effects, carcinogenicity, other toxicological effects, chemistry and fate, ecosystem effects, occupational exposure, environmental	Scoring system components developed separately for each category of criteria Sometimes ITC's current method endorsed, sometimes changes proposed Scores assigned within each category	6 chemicals	*14*

(*continued*)

TABLE I (*Continued*)

Title/purpose of scoring method	Developer	User	Criteria used for scoring	Algorithms used	Substances scored	References
for which testing is required for ITC			release, general population exposure			
An Approach to Prioritization of Environmental Pollutants: The Action Alert System/to help the OWRS to set priorities regarding chemicals identified in water	A. D. Little	EPA/OWRS	Concentration in surface and drinking water, concentration in food, acute effects data, chronic effects, aquatic effects data	Six auxiliary modules to supply data if not available (1) Environmental release (2) Environmental fate (3) Food and water contamination (4) Alternative exposure routes (5) Subordinate adverse effects (6) Subpopulations exposed Exposure and effects data are scored and the results placed on a grid to identify a chemical's location in one of three action zones	129 priority pollutants	*15*
Scoring of Organic Air Pollutants/to select organic air pollutants for more indepth study/monitoring	MITRE Corp.	EPA/OAQPS	United States production volume, fraction of production lost, volatility, acute toxicity, nonlethal effects, carcinogenicity, mutagenicity, teratogenicity, occupational standards	First, each variable was scored on a scale of 1 to 5 $P \times F_{PL} \times V \times \frac{T_{total}}{T_{possible}}$ = Score, where P = production F_{PL} = fraction of production lost V = vapor pressure T = toxicity values	637 chemicals	*16*

Ranking of Environmental Contaminants for Bioassay Priority/to select chemicals for NCI bioassay	SRI	NCI	Production, population exposed, annual intake per person by each route, probability of chemical's carcinogenicity, relative potency	$R = \sum_{ijk} S_i \times Q_{ik} \times F_{jk} \times P_j(c) \times P_j$ where S_i = number of individuals in population group i; Q_{ik} = average quantity available through use k to each member of population group i; F_{jk} = intake attenuation factor by route j and use k; $P_j(c)$ = probability of carcinogenicity by intake route j; P_j = potency by intake route j	*17*
PHL Model/to identify landfill components likely to represent human health hazards			Human toxicity, LOAEL for water, persistence, mobility in environment	Linear additive combination model in which scores are assigned, multiplied by weights, and summed	*18*
Hazard Evaluation Procedure for Potentially Toxic Chemicals/screening procedure to identify high-risk chemicals	MARC (Monitoring and Assessment Research Centre)	UNEP	Production volume, use patterns, magnitude of exposure; transformations, environmental fate; target organisms, toxicity, probability of accidental exposure	Ranking of five factors on a scale of high, medium, low by responding to specific questions	*19*

(*continued*)

TABLE I (*Continued*)

Title/purpose of scoring method	Developer	User	Criteria used for scoring	Algorithms used	Substances scored	References
Selection of Chemicals for Inclusion in a Trend Monitoring/to select chemicals and chemical classes to include in a monitoring program to follow trends	MITRE	Federal Republic of Germany	Bioaccumulation, persistence, acute toxicity, chronic toxicity	Scores assigned to each criterion Priorities defined based on scores (P_1 to P_6), e.g., P_1 = positive score on all criteria P_2 = high total score, positive persistence, bioaccumulation, and toxicity Priorities were assigned to each chemical group	700 chemicals	*22*
RCRA Risk/Cost Policy Model/to identify relative risks from exposure to chemicals in wastes	ICF Inc. Clement SCS Engineers	EPA/OSW	Production/use quantities, releases, minimum effective dose (mg/kg/day), exposure medium	Risk = population at risk × exposure × inherent hazard Exposure = $Q \times F \times D_i \times I_n \times A$ where Q = quantity of material F = fraction released D_i = dispersion factors I_n = intake factor A = absorption coefficient Unit risk = MED/10 (human) MED/100 (chronic animal)	140 compounds	*23*

				MED/300 (subchronic animal) Inherent hazard/mg/kg/day = 0.01/human MED/10 = a score		
ITC Scoring for Biological Effects	Clement Associates	EPA/ITC	Carcinogenicity, mutagenicity, teratogenicity, acute toxicity, other toxic effects, ecological effects, bioaccumulation	Each factor (f) scored on a scale of 0 to 3, a negative value indicates need for testing Weights (w) are assigned: $R = \sum_{i=1}^{5} w_i \times \frac{f_i}{S_i}$ S = possible score for that factor	1st scoring exercise (1977), 250 chemicals 2nd scoring exercise (1979), 235 chemicals 3rd scoring exercise (1981), 214 chemicals 4th scoring exercise (1982), 218 chemicals	*24*
Ranking of Food Contaminants/to identify for OTA organics, inorganics, and radionuclides that are possible food contaminants	Clement Associates	OTA	Bioaccumulation, persistence, occurrence in water, production, use pattern, occurrence in foods, potential for post-sale contamination, volatility, exposure of susceptible populations, acute toxicity, carcinogenicity, mutagenicity, teratogenicity, other toxic effects	Scores assigned to each factor Weights assigned for importance of effects Scores multiplied by weights and added Scores normalized	143 chemicals	*27*

(continued)

TABLE I (*Continued*)

Title/purpose of scoring method	Developer	User	Criteria used for scoring	Algorithms used	Substances scored	References
Rapid Screening and Identification of Airborne Carcinogens of Greatest Concern	SAI	Air Resources Board, State of California	Present use in California, growth in California use, emission potential, stability in ambient air, dispersion potential, evidence of carcinogenicity	Each factor scored on scale of 1 to 5 for multiplicative applications and 0 to 5 for additive applications In the additive approach, the scores are multiplied by weightings and then added In the multiplicative approach, the ratings are merely multiplied A panel of experts was then used to reach a consensus between the results from the two approaches	47 chemicals with known carcinogenicity from MITRE list of organic air pollutants (ref. *15*)	*28*
Critical Materials Register/to construct a register of chemicals of concern	State of Michigan	Michigan Department of Natural Resources	Acute toxicity, carcinogenicity, mutagenicity, teratogenicity, persistence, bioaccumulation, other adverse toxicity	Each factor scored on a scale of 0 to 7 Scores added A score of 7 for any factor or a cumulative score of 7 resulted in the listing of that chemical in the register	178 chemicals or chemical classes	*29*
National Occupational Hazard Survey/to rank hazards according to the amount of occupational exposure	NIOSH	NIOSH	Size of facility, SIC code, number of employees, exposure to chemicals by job title: duration, intensity, form, control	Ranked exposures by job title by industry Chemicals assigned a hazard code	7145 chemicals	*30*

Assessment of Oncogenic Potential/to identify carcinogens and to rank them relative to the evidence	Hooker Chemical		Length of test (PS), metabolism same in test animal and man (M), route (R), confidence factor (C), dose-level factor (D), time to tumor (T), tumor type, suitability of control, dose–response, exposure levels, epidemiology studies	$AP = PS \times M \times R \times C$ $WS = (D \times AP) + (T \times AP)$ AP = adjusted primary score WS = weighted score Additional equations cover exposure potential		*31*
ITC Scoring for Exposure/to rank chemicals on the basis of potential for human exposure and environmental release	Clement Associates	EPA/ITC	Production quantity, quantity released to environment, persistence, occupational exposure, general population exposure (quantity/frequency/intensity/penetrability)	$R_j = \sum_{i=1}^{x} W_i \frac{f_{ij}}{S_i}$ W_i = weight assigned ith factor f_{ij} = ith factor score of jth chemical S_i = scaling factor chosen to normalize scores $S_1 = 4.3191$, $S_2 = 6$, $S_3 = 3.3680$, $S_4 = 12$ Each factor is scored on a scale of 0–3	1st scoring exercise (1977), 528 chemicals 2nd scoring exercise (1979), 412 chemicals 3rd scoring exercise (1981), 481 chemicals 4th scoring exercise (1982), 413 chemicals	*35*
Ordering of Commercial Chemicals on NIOSH's Suspected Carcinogens List/to determine which suspected	EPA/OPTS		Species in which neoplastic or carcinoma response found (score), number of species, route of exposure (score), total number of	Scores not combined but 4 lists prepared (1) Chemicals needing further delineation of hazard potential (2) Chemicals under other	75 chemicals on NIOSH Suspected Carcinogen List 1693 chemicals for EPA/OPTS	*36*

(*continued*)

TABLE I (*Continued*)

Title/purpose of scoring method	Developer	User	Criteria used for scoring	Algorithms used	Substances scored	References
carcinogens are of concern to OPTS			neoplastic or carcinoma responses reported, production, use	regulatory jurisdictions (3) Chemicals with insufficient information (4) Chemicals carcinogenic in lab animals by non-environmentally significant routes; ranked by production		
Identification of High-Risk Occupational Groups and Industrial Processes Using RTECS/NOHS Data/tool to objectively assess potential health risk from workplace exposures	Tracor Jitco	NIOSH DCCP/NCI	Toxicity (RTECS), exposure (NOHS), census data, SIC codes, test species, endpoint, number of species, route of administration, number of chemicals in test class, form of chemical, number of workers	Dose data normalized Combine species and route data into test classes Calculate hazard risk index for each chemical Compute occupational risk index Compute industry risk index Intersections of lists	28,000 chemicals in RTECS file	*38*
OECD Ecotoxicology Testing Scheme/to test how well aquatic tests predict hazard potential	Battelle	EPA/OPTS	Type of organism, type of data, length of test, type of test, test conditions	Scores assigned to values for each criterion Toxicity values established in different species and compared Tests scored for how well they predicted toxicity	53 chemicals	*39*

Chemical Scoring System Development/to select chemicals for more indepth evaluation by OPTS	ORNL	EPA/OPTS	Carcinogenicity, mutagenicity, embryotoxicity and fetotoxicity, reproductive effects, chronic toxicity, acute toxicity, production volume, environmental exposure, occupational exposure, consumer exposure	Criteria established for each category of data Scores assigned based on these criteria: 1–9 for effects, 1–10 for exposure Scores weighted for severity and summed, then normalized Scores then totaled for each chemical High scores in a given area call for particular attention	6 chemicals	*40*
Environmental Scoring of Chemicals/to select chemicals presenting an environmental risk under TSCA and for use by ITC to identify chemicals for additional environmental testing	ORNL	EPA/OTS	Production quantity, F_D = fraction released to environment, release distribution (D), medium (i), mobility (M), persistence (P), bioaccumulation (B), acute toxicity (terrestrial/aquatic), chronic toxicity (terrestrial/aquatic)	Scores assigned Mobility × persistence by medium, $V + D + \sum_{i=1}^{3} M \times P + B$ = Exposure Toxicity multipliers used $\sum_{ij}$ toxicity scores = Toxicity rank i = medium j = acute/chronic	10 chemicals	*41*
Ranking Animal Carcinogens/to classify animal carcinogens to permit the use of different regulatory options	Squire		Number of species showing carcinoma or neoplastic response, number of types of neoplasms, spontaneous incidence, dose–response	Scores assigned to each factor and all scores added Total score then used to determine regulatory option	10 chemicals	*42*

(*continued*)

TABLE I ***(Continued)***

Title/purpose of scoring method	Developer	User	Criteria used for scoring	Algorithms used	Substances scored	References
			relationships, malignancy of induced neoplasms, genotoxicity			
Hazard Assessment by a Qualitative System/to determine whether a new chemical represents a hazard based on MPD data	Assn. Chimie et Ecologie	French Ministère de l'Environnement	Production, market dispersion, persistence, initial distribution, preferential distribution, bioconcentration, toxicity, mutagenicity, sensitization, irritation Above factors are scored using a three-level scale Penalties then assigned based on interactions	Scores are added and penalty points assigned based on the interactions; new chemicals are then compared to existing chemicals using a graphical grid system	47 known chemicals	*25*

A. Needs of the Program

An effective scoring system must meet the needs of the program for which it will be implemented, and these needs must be clearly defined. Scoring for hazard assessment is most often used to set priorities for research or regulatory programs; the methods used for these purposes vary markedly. A regulatory program usually needs to identify those chemical substances for which there is strong evidence of a high degree of hazard (exposure and effects) to the population. A testing program needs to identify those chemicals that are likely to present a high degree of hazard to the population but for which good evidence of hazard does not yet exist. For this reason, it is rarely possible to select, without modification, a system designed for another program.

B. Availability of Data

Because the scoring exercise is heavily dependent upon the completeness of the data base, the search for data should be as exhaustive as practically possible within the limitations of the available resources. Inadequacy of the data base is not always due to limitations of resources for retrieval and evaluation. Very often, the information desired for hazard assessment is simply unavailable. In fact, availability of information is usually the major limiting factor in scoring for hazard assessment. Most parameters used in a scoring system are actually surrogates for the criteria of exposure and toxicity.

It is often desirable to approach scoring in a stepwise fashion with increasing numbers of data being considered at each step. In this way, the resources available for data retrieval can be maximized. The Interagency Testing Committee (ITC) scoring approach follows this form with many more chemicals being scored for exposure than are scored for biological effects (*24, 35*).

Any scoring system must confront the problem of data gaps. After the scoring criteria that meet the needs of the program are defined, it often becomes apparent that the data base needed to evaluate chemicals based on these criteria is not complete. The criteria may have to be redefined in terms of parameters for which data are available. Alternative methods can be used to estimate missing data. Depending on the purpose of the method, default options that rely on worst-case values, average values, or values for structurally similar chemicals can be used.

C. Role of Expert Judgment

Expert judgment plays a role in all scoring systems. It is used in selection of criteria and design of the scoring system. Even the most highly quantified scoring process relies on human judgment to interpret and weigh raw data and convert them into a score. Both exposure and toxicology evaluations require a high degree of expertise. Training and experience are especially important where inferences must be drawn from inadequate data.

Some scoring systems, such as that developed by Jouany *et al.* (*25*) or that used by the Toxic Substances Control Act (TSCA) Interagency Testing Committee, permit intervention by experts at several points in the process to raise or lower chemicals in priority or to add chemicals that clearly deserve priority action but for some reason have not been selected by the scoring system. After the scoring phase, the high-scoring chemicals are often individually evaluated and expert judgment is used to make the final decisions for action.

D. Bias

There is a bias toward high scores for those chemicals that have received the greatest amount of previous attention. For such chemicals, there are likely to be a large number of case reports implicating the compound as the cause of adverse effects on man and the environment. There are also likely to be reports of its occurrence in air, water, the food supply, or body tissues, and toxicity tests are likely to have been performed. The more tests performed on a specific compound, the greater are the chances for detecting adverse effects. On the other hand, a lack of information minimizes the chance of a chemical, even a potentially hazardous one, being selected on the basis of scoring.

Methods have been developed to minimize bias due to differences in the amount of testing performed on various chemicals. An example is the normalization procedure used by the MITRE Corporation in scoring organic air pollutants (*16*). Toxicity scores were determined by adding scores for various categories of toxicity. Normalization was accomplished by dividing this total score by the maximum score possible for those types of toxicity for which data were available.

IV. STEPS INVOLVED IN SCORING

Though scoring is done for different purposes, there are a number of steps that are common to application of most scoring methods.

A. Compiling a Candidate List

The first step in scoring is to compile a comprehensive list defining the universe of chemicals under consideration. The establishment of the initial universe in itself constitutes a significant and error-prone step in the process of scoring to set priorities for research or regulatory programs. Most often, the sources of chemicals for this comprehensive list are previously existing lists. It is interesting to note that the base set of chemicals serving as input to the state of California's rapid screening method is the output from the MITRE scoring exercise on air pollutants (*16, 28*). The MITRE exercise scored 637 compounds compiled from eight previously existing lists (*16*). Errors of omission are likely with this method; since the same source lists tend to be used for scoring by various programs, the same chemicals tend to be repeatedly overlooked.

A reasonable attempt to define the universe of chemical substances to which humans are exposed in the United States can be made by starting with the Inventory of Chemicals in Commerce in the United States compiled under Section 8(b) of the Toxic Substances Control Act and then systematically adding lists of those chemicals explicitly excluded from the inventory. Substances omitted from the inventory because they are regulated under other laws include foods, drugs, cosmetics, and pesticides. One could, therefore, add to the inventory the list of food additives approved for use by the Food and Drug Administration (FDA), the list of drugs and ingredients approved for use by the FDA, the Cosmetics, Toiletries, and Fragrances Act (CTFA) cosmetics ingredients list, and the list of pesticides registered for use by EPA. The above lists and other commonly used source lists such as annual reports of the International Trade Commission focus on chemicals in commercial production. Groups of chemicals that tend to be omitted from these source lists include (1) impurities in commercial chemical products, (2) captive intermediates in industrial processes, (3) chemicals produced from incomplete combustion of fossil fuels or from other energy-producing processes, (4) by-products or wastes of industrial processes, (5) naturally occurring chemicals, and (6) environmental degradation products. Identification of substances in these categories remains a challenging problem for those compiling candidate lists for scoring.

Not every scoring effort begins with an attempt to list the entire universe of chemicals to which humans are exposed. Often, only a subset is required, such as air pollutants or chemicals with consumer exposure, and the source lists can be chosen accordingly. Another way to identify chemicals is to have chemicals selected for consideration by panels of experts. When either of these methods is utilized, the judgment of experts will already have been employed to prescreen chemicals.

Errors of omission are likely whether chemicals are selected from previous lists or nominated by experts. In using existing lists, the chances for error increase with the extent to which the purposes of the earlier program differ from those of the present one. Having a panel of experts select chemicals also is likely to result in errors of omission since experts will tend to select chemicals with which they are already familiar. This method is particularly inappropriate when selecting chemicals for research or testing since the bias against chemicals for which little information is available tends to exclude the very chemicals most in need of further research.

B. Truncating

It is almost always necessary to reduce the size of the comprehensive list before ranking. Errors of omission are very likely to occur during this truncating step because, for practical reasons, decisions at this step usually are based on less information than will be compiled and analyzed in subsequent phases of the scoring effort. In a multiphase scoring effort such as that used by the Interagency Testing Committee, each step is followed by a more intensive effort to obtain information on a reduced set of chemicals.

A multistage screening system balances the costs of generating information on a large number of chemicals in earlier stages against the costs of generating more detailed information on fewer chemicals at later stages. The advantage of such a system is that it uses simple, readily retrieved data to eliminate low-priority chemicals quickly. The disadvantage is that some chemicals might be erroneously eliminated based on crude criteria applied in the early phase or because only insufficient data could be found on them.

C. Grouping

Grouping is an optional and often controversial step. Individual chemicals are often combined into groups or classes in an attempt to reduce the size of the list of chemicals. Grouping is based usually on chemical or physical properties or on chemical structure. One system that used grouping was that developed by MITRE to select chemicals for inclusion in a trend monitoring program for the Federal Republic of Germany (*22*). It was the goal of this program to select 50 representative persistent chemicals; it was desirable that as many structurally diverse classes be represented as possible. Candidate chemicals were identified then assigned to structural categories. Scores were obtained and those categories with the highest average scores were chosen. From within these categories, representative chemicals were then selected.

There is often a legitimate objection to assigning scores to a group rather than to individual chemicals. Members of a group may vary widely with respect to the factor being scored making it difficult or meaningless to assign a single score to the group. As in the case above, however, grouping can be beneficial, and its use should be considered in designing the scoring strategy.

D. Collecting Data

The effectiveness of the scoring system is highly dependent on the quality and completeness of the data base. The data base on toxicity should contain enough information to identify the chemicals' adverse health effects and to permit assessment of the risks associated with anticipated human exposure. Since reliable toxicity information based on experience of exposed humans is often not available, information for scoring is usually based primarily on laboratory animal data.

Table II provides a list of secondary sources in which toxicological data have been compiled and Table III provides a list of computerized bibliographical data banks that can be used to locate primary studies of the effects of chemical substances on humans, laboratory animals, and other biota. The data base compiled for exposure scoring should adequately reflect the complexities of real-life exposure situations. A guide to sources of monitoring and exposure data is provided by Hushon and Clerman (*20*).

In the interest of cost effectiveness, compromises are often made between completeness of the data base and efficiency. Investment of large amounts of

TABLE II

Secondary Sources of Toxicology Data

Water Quality Criteria, 1972	Water Quality Criteria Documents (EPA)
Farm Chemicals Handbook	IARC Monographs
NIOSH Criteria Documents	Metabolism of Pesticides (Menzie)
NCI/NTP Bioassay Reports	CRC Handbook of Food Additives (Furia)
Handbook of Toxicology	Foreign Compound Metabolism in Mammals
Handbook of Industrial Toxicology	Handbook of Flavor Ingredients (Fenaroli)
Dangerous Properties of Industrial Materials (Sax)	Toxicology of Pesticides (Hayes)
Merck Index	Handbook of Toxicology of Pesticides to Wildlife (DOI)
CRC Handbook of Analytical Toxicology	Chemical Carcinogens (Searle)
Toxic and Hazardous Industrial Chemicals Safety Manual	Practical Toxicology of Plastics
Chemical Carcinogens: Survey of Compounds Tested for Carcinogenic Activity (NCI)	Drinking Water and Health (NAS/NRC)
Catalog of Teratogenic Agents (Shepard)	Metal Toxicity in Mammals (Luckey and Venugopal)
Clinical Toxicology of Commercial Products (Gosselin *et al.*)	Toxicology (Cassarett and Doull)
Industrial Hygiene and Toxicology (Patty)	Handbook on the Toxicology of Metals (Friberg *et al.*)
Potential Industrial Carcinogens and Mutagens (Fishbein)	Handbook of Acute Toxicity of Chemicals to Fish and Aquatic Invertebrates (DOI)
Toxicity and Metabolism of Industrial Solvents (Browning)	Lethal Dietary Toxicities of Environmental Pollutants to Birds (DOI)
Chemical Mutagens (Legator)	Registry for Toxic Effects of Chemical Substances (NIOSH)
Toxicity of Industrial Metals (Browning)	Pesticides Studied in Man (Wilkins)
Handbook of Environmental Data on Organic Chemicals (Verschueren)	Metabolic Maps of Pesticides (Aizawa)
Pharmacological Basis of Therapeutics (Gilman *et al.*)	Environmental Contaminants in Food (OTA)
Encyclopedia of Occupational Health and Safety	Hazard Assessment Documents (EPA)
Documentation of the Threshold Limit Values for Substances in Workroom Air	Hazard Information Reviews (EPA)
Hygienic Guide Series	Health Effects of Environmental Pollutants (Waldbott)
	Chemical Hazard Information Profiles (CHIPS) (EPA)

TABLE III

Computerized Abstracts of Health and Environmental Effects Literature

Acronym	Expanded Name	Offeror(s)	Coverage
AFEE	Association Francaise pour l'Etude des Eaux	SPIDEL[a]	Articles, reports, and monographs related to water pollution and toxicology of water pollutants
APILIT	Index to American Petroleum Institute Abstracts of Refining Literature	SDC	Citations to toxicology studies of petroleum-related materials
ASFA	Aquatic Sciences and Fisheries Abstracts	DIS, QL[b]	References to all aspects of the aquatic environment including human and aquatic biota effects
AVLINE	Audio Visual Line	NLM, DIMDI[a]	References to health and environmental nonprint materials
BIOSIS	BIOSIS Previews	BRS, CISTI,[b] DATA-STAR,[a] DIMDI,[a] DIS, ESA–IRS, SDC	Citations to international life sciences literature
CA Search	Chemical Abstracts Service Search	BRS, CISTI,[b] DIS, ESA–IRS, SDC, TIQ	Coverage of chemical science literature including pharmacology and toxicology
CANCERLIT	Cancer Literature Information On-Line	NLM, DIMDI[a]	Cancer literature worldwide
CANCERNET	CANCERNET	T/Q	Coverage of world cancer literature in French and English
CDI	Comprehensive Dissertation Index	BRS, DIS	Index of all United States and 210 non-United States countries' doctoral dissertations
CTCP	Clinical Toxicology of Commercial Products	CIS	Compilation of ingredients of commercial products and their respective toxicities
EMIC	Environmental Mutagen Information Center File	RECON	References to mutagenicity literature
ENVIROLINE	Environmental Abstracts On-Line	BRS, DIS, ESA–IRS, SDC	Literature abstracts on all aspects of the environment including toxicology

ETIC	Environmental Teratogen Information Center	RECON	References to teratology literature
EXCERPTA MEDICA	Excerpta Medica	DIMDI,[a] DIS	Coverage of world literature related to medical science
Foods Adlibra	Foods Adlibra	DIS	Covers information on the food industry including nutrition and toxicology
IPA	International Pharmaceutical Abstracts Information System	DIS	International pharmaceutical abstracts
IRL Life Sciences	IRL Life Sciences Collection	DIS	Worldwide literature in major areas of biology, medical science, biochemistry, and ecology
MEDBACK	Backfiles of MEDLARS On-Line	NLM, DIS, DIMDI,[a] CISTI,[b] BRS, BLAISE[a]	Comprehensive coverage of the biomedical literature of the United States and 70 foreign countries from 1966–1980
MEDLINE	MEDLARS On-Line	NLM, DIS, DIMDI,[a] CISTI,[b] BRS, BLAISE[a]	Comprehensive coverage of the biomedical literature of the United States and 70 foreign countries for 1981–1982
NIOSHTIC	National Institute for Occupational Safety and Health Technical Information Center	NIOSH	References to industrial hygiene and toxicology
NTIS	National Technical Information Service Bibliographic Data File	BRS, DATA-STAR,[a] DIS, ESA–IRS, INKA,[a] SDC	United States government-sponsored studies in areas including biology, medical sciences, and environmental effects
OCABS	Oceanic Abstracts	DIS, BNDO[a]	Effects on the marine environment
PESTDOC	Pesticide Control Literature Documentation	SDC	Includes references to toxicology of insecticides, rodenticides, fungicides, herbicides, and molluscicides
POLLUTION	Pollution Abstracts	DIS, BRS	Effects of pollutants on the environment
Psyc INFO	Psychological Abstracts Information Services	BRS, DATA-STAR,[a] DIS, DIMDI,[a] SDC	World literature abstracts related to behavioral toxicology

(*continued*)

TABLE III (*Continued*)

Acronym	Expanded Name	Offeror(s)	Coverage
RAPRA	Rubber and Plastics Research Association Abstracts	DIS, ESA–IRS	Citations to toxicology and industrial hygiene related to rubber, polymers, plastics
RTECS	Registry of Toxic Effects of Chemical Substances	BLAISE,[a] CIS, DIMDI,[a] NLM	Toxicity data and references for specific chemicals
SAFETY	Safety Science Abstracts Journal	SDC	Industrial hygiene and public health abstracts
SCI	SCISEARCH	BLAISE,[a] BRS, DIMDI,[a] DIS	Survey of science and technology literature
SDILINE	Selective Dissemination of Information On-Line	BLAISE,[a] BRS, CISTI,[b] DIMDI[a]	Items input to MEDLINE for most recent month
TITUS	Textile Information Treatment Users' Service	SDC, T/Q	Environmental and health effects related to textiles
TOXBACK	Backfile of Toxicology Information On-Line	BLAISE,[a] NLM	Comprehensive literature of all areas of toxicology from 1965–1977
TOXLINE	Toxicology Information On-Line	BLAISE,[a] NLM	Comprehensive literature in all areas of toxicology from 1978–present

[a] Available in Europe only.
[b] Available in Canada only.

resources in completing the data base defeats the purpose of the screening process. These compromises often mean that secondary sources, such as the National Institute for Occupational Safety and Health (NIOSH) *Registry for Toxic Effects of Chemical Substances* or the International Agency for Research on Cancer (*IARC*) *Monographs,* are used rather than the original research reports. Evaluation of the quality of data used for scoring takes a much larger investment of resources than merely using the conclusions of the original author. Compromises in data evaluation as well as in completeness of the literature research are particularly likely to be made in the earliest stages of a multistep screening process when data might be compiled for hundreds or thousands of chemicals. In later stages, when important decisions on selecting chemicals for priority consideration are being made, critical review of key studies becomes a necessity.

A scoring system should provide a mechanism for dealing with data gaps. One such mechanism is consideration of the properties of structurally related compounds. For some factors, systems of structure–activity relationships have been developed. Structure–activity relationships are not often formally incorporated into the scoring system. These relationships often do, however, form the basis for expert judgment.

In scoring for research or testing, the data gap itself may be exactly what one is attempting to identify. In scoring for regulatory action, a decision could be made to postpone a final decision until the missing data are obtained through further monitoring or testing.

E. Scoring

Some of the knowledge on which the score is based will be objective while some will be subjective; some will be quantitative while some will be qualitative. By assigning scores to a number of factors or criteria relevant to hazard assessment, knowledge of these factors is combined into common quantitative units that can be further combined to obtain a single number representing the severity of the hazards attributable to a chemical substance.

Scoring is most often accomplished by describing each factor by a series of statements defining the grades of severity, intensity, strength of evidence, etc. A number is then assigned to each graded response. Judges then select the appropriate response with its corresponding score for each factor.

It is often desirable to use a team approach for assigning scores since expertise is needed in a number of disciplines. Members of the team should have expertise in complementary disciplines. Where a number of chemicals are being compared, all chemicals should be scored for the same criterion by the same expert or experts.

Judges may sometimes have to decide among a number of data sets. For example, toxicology tests using different protocols or epidemiology surveys using different cohorts may yield different results. Three alternative approaches to this problem are to determine a mean score, to base the score on one data set that is determined to be more valid than the other, or to base the score on the data indicating the greatest potential hazard of the chemical. This last method can be considered the most conservative approach and is likely to be used by regulatory agencies when there is no obvious difference in validity among the available data sets.

Mathematical modeling can be an aid to scoring. Stanford Research Institute (SRI) International developed a scoring method to assist the United States Army Medical Research and Development Command (USAMRDC) in allocating resources among competing research programs on health hazards of chemicals (*10*). A mathematical model was developed for the process leading from the levels of pollution of air, water, or land to the essential environmental effects of the chemicals in question. The model estimates a total hazard value, weighted among various human and ecological effects, with a corresponding uncertainty due to lack of knowledge.

Some scoring methods assign chemicals to one of several priority categories. The Flavor and Extract Manufacturers' Association developed a decision tree approach for dividing chemicals into one of three classes reflecting a presumption of low, moderate, or serious toxicity (*13*). The decision tree consists of a series of 33 questions, most of which relate to molecular structure, and each of which can be answered by "yes" or "no." The logic of the tree depends primarily on known data on metabolism and toxicity.

Another priority category scoring system is that developed by A. D. Little for the EPA Office of Water Regulations and Standards (OWRS) (*15*). Risk estimates (probabilities of suffering adverse effects) were compared with action levels (scientifically based reference values) to assign environmental pollutants to one of three action alert categories.

Scoring for monitoring or regulation is usually based primarily on strength of evidence and severity of adverse effects. Scoring for research or testing, on the other hand, is often based on uncertainty or suspicion of adverse effects. Structure–activity relationships can be extremely useful when selecting chemicals for testing, but are not sufficiently reliable to form a basis for regulation. Structure–activity methods permit the estimation of biological activity from chemical structure through mathematical correlations and pattern-recognition techniques. Veith (*43*) has used octanol/water partition coefficients along with chemical structure to estimate bioconcentration factors and predict toxicity of chemicals to fish by summarizing toxicity data for semihomologous series of chemicals and then correlating the data with major structural features.

Rather than using structure–activity correlations to estimate activity based on the original chemical, Wipke (*44*) developed a system that acknowledges the importance of metabolism and analyzes activity on the basis of metabolite structures. Wipke's system first generates plausible metabolites for a given compound and then identifies those with possible biological activity.

Jurs (*26*) used computerized information-handling methods for chemical structures to organize large files of molecular structures and to generate molecular structure descriptors. He then used pattern-recognition techniques to separate chemicals into active and inactive classes for various toxic effects.

Special techniques have been developed to score for uncertainty when selecting chemicals for research or testing. In developing a system for the Interagency Testing Committee to use in selecting chemicals for testing under Section 4 of the Toxic Substances Control Act, Clement Associates used two independent components to score for biological activity (*24*). For each factor, a chemical was given a numerical score reflecting a judgment of the severity of its potential effects in that area based on existing data, or a letter score reflecting a judgment that it had not been adequately tested and therefore needed testing. The letter scores also indicated a judgment as to the seriousness of the biological effects.

A very different scoring system for testing priorities was developed by SRI International for the United States Army Medical Research and Development Command (*10*). A mathematical model was used to estimate the uncertainty due to lack of knowledge for each biological effect being scored. Comparisons were based on the ratio of the reduction in hazard uncertainty expected to be achieved after a research study to the cost of the study. Resources could then be allocated to those studies expected to result in the greatest reduction of uncertainty for the least expenditure of research funds.

F. Weighting and Combining Scores

Most methods of scoring for hazard assessment involve assigning subscores to a number of exposure and toxicity factors and then combining the subscores to get a single score representing overall hazard assessments. Subscores may or may not be weighted before combining.

Weighting is used to give factors judged to be most important for the needs of the program the greatest influence in determining the final score. To provide a standard for comparison among effects of different kinds, each effect is assigned a value relative to the others. For example, chemicals found in the environment are more likely to occur at levels having chronic effects than at the very high levels usually required to produce acute effects. Therefore, when the MITRE Corporation was developing a scoring system to select hazardous substances for a German environmental monitoring program, less weight was assigned to acute effects than to chronic effects (*22*). The weighting scheme for acute toxicity: chronic toxicity:bioaccumulation:persistence was 1/3:1:1:1.

Different weighting schemes may be appropriate for different purposes. For example, a positive result in a poorly conducted study may be given little weight when scoring for regulatory purposes, but might be given considerable weight as justification of the need for further testing. Because the success of priority setting is usually limited by an inadequate data base, there is little gain from development of elaborate algorithms for weighting and combining scores. Clement Associates performed a sensitivity analysis to determine the effect of weighting on scores obtained by combining four factors in exposure scoring (general population exposure, quantity released into the environment, production volume, and occupational exposure) (*34*). In the first run, they gave each factor equal weights. In each of four subsequent runs, they gave a different factor a weight of three times that of the remaining factors. They found that the overall ranked order of the approximately 900 chemicals being scored was not sensitive to the different factor weights. They therefore decided to give each factor equal weight on subsequent scoring exercises.

There are several ways in which subscores can be combined. There are also legitimate arguments for not combining subscores. Jouany *et al.* (*25*) stressed the importance of maintaining exposure scores separate from biological effect scores. They noted that, for the general public, an event that occurs frequently but with minimal consequences is not at all equivalent to an infrequent event with severe consequences. On the other hand, Fiksel and Segal (*15*) expressed the opinion that exposure and biological effect factors should be combined so that decisions can be made on the basis of risk/hazard.

Whether or not subscores are eventually combined, scores for individual factors provide valuable information for decision making and should not be lost in the process of determining a single comprehensive score. Ross and Lu (*40*) note that a single combined score tends to hide information and could lead to incorrect conclusions when scores of several chemicals are compared. For example, a chemical receiving a moderately low score for all components could have essentially the same score as a chemical receiving a high score in one or two components and low scores for the remaining components. The former compound may present no serious health risk whereas the latter compound may present a risk of, for example, severe central nervous system damage to those that are exposed.

Various display formats such as grids or matrices are available for displaying exposure and toxicity scores (*16, 25, 41*).

Some subfactors obviously interact multiplicatively. For example, quantity released will equal the production volume times the fraction released. The amount of material to which a person is exposed equals the frequency of exposure multiplied by the extent to which the person is exposed per event multiplied by the fraction that enters the body (*34*). Other factors interact additively. For example, the population exposed to a chemical in manufacturing can be added to the population exposed to the chemical by consumer use. Most factors, however, have no obvious additive or multiplicative relationship. Similar factors may be added in some scoring systems and multiplied in others. Toxicity scores for various toxic effects are often added together to obtain an overall toxicity score.

Multiplication is usually used to combine exposure scores with toxicity scores. For example, to rank chemical components of consumer products for health hazard, the system developed by Illinois Institute of Technology Research Institute (IITRI) for the Consumer Product Safety Commission (CPSC) multiplies the number of people exposed by the mean dose. This exposure factor is then multiplied by the probability of toxic effect. A reason for multiplying subscores is that multiplication gives a wider spread of scores, which facilitates comparisons when ranking large numbers of chemicals.

One effect of multiplying toxicity and exposure scores is that, as either toxicity or exposure approaches zero, the overall hazard score also approaches zero. This approximates the real world situation for as toxicity and exposures increase the increase in hazard is greater than the additive rate (*32*).

When toxicity is expressed in terms of probability of response, it is multiplied by exposure factors. The system developed by Gori to set priorities for carcinogenesis testing uses multiplication to combine four elements: the number of people exposed, estimated annual intake per person by each route of intake, probability that the chemical is carcinogenic for each intake route, and estimated potency for each intake route (*17*). A similar algorithm was used by ICF, Inc. in scoring for human health risks from waste streams (*23*). This system defined overall risk as the expected number of people affected. This overall risk was determined by multiplying three factors: the population at risk, the average intake per person per day, and the probability of response per unit intake. Multiplying by weights for severity of effect was suggested as an option.

This safety factor approach integrates toxicity and exposure information by determining the ratio of the lowest concentration at which an adverse effect is observed to the concentration of the chemical in the environment. This system is valuable in ranking or identifying chemicals in need of regulation or control (*19, 40, 41*). If the safety factor concept is used together with sequential testing, the contribution of uncertainty to the safety factor decreases (*12*). As testing proceeds through succeeding tiers of tests, estimates of expected environmental concentrations and of the concentration producing biological effects can be made with an increasing degree of accuracy and confidence.

V. CRITERIA FOR SCORING

The criteria used for scoring are those one would use in developing a preliminary assessment of probable risk associated with chemical exposure. These criteria must be defined in terms of parameters for which data are readily available for large numbers of chemicals.

A. Criteria for Assessment of Exposure

The three aspects of exposure that are of greatest importance in hazard ranking are the size of the population exposed, the frequency and duration of exposure, and the concentrations/quantities of the chemical to which organisms are exposed. When direct information is available for these exposure factors (e.g., from monitoring chemical levels in the ambient environment or in human tissues), it is obviously desirable to use this information. In most cases, however, direct information regarding the actual quantities of a chemical taken in by humans over a specific time or even the amount of a chemical to which members of a population are exposed is not available, so one must rely on indirect information. Several factors have been used as surrogates for direct measures of exposure.

1. Production

The surrogate most commonly used as an index of exposure is annual production volume. Since this information is quantitative and has been compiled, it is easy to use for scoring or ranking. In most cases, production volume provides a rough indication of the amount of a chemical substance potentially available for release into the environment. The fraction of production actually released to the environment may, however, vary over a wide range. A more serious indictment of production volume as an index of exposure is, however, that it provides information only on chemicals intended for entry into commerce. Production volume as an index of exposure is misleading for substances found in emissions, effluents, or other waste streams, which have a high potential for polluting the environment. Production volume also fails to provide useful information on exposure to impurities, captive intermediates, naturally occurring chemicals, or degradation products.

2. Use

The use pattern can be a major determinant of the type, frequency, and amount of human contact with a chemical substance. Experience of the authors with several of the scoring exercises in the survey has indicated that accurate, complete, and up-to-date use information is very difficult to acquire. Since some uses are more likely than others to lead to human exposure, a quantitative breakdown of uses is particularly valuable but rarely attainable. General categories of use can sometimes be estimated based on the chemical's structure, properties, and similarity to other chemicals for which more complete data are available.

Under Section 8(a) of TSCA, the EPA is currently obtaining use information for a number of high-volume chemicals. This should help to augment the base of available data. It is particularly important to differentiate uses that lead to consumer, occupational, and environmental exposures.

3. Release

Releases to the environment occur as a result of production losses, uses, and disposal. Production losses include the loss of the product chemical in effluents, emissions, and solid wastes and average about 15%, depending on the process used and the properties of the chemical. Releases as a function of use have been reviewed and summary release rates for a number of use categories have been estimated by Becker *et al.* (*4*). It also is important to estimate release rates from disposal as is shown by the following example. In 1971, the sole United States producer of polychlorinated biphenyls (PCBs) voluntarily restricted sales to closed-system applications, but environmental contamination by PCBs continued because most environmental release of PCBs occurs not through use, but rather through disposal.

Release rates, when determined, are often used as multipliers of production, use, or disposal volumes during scoring. For examples, see the SRI systems developed for the National Science Foundation (NSF) (*8, 33*) or the MITRE system for the EPA Office of Air Quality Planning and Standards (OAQPS) (*16*).

4. Environmental Fate

a. Persistence. In hazard assessment systems, persistent chemicals are often given priority over those that are easily degraded or transferred. It is assumed that persistent chemicals are more likely to spread through the environment and that the possibility of exposure is increased proportionally to environmental residence time. A chemical that is easily degraded or transferred should not, however, automatically be given low priority. There are cases in which degradation products or secondary pollutants are more toxic than the chemical substances initially released.

Some examples of systems that consider persistence include that designed by MITRE for Germany's Trend Monitoring Program (*22*) and the Clement system to rank food contaminants for the Office of Technology Assessment (OTA) (*27*).

b. Distribution. To evaluate the distribution of a chemical, initial environmental loading and transport and fate should be considered. There are a number of factors involved at each stage in a chemical's life cycle. The first stage is manufacturing. For this stage, data are needed on the number and geographical distribution of the manufacturing facilities, the production volumes of plants, and the fraction of production lost to each of the environmental media. Monitoring data are sometimes available and can be used to define the fractions lost. Similar evaluations are needed for uses and disposal. These are then combined to give an overall picture of the initial environmental loading.

Transport and fate of the chemicals also should be considered in the scoring system. These require data on the chemical's reactivity, and its likelihood of undergoing transformation through such processes as biodegradation, photolysis, OH-radical reactions, and hydrolysis. There are several models that can be used to predict the ultimate fate of a chemical given its initial loading pattern and transformation data. These are reviewed by Hushon *et al.* (*21*). The models can give an estimation of a chemical's distribution under equilibrium-type conditions and are a useful supplement to monitoring data.

Additional distribution factors that are often considered are persistence and bioaccumulation or bioconcentration potential. These may be indicated by laboratory data or, in the case of the latter, by surrogate data, the log octanol/water partition coefficient. Bioconcentration and persistence are occasionally included in exposure scoring as weighting factors or multipliers instead of additive terms because of their broad impact on exposure.

5. *Population at Risk*

Some scoring systems do consider the size and type of populations exposed as a factor in ranking toxic chemicals. Types of populations considered at risk in scoring systems include (1) general human population, (2) population of a geographical region, (3) population in the immediate vicinity of industrial plants, (4) people whose diets include large quantities of particular foods, (5) occupational groups, (6) groups highly susceptible to specific health effects, and (7) users of common products.

Population groups can be further classified as to whether exposure is voluntary or involuntary. Scores indicating priority consideration would be given to chemicals to which there is extensive involuntary public exposure or to which susceptible segments of the population are exposed. For example, the scoring system developed by the ITC Scoring System Workshop separately considers occupational and general population exposures (*14*). Similarly, the ranking of food contaminants developed by Clement for OTA considers exposure to susceptible subpopulations as a scoring factor (*24*).

6. *Dose*

In order to determine the actual quantity of a substance taken in by humans over a specified time and route, the frequency and duration of exposure should be known. This information is not readily available for most chemicals (drugs are an obvious exception). Although rarely used in scoring, frequency and duration of exposure were included as criteria in the exposure scoring system developed for EPA by Auerbach Associates (*3*). More commonly, dose is expressed as milligrams/kilogram/day as was done in the system developed by Nees (*31*).

Biological monitoring reflects the dose that an organism has absorbed and accumulated. Where measurements of a chemical or its metabolites in biological tissues are available, this information can be used in scoring.

B. Criteria for Assessment of Biological Effects

1. Metabolism and Pharmacokinetics

Although metabolism and pharmacokinetics are relevant to both exposure and health effects, metabolic factors were not often used as criteria in the scoring systems surveyed. Bioaccumulation or ecological magnification were occasionally used. Absorption, bioconversion, and elimination are complex factors and are difficult to handle for scoring purposes. Also, information is not readily available for many chemicals and is rarely compiled in easy-to-use secondary sources.

Relating metabolic information to adverse health effects often requires a high level of expertise in toxicology. Despite these problems, a scoring system that takes these factors into account is generally superior to one that does not. One system among those reviewed that considers metabolism is that developed by the Flavor and Extract Manufacturers' Association to evaluate the hazard from food contaminants (*13*). Another is Hooker's assessment system for oncogenic potential which considers metabolic similarity between test animals and man (*31*).

2. Toxicity

Although toxicity is a criterion for setting priorities for nearly all the scoring systems surveyed, systems varied in the detail with which they specified precisely which effects they were rating. Acute toxicity, carcinogenicity, mutagenicity, and teratogenicity were frequently scored criteria, but other toxic effects also have been considered.

a. Acute Toxicity. Lethal dosage levels are the most frequently used index of acute toxicity, the most common parameters being the LD_{50} and LC_{50} (median lethal dose and concentration, respectively). Because these parameters are quantitative and are summarized in secondary literature (e.g., the NIOSH *Registry for Toxic Effects of Chemical Substances*), they are easy to use for scoring. Since acute lethal doses are much higher than levels usually found in the environment, one could object to the use of acute lethality as a basis for ranking environmental contaminants. Use of this parameter can be justified, however, on the basis that LD_{50} is at least a crude index of biological activity.

Data on acute toxic effects other than lethality are less frequently compiled and are more difficult to use for scoring purposes. Nevertheless, such data, especially information on target organ toxicity, can provide extremely valuable information.

b. Carcinogenicity. Scoring systems for setting priorities among known carcinogens can be based on the results from well-designed and -conducted laboratory animal or epidemiological studies. Scoring systems for selecting suspect carcinogens for research or testing programs are often based on structure–activity relationships, metabolic pathways leading to known carcinogens, *in vitro* tests, short-term *in vivo* bioassays, and individual human case reports. Valuable information for setting priorities among suspect carcinogens can even be obtained from chronic animal bioassays or epidemiological studies that are considered inadequate or inconclusive.

None of the lists surveyed used quantitative risk assessment as a means of setting priorities among carcinogens, although this method has many strong advocates. Scoring systems can consider more factors relevant to strength of evidence of carcinogenicity and to carcinogenic potency than dose–response extrapolation methods. The incidence of a specific tumor type and the lowest dose at which a statistically significant increase in tumor types occurred are factors used in scoring that would also be reflected in quantitative risk assessment, but some additional factors that would be overlooked if chemicals were ranked solely by unit risk include (1) rarity of the elevated tumor type, (2) degree of malignancy, (3) multiplicity of primary tumor types, (4) number of species in which the chemical has been demonstrated positive or tested or found to be negative, and (5) time from the initial administration of the chemical until observation of tumor. Nees (*31*) and Squire (*42*) have developed multifactorial scoring systems that take some of these additional factors into consideration for carcinogenicity.

c. Mutagenicity. Before scoring chemicals for mutagenicity, it is desirable to clear up any ambiguity as to whether the score is to be based on mutagenicity as an indicator of potential carcinogenicity or mutagenicity as the cause of serious heritable disorders. The TSCA–ITC workshop recommended that results of short-term *in vitro* tests be scored separately for mutagenic activity and for carcinogenic potential (*14*). Most of the lists surveyed used very simple scoring systems for mutagenicity. In general, chemicals are scored based on whether or not there was evidence of mutagenicity or on the number of test systems producing positive results for mutagenicity.

d. Teratogenicity and Reproductive Effects. It also is difficult to score chemicals on the basis of teratogenic potency. The data base for predicting human health effects is extremely limited in the areas of reproductive effects, teratogenicity, and embryotoxicity; the types of dose–response data needed for quantification of teratogenic risk are seldom available. It also is often difficult to differentiate between maternal toxicity and fetotoxicity. In addition, interspecies extrapolation is poorly understood and there is a wide range of variability of teratological endpoints (*14*).

Those scoring systems that consider teratogenicity as a criterion for ranking chemicals do so on the basis of the presence or absence of evidence for teratogenicity in animals or humans, though whether or not chick embryo data are considered varies from system to system. Effects observed in nonmammalian model systems, such as chick embryos, are generally given less weight. Some systems combined embryotoxicity with teratogenicity for scoring purposes; other systems combined teratogenicity with other reproductive effects. Epidemiological evidence is usually weighted more heavily than data in other mammalian species.

e. Chronic Systemic Toxicity. Chronic exposure test results are more relevant than are acute test results to the types of exposure due to environmental contaminants. It is, however, difficult to design a single scoring system to cover the wide range of endpoints observed in chronic studies; comparisons between different types of chronic effects are difficult to make. Since comparing one type of chronic effect with another on the basis of severity is highly subjective, scoring for chronic toxicity requires a high level of scientific expertise.

The TSCA–ITC workshop developed a scoring system for toxicological effects in which a chemical's score is based on the dose range at which toxic effects are observed multiplied by a severity coefficient. Severity of the toxic effect is classified as being life threatening, incapacitating, irreversible, or causing minor impairment of function.

C. Criteria for Assessment of Environmental Effects

Scoring for adverse effects on ecological species other than man is seriously hampered by lack of adequate data for chemicals other than pesticides, PCBs, and heavy metals. Scoring for environmental and ecological effects was given serious consideration at the TSCA–ITC workshop (*14*), and a follow-up EPA Workshop on the Environmental Scoring of Chemicals (*41*). The TSCA–ITC workshop developed a matrix by which scores for five factors (lethality, growth and development, reproduction, bioaccumulation, and other toxicological effects) were assigned to each of four ecological categories: (1) microbes, algae, and plants, (2) invertebrates, (3) fish, and (4) birds and mammals. The EPA workshop developed a scoring system in which toxicity multipliers based on threshold levels of concern for various biota would be applied to environmental exposure scores. This is similar to the severity index developed for chronic health effects.

VI. COMPARISON OF SCORING SYSTEM CAPABILITIES

The scoring systems presented in Table I were designed to evaluate chemicals under different prescribed sets of conditions for a variety of purposes. In the following section, systems that were designed with similar endpoints in mind are compared and, where possible, conclusions are drawn. For the purposes of grouping the systems, those trying to evaluate chemicals in particular media are grouped as are those that concentrate predominantly on exposure or toxicity. Finally, those systems that try to combine a number of media, exposure routes, and effects to provide a comprehensive integrated assessment are considered.

A. Systems Focused on a Particular Medium

1. Air

Three of the systems described in Table I were designed for scoring air pollutants (*1, 16, 28*). The Pesticide Manufacturing Air Prioritization developed for EPA by Monsanto is the most mathematically complex of the three since it involves the use of a simplified air dispersion model as part of the scoring (*1*). It also requires the input of data on air concentrations and population densities which are time consuming and expensive to obtain. This would make this model an unlikely candidate for broad application by a variety of users.

The other two scoring methodologies make use of the same data base, that collected by MITRE for application of their scoring method (*16*). The MITRE method has the advantages that it uses data that are readily available on a wide range of chemicals and that it provides a set of default values and ways of compensating for missing data. Being based largely on objective criteria, scoring requires a minimum of expertise. The Science Applications, Inc. (SAI) Rapid Screening and Identification of Airborne Carcinogens of Greatest Concern (*28*) added some additional data to the MITRE data set on stability in the atmosphere and used California production and use data in place of the national data employed by MITRE. The SAI method was concerned only with evaluating the 47 chemicals identified by MITRE as being carcinogenic. The SAI approach also required the use of experts to evaluate the differences in results obtained by combining the data multiplicatively or additively. The authors justified their use of experts as a guard against assigning importance to false positives.

With regard to scoring air pollutants, one therefore has a choice to make between the Monsanto pesticides approach (*1*), which requires the input of monitoring and site-specific data, and the more generalized MITRE approach (*16*). In most cases, the MITRE approach will probably suffice.

2. Water

A number of systems have been developed to evaluate the transport and effects of a broad variety of substances in an aquatic environment (*6, 9, 12, 15, 39*). The major differences among the systems relate to whether they consider human as well as aquatic biota effects and whether they require the input of concentration data for the model to work. The systems developed by the American Society for Testing and Materials (ASTM) (*12*) and Battelle Laboratories (*39*) both concentrate entirely on effects on aquatic ecosystems. Both are designed to use aquatic toxicity test results to predict aquatic hazards from a set of compounds and to indicate areas in which additional testing may be required.

The methods developed by SRI for the Commission of the European Communities (CEC) (*9*) and by A. D. Little for EPA (*15*) also seem similar in approach. Of the two methods, the SRI method is much easier to use. The A. D. Little method is more complex, but it was designed to be applied to a limited set of chemicals, the set of 129 priority pollutants under the Clean Water Act, about which relatively much is known. The SRI method, on the other hand, was designed for application to the Biokon List which consists of 1500 substances that have been identified in European surface waters. Both of these methods differ from the first methods described in this section in that they utilize human as well as aquatic effects data and consider production and use quantities that may reach surface waters.

The final water-related scoring method is that developed by the National Academy of Sciences (NAS) for the United States Coast Guard (*6*). This method is unique in that it is designed to assess the hazard of a chemical that is spilled into a river or other waterway. In this system, the major concern is for human health with wildlife effects being secondary.

The water pollutant evaluation methods therefore fall into three categories: those evaluating aquatic testing results, those concerned with both human and aquatic effects data for ranking and hazard assessment, and the system for evaluating the impact of accidental chemical releases in surface waters. Of the two aquatic test evaluation methods (*12, 39*), the one by Battelle seems to be easier to use and more comprehensive than the ASTM method. Similarly, of the two methods for considering both aquatic and human effects data (*9, 15*), the SRI approach for CEC probably has broader application than that developed by A. D. Little for EPA since it requires less detailed measurements as inputs. The SRI model also has been successfully applied to a far larger set of chemicals.

3. *Dumpsites*

Three of the models presented in Table I are designed to estimate the hazard of chemicals disposed of in dumpsites (*4, 18, 23*). Of the three, only the PHL model takes into account the chemical's persistence and its environmental mobility (*18*). The other two (*7, 23*) are based predominantly on toxicity and volume of the chemical disposed. The Barring model by Booz–Allen also considers explosion potential and its toxicity relates to both humans and the ecosystem (*7*). The Resource Conservation and Recovery Act (RCRA) Risk/Cost Policy model by ICF and Clement (*23*) is somewhat more complex than the Booz–Allen model and is multiplicative, which can tend to exaggerate the impact of a single high value.

Of the three, the PHL model has the broadest potential for application. It is easy to use since it is a linear additive model with weights. It also has the advantage of taking into account the mobility of the chemical among the environmental compartments as well as volumes disposed and toxicity.

4. Food and Food Additives

Two very different methods have been developed for estimating hazards associated with potentially toxic substances in the food supply. The "Decision Tree Approach," developed by the Flavor and Extract Manufacturers' Association (*13*), is extremely useful for substances for which little or no toxicity testing has been done and for which little information is available. Presumable risk estimates are based primarily on structure–activity relationships. The decision tree uses information on known associations between various functional groups and metabolic or toxic properties to assign substances to risk categories based on their molecular structures. This method provides only predictions of hazard. Results of animal tests or human epidemiological studies would take precedence over predictions based on structure–activity relations. If information on metabolism and toxicity is available, a more traditional scoring approach such as that developed by Clement Associates (*27*) to score chemicals contaminating the food supply is preferable.

With Clement's method, scores were assigned for a number of exposure and biological factors (see Table I), and weighted sums of these scores were used for ranking. Criteria were defined so as to make maximum use of readily available objective data. This system is applicable for a broad range of purposes for which one might need to score toxic chemicals in the food supply.

B. Systems Focused on Only One Aspect of Hazard Assessment

Though hazard assessment generally involves both exposure assessment and effects assessment, separate systems have in many cases been developed for each of these areas. In addition, the category of effects is often divided into toxicity and environmental effects. In this section, the systems developed to address each of these areas will be considered. In addition, several systems developed to identify and rank carcinogens will be discussed.

1. Exposure

Three schemes have been developed that attempt to compare chemicals on the basis of exposure (*3, 5, 35*). The system developed by IIT Research Institute for the Consumer Product Safety Commission limits itself to evaluating consumer exposure to chemical substances used in commercially fabricated products (*5*). This system has not been applied by the agency so it remains to be seen whether the proposed scheme is effective. As described, the system requires the making of a number of assumptions by the user as to the likely exposure concentrations and how often the exposure would occur.

The other two exposure schemes were both developed for the EPA in connection with the implementation of the Toxic Substances Control Act. The first was developed by Auerbach Associates to indicate the relative exposure associated with a given use of a chemical for the purpose of identifying chemicals with extremely high exposures for closer examination (*3*). It requires information on the numbers of people exposed, the exposure dose, and the levels of background exposure. Each of these variables is then assigned a factor and weights are applied.

The second scheme is that developed by Clement Associates for the Interagency Testing Committee to provide an indication of the exposure potential to existing chemicals that are being considered as candidates for testing (*35*). This scheme requires less detailed information than the Auerbach model, and it also explicitly includes information on persistence in the environment and the extent of occupational exposure as well as general population and environmental exposure. This scheme likewise involves the assigning of factor scores instead of trying to obtain "hard" numbers. Because of its use with over 1800 chemicals in four scoring exercises, the Clement/ITC system has gained a great deal of credibility. It also has proven to be relatively easy to use and does not require the services of experts to evaluate the data. Though the Auerbach and Clement systems may well give similar answers in a number of cases, the broad base of practical application of the Clement approach would tend to bias future users.

Several additional systems have been developed to evaluate specific categories of exposures including occupational and long-term exposure. These separate subclasses of exposure will be considered below.

a. Occupational Exposure. Three systems have been developed that attempt to rank chemicals on the basis of occupational exposure potential (*10, 30, 38*). The first, and probably the most famous, was developed by NIOSH as a means of evaluating the data from the National Occupational Hazard Survey (NOHS) (*30*). This approach ranked chemicals on the basis of the amount present in various workplaces and on the numbers of persons employed. This was the first time that a comprehensive and systematic approach to identifying and evaluating chemicals in the workplace had been undertaken.

The system developed by Tracor Jitco for the United States Department of Health and Human Services and NIOSH takes the above survey a bit further and tries to use it to identify high-risk occupational groups (*38*). This scheme combines data on chemicals that exist in certain industries and industry operations to try to identify those that pose the greatest risk. The exposure variables come from the NOHS described above, but in this scheme, the data are combined differently to try to reach different conclusions. In many ways, this is a logical extension of the NOHS study described first.

The final occupational exposure study to be evaluated is that performed by SRI for the United States Army to help them evaluate the potential for exposure of persons working in ammunition plants and arsenals to Army chemicals. In considering the concentrations to which the workers are exposed, the model considers the sources, transport, transformation, and intercompartmental transfer of the pollutants. It makes use of separate submodels for air pollutants (a dispersion model), surface water pollutants (a dilution model), and groundwater pollutants (a diffusion model). This type of approach to exposure evaluation is very comprehensive, but requires a significant degree of experience with these various types of models. It is also relatively costly to implement since it requires software development and tailoring as well as the collection of detailed data on the chemicals of concern.

Therefore, of the occupational exposure scoring methodologies reviewed here, that developed by Tracor Jitco probably has the widest application though it makes use of the data base collected by the NOHS. The outputs of the SRI model are probably more accurate than those of the Tracor model, but most workplace exposure takes place within a manufacturing facility and it is unusual to have to consider environmental behavior of chemicals as part of evaluating occupational exposure.

b. Long-Term Exposure. One of the scoring approaches evaluated in this study attempted to identify chemicals that are highly persistent that could be used to monitor long-term exposure trends. In this approach developed by MITRE for the Federal Republic of Germany, basic chemical data on persistence and bioaccumulation were used as estimators of exposure since the likelihood of exposure to a chemical increases directly with its presence in the environment (*22*). This study also used inputs on structural relationships to identify as many different persistent chemical classes as possible. This study provides a unique approach to exposure evaluation, though its potential useful application to areas other than those for which it was intended is unlikely.

2. *Environmental Effects*

Only two of the scoring methods considered were completely concerned with evaluating environmental effects (*39, 41*). The first method was developed by Battelle Laboratories for EPA and the Organization for Economic Cooperation and Development (OECD) (*39*). This method was an attempt to use ecotox test data to predict the toxicity of a chemical in the environment relative to the toxicity of other known substances. Scores were assigned to the various test results and combined to give an overall estimate of ecotoxicological hazard.

The second method evolved at a workshop sponsored by Oak Ridge National Laboratory (ORNL) for EPA's Interagency Testing Committee to identify chemicals in need of additional ecological testing (*41*). This methodology is more comprehensive than the previously described method in that it considers acute data separately from chronic data. In addition, special weights are assigned for chemicals that are persistent or bioaccumulate since there is an increased likelihood of exposure and effects. Though this methodology was applied only to the sample chemicals, it appears to have a significant application potential for other situations in which ranking of chemicals on the basis of environmental effects is required.

3. *Toxicity*

A good example of a scoring system based on toxicity is the system developed for the TSCA Interagency Testing Committee by Clement Associates (*24*). This scoring system was applied to a preliminary list of approximately 300 substances previously scored and selected on the basis of exposure.

For each of the seven factors shown in Table I, scoring criteria are carefully defined to facilitate uniformity of scores assigned by different experts. Since the purpose of this scoring exercise is to identify chemicals needing testing, the scoring system has two independent scoring systems. Numerical scores are used to score chemicals on known biological activity. Letter scores are used to score on need for further testing (these letter scores were converted to negative scores for entry on the computer). The positive numerical scoring system can be used on its own when scoring chemicals of known toxicity for regulatory purposes.

4. *Carcinogen Evaluation*

Scoring for carcinogenicity allows consideration of a multiplicity of factors associated with carcinogenic risk. This gives scoring systems an advantage in risk extrapolation over mathematical models, which neglect a great deal of relevant biological information.

Three scoring systems presented in Table I use results of carcinogenicity studies to score known carcinogenicity for risk to the human population (*17, 31, 42*). Nees (*31*) and Squire (*42*) both express the opinion that the extent of risk to humans should be considered in determining appropriate regulatory options for chemical carcinogens. They have each proposed multifactorial scoring systems to categorize carcinogens for this purpose. Gori (*17*) proposed a scoring system to select chemicals for testing based on suspicion of carcinogenic hazard.

Nees (*31*) developed a scoring matrix in which scores from animal studies, epidemiology studies, and short-term tests are adjusted for relevant factors and then combined to provide an index of oncogenic potential. Factors by which animal test scores are adjusted include metabolic similarity between test species and man, appropriateness of the route of exposure, quality of the study, margin of safety between lowest dose producing animal response and highest human exposure level, and time to tumor. Factors by which epidemiology scores are adjusted include tumor type, suitability of controls, incidence in exposed population relative to controls, exposure of cohort to other suspect chemicals, exposure level, dose–response, and power or sensitivity of the study. Short-term studies also are scored and weighted. The sum of weighted scores for all animal studies, epidemiology studies, and short-term tests is the index of oncogenic potential.

Squire (*42*) proposed a scoring system to rank animal carcinogens according to the most relevant toxicological evidence derived from animal and genotoxicity studies. This method was designed to score animal carcinogens that have been identified by adequate testing at multiple doses in at least two species. The proposed system scores for each of six factors: number of species showing carcinogenic or neoplastic response, number of types of neoplasms, spontaneous incidence of tumors, dose–response relationships, malignancy of induced neoplasms, and whether the mechanism is genotoxic. These factors provide a comprehensive coverage of the carcinogenicity subject area.

The Gori approach (*17*) relies more on noncarcinogenic factors such as production, population exposed, and exposure levels, but it does have a factor for the probability that a chemical is carcinogenic by the various exposure routes and a score for potency. These latter two scores are highly subjective and require considerable familiarity with the whole carcinogenicity data base to apply accurately.

Both the Nees (*31*) and Squire (*42*) approaches represent advantages over the standard unit-risk approach in that they consider numerous factors, including carcinogenic potency and genotoxicity, that are ignored in the unit-risk approach. These two methods are roughly equivalent in ease of use.

C. Comprehensive Hazard Evaluation Systems

A number of systems have been developed that attempt to combine exposure and effects data in an attempt to determine the relative hazard posed by a group of chemicals (*8, 11, 14, 19, 25, 29, 40*). The systems, however, differ markedly in their degree of sophistication and complexity.

1. Less Complex Models

Probably the least complex multiple-component system is that developed by SRI for the EPA Office of Research and Development (ORD) to select chemicals for more in-depth study (*11*). Instead of requiring the collection and evaluation of large numbers of data, this system relies on the judgment of a group of experts to assign scores of 0 to 10 in a wide variety of categories. The quality of the evaluation, therefore, is directly proportional to the knowledge and experience of the experts.

Another straightforward evaluation scheme is that developed by SRI for NSF to identify chemicals of potential hazard to the environment (*8*). Based on a cursory overview of the scientific literature, the experts were asked to assign scores for environmental and health impacts. These were then combined with production and release-rate data to give a hazard estimate. This method also relies on the use of experts and is very similar to that described above (*11*).

Finally, the State of Michigan developed a simple model to identify chemicals that should be of potential concern (*29*). It combines factored toxicity data with data on bioaccumulation and persistence. This model is unusual in that a high score in any category will automatically flag a chemical for closer examination, as will a high total score.

Of these three models, that developed by the State of Michigan is the easiest to apply, but it lacks a real consideration of exposure or environmental effects (*29*). Of the two models by SRI that rely on expert judgment, the one for NSF is probably the easiest to apply (*8*). None of these models has been used beyond its initial application.

2. More Comprehensive Approaches

Of the more comprehensive models, that developed by the Monitoring and Assessment Research Centre (MARC) of the United Kingdom is unique in that it is qualitative rather than quantitative (*19*). Evaluators are asked to rank exposure from production and use, environmental fate, target organisms, and toxicity on a three-value score (high, medium, low). Several options for combining scores are offered. They suggest simply adding the scores, using a pattern-analysis scheme, or using more novel techniques such as Delphi analysis or cross-impact analysis. The description of this method was meant to provide general guidance and not to provide a definitive methodology. It is unclear whether this model, as documented, has ever been applied, but it takes into account a broad range of potential impacts and is thus of interest.

The model developed by M. Jouany *et al.* (*25*) for the French Ministry of Environment is designed to evaluate the potential hazard from exposure to new chemical substances as required by the Commission of the European Communities sixth amendment to the Council Directive of 27 June 1967 on the Approximation of the Laws of Member States Relating to the Classification, Packaging, and Labeling of Dangerous Substances. This model is forced to rely upon only those data on the chemical that are supplied by the manufacturer as part of his notification package. These data are known as the minimum premarket data set or MPD. This model, therefore, relies on predictions of production, distribution, and dispersion plus test results on peristence, bioconcentration, animal and aquatic acute toxicity, mutagenicity, and skin irritation and sensitivity. Scores are assigned by expert reviewers in each of these categories. In addition, penalty points are assigned for interactions if the result is unfavorable. These scores are then combined algebraically and scores obtained for each new chemical for exposure and effects. Finally, these scores are compared to those for a number of existing chemicals using a graphical grid. Those appearing to be similar in hazard potential to existing hazardous chemicals are subsequently examined in closer detail and potentially regulated. This approach is only applicable for new chemicals in that it does not take into account any knowledge concerning a chemical's potential to cause chronic effects. Unlike some other methods, it does not try to predict these chronic effects based on structural similarity to known "bad actors."

The final two methods to be considered here, that by Enviro Control, Inc. for the Interagency Testing Committee (*14*) and that by Oak Ridge National Laboratories for EPA (*40*), are both very comprehensive and allow for the use of many kinds of data on existing chemicals to reach a single ranking score. The Enviro method is probably slightly less useful in that the segments of the model do not fit well together. This is because the various segments were developed independently by nine groups of experts during a 3-day workshop. The segments, however, when considered separately, are extremely well-conceived and demonstrate a good understanding of the available data in each area. Specific model segments focus on environmental release, environmental chemistry and fate, occupational exposure, general population exposure, ecosystem effects, mutagenicity, carcinogenicity, reproductive effects, and other toxic effects.

The chemical scoring system developed by ORNL (*40*) relies heavily on the earlier Enviro effort described above (*14*). The categories for scoring are almost identical to those above except that fetotoxicity is split out from reproduction effects and separate categories are provided for acute and chronic toxicity. The big advantage of this approach is its having developed a way of combining scores in all of these areas. This is done by first scoring then multiplying these scores by a set of preestablished weights. These weights relate to the relative severity of the various exposures and effects. The products of scores and weights are then summed to obtain a chemical score. This scoring system is very comprehensive, but it has not been applied to a sufficient number of chemicals to determine its effectiveness. Because it relies heavily on several previous efforts that were validated, it stands a good likelihood of giving appropriate results upon broader practical application.

VII. APPLICATIONS OF SCORING SYSTEMS

Scoring for hazard assessment must be based on sound scientific judgment applied to a data base of information concerning exposure levels and adverse health and environmental effects. The survey of scoring systems described in Table I shows that scoring can be used for a wide variety of specific purposes.

Scoring systems tend, however, to fall into two major categories: those used to select chemicals for testing, and those used to select chemicals for regulation. In the first case, scoring is based on estimates or projections from the available data. In these cases, the scoring system not only identifies chemicals, it also identifies those areas of concern requiring further testing.

In selecting chemicals for regulatory action, however, the scoring system is applied for the purpose of ranking chemicals on the basis of clearly demonstrated adverse effects. Scoring, in this case, represents the first step in the regulatory review process. Once a chemical is identified as a candidate for regulation or control, the regulatory agency will perform a thorough investigation of exposure, health effects, and regulatory options.

By varying the specific criteria, the parameters on which scoring is based and the weights given to specific factors, scoring systems can be adapted to meet the needs of a variety of programs. The systems surveyed were used to score pesticides, new chemicals, food contaminants, synthetic organic chemicals, and hazardous wastes. Some scoring systems were also developed to deal with specific environmental compartments such as the atmosphere or aquatic life. Systems were developed for such purposes as assessing risks from chemicals in waste streams, dumpsites, or chemicals being transported by water. Some scoring systems assessed risks to specific populations such as employees of certain industries, users of consumer products, or residents in the vicinity of a landfill.

Scoring can thus function as a flexible tool adaptable to many purposes by a variety of users.

References

1. Archer, S. R., McCurley, W. R., and Rawlings, G. D. (1978). ''Source Assessment: Pesticide Manufacturing Air Prioritization,'' EPA-600/2-78/004d. EPA-IERL, Research Triangle Park, North Carolina.
2. Astill, B. D., Lockhart, H. B., Jr., and Moses, J. B. (1980). ''Sequential Testing for Chemical Risk Assessment.'' Paper presented at the Second International Congress on Toxicology, Brussels, July 6–11.
3. Auerbach Associates, Inc. (1977). ''Test of EPA Index of Exposure.'' Auerbach Assoc., Philadelphia.
4. Becker, D., Fochtman, E., Gray, A., and Jacobuis, T. (1979). ''Methodology for Estimating Direct Exposure to New Chemical Substances,'' EPA 560/13-79-008. IIT Res. Inst., Chicago.
5. Becker, D. S. (1978). ''Design of a Chemical Hazard Ranking System,'' Final Rep., Contract No. CPSL-R-77-0068. IIT Res. Inst., Chicago.
6. Beckman, R. B. (1974). ''System for Evaluation of the Hazards of Bulk Water Transportation of Industrial Chemicals.'' Nat. Acad. Sci., Washington, D.C.
7. Booz–Allen Applied Research, Inc. (1975). ''A Study of Hazardous Waste Materials, Hazardous Effects, and Disposal Methods,'' Vol. 1, BARRING Rep. No. 9075-003-001. Booz–Allen, Bethesda, Maryland.
8. Brown, S. L., Chan, F. Y., Jones, J. L., Liu, D. H., McCaleb, K. E., Mill, T., Sapios, K. N.,

and Schendel, D. E. (1975). "Research Program on Hazard Priority Ranking of Manufactured Chemicals." NTIS PB-263-161/2ST; PB-263-162/(ST; PB-263-163/8ST; PB-263-164/6ST; PB-263-165/3ST). Stanford Res. Inst., Menlo Park, California.

9. Brown, S. L., Cofer, R. L., Eger, T., Liu, D. H. W., Mabey, W. R., Suttinger, K., and Tuse, D. (1980). "Ranking Algorithm for the EEC Water Pollutants," Final Rep., Contract No. ENV/223/74-EN. Stanford Res. Inst., Menlo Park, California.
10. Brown, S. L., Cohen, J. M., Macrea, N., and Small, M. J. (1978). "Setting Priorities for R&D on Army Chemicals," EGU-4479. Stanford Res. Inst., Menlo Park, California.
11. Brown, S. L., Holt, B. R., and McCaleb, K. E. (1976). "Systems for Rapid Ranking of Environmental Pollutants—Selection of Subjects for Scientific and Technical Assessment Reports," EPA Contract No. 68-01-2940. Stanford Res. Inst., Menlo Park, California.
12. Cairns, J., Jr., Dickson, K. L., and Maki, A. W. (1979). "Estimating the Hazard of Chemical Substances to Aquatic Life." Am. Soc. Test. Mater., Philadelphia.
13. Cramer, G. M., Ford, R. A., and Hall, R. L. (1978). *Food Cosmet. Toxicol.* **16,** 255–276.
14. Enviro Control, Inc. (1979). "Scoring Chemicals for Health and Ecological Effects Testing," Proceedings of TSCA-ITC Chemical Scoring System Workshop, San Antonio, Feb. 25–28, 1979. Enviro Control, Rockville, Maryland.
15. Fiksel, J., and Segal, M. (1982). "An Approach to Prioritization of Environmental Pollutants: The Action Alert System," Final Draft Rep. (Rev.), EPA Contract 68-01-3857. Arthur D. Little, Cambridge, Massachusetts.
16. Fuller, B., Hushon, J., Kornreich, M., Ouellette, R., Thomas, L., and Walker, P. (1976). "Scoring of Organic Air Pollutants," MTR-7248. MITRE Corp., McLean, Virginia.
17. Gori, G. B. (1977). *In* "Contaminants for Bioassay Priority in Air Pollution and Cancer in Man" (V. Mohr, D. Schmal, L. Tomatis, and W. Davis, eds.), pp. 99–111. Int. Agency Res. Cancer, Lyon, France.
18. Hagerty, J. D., Pavoni, J. L., and Heer, J. E. (1973). "Solid Waste Management." Van Nostrand Reinhold, New York.
19. Harriss, R. C. (1976). "Suggestions for the Development of a Hazard Evaluation Procedure for Potentially Toxic Chemicals," MARC Rep. No. 3. Monitoring and Assessment Research Centre of the Scientific Committee on Problems of the Environment, Chelsea College, U.K.
20. Hushon, J. M., and Clerman, R. J. (1981). *Hazard Assess. Chem.: Curr. Dev.* **1,** 323–388.
21. Hushon, J. M., Klein, A. W., Strachan, W. J. M., and Schmidt-Bleek, F. (1983). *Chemosphere* **12**(6), 887.
22. Hushon, J., Saari, S., Small, R., Thoman, D., Clerman, R., and Zimmerman, T. (1978). "Baseline Plan for Design of a Hazardous Substances Monitoring Program," MTR-7918. MITRE Corp., McLean, Virginia.
23. ICF, Inc. (1982). "RCRA Risk/Cost Policy Model Project," Phase 2 Rep. Office of Solid Waste, U.S. Environ. Prot. Agency, Washington, D.C.
24. Interagency Testing Committee. (1977). "Initial Report to the Administrator Under the Toxic Substances Control Act." *Fed. Regist.* **42**(197), 55026–55080.
25. Jouany, J. M., Vaillant, M., Blarez, B., Cabridenc, R., Ducloux, M., and Schmitt, S. (1982). "Approach to Hazard Assessment by a Qualitative System Based on an Interaction Concept between Variables." Paper presented at Symposium on Chemicals in the Environment, Lyngby-Copenhagen, Denmark, Oct. 18–20.
26. Jurs, P. C. (1979). Computer-assisted structure/activity. Studies of chemical carcinogens using pattern recognition. *In* "Scoring Chemicals for Health and Ecological Effects Testing," Proceedings of TSCA-ITC Chemical Scoring System Workshop, San Antonio, Feb. 25–28, 1979. Enviro Control, Rockville, Maryland.
27. Kornreich, M. R., Nisbet, I. C. T., Fernsterheim, R., Beroza, M., Shah, M., Bradley, D., Turim, J., Pinkney, A., and Smith, D. (1979). "Priority Setting of Toxic Substances for

Guiding Monitoring Programs,'' Contract No. OTA-C-78-372. Clement Assoc., Washington, D.C.
28. Margler, L. W., Rogozen, H. B., Ziskind, R. A., and Reynolds, R. (1979). *J. Air Pollut. Control Assoc.* **29,** 1153–1157.
29. Michigan Department of Natural Resources. (1979). ''Critical Materials Register 1979,'' Publ. No. 4833-5323. Mich. Dept. of Nat. Resources, Lansing.
30. National Institute for Occupational Safety and Health. (1977). ''National Occupational Hazard Survey,'' 3 vols., DHHS Publ. No. 78-114. Depart. of Health Human Serv., Washington, D.C.
31. Nees, P. O. (1979). *In* ''Toxic Substances Control'' (M. L. Miller, ed.). Government Inst., Inc., Washington, D.C.
32. Nees, P. O. (1979). The Hooker Chemical Corporation's scoring system. *In* ''Scoring Chemicals for Health and Ecological Effects Testing,'' Proceedings of TSCA-ITC Chemical Scoring System Workshop, San Antonio, Feb. 25–28, 1979. Enviro Control, Rockville, Maryland.
33. Nelson, N., Van Duuren, B., and Goldschmidt, B. M. (1975). ''Final Report of the NSF Workshop Panel to Select Organic Compounds Hazardous to the Environment.'' Natl. Sci. Found., Washington, D.C.
34. Nisbet, I. C. T. (1979). Ranking chemicals for testing: A priority-setting exercise under the Toxic Substances Control Act. *In* ''Scoring Chemicals for Health and Ecological Effects Testing,'' Proceedings of TSCA-ITC Chemical Scoring System Workshop, San Antonio, Feb. 25–28, 1979. Enviro Control, Rockville, Maryland.
35. Office of Toxic Substances/EPA. (1977). ''Preliminary List of Chemical Substances for Further Evaluation by the TSCA Interagency Testing Committee.'' U.S. Environ. Prot. Agency, Washington, D.C.
36. Office of Toxic Substances/EPA. (1978). ''An Ordering of the NIOSH Suspected Carcinogens List Based on Production and Use Data,'' EPA-560/1-78-001. U.S. Environ. Prot. Agency, Washington, D.C.
37. Office of Toxic Substances/EPA. (1980). ''Chemical Selection Methods: An Annotated Bibliography.'' U.S. Environ. Prot. Agency, Washington, D.C.
38. Pielmeier, G. R. (1981). ''Identification of High Risk Occupational Groups and Industrial Processes Using RTECS/NOHS Data,'' Final Rep., Contract No. 210-78-0076. Tracor Jitco, Rockville, Maryland.
39. Pommeroy, S. E., Brauning, S. E., and Kidd, G. H. (1980). ''Validation of the OECD Ecotoxicology Testing Scheme Base Set.'' Battelle Laboratories, Columbus, Ohio.
40. Ross, R. H., and Lu, P. (1981). ''Chemical Scoring System Development,'' Contract No. W-7405-eng-26. Oak Ridge Natl. Lab., Oak Ridge, Tennessee.
41. Ross, R. H., and Welch, J. (1980). ''Proceedings of the EPA Workshop on the Environmental Scoring of Chemicals (Aug. 13–15, 1979),'' ORNL/EIS-158, EPA-560/11-80/010. Oak Ridge Natl. Lab., Oak Ridge, Tennessee.
42. Squire, R. (1981). *Science* **214,** 877–880.
43. Veith, G. D. (1979). Structure/activity research for aquatic toxicity testing. *In* ''Scoring Chemicals for Health and Ecological Effects Testing,'' Proceedings of TSCA-ITC Chemical Scoring System Workshop, San Antonio, Feb. 25–28, 1979. Enviro Control, Rockville, Maryland.
44. Wipke, N. T. (1979). XENO: Computer assisted prediction of plausible metabolites from xenobiotic compounds. *In* ''Scoring Chemicals for Health and Ecological Effects Testing,'' Proceedings of TSCA-ITC Chemical Scoring System Workshop, San Antonio, Feb. 25–28, 1979. Enviro Control, Rockville, Maryland.

The Role of Medical Records in Evaluating Hazardous Chemical Exposures

David A. Savitz

Department of Preventive Medicine and Biometrics
University of Colorado School of Medicine
Denver, Colorado

I. Introduction ... 111
II. Potential Uses of Medical Record Systems ... 113
III. Characteristics of Medical Records as a Data Source ... 115
IV. Attributes of Medical Records Determining Their Effectiveness in Evaluating Hazardous Chemical Exposures ... 119
  A. Sensitivity and Specificity ... 119
  B. Speed of Detection ... 121
  C. Feasibility ... 123
  D. Comprehensive Coverage of Health Outcomes ... 123
V. Methodological Considerations in Implementing a Record-Based Hazard Evaluation System ... 124
  A. Cross-Sectional versus Cohort Design ... 124
  B. Objective Assessment ... 125
  C. Data Validation ... 127
  D. Consideration of Potential Confounding Factors ... 127
VI. Case Studies ... 128
  A. Pesticide Contamination of Drinking Water ... 128
  B. Carbon Disulfide and Coronary Heart Disease ... 131
  C. Environmental Dioxin Exposure and Adverse Reproductive Outcomes ... 134
VII. Recommendations ... 136
  References ... 138

I. INTRODUCTION

In recent years, societal concern over the potential adverse health effects of environmental chemicals has increased. This has engendered a variety of tech-

HAZARD ASSESSMENT OF CHEMICALS:
Current Developments, Vol. 3

ISBN 0-12-312403-4

nologies for anticipating and identifying the links between chemical exposure and health (*26*). Examples include animal modeling, studies of controlled exposure to humans, and epidemiologic assessment of human health as it occurs in the community. Epidemiologic studies have inherent strengths and weaknesses compared to other approaches. The principal advantage is the "real world" basis: exposures and effects are the items of ultimate interest rather than surrogates to be extrapolated across animal species, chemicals, or levels of exposure. The key limitation of an epidemiologic approach to hazard assessment is that it is usually an observational rather than experimental science. As such, components of the observed natural "experiment" underlying the community's health experience must be deciphered, since the exposure conditions cannot be manipulated. Measurements of exposure and health outcomes and careful consideration of research design can, however, lead to causal inferences from epidemiologic evidence (*21*).

Within the field of epidemiology, there is a variety of approaches to address adverse chemical effects. The most common is the implementation of a specific research protocol to document exposure and health status to then determine whether an association between the two is present. For example, Kreiss *et al.* (*19*) constructed a DDT exposure history for a number of the residents of Triana, Alabama and assessed the link to a variety of biochemical, physiological, and clinical health endpoints. Often, epidemiologists opportunistically acquire data that had been collected for other purposes and use them to measure exposure or health. For example, vital statistics on birth and death are maintained for governmental administrative purposes but lend themselves to epidemiologic interpretation.

Similarly, medical records constitute a potentially rich epidemiologic data source collected primarily for the purposes of clinical patient care. In some special situations, the possibility of epidemiologic exploitation is built in (*20, 27*), but more often those who design and work with the data collection system have other goals. Nonetheless, the wealth of health data collected on patients that is recorded in medical records provides information on health status that potentially can be linked to environmental exposures. This article provides recommendations for the optimal use of medical records for the purpose of making inferences about potential hazardous health effects of chemical exposures. Options other than medical records, such as special data collection efforts, should be viewed throughout this discussion as implicit alternatives.

This article is addressed principally to persons who have chosen to use health records to examine a potentially adverse chemical effect. The principles, however, are also useful for designers of such record systems since a recognition of the needs of the data users will lead to more optimal methods of data collection. In addition, those who must interpret and then take action based on the results of research derived from medical records should understand the strengths and weaknesses of such research.

This article first delineates the purposes for which medical record systems can be used. A distinction is made between surveillance functions and the use of records to address specific hypotheses. Then, the sources and types of records are described including both their potential and actual contents. A differentiation between records in a general community setting and an occupational setting is made. The characteristics of a medical record system that are necessary for it to be useful are described. These include the system's sensitivity and specificity, rapidity of identification, feasibility, and comprehensiveness. Epidemiologic concerns about the implementation of a data system to evaluate chemical hazards focus on the need for a defined cohort, objective assessment of health status, data validation, and consideration of potential confounding factors.

Three detailed case studies are presented to illustrate these principles. The first is a hypothetical investigation of neurotoxicity in relation to a pesticide in drinking water. Next, the cardiovascular effect of occupational exposure to carbon disulfide as determined through medical records is discussed. Third, the adverse reproductive impact of community dioxin exposure illustrates the complexities of addressing this facet of health. Finally, recommendations are offered to persons designing, using, or interpreting results from medical record systems in relation to hazardous chemical exposures.

II. POTENTIAL USES OF MEDICAL RECORD SYSTEMS

Medical record systems offer the potential to document comprehensively the health experience of a community or company. The patient record system of the Mayo Clinic in Rochester, Minnesota illustrates the uses of medical records as an epidemiologic resource (*20*). Rates of disease occurrence calculated using this record system have provided a description of the community's health status to allow for planning of health services, have monitored temporal trends in disease patterns, and have allowed comparison to disease rates in other areas. More important is the value of these data as a starting point for analytic (rather than descriptive) epidemiologic studies which seek causal factors in disease occurrence. Prospective studies examining the health of persons with and without a given exposure can use such records to determine various health endpoints of interest. Case-control studies, which contrast the exposure histores of persons with and without illness, can rely on patient records as the source of cases. The careful documentation of the health status of any defined group facilitates analytic studies linking environmental exposures to health.

Within the area of environmental epidemiology, medical records can serve as part of an evaluation system concerned with possible chemical influences on health. In this particular use, illness rates derived from such records are linked in some manner to a measure of the patients' exposure. The interpretations of the resulting environment–health association (or lack of an association) can be di-

TABLE I

Characteristics of Hypothesis-Testing and Hypothesis-Generating Research

Hypothesis-testing	Hypothesis-generating
Background literature indicates specific questions to be addressed	Absence of literature to focus the research
Hypothesis stated a priori	No hypothesis formulated a priori
Few exposures and health endpoints considered	Many exposures and/or health endpoints considered
Observation of purported link supports causal inference	Observation of association supports only further study

chotomized into those that address a specific, well-defined hypothesis (hypothesis testing) versus those resulting from a more exploratory surveillance system (hypothesis generating) (Table I).

Hypothesis testing in the context of environmental exposures refers to establishing, a priori, a health outcome thought to be related to hazardous chemicals and then assessing whether that association is observed. The impetus for such a study might be toxicological studies, other epidemiologic investigations, or inferences based on chemical similarities to toxins with a known effect. Nonetheless, the investigator could state at the outset of the study what exposure–health relationship is postulated and give some rationale for his or her hypothesis.

The counterpart to this is surveillance in which there is a general interest in the group's health experience with no specific hypothesis to be addressed. The health of a population is monitored to generate plausible hypotheses about exposure and health. Specific health problems found to be in excess would then serve as the starting point for a hypothesis-testing investigation.

Both hypothesis testing and generation are valid uses of medical records in the context of chemical exposures, but the inferences that can be made differ depending on the type of approach used. Assume an association has been found between a chemical X and health outcome Y. If this hypothesis were stated prior to the study with a firm theoretical or empirical basis, this finding carries substantial weight in making an etiologic inference that X causes Y. If, however, a number of different chemicals (X_1, X_2, . . . X_n) and health outcomes (Y_1, Y_2, . . . Y_n) were considered because of a general interest in chemical–health associations, the meaning of any particular association is weakened. The X–Y association could have easily been a chance finding. In the next investigation this unanticipated association could be specified as a hypothesis to be tested.

Although this dichotomy between hypothesis-testing and hypothesis-generating studies is in fact more of a continuum, it serves to characterize the different uses and interpretations of data derived from medical records. There is often a strong temptation to discover an association fortuitously and then assemble evidence to make the link plausible, claiming that evidence of a causal association was identified. This pitfall occurs because post hoc and a priori explanations are confused.

III. CHARACTERISTICS OF MEDICAL RECORDS AS A DATA SOURCE

The scope of health outcomes that can be investigated in medical records is defined by the contents of such records. The content is a function of both the information requested in the system, as outlined in Table II, and what *actually* is recorded in a readable and retrievable manner.

The potential scope of medical record contents is extensive (Table II). Background characteristics, including the individual's medical and social history and specific health risks, are commonly recorded by the health care provider. The procedures used to arrive at a diagnosis and monitor the course of a disease offer potential as health outcomes in studies of chemical exposures. Medications, surgery, and other medical treatments are another major body of information. Finally, and most importantly, the diagnoses themselves are recorded. These are

TABLE II

Typical Contents of Medical Records from Hospital, Outpatient Clinic, and Occupational Settings

	Hospital[a]	Outpatient clinic[b]	Occupational setting[c]
Background data	X	X	X
Medical history	X	X	X
Physical examination	X	X	X
Clinical notes	X	X	X
Diagnostic tests	X	X	X
Diagnoses	X	X	X
Therapeutic program	X	X	
Functional abilities[d]	X		
Medications	X	X	X
Reproductive history		X	
Pulmonary function tests			X
Audiometry			X
Vision tests			X
Workmen's compensation information			X

[a] University Hospital, University of Colorado, Denver, Colorado.
[b] A. F. Williams Family Medicine Center, Denver, Colorado.
[c] General Foods Corporation, White Plains, New York, and Mountain Bell, Denver, Colorado.
[d] Patients' capability to wash, dress, eat, etc.

the items most likely to be of use in assessing adverse chemical influences on health because of their value in summarizing health status.

As data resources, medical records offer the possibility of assessing many health affects, but there are important exceptions. For example, an interest in the arsenic content of hair, level of lead in the urine, or serum cholinesterase levels will not be addressed in routine records. Occupational records, however, which contain data on more defined populations, have begun to incorporate such special studies. For example, spirometry and audiometry have become part of routine medical examinations in settings where respiratory toxicants and noise are of concern (*17*). Since medical records have great potential usefulness, the next concern is whether their potential can be realized. Unfortunately, experience has shown that it is notoriously difficult to extract usable information from medical records as they are routinely collected. Fries (*9*) noted that ''the medical record has been described as bulky, disorganized, unstructured, and redundant'' (p. 871). Murnaghan and White (*24*) described medical records as ''disorganized, illegible, incomplete, or not to be found in the first place'' (p. 825). They are compiled in a haphazard fashion with little logical organization to assist in data retrieval. Cryptic notes from busy clinicians are often indecipherable (Fig. 1).

The question of whether the information listed in Table II is actually present cannot always be answered in the affirmative. For example, Monson and Bond (*23*) considered the completeness with which medical records recorded medication prescriptions to outpatients. This is a data item that is crucial to clinical care and thus there is a strong incentive for accurate information. Over 20% of the prescriptions were not recorded at all, and many more contained inaccuracies regarding refills and dosage. Given these results, such items as smoking history or nonspecific symptoms cannot be expected to be recorded accurately in a high proportion of records. Accurate abstraction of data from records requires that the information be both present and readable. Legibility must be a key concern in deciding whether medical records will be useful for investigative (and clinical) purposes.

The increasing use of computerized records, illustrated by the General Foods Corporation approach (Fig. 2), helps to overcome the above concerns with completeness and legibility (*8*). First, the format required for computerization consists of fixed choices (e.g., Yes/No, Past/Now), thus making missing data on a particular item apparent. Second, those answers that are provided are unambiguous to the abstractor. [It is worth noting that these attributes are actually artifacts of computerization that can also be accomplished through development of appropriate forms and conscientious efforts to complete them as are done at the Mayo Clinic (*20*).] Third, the data can be retrieved and analyzed without incurring substantial clerical time in locating and abstracting the records. Movement toward computerization of medical records should be welcomed by all users of such data, including investigators interested in toxic chemical effects.

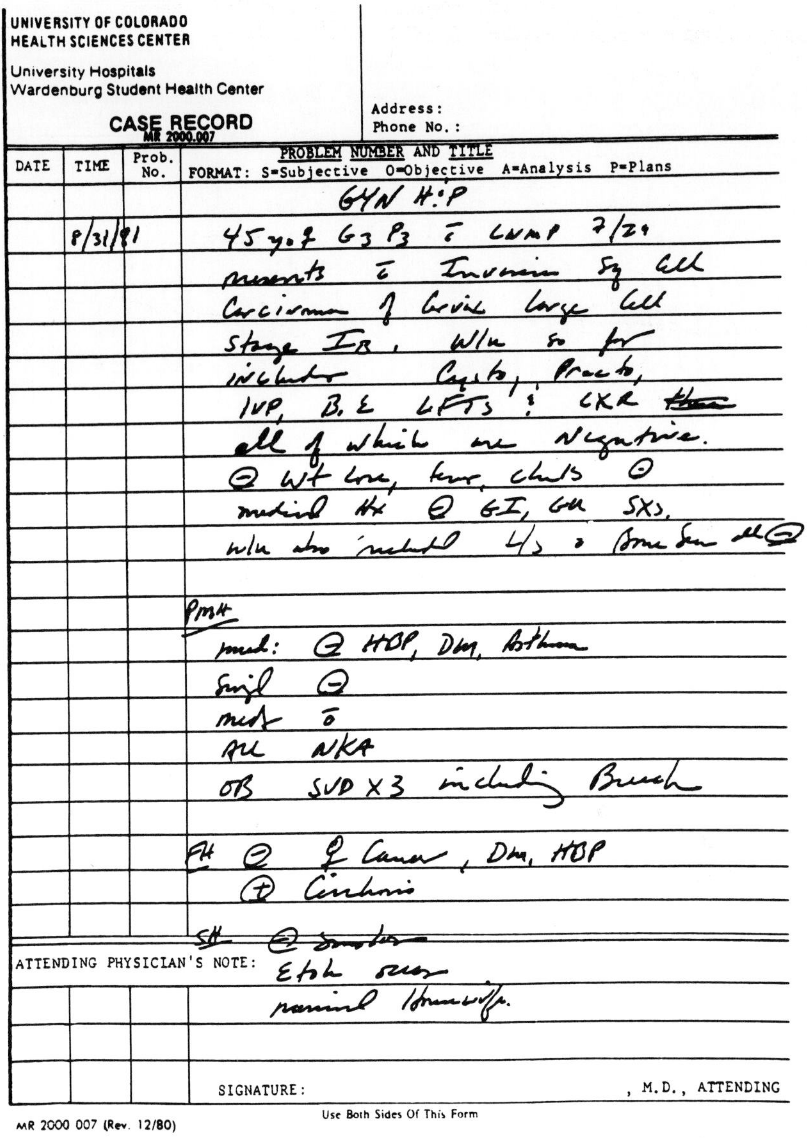

UNIVERSITY OF COLORADO
HEALTH SCIENCES CENTER

University Hospitals
Wardenburg Student Health Center

CASE RECORD
MR 2000.007

Address:
Phone No.:

DATE	TIME	Prob. No.	PROBLEM NUMBER AND TITLE FORMAT: S=Subjective O=Objective A=Analysis P=Plans
			GYN H&P
8/31/81			45 y.o.f G3 P3 c LNMP 7/29
			presents c Invasive Sq Cell
			Carcinoma of Cervix large cell
			Stage IB, W/u so far
			includes Cysto, Procto,
			IVP, B.E LFTs & CXR ~~then~~
			all of which are Negative.
			⊖ Wt loss, fever, chills ⊖
			medical Hx ⊖ GI, GU SXs.
			w/u also included L/S & Bone Scan all ⊖
			PMH
			med: ⊖ HBP, DM, Asthma
			Surg ⊖
			meds ō
			All NKA
			OB SVD X3 including Breech
			FH ⊖ of Cancer, DM, HBP
			⊕ Cirrhosis
			~~SH ⊖ Smoker~~
ATTENDING PHYSICIAN'S NOTE:			EtOH occas
			married Housewife.
			SIGNATURE: , M.D., ATTENDING

MR 2000 007 (Rev. 12/80)

Use Both Sides Of This Form

Fig. 1. Example of case notes from hospital medical records illustrating the difficulty in extracting information from such a resource.

GENERAL FOODS CORPORATION
MEDICAL AND ENVIRONMENTAL HEALTH SYSTEM

PAGE 5

FORM CODE — CASE NUMBER

ILLNESSES, DISEASES AND SYMPTOMS

Social Security Number ____________

CHECK ITEMS THAT REQUIRED MEDICAL CARE OR HOSPITALIZATION, OR WERE OF SEVERE CONCERN TO YOU.
FIRST COLUMN FOR THOSE IN THE PAST, SECOND COLUMN FOR THOSE AT THE PRESENT TIME.

PAST NOW ITEM

GENERAL
Under stress
Fever
Chills
Tire easily
Night sweats
Run-down
Dizziness

INFECTIONS
Polio
Measles
Tuberculosis
Mumps
Chicken pox
Whooping cough
Rheumatic fever
Scarlet fever
Amebiasis
Malaria
Parasites
Meningitis or Encephalitis
Other General Infections

EYES
Cataracts
Glaucoma
Watering of eyes
Excessive eye prominence
Infections (besides conjunctivitis)
Crossed eyes
Redness
Discharge
Other Eye Illnesses
Blurring
Double vision
Loss of vision (either eye)
Severe decrease in vision (either eye)
Visual field defects
Pain
Night blindness

EARS
Middle ear infection
Otosclerosis
Menieres Disease (dizziness & hearing loss)

PAST NOW ITEM

EARS (continued)
Severe motion sickness
Ringing or buzzing
Draining/perforation
Loss of hearing, R or L
Loss of hearing, both
Pain in ears

NOSE
More than 2 colds/year
Stuffiness
Discharge, clear
Nasal polyps
Loss - sense of smell
Post-nasal discharge (constantly)
Frequent nosebleeds

MOUTH
Sore or irritation of lips
Sores in mouth
Teeth in need of repair
Partial or complete dentures
Bleeding gums
Dental pain
Tongue - sores or lumps

THROAT
Tonsils are out
Pain in throat
Sore throat with fever
Positive culture for streptococcus (strep throat)
Hoarseness
Difficulty in breathing in
Tonsilitis
Polyps or growths
Infectious Mono

NECK
Lumps tenderness
Enlarged or tender thyroid gland
Pain, stiffness
Other

RESPIRATORY
Asthma
Bronchitis
Emphysema
Pneumonia
Tuberculosis

PAST NOW ITEM

RESPIRATORY (continued)
Fungus infection
Lung tumor
Positive skin test for Tuberculosis
Pleuritis (Pleurisy)
Embolism (blood clot)
Abnormal Chest X-Ray

CHEST
Cough - dry
Cough up, phlegm
Cough up, blood
Cough wakes me up at night
Morning cough
Pain in breathing
Wheezing-asthma

WOMEN ONLY
Fibroids of uterus
Cancer of uterus
Cancer of cervix
Infection of uterine tube
Ovarian cyst
Ovarian cancer
Breast cancer
Breast - benign tumor
Breast - Other Diseases without tumor
Other Women's Disease

Menstruation
Onset 12 or younger
Onset 13 or older
Cessation (menopause) Age 42 or less
Cessation (menopause) Age 43 to 49
Cessation (menopause) Over age 50

If you are menstruating do you have:
Excessive bleeding (unusually heavy periods)
Irregular periods
Excessively prolonged periods
Bleeding between periods

PAST NOW ITEM

If periods have stopped
Last period was more than 2 years ago
Have had bleeding or spotting since
Was due to natural causes (menopause)
Was due to surgery
Was due to X-Ray treatment
Was due to Other

Pregnancies
Never been pregnant
One
Two to five
Five or more
Am now pregnant
Birth weight of child over 10 lbs./in past

Miscarriages and Abortions
None
One
Two or more
D & C required for one or more

Births
All live births
Still births one or more
Died shortly after birth one or more
Multiple births (twins, triplets)
Birth defects one or more births

Complications of Pregnancy
High blood pressure
Kidney or bladder infections
Bleeding
Caesarian birth
Tubal pregnancy
Pelvic infections
Thrombophlebitis (clots in legs)
Do you now have or did you have pain in the pelvis
Is it associated with fever
Is it associated with periods
Is it associated with intercourse
Is it between periods

Vaginal Discharge
None
Moderate
Heavy

FORM 53-202 3/77 ©AMOCO COMPUTER SERVICES COMPANY 1977

Fig. 2. Page from General Foods Corporation computerized medical record form.

IV. ATTRIBUTES OF MEDICAL RECORDS DETERMINING THEIR EFFECTIVENESS IN EVALUATING HAZARDOUS CHEMICAL EXPOSURES

The presumed goal of the evaluation of hazardous chemical exposures is the rapid, accurate characterization of any health risks so that appropriate corrective actions (if any are indicated) can be taken. This characterization can include environmental monitoring, use of human or nonhuman bioindicators, and studies of human health outcomes. Medical records constitute one form of human health assessment which in turn is only one of several approaches to the identification of environmental hazards.

The effectiveness of medical records must be considered in light of the broad goals of hazard recognition and abatement. Simply stated, the demand placed upon this resource is the accurate, rapid, efficient, comprehensive characterization of health and disease. Recognition of how such a data source can fail to meet these demands leads to suggestions for improvements and to more accurate interpretations of findings that are derived from this resource.

The five properties of principal concern in medical records use are sensitivity, specificity, rapidity, feasibility, and comprehensiveness. These properties must be viewed with the understanding that the relative importance of each characteristic depends on the particular exposure and outcome of interest, with no possible assignment of intrinsic value to each. There are clear trade-offs among these attributes (e.g., rapidity versus sensitivity, comprehensiveness versus cost), and the optimal balance among them cannot be specified for all possible situations. Discussion of these properties of medical records is included to provide a list of considerations that should be of concern to those who design, use, or interpret the results from such data systems.

A. Sensitivity and Specificity

In the context of a screening test, the terms sensitivity and specificity characterize how well a test identifies persons with and without disease. Medical records can be thought of as screening in that they also imperfectly identify persons with and without a given disease. Assuming the true presence or absence of a disease is known, the analogy to a screening test can be applied to the accuracy of medical records.

Sensitivity refers to the proportion of truly diseased persons that is correctly identified by the records. Specificity is defined as the proportion of nondiseased persons accurately identified in the records. These terms are identified graphically in Fig. 3. One method of increasing the sensitivity is to broaden the criteria by which a given disease is defined. This has the effect of shifting persons from cell C to cell A in Fig. 3. However, one will simultaneously label more nondiseased

		Disease Present	Disease Absent
Medical Records	Indication of Disease Presence	A	B
	Indication of Disease Absence	C	D

Sensitivity: $\frac{A}{A + C}$

Specificity: $\frac{D}{B + D}$

False Positives: B

False Negatives: C

Fig. 3. Medical records as screening test for disease.

people (D) as diseased (B), i.e., shift persons from D to B. As is true for screening tests, the degree to which false positives versus false negatives can be accepted determines the appropriate criteria for labeling diseased persons.

When medical records are used to evaluate hazardous chemical exposures, it is important to realize that records are an imperfect reflection of disease status. The specificity and sensitivity of these records will never be 100%. Quantification of the magnitude of these imperfections is highly desirable, as will be discussed in Section V,C.

All possible actions should be taken to minimize the errors of misclassification. Beyond that, explorations of the trade-off between sensitivity and specificity can provide useful information. For example, the categories of disease outcome might be divided into three levels of certainty: definite, probable, and possible. The sensitivity would increase and the specificity would decrease across this gradient of decreasing certainty. Relating each disease endpoint (definite, probable, and possible) to the exposure may provide an indication of which is most useful in the context of a particular study. The definition producing the closest association between health and exposure, for example, suggests that that particular entity receive further study (in a hypothesis-generating mode). Most important, using several possible disease criteria allows readers to examine the pattern of results and reach their own conclusions.

B. Speed of Detection

The temporal relationships among exposures, health effects, and the observation and recording of those effects are highly variable. Some adverse health effects such as headache in response to solvent exposure are essentially simultaneous with the exposure. In fact, in occupational settings the close temporal relationship of known symptoms to exposure often serves to indicate that the toxic agent is present in the workplace. At the other extreme, some cancers are thought to have 20- to 40-year latency periods between the exposure and the clinical recognition of disease. The effectiveness of medical records is partially determined by the point in the temporal course of exposure and disease at which records effectively identify a hazard.

Beginning with the introduction of a new chemical in either the occupational or general environment, a typical sequence of events can be schematized as in Fig. 4. The first effect is likely to be a measurable change in the environment, e.g., drinking water, soil, air. Medical records are of no value in detecting such problems. If the chemical's toxicity is known and environmental monitoring detects unacceptable levels, then corrective actions are needed even in the absence of overt health problems. Besides direct human exposures, ingestion of contaminated plants or animals may produce indirect human exposures. Next, there may be evidence of human exposures in the form of biological indicators. These endpoints reflect exposure but are not clinically undesirable in their own right (*4*). Bioindicators of lead exposure include blood test levels, urinary lead level, zinc protoporphyrin, and free erythrocyte protoporphyrin. The early recognition of unacceptably high exposures to humans should lead to changes that will reduce the risk of future illness in the index case and alter the environment to reduce risk to others.

The current usefulness of medical records as reflections of these bioindicators is minimal since the clinical practitioners do not routinely apply such tests. Nonetheless, the potential value of exposure monitoring through analyses of blood or urine could be great when special concerns merit such studies. These tools are used regularly in research protocols (*2*), and the repeated demonstration of their value as preclinical warnings can lead to their implementation in routine practice when warranted by the likelihood of exposure. Occupational settings offer the most potential for routine biological monitoring since the exposures are relatively high compared to those in the general environment and medical care providers should be aware of the chemicals to which workers are potentially exposed. A proposal to define exposure limits in terms of bioindicators offers intriguing potential for more effectively limiting health risks (*13*). The present lack of bioindicator data in routine medical records makes most chemically related problems undetectable prior to some clinically evident symptomatology.

Early manifestations of ill health are probably the outcome for which medical records have the greatest value. Such symptoms as minor respiratory or neurological complaints lead to many outpatient visits and, indeed, those visits to physicians may often be the only documentation of a health problem. Addressing those minor complaints can directly prevent the inherent discomfort they reflect and avoid diseases of long latency (e.g., emphysema, lung cancer) related to the same exposures. In a sense, those early symptoms can be viewed as bioindicators of the chronic health effects that might ensue.

The final outcomes in Fig. 4 are delayed health effects and death, the outcomes a hazard identification system seeks to avoid. Environmental changes offer little help to the patient with late, chronic effects, e.g., reducing asbestos exposure after lung cancer was produced. Nonetheless, accurate identification of these adverse health effects can lead to environmental changes that will protect others. Such events have traditionally served as the primary means of hazard identification, even though they come many years after the exposure. For example, recognition that an excess number of coke oven workers were dying of lung cancer (*5*) ultimately led to a revised standard for coke oven emissions.

In a steady-state ecosystem in which exposure has existed for a period of years, continuing pollution, human exposure, early symptoms, later illness, and death are all occurring simultaneously. Since all aspects of adverse effects are present, the speed with which the manifestations occur is not relevant in that all are accessible to the investigator. For many chemical exposures such as silica, benzene, and formaldehyde, this is the case. Early recognition of individual exposures and early health effects often provides clinically useful data for the treatment of that patient, but offers little toward characterizing and ameliorating the public health problem.

The rapid rate at which new chemicals are introduced, however, makes the

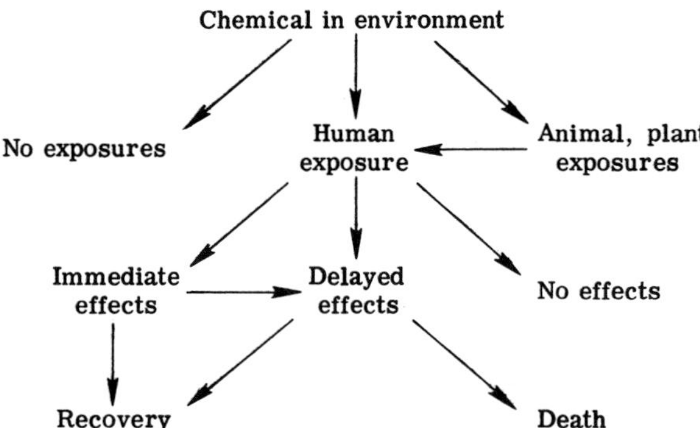

Fig. 4. Possible temporal courses of events following introduction of a chemical into the environment.

earlier identification of newly introduced hazards quite valuable. Medical records offer the potential of moving the time of detection from delayed chronic illness to early symptomatology. Given the severity and irreversibility of many cancers and other chronic diseases, this would be a significant achievement in reducing the effects of environmental chemicals.

C. Feasibility

Costs, confidentiality, the work load placed on respondents, and other logistic issues must be considered when evaluating medical records as a hazard characterization system. When medical record data are adequate to answer the question of interest, these practical issues should be given serious consideration. The question of whether the records are adequate, however, must be addressed and answered affirmatively prior to considering these issues.

Finances generally favor use of an existing record system over collection of new data from respondents. Studies based on medical records require data retrieval, analysis, and interpretation. In contrast, an interview study requires development and administration of a questionnaire in addition to data coding, analysis, and interpretation. Developing an instrument and training and supporting interviewers can be quite costly.

Confidentiality and the rights of privacy have become important considerations in the conduct of epidemiologic research based on medical records (*11*). Patients' permission may be required for examination of records, introducing the potentially expensive and time-consuming process of identifying patients and soliciting their cooperation prior to examination of their records.

The work load of respondents tends to favor record-based studies over personal interviews. The time consumed in the interview is a potential inconvenience. In addition, querying sensitive and sometimes painful topics results in a burden to the respondent that is difficult to quantify but which cannot be ignored. For example, reviewing the circumstances surrounding the birth of a child with a congenital anomaly has a high potential for causing the respondents to suffer.

Most logistic considerations favor medical records over alternative data collection methods. Since medical records contain only a limited range of information with limited accuracy, however, this strength must be evaluated in the context of a particular research question.

D. Comprehensive Coverage of Health Outcomes

The amount of information potentially available in health records is one of their most favorable attributes. In contrast to focused systems such as cancer registries, medical records allow examination of a broad array of health outcomes. This facilitates the identification of unanticipated health problems associated with environmental exposures, which is not possible with studies of single diseases.

The disadvantage of medical records is a lack of real detail in regard to any specific outcome. The virtue of comprehensiveness is most significant in hypothesis-generating efforts rather than in the testing of a focused biological hypothesis.

V. METHODOLOGICAL CONSIDERATIONS IN IMPLEMENTING A RECORD-BASED HAZARD EVALUATION SYSTEM

The application of medical records to chemical hazard assessment falls naturally within the scope of epidemiology. Epidemiologic methods and principles especially applicable to using medical records in this manner will be reviewed.

A. Cross-Sectional versus Cohort Design

Cross-sectional investigations examine exposure and health status at a given point in time, whereas cohort studies identify a group of persons with a common exposure (i.e., a cohort) and ascertain their health status at a later time. Cohort studies have distinct methodologic advantages in making etiologic inferences which will be illustrated by the contrast to cross-sectional designs.

One assumption in a cross-sectional study is that currently exposed persons adequately represent all persons who were ever exposed. There is the possibility for selective exclusion leading to bias and ultimately incorrect inferences. For example, in occupational studies, the health experience of retirees does not reflect the experience of active workers (*5, 10*). Similarly, persons who currently live near a chemical waste disposal site may poorly represent the cohort of all persons who ever lived in the area. They may be poorer, unable to move, older, or less concerned about chemical exposures. That is, they may be different in some way that also is associated with poorer health status, independent of residence near the waste site. If a cohort design were used, with the cohort defined as "persons with *X* years of residence near the site," then the investigator would know whether the currently accessible persons adequately represent all persons who have lived in the area.

When the effect of interest is a product of long-term exposure, then this potential problem of selective retention of study members is enhanced. Those currently exposed may have a brief history of exposure and persons who moved away may have lengthy exposures. This migration, even if it is not related to health status, can make associations between the environment and health difficult to detect (*28*). Defining cohort eligibility as "*X* or more years of exposure" and ascertaining health outcomes for all cohort members circumvent this problem. Another implicit assumption in cross-sectional studies is that any latency period between exposure and manifestation of health effects has passed. For some diseases such as cancer, which may have a latency period of decades, this is a very tenuous assumption. The community's current cancer risks may reflect exposures 20–40 years ago, whereas the investigator is assessing only very recent exposures. In a cohort design, the follow-up or observation period is explicitly addressed. The sequence of exposure, latency period, and disease expression period can be examined by calculating disease rates at varying intervals after exposure (*29*). Knowledge of the disease processes of interest indicates the necessary follow-up period: for respiratory irritation from chlorine gas, 1 h may suffice whereas 20 or more years of observation would be essential to assess the carcinogenic effects of asbestos. Pressures to provide quick answers to public health concerns may prompt studies before a sufficient amount of time has elapsed for disease risks to possibly appear. Such efforts are worthless in characterizing environmental hazards and can give only false assurance that no health dangers have occured.

A cohort investigation begins with a defined group of persons receiving an exposure and then ascertains its subsequent health experience. A cross-sectional study may provide results similar to those of a cohort study if there is not substantial migration and if sufficient time has elapsed for health risks to appear. This may often be the case, but those who conduct and interpret cross-sectional studies should be aware of these assumptions and make some effort to confirm their validity.

B. Objective Assessment

Medical records reflect in part the judgments of both the patient who reports a problem and the clinician who interprets and records that information. Figure 5 illustrates this process. Since the data are not gathered within a defined research protocol that stresses objectivity, the translations from objectively defined illness to the medical record notation are potentially biased. Whereas random errors predictably decrease the likelihood of detecting an association, systematic error or bias can spuriously create or mask an association between the exposure and the health outcome (*6*). The discussion of sensitivity and specificity (Section IV,A) addressed erroneous medical record data that were assumed to be unrelated to exposure status. Here the concern is with misclassification of health that is related to exposure. Errors can be introduced at every step in the process illustrated in Fig. 5. First, the patients may respond selectively to their own symptoms based on whether they think the symptoms resulted from exposure. Individuals vary in the way they interpret and respond to perceived ill health (*1*), and an awareness of a potentially hazardous exposure could increase the propensity

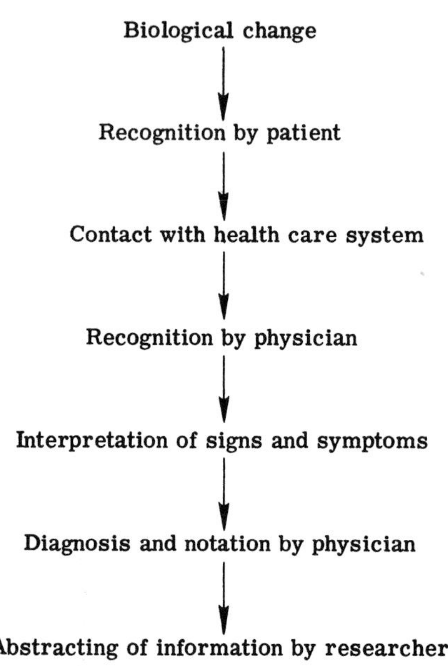

Fig. 5. Sequence of events linking biological disease process to identification in medical records.

to perceive minor symptoms (e.g., respiratory discomfort) as indicators of more serious illness. This would then lead the person to contact a physician, providing an opportunity for the diagnosis of illness that would not have been made otherwise.

Skegg *et al.* (*31*) provided an interesting empirical example of contact with the health care system that gave a false impression of ill health. In examining the records of general practice physicians for adverse drug reactions, oral contraceptives were found to be associated with the common cold. The authors felt this could be accounted for by the more frequent physician visits among women taking oral contraceptives, which provided the opportunity to have colds noted in the record. Obviously, there is a pool of persons with undiagnosed colds in the general population that would be identified if they sought medical care.

After contact with the health care provider, additional biases can occur. Since diagnoses are always made using complex judgment patterns and perhaps some element of art, awareness of a potentially toxic chemical exposure could readily influence the physician's decision. Besides affecting the labeling of the disease, this concern with the chemical environment could alter the information ultimately recorded in the case notes. A complaint of headache may arouse greater interest if it comes from a painter than from a clerical worker, since an astute clinician would be sensitized to the possibility of a chemical etiology for the headache.

The final step in Fig. 5, data abstraction by the investigator, also is a potential source of biased recording. Given the nature of medical record information as depicted in Fig. 1, an abstractor with preconceived notions of who is likely to manifest a health problem could quite plausibly interpret ambiguous notes accordingly. Unlike the other concerns discussed above, this one is under the control of the investigator. The optimal approach is to review the records without knowledge of the exposure status. If this is not possible, explicitly defined criteria for what constitutes the outcome of interest leave little room for subjectivity and hence, bias.

C. Data Validation

Given the numerous opportunities for random and systematic error, validation of the records of a sample of patients is extremely desirable. The goal is to trace information back through the sequence of events in Fig. 5 as far as possible. Direct confirmation of the patient's health condition would be the most useful, but intermediate levels of confirmation (e.g., confirming the physician's diagnosis) also have value. By selecting a sample of exposed, nonexposed, diseased, and nondiseased patients for questioning or possibly for repeat examination, the magnitude of systematic error can be quantified and the results adjusted for the presence of such bias (*12*).

D. Consideration of Potential Confounding Factors

Special attention must be paid to factors that may distort the results of an observational study. Factors that are independently related to both an exposure and a disease can distort the measured association between the exposure and the disease (*22*). Weinberg *et al.* (*35*) provided an excellent example of confounding in environmental epidemiology. Communities with high levels of air pollution in the Pittsburgh metropolitan area had been found to have higher rates of lung cancer than less polluted areas. Upon closer examination, however, cigarette smoking rates were found to be the explanatory factor. Areas of high pollution (lower social class) had a higher proportion of cigarette smokers, thus linking smoking to the exposure variable in this study. In addition, smoking is a cause of lung cancer, i.e., independently linked to disease. Failure to adjust for cigarette smoking would have thus led to incorrect inferences regarding air pollution and lung cancer.

Special concerns with confounding variables arise from the use of medical records. All other known causes of the outcome under investigation must be considered as potentially confounding the exposure–disease relationship of interest. The problem in many record-based studies is the general lack of information on such risk factors. Although dietary influences might confound a study of the adverse influence of the general environment, sufficient information on diet is generally not available from medical records. Information on other aspects of lifestyle and genetic background, for example, is likewise not available in such sources, making it impossible to address these factors. Knowledge of demographic and social characteristics of the groups under study (e.g., occupational profile, ethnicity, income level) is of some value in avoiding confounding, but its presence does not provide assurance that specific risk factors like smoking or alcohol consumption are also equivalent across the exposure groups. Although these unaddressed concerns do not invalidate the method, investigators can strengthen their results by discussing and speculating intelligently about the role of unmeasured potential confounding variables. Techniques have been developed (*3, 32*) to actually estimate the magnitude of distortion for which an unmeasured confounding variable could account. This aids in making valid interpretations of the findings of studies based on medical records.

VI. CASE STUDIES

The application of the principles developed in earlier sections of this article must be done in the context of a particular situation. Both the importance of these specific guidelines and the feasibility with which they can be implemented depend on such factors as the quality of exposure information, the nature of the disease outcome (severity, symptomatology), and the temporal course of exposure and disease. Though a comprehensive or even representative review of all possible applications is impossible, selected case examples are provided to illustrate the usefulness of addressing the issues outlined in this article.

A. Pesticide Contamination of Drinking Water

A recent report by Zaki *et al.* (*36*) documents an episode of pesticide contamination of groundwater in Suffolk County, New York. Aldicarb, a carbamate pesticide, is known from toxicological studies to cause cholinesterase depression. It was used extensively on potato farms in Suffolk County and was found in 1979 to be present in drinking water supplies, with no public knowledge of the contamination prior to that time. A survey of over 8000 wells indicated that 13.5% contained Aldicarb in excess of the recommended maximum of 7 ppb. The situation was ameliorated through the use of alternative water sources and, where that was not possible, with carbon filters supplied by the pesticide manufacturer, Union Carbide. This is an environmental hazard with an unusually well-documented exposure in that the source (contaminated groundwater) and route (ingestion) are well understood.

The authors (*36*) noted a need for controlled epidemiologic studies of the effects of long-term exposure to low concentrations of pesticides. For illustrative purposes, a simple approach is to stratify current residents into 3 levels of exposure: no detectable Aldicarb in the water (5896 wells), 1–7 ppb Aldicarb (1068 wells), 7+ ppb Aldicarb (1087 wells). The health effects known to be related to acute Aldicarb poisoning include nausea, vomiting, blurred vision, dyspnea, perspiration, headache, and temporary paralysis of extremities lasting 4–12 h (*36*).

A review of medical records from health care providers in the area could be used to assess the prevalence of these symptoms in each exposure group, especially if conducted over a period prior to public knowledge of the contamination. One might contact each of the households whose water supply was surveyed, determine their source of medical care, and seek permission to review their medical records. An abstractor could then be sent to physicians' offices, clinics, and hospitals to review records for indications of the gastrointestinal and neurological symptoms known to be associated with Aldicarb exposure. A review of the concerns with this approach leads directly toward more useful methods of addressing the same question.

The contents of the health record and results of medical examinations would, in practice, consist of whatever physicians, nurses, and other health practitioners noted in the hospital, clinic, or office records. No special studies of cholinesterase levels would have been done routinely prior to the knowledge of contamination, and, therefore, such data would not be present in medical records.

It is useful to examine the proposed approach in light of the previously discussed principles. It is important to ask first if the records are adequately sensitive since the outcomes of interest are relatively minor. Temporary partial or complete paralysis would probably lead to contact with the health care system. However, a lower order effect such as transient muscle fatigue, weakness, or incoordination would, like headache and nausea, be less likely to lead to a recorded medical evaluation because either the patient or physician dismissed the complaint. A small proportion of these affected persons would therefore be accurately identified, and the records would fail to indicate their condition. The accuracy of identification of truly unaffected persons by the health care provider is less of a problem. It is, of course, possible for persons to have symptoms or complaints incorrectly attributed to them in the health record (false positives), but specificity is usually adequate since errors of omission are more likely than errors of commission.

Evaluation of the rapidity of this type of detection system illustrates some interesting concerns. Given the current knowledge regarding the health effects of this particular pesticide, there are no health effects known to occur earlier than these symptoms. There are, however, bioindicators of depressed cholinesterase activity that, in a research setting, could be linked to exposure. Ultimately, the earliest event that warrants corrective action is the presence of Aldicarb in the well water. In the sequence of events from environmental exposure to health alterations these symptoms are plausible as the first actual health effects. Thus, the method is rapid relative to chronic diseases, but slow relative to environmental or biological indicators of exposure.

Feasibility and cost are pertinent even in this simple record review study. Depending on the characteristics of health care delivery in the area, numerous hospitals, clinics, and physicians within and outside the region would have to be contacted. Even with the patient's permission to review records, they are not always easily located. The other logistic issue is how to obtain the respondent's permission to review his or her records. Finally, the range of health outcomes that can be ascertained in this study is limited largely by the detail with which the record review is conducted. A systematic data collection effort covering gastrointestinal, neurological, and neuropsychological functions merits exploration. Since low-dose chronic exposures may produce a totally different symptom profile than acute exposures, it would be useful to review symptoms not known to be associated with Aldicarb.

Specific research design issues discussed earlier are applicable in this situation. Although the establishment of a cohort based on current exposure may adequately characterize all persons who were ever exposed, this is an empirical question. Pertinent considerations are (1) how long the exposure is likely to have existed and (2) how many persons were exposed who have since moved. If migration is low overall (and the brief period of recognized exposure suggests it might be), the design is appropriate. If, however, a large proportion of ever-exposed persons no longer reside in the area, a study that follows former residents would be necessary.

The temporal course of the exposure and effect under study is open to debate. If the effects take 20 years to emerge through repeated, low-dose insults, then the proposed study is worthless. If symptoms develop after only a few years of exposure, then the study is suited to the time course of the phenomenon.

Objective assessment of these outcomes is crucial since they are likely to be substantially underreported. If it were possible to review records from a time period before exposure status was known to patients or physicians, i.e., before the publicity began in 1979, symptoms like nausea, headache, and paralysis would be likely to be recorded without bias. After the publicity, they could be artificially elevated in a psychologically sensitized group. Assuming the usual media and word of mouth coverage of such environmental hazards, the potential for bias is apparent. What would have been dismissed as a transient tension headache now leads the victim to contact a physician if he or she believes it was chemically induced. That physician, being aware of the concern, dutifully records the headache in the chart.

Referring back to the original study design, several additions should be considered. Migration should be assessed to determine whether a study of current residents adequately represents all exposed persons. Development of a program of objective testing for health endpoints would be critical to complement the record results. The records have the advantage of being historical, allowing reference to events that occurred before knowledge of exposure. They are so fallible in other ways, however, that they alone cannot produce convincing results. Neurological examinations with neurophysiological and neuropsychological testing would add to the understanding of human toxicity of Aldicarb and help in the interpretation of the health record results. Even if resources allowed only a small sample of persons to be tested, validation of persons identified as affected or unaffected based on medical records would help to characterize the adequacy of the proposed approach.

B. Carbon Disulfide and Coronary Heart Disease

There is substantial evidence that chronic carbon disulfide exposure increases cardiovascular disease mortality and morbidity (*14, 15, 25, 33, 34*). This literature establishes the criteria against which the study by Sakurai (*30*), based on health records, can be evaluated. This allows an assessment of what information medical records can provide regarding carbon disulfide's cardiotoxicity in the presence of strong independent evidence for a causal association.

Sakurai conducted an investigation of viscose rayon workers in Japan that lends itself to an assessment of whether health records alone could have identified the coronary heart disease risk due to occupational exposure. The overall goal of the study was to assess the morbidity associated with carbon disulfide exposure. Records of all medical care provided to employees of a rayon staple plant and neighboring spinning mill over a 56-month period were available through a health insurance office, including records of treatment provided by general practitioners, hospitals, and clinics. Company records were used to construct a carbon disulfide exposure history based on exposure level (none, lower, higher) and duration (<10 years, 10+ years).

The health outcome measures were period prevalence rates of various health conditions, i.e., the proportion of workers who were noted in the medical records to have had the problem at some time during the 56-month study period. The outcomes of primary interest (for this illustration) were ischemic heart disease and hypertensive heart disease. The ischemic heart disease prevalence rate was 8.2% among nonexposed workers, and climbed from 3.8% in those with lowest exposure to 15.3% in persons with the longest, most intense exposure, an elevation that was not, however, statistically significant. The prevalence of ischemic heart disease among exposed workers in the aggregate was 10.4%, not notably different from the 8.2% prevalence among nonexposed workers. The pattern of results for hypertensive disease was similar. The rates among the aggregate of exposed versus nonexposed workers were not significantly discrepant (29.0 and 25.7%, respectively), but the most highly exposed group had a notably and significantly higher (44.4%) prevalence.

The methodological considerations pertaining to medical records are applicable to Sakurai's study. Coronary heart disease is a health endpoint that merits careful investigation even in the absence of specific chemical risks. The frequency of this disease in the population and its seriousness make it a prominent concern of health care providers. Unfortunately, the diagnoses are much more error prone than would be desirable due to such factors as labile blood pressure, subjective reporting of anginal pain, and interindividual variability in reading electrocardiograms. In addition, the underlying pathological process of atherosclerosis of the coronary vessels must generally be inferred from indirect evidence. A detailed characterization would require visualization of the coronary vessels through angiography, an invasive and very hazardous procedure.

The sensitivity and specificity of these medical records as indicators for the presence of coronary disease raise several issues. Severe outcomes like myocardial infarction would likely be accurately identified. Few false positives or false negatives would be expected because of the magnitude of the event. Mild angina, in contrast, might be perceived differently by individual patients and physicians. False positives and false negatives would be likely in the identification of angina, diminishing the specificity and sensitivity, respectively. Hypertension is even more difficult to determine unequivocally in spite of an objective measuring technique. Assuming the concept of a hypertensive/normotensive dichotomy is valid, then the typical blood pressure reading taken in a physician's office can be viewed as a screening test for the disease. But research clearly indicates the complexity in definitively classifying patients with nonsystematic blood pressure measurements (*7*).

The medical record review undertaken by Sakurai (*30*) has the potential to detect adverse effects much more rapidly than studies of mortality if the exposure had begun only recently, since manifestations of the disease usually (but not always) precede death. Since the factory had operated for some time prior to the study, however, the steady state would include late as well as early manifestations of carbon disulfide toxicity.

With respect to feasibility and cost, Sakurai had the very fortunate opportunity to begin with a listing of the source of all medical care for all persons in the study, compiled by the health insurance office. Without such a system, the workers themselves would have to be queried for such information, a process both difficult and prone to underreporting. Alternatively, the records of all health care providers in the region could be searched. There would be some difficulty in obtaining the needed broad access to records and substantial expense in perusing thousands of mostly uninformative records.

The comprehensiveness of health outcomes available in this study is the major strength of such a broad medical record review. Unlike mortality studies, diagnoses such as hypertension, diabetes, and nephritis were available for analysis. Sakurai's methodology offers a very wide ascertainment net in identifying anticipated and unexpected health problems among these workers.

The major sacrifice in this type of study is the failure to study a defined cohort. Only the workers who were alive and working during the 56-month period were included. Persons who left the company due to a special sensitivity to carbon disulfide exposure or due to death would have been excluded. A preferable approach would have been to define a cohort based on exposure at some time in the past and examine its morbidity experience over a defined time interval. Without this cohort approach, an occupational risk may be overlooked as exposed persons die and are excluded from the study. In fact, if the occupational chemical exposure increased the likelihood of fatality among diseased persons, the measured prevalence of coronary heart disease might be substantially reduced among the exposed due to depletion of such cases through death.

Another weakness in this is the inability to make any inferences about latency. The health experience subsequent to a terminated exposure period is unavailable in a study limited to active workers.

Objectivity of assessment is another area of concern. To what extent do the final tabulations of prevalence rates for coronary heart disease and hypertension reflect the health of these workers? The most significant concern is the link between symptomatology and contact with the health care system. Although this is not pertinent to severe events such as myocardial infarction, it is very important for hypertension. It is plausible that persons exposed to carbon disulfide experience a wide variety of minor neuropsychological symptoms that motivates them to see a physician. At that time, a complete medical examination reveals hypertensive disease. This ascertainment bias conceivably accounts for the observed risk differences in hypertension reported in Sakurai's study since not all workers were screened. Another reason for more effective disease identification among exposed persons might be the worker's awareness of the toxic effects of carbon disulfide.

As a complement to the literature on mortality associated with carbon disulfide, Sakurai's study was quite useful. The unique opportunity to examine the complete medical experience of a group of workers was effectively exploited, and such a research approach would be worthwhile even in the absence of other supportive data. In fact, if such insurance records were widely available to index health care sources and events, this could be a very useful exploratory approach in the assessment of hazardous chemical effects.

C. Environmental Dioxin Exposure and Adverse Reproductive Outcomes

A recent episode of environmental contamination in Seveso, Italy (*16*) provides an example of a well-documented exposure for which the health outcome of fetal loss was rather poorly examined. The more general problems of assessing reproductive hazards through medical records are also illustrated by this study. Homberger *et al.* (*16*) reported the details of a dioxin release that occurred recently and a summary of the extensive series of health and ecological studies that followed to estimate the impact. On July 10, 1976, the ICMESA plant in Seveso, Italy accidentally released dioxin-contaminated material over a 700-acre area. Soil samples taken in the immediate area of the factory contained over 15 ppm dioxin with lower concentrations at greater distances from the plant. The extent of environmental exposures was also characterized by plant toxicity, animal (especially rabbit) illness and death, and dioxin content in vegetation.

A rather thorough investigation addressed the incidence of chloracne, a dermatological condition in humans specifically related to aromatic chlorinated hydrocarbons. Though the disease can cause substantial discomfort and disfigurement, the present focus on reproduction places chloracne in the role of a biological indicator of exposure. The geographical distribution of chloracne cases indicates where significant human dioxin exposures occurred.

Communities were identified as exposed or unexposed, based upon the pattern of dioxin release. The chloracne prevalence rate was 2.5% among school children in exposed communities and 0.6% among children in unexposed communities, giving a pronounced relative risk of 4.58. This demonstrates that the population near the factory was, in fact, exposed to dioxin. This is much more direct evidence than monitoring of air, water, and plants can provide since the latter measures indicate only the *potential* for exposure.

Homberger *et al.* were not explicit regarding their methods of ascertaining spontaneous abortions, but health records appear to have been the primary source of data since only area rates could be calculated. (For the outcomes derived by special surveys, e.g., chloracne, the results indicated the individual risk levels.) The ratio of stillbirths, spontaneous abortions, and induced abortions to total pregnancies was calculated for the exposed and unexposed areas over the period 1973–1977. There were no differences in any of these outcomes related to exposure, either spatially or temporally. A finer breakdown by 3-month intervals from the third quarter of 1976 through the fourth quarter of 1977 again provided no evidence of excess risk with spatial or temporal proximity to the dioxin release. Rates of fetal loss (pregnancies not resulting in a live birth) were on the order of 9–15%, comparing favorably to an expected rate of 15–20%.

The sensitivity and specificity of medical records as identifiers of fetal losses are important considerations. Both would be expected to vary in relation to the time in pregnancy that the loss occurred. Early miscarriages are difficult to identify and would be much less likely to be reported to health care providers than later miscarriages or stillbirths. Conversely, uncertainty regarding early pregnancy also allows for possible false positive reporting due to menstrual irregularities, thus lowering specificity. Separate tabulations of the reproductive events by the timing in pregnancy (a proxy for reporting accuracy) would provide useful clues to determine the likelihood that a chemically induced adverse effect was present.

The rapidity of these events as reflections of toxic chemical exposure is thought to be favorable relative to most chronic diseases. For example, genetically toxic agents might cause both genetically altered fetuses, which are aborted, and cancer. The limited duration of pregnancy makes reproductive effects among the earliest clinical indicators of genetic damage. Nonetheless, environmental evidence of exposure and biological evidence of human exposure in the form of chloracne preceded these measures.

The system utilized to evaluate adverse reproductive effects is low in cost relative to studies requiring interviews of individual respondents. Also, since reproduction is a potentially sensitive topic, reliance on a record system that avoids direct contact with respondents is favorable in averting unnecessary discomfort.

The need for a defined cohort is not especially strong here due to the relatively short time between exposure, manifestation of adverse effects, and the assessment of those effects. Little random or health-based migration could have occurred over a 2-year time interval, making the cross-sectional results very similar to what would be expected from a cohort defined as those living in the area at the time exposure began.

Objectivity of assessment is a significant consideration in evaluating the applicability of medical records in this situation. Unlike the first example in which the Aldicarb exposure in drinking water was not known to the public for some time, the Seveso accident was of great public interest from the beginning. Furthermore, the general public has a strong emotional concern about the health of its offspring. Patient's reports, physicians' perceptions, and ultimately the data entered into medical records are vulnerable to bias due to the public awareness following such an accident. Homberger *et al.* noted the general underreporting of these events, which allows for the possibility of a reporting bias.

The array of potential sources of error contains elements that bias the results toward the null hypothesis, i.e., no association between dioxin exposure and fetal loss, and other elements that would tend to spuriously create such an association. The random inaccuracy in reporting miscarriages and whatever migration occurred into and out of the area diminish the study's capability of finding an association if one is in fact present. The publicity and consequent sensitization of the exposed residents could have created a bias toward overreporting problems in that area. The lack of any validation of these data renders these results inconclusive. Although no adverse effect was found, the study does not allow the reader to assess the level of confidence that can be placed in these findings.

VII. RECOMMENDATIONS

The considerations presented in this article on medical records as a data resource for the evaluation of chemical hazards allow several recommendations to be made. Like most guidelines in the conduct of epidemiologic research, these are not to be interpreted as unequivocal rules. Failure to conduct a research program exactly as outlined in textbooks does not completely invalidate the results, though it generally weakens them. Conversely, conducting a study with all the recommended standards does not guarantee valid results and conclusions, but it does increase the likelihood.

The first recommendation is that the research question should be stated explicitly, because the appropriateness of records can be examined only in the context of a specific question. The utility of medical records is so dependent upon the specific outcome of interest that the general question "Are records useful?" cannot be answered. A more useful formulation is to ask whether records are useful to detect excess hypertension within an occupational cohort or to detect an increased rate of miscarriages among residents in the vicinity of chemical contamination. Even in a hypothesis-generating study, the potential outcomes of interest must be defined before addressing the adequacy of medical records.

The second recommendation is to carefully examine whether the information available from medical records is relevant to the question of primary interest. Opportunism in exploiting data collected for other purposes is desirable, but only so long as it is of some use in addressing the real concern. There is an unfortunate tendency to reformulate the research question into one that the available data can answer, even if it is not really of major interest. Many (perhaps most) concerns in occupational and environmental epidemiology require specially designed research efforts. Questionnaires, biochemical measurements, and clinical examinations are frequently required to effectively evaluate the health consequences of a chemical exposure. Depending on the relative cost and information gained, a detailed study of fewer persons may be more cost-effective than a record survey of a larger number. Alternative methods of answering the primary question should be given very serious consideration and medical records should be used only when they are the best approach.

Third, attention should be given to the validity of the data collection system when medical records are chosen as the assessment method. Ideally, the program should include methods for empirically addressing the validity of the methods used. These include overall sensitivity and specificity as well as response biases related to exposure. Even with limited resources, establishing the validity for a subsample of respondents as a check on the accuracy of records is very important. In the absence of empirical examination of validity, the literature to address data quality concerns should be examined and used in the interpretation of the study's results.

A final global recommendation is to apply sound principles of epidemiologic study design in using health records. Evaluating hazardous chemical exposures through medical records is, after all, a specific method within environmental epidemiology which, in turn, is a subset of epidemiology. The principles of designing observational investigations of health in human populations are the essence of epidemiology. This article touched on a number of especially pertinent issues, but all of the conceptual and statistical tools of epidemiology are applicable (*18*). This may require consultation with an epidemiologist both to evaluate the usefulness of a study based on medical records and then to design the study appropriately.

Health records can be of real value in assessing hazardous chemical exposures, but caution is required. If there has been a dramatic adverse health effect, this method would probably be capable of identifying it. However, such dramatic events rarely occur. Society's concern has become increasingly focused on subtle effects, in which small sources of inaccuracy and bias can produce erroneous conclusions.

When a special opportunity exists to favor this methodology, such as the centralized record-keeping system identified by Sakurai (*30*), medical record studies can provide an economical screening method. There are two primary drawbacks: (1) The overall inaccuracy of medical records can make them an ineffective tool in hazard detection, giving false security to those who find no differences related to exposure; and (2) Reporting biases can overwhelm other health determinants and falsely create an apparent problem. Much greater care is needed in the design of health data collection systems that avoid both random and systematic inaccuracies. None of the pitfalls is inherent to the method, but are common in records as they currently exist. The design of medical record systems with due consideration to their value as an epidemiologic data resource would eliminate many of the problems reviewed here.

Acknowledgments

I thank Dr. Richard F. Hamman for his extensive assistance in organizing and clarifying the contents of this article. His epidemiologic and editorial skills were indispensable. Mr. Bruce Perry and Dr. Miriam Orleans also provided valuable assistance in the development of this article.

References

1. Anderson, R., Anderson, O. W., and Smedby, B. (1968). *Med. Care* **6,** 18–30.
2. Brilliant, L. B., Wilcox, K., Van Amburg, G., Eyster, J., Isbister, J., Bloomer, A. W., Humphrey, H., and Price, H. (1978). *Lancet* **2,** 643–646.
3. Bross, I. D. J. (1966). *J. Chronic Dis.* **19,** 637–647.
4. Calabrese, E. J. (1978). "Methodological Approaches to Deriving Environmental and Occupational Health Standards." Wiley, New York.
5. Collins, J. F., and Redmond, C. K. (1976). *J. Occup. Med.* **18,** 595–602.
6. Copeland, K. T., Checkoway, H., McMichael, A. J., and Holbrook, R. H. (1977). *Am. J. Epidemiol.* **105,** 488–495.
7. Donner, A., and Bull, S. (1981). *J. Chronic Dis.* **34,** 527–531.
8. Finucane, R. D. (1982). *J. Occup. Med.* **24,** 794–798.
9. Fries, J. F. (1974). *Med. Care* **12,** 871–881.
10. Gilbert, E. S. (1982). *Am. J. Epidemiol.* **116,** 177–188.
11. Gordis, L., and Gold, E. (1980). *Science* **207,** 153–156.
12. Greenland, S. (1982). *Am. J. Epidemiol.* **116,** 402–406.
13. Grunder, F. I., and Moffitt, A. E. (1982). *Am. Ind. Hyg. Assoc. J.* **43,** 271–274.
14. Hernberg, S., Partanen, T., Nordman, C.-H., and Sumari, P. (1970). *Br. J. Ind. Med.* **27,** 313–325.
15. Hernberg, S., Tolonen, M., and Nurminen, M. (1976). *Scand. J. Work, Environ. Health* **2,** 27–30.
16. Homberger, E., Reggiani, G., Sambeth, J., and Wipf, H. K. (1979). *Ann. Occup. Hyg.* **22,** 327–370.
17. Joiner, R. L. (1982). *J. Occup. Med.* **24,** 863–866.
18. Kleinbaum, D. G., Kupper, L. L., and Morgenstern, H. (1982). "Epidemiologic Research. Principles and Quantitative Methods." Wadsworth, Belmont, California.
19. Kreiss, K., Zack, M. M., Kimbrough, R. D., Needham, L. L., Smrek, A. L., and Jones, B. T. (1981). *J. Am. Med. Assoc.* **245,** 1926–1930.
20. Kurland, L. T., and Molgaard, C. A. (1981). *Sci. Am.* **245,** 54–63.
21. MacMahon, B., and Pugh, T. F. (1970). "Epidemiology. Principles and Methods." Little, Brown, Boston.
22. Miettinen, O. S., and Cook, E. F. (1981). *Am. J. Epidemiol.* **114,** 593–603.
23. Monson, R. A., and Bond, C. A. (1978). *J. Am. Med. Assoc.* **240,** 2182–2184.
24. Murnaghan, J. H., and White, K. L. (1971). *N. Engl. J. Med.* **284,** 822–828.
25. Nurminen, M. (1976). *Int. J. Epidemiol.* **5,** 179–185.
26. Office of Technology Assessment (1981). "Assessment of Technologies for Determining Cancer Risks from the Environment." U.S. Govt. Printing Office, Washington, D.C.
27. Ott, M. G., Hoyle, H. R., Langner, R. R., and Scharnweber, H. C. (1975). *Am. Ind. Hyg. Assoc. J.* **36,** 760–766.
28. Polissar, L. (1980). *Am. J. Epidemiol.* **111,** 175–182.
29. Rothman, K. J. (1981). *Am. J. Epidemiol.* **114,** 253–259.

30. Sakurai, H. (1982). *Br. J. Ind. Med.* **39,** 39–44.
31. Skegg, D. C. G., Richards, S. M., and Doll, R. (1981). *J. Epidemiol. Community Health* **35,** 32–34.
32. Schlesselman, J. J. (1978). *Am. J. Epidemiol.* **108,** 3–8.
33. Tiller, J. R., Schilling, R. S. F., and Morris, J. N. (1968). *Br. Med. J.* **4,** 407–411.
34. Tolonen, M., Hernberg, S., Nurminen, M., and Tiitola, K. (1975). *Br. J. Ind. Med.* **32,** 1–10.
35. Weinberg, G. B., Kuller, L. H., and Redmond, C. K. (1982). *Am. J. Epidemiol.* **115,** 40–58.
36. Zaki, M. H., Moran, D., and Harris, D. (1982). *Am. J. Public Health* **72,** 1391–1395.

Mediation of Toxicological Properties of Chemicals by Particulate Matter

David R. Bevan

Department of Biochemistry and Nutrition
Virginia Polytechnic Institute and State University
Blacksburg, Virginia

Joseph R. Lakowicz

Department of Biological Chemistry
University of Maryland School of Medicine
Baltimore, Maryland

I.	Introduction	142
II.	Tobacco Smoke	143
	A. Composition of Tobacco Smoke	143
	B. Perspective	143
	C. Asbestos and Smoking	144
	D. Other Mineral Particulates and Smoking	145
	E. Animal Studies	146
	F. Summary	147
III.	Polynuclear Aromatic Hydrocarbons	147
	A. Environmental Occurrence of PAH	147
	B. Perspective	149
	C. Particle-Enhanced Uptake of PAH into Membranes	150
	D. *In Vitro* Methods	152
	E. Animal Studies	154
	F. Summary	164
IV.	N-Nitroso Compounds	164
V.	Pesticides	167
VI.	Gases and Vapors	169
VII.	Metal Compounds	172
VIII.	General Summary	173
IX.	Prospectus	173
	References	174

HAZARD ASSESSMENT OF CHEMICALS:
Current Developments, Vol. 3

ISBN 0-12-312403-4

I. INTRODUCTION

Man and other living organisms are exposed to a complex array of chemical compounds and other agents that may be present at concentrations sufficient to exert adverse effects such as toxicity, carcinogenicity, and mutagenicity. The deleterious health effects of individual compounds may be inferred from human epidemiological studies or demonstrated in animal testing. However, it is more difficult to assess the effect of exposure to multiple compounds. A given compound or agent may diminish the effect of another, the effect of the two may be a simple sum of their individual effects, or the effect of the two may be multiplicative, that is, greater than their sum, in which case a synergism is said to occur.

In this article we will discuss the mediation of toxicological properties of chemicals by particulate matter. Initially, we will discuss the increased health risk that results from exposure to particulates and tobacco smoke, which contains numerous hazardous compounds. In subsequent sections, the effects of particulates on the toxic manifestations of specific classes of chemical compounds will be considered. These classes of chemicals include polynuclear aromatic hydrocarbons (PAH), N-nitroso compounds, pesticides, gases and vapors, and metal compounds. We will discuss the environmental occurrences of the chemicals, their association with particulates, the various studies in which particulates have been found to mediate the toxicities of the chemicals, and the possible mechanisms by which particulates may alter the potency of toxic substances.

With some classes of chemicals a variety of testing methods has been applied to evaluate the effects of particulates. By using various experimental approaches, it is possible to assess better the interactions that occur and to gain insight into the mechanisms of the interactions. Consequently, the discussion regarding the classes of chemicals will be divided, where possible, according to the types of testing methods employed. In considering the various testing methods, it is instructive to outline the types of tests that have been applied to studies of particulate–chemical interactions. Note that not all of the methods have been applied to studies with each class of chemicals. The testing methods are (1) human epidemiology, (2) animal studies, (3) organ tissue cultures, (4) cell cultures, and (5) subcellular and model systems.

Epidemiological studies are essentially the only approach available to assess the toxicological properties of chemicals and/or particulates in humans. However, these studies are difficult to perform because of the difficulty in collecting data and because humans are exposed to a wide variety of environmental pollutants, each of which may pose a distinct health hazard. The nonepidemiological methods are useful for analyzing the toxicity of a particular chemical compound or combinations of compounds. Unfortunately, extrapolating the results obtained in studies using these systems to the potential health effects in the human population is fraught with difficulties. Nonetheless, all of these testing methods are potentially useful in studies of the effect of particulates on the toxicity of chemicals, and, taken together, provide information on the occurrence of these interactions and the possible mechanisms involved.

II. TOBACCO SMOKE

A. Composition of Tobacco Smoke

Tobacco smoke is an aerosol composed of gases, organic vapors, and a particulate phase, sometimes referred to as the smoke condensate (*186*). The relative proportion by mass of the gaseous and vapor phases is 92% and that for the particulate phase is 8% (*84, 156*). Of the more than 1200 components identified in tobacco smoke, more than 95% are present in the particulate phase (*186*). Toxic compounds found in the gaseous, vapor, and particulate phases include carbon monoxide, nitrous and nitric oxides, formaldehyde, PAH, aromatic amines, and nitrosamines (*1, 39, 77, 186*).

Because of the large number of compounds present in tobacco smoke, it is difficult to determine the contribution of each of the constituents to the adverse health effects associated with tobacco smoking. Among these health effects are lung cancer, chronic obstructive pulmonary disease, and coronary–artery disease (*174*). It is postulated that these health effects of tobacco smoking may originate from the impaired pulmonary function associated with smoking (*20*).

As mentioned above, tobacco smoke contains a particulate phase as well as gaseous and vapor phases. In reviewing the effects of particulates on the toxicity of tobacco smoke, the particulates to which we refer are not those normally associated with the smoke. In fact, the role of the particulate phase of tobacco smoke in producing the adverse health effects associated with the smoke has not been delineated. However, it has been observed that certain mineral particulates may exacerbate the health effects associated with smoking. It is these studies that we will now address.

B. Perspective

Human epidemiological studies, which will be described below in detail, have indicated that asbestos workers who also smoke cigarettes are at much greater risk of dying from lung cancer than are nonsmoking asbestos workers or smokers and nonsmokers who are not occupationally exposed to asbestos. However, a synergism between particulate exposure and cigarette smoking in the incidence of disease is not seen among workers exposed to particulates other than asbestos or in animals exposed to cigarette smoke and asbestos. Nevertheless, the data pertaining to a synergism between asbestos inhalation and cigarette smoking in the production of lung cancer are compelling, so it is of value to consider mechanisms by which particulates, in combination with cigarette smoking, may enhance lung tumor formation. Several factors may contribute to this process: (1) smoke and/or particulates may interfere with pulmonary function, (2) smoke and/or particulates may alter the activities of enzymes involved in xenobiotic metabolism, and (3) particulates may produce scarification of the lung.

Studies have shown that both cigarette smoking and particulates reduce the rate of pulmonary clearance so that inhaled carcinogens will be retained in the lungs for longer periods of time (*43, 174*). In addition, components of cigarette smoke and particulate matter each alter the activities of some of the enzymes involved in xenobiotic metabolism (*24, 79, 176*). It is now well recognized that certain carcinogens, such as the polynuclear aromatic hydrocarbons, must be metabolized to reactive forms to produce carcinogenesis. An increased production of the ultimate carcinogenic forms of the polynuclear aromatic hydrocarbons would presumably result in enhanced carcinogenesis. Particulates, such as asbestos, produce scarification of the lungs. It has been shown in animal studies that the scars produced as a result of an injury repair are the sites from which tumors arise upon application of a carcinogen (*12, 34*). In addition, lung scars in humans are frequently the sites from which tumors are seen to originate (*128, 129*). Thus, particulates could enhance the carcinogenicity of compounds in cigarette smoke by causing scarification of the lungs.

C. Asbestos and Smoking

A series of epidemiological studies have been performed by Saffiotti and coworkers regarding the incidence of lung cancer among workers who are occupationally exposed to asbestos. In a study of a cohort of 632 asbestos insulation workers, it was observed that 42 died of lung cancer, whereas only 6.6 deaths from lung cancer were expected based on calculations involving the general population (*141*). Clearly, the asbestos insulation workers were at increased risk of dying from lung cancer. In a subsequent study, the incidence of bronchogenic carcinoma in asbestos insulation workers was determined with consideration given to smoking histories (*142*). Of 283 workers who were cigarette smokers, 24 died of bronchogenic carcinoma, whereas no deaths from bronchogenic carcinoma were observed among the nonsmokers. Calculations based on age and smoking habits indicated that 2.98 deaths due to bronchogenic carcinoma were expected among the cigarette smokers, and 0.16 deaths were expected among nonsmokers who were not exposed to asbestos. From these data, it was calculated that asbestos increases the risk of death from bronchogenic carcinoma in cigarette smokers by 8-fold and that asbestos workers who smoke have a 92-fold increased incidence of this disease as compared to nonsmoking individuals without occupational exposure to asbestos.

Unexpectedly, 10 mesotheliomas also were observed in this group of asbestos insulation workers (*142*). Although the majority of men who died of mesotheliomas were also regular smokers, the number of individuals involved was too small to allow an analysis of possible interactions between asbestos and smoking in the production of mesotheliomas. Subsequent studies using larger cohorts of asbestos insulation workers and using amosite asbestos factory workers supported the earlier epidemiological studies (*58, 143*). It also should be noted that in the above studies, increased death rates from cancer at sites other than the lung also were observed. However, the number of cancers at any of these sites was smaller than the number of lung cancers, so the relative contributions of cigarette smoking and asbestos exposure to the production of these cancers could not be determined (*143*).

The effects of asbestos exposure and cigarette smoking in the incidence of asbestosis also have been considered. A threefold increased incidence of death from asbestosis was observed in asbestos insulation workers who smoked, but no such increase was observed in amosite asbestos factory workers (*58, 143*). In another study, the combined effects of cigarette smoking and asbestos exposure in asbestosis were found to be additive, but not multiplicative (*138*). Thus, if a synergism between asbestos inhalation and cigarette smoking exists in producing asbestosis, it is not of the same magnitude as that for the production of lung cancer. Similarly, a study of the incidence of pulmonary fibrosis in smoking and nonsmoking asbestos textile workers revealed that the smokers had higher incidences of pulmonary fibrosis though it could not be determined if the effect was additive or multiplicative (*185*).

D. Other Mineral Particulates and Smoking

It also is of interest to consider whether or not the synergistic effect observed with cigarette smoking and exposure to asbestos particulates occurs with other types of particulates. In general, detailed analyses similar to those performed on asbestos exposure and smoking are not available. It was found that ferrous foundry workers have a 1.4- to 2.5-fold increased incidence of death from bronchogenic carcinoma as compared to nonfoundry workers (*38, 54, 171*). The foundry workers are occupationally exposed to airborne particulates such as metal oxides, silica, carbon, and condensed organic matter, which could contribute to the observed increase. Sufficient data are not available to determine if cigarette smoking also is involved. A more complex situation exists when the excess of lung cancers among uranium miners is considered. These workers are occupationally exposed to particulates containing radon and radon daughters as well as to silica particles. The contribution of inhaled particulates to carcinogenesis, as opposed to the radiation associated with the particulates, is not easy to assess. However, exposure to radiation has been shown to increase the risk of cancer, so in this case the radiation effect probably exceeds that of the particulates. Studies of a cohort of workers in the Colorado Plateau area indicated that cigarette smoking did not contribute significantly to the increased incidence of lung cancer among uranium miners (*100, 175*). Follow-up studies of the same cohort of workers indicated that smokers represented a disproportionate number of miners who developed lung cancer and that cigarette smoking reduced the length of time required for lung cancer to develop in the uranium miners (*9*). However, it is believed that radiation is the primary cause of lung cancers in the miners and that smoking may account for only a slight increase in the lung cancer incidence (*9, 100, 175*).

E. Animal Studies

In addition to the human epidemiological studies discussed above, animal experiments also have been done to determine if particulates and cigarette smoke may interact in some manner to produce incidences of lung cancer that are greater than would be expected from exposure to either particulates or cigarette smoke alone. Wehner *et al.* (*179–183*) have investigated the effect of cigarette smoke in combination with chrysotile asbestos, nickel oxide, and cobalt oxide on the incidence of tumors in Syrian golden hamsters. From their data, an increase in lung tumors is associated with inhalation exposure to cigarette smoke, but the incidence of tumors was not increased with concurrent exposure to any of the particulates. In fact, the tumor incidence was decreased in the animals exposed to smoke and asbestos and unchanged in the animals exposed to smoke and either nickel oxide or cobalt oxide. These data probably result from the effects of cigarette smoking and exposure to particulates on the life span of the hamsters. That is, smoke inhalation results in a significantly increased life span as compared to sham smoke-exposed controls that could lead to an increased incidence of tumor formation. Asbestos exposure produced asbestosis, which significantly reduced the life span of the animals thus producing a lower expected incidence of tumor formation. Nickel oxide and cobalt oxide did not alter the life span, thus the observed effects were due to exposure to the cigarette smoke alone. Studies also have been done in beagle dogs in which the effect of uranium ore dust and radon daughters in combination with cigarette smoke was investigated. It was concluded that the respiratory pathology that occurred was due to exposure to the uranium ore dust and radon daughters with no additional effect contributed by the cigarette smoke (*48*). Studies of the effect of coal dust and cigarette smoke in Sprague–Dawley rats revealed that the pulmonary lesions could be accounted for by exposure to smoke alone (*95*). Thus, from the cited animal studies, evidence for an interaction between cigarette smoke and particulates to produce an enhanced incidence of lung tumors is not available.

F. Summary

The evidence supporting a synergism between particulate exposure and cigarette smoking in the production of disease comes from the epidemiological studies of Selikoff and co-workers (*141–143*). That is, asbestos workers who also smoke cigarettes have a much higher incidence of death from bronchogenic carcinoma than do nonsmoking asbestos workers. A synergism was not observed among workers exposed to other particulates, though this may reflect the lack of appropriate data. Moreover, a synergism was not observed when animals were exposed to cigarette smoke and asbestos (*179–183*).

The ambiguity observed among these various studies probably reflects the complex situation under consideration. That is, cigarette smoke contains numerous components that could contribute to the production of disease, the particulate exposures in humans are likely to vary among work environments, and the human population is genetically heterogeneous. The lack of support from animal studies for a synergism may be a further indication of the complexity of the situation.

It is unfortunate that data are not available to determine the mechanisms by which particulate exposure and cigarette smoking exert a synergistic effect. It is inherently difficult to obtain mechanistic information from human epidemiological studies. However, studies with animals or animal tissues can be designed to evaluate possible mechanisms. The factors mentioned above that might be involved are lung scarification and alterations in pulmonary function and metabolism. The individual effects, but not the combined effects, of particulates and cigarette smoke on these factors have been considered (*12, 24, 34, 43, 79, 174, 176*). To gain further insight into the synergism between particulates, such as asbestos, and cigarette smoking, future studies using experimental animals and animal tissues must evaluate the combined effect of particulates and cigarette smoke on the parameters that may be altered by these exposures. Examples of studies of this kind are those in which the combined effects of particulates and PAH, which are components of cigarette smoke, were examined. These studies will be considered in Section III.

III. POLYNUCLEAR AROMATIC HYDROCARBONS

A. Environmental Occurrence of PAH

PAH, which are produced in incomplete combustion processes, are ubiquitous in the environment (*29, 140*). Benzo[*a*]pyrene (B[*a*]P) is frequently used as an index of the amounts of total PAH produced by various combustion processes primarily because it is a highly carcinogenic PAH that has been studied extensively. It is estimated that 800 tons of B[*a*]P are emitted into the atmosphere per year (*11, 56*). Much of this emission results from stationary sources such as the burning of coal, oil, wood, and gas in heat and power production, and industrial processes such as coke production and refuse burning. In fact, coal burning is responsible for more B[*a*]P production than any other stationary source. Mobile sources, which include vehicles that burn petroleum products, also are important environmental sources of B[*a*]P and other PAH. As mentioned above, PAH are produced also in tobacco combustion and thus are present in cigarette, cigar, and pipe smoke.

Because of the carcinogenic properties of the PAH, numerous studies have been done to determine the levels of these compounds in the environment. The widespread production of PAH has resulted in their being found in the air, soil, and water with the highest levels being found in the vicinity of urban areas (*7, 139, 150–152*). As might be expected, the incidence of lung cancer is higher among individuals living in urban areas as compared to those living in rural areas (*36*).

Of particular importance to this discussion is the nature of PAH in the atmosphere. PAH are found in association with particulate matter (*81, 122*). The particulates with which the PAH are normally associated are those that are also byproducts of the combustion processes by which the PAH are produced, and thus are particulates of a nonuniform size distribution composed of a variety of organic and inorganic constituents. In a study of PAH in airborne particulates in a large Canadian city, it was found that 85% of the PAH was associated with particles less than 5 μm in diameter (*4*). Similarly, a study of coke oven emissions revealed that 94% of the PAH was associated with particles less than 3 μm in diameter (*17*). Particles less than 5 μm in diameter are defined to be ''respirable'' particles because they can penetrate through the upper respiratory airways to the lower airways and even to the alveoli (*64*). Thus, it is seen that the majority of PAH in the environment is associated with respirable particles, which could drastically increase the health risk simply by facilitating deeper penetration of the PAH into the lungs. From a theoretical study of the adsorption of PAH to particulates, it was proposed that the PAH preferentially adsorb onto the smaller particles because they represent a larger proportion of the total surface area, and that the extent of adsorption is proportional to the frequency of collision with the available surface area (*113*). Another example of the association of PAH with particulates in the environment involves a particulate that was previously discussed in conjunction with the adverse effects of cigarette smoking, namely, asbestos. It has been shown that B[*a*]P is found in the oils associated with the amosite and crocidolite types of asbestos and also that B[*a*]P can be adsorbed onto amosite, crocidolite, and chrysotile asbestos from the jute bags in which they are stored (*59*).

The role of PAH, particularly in association with particulates, in producing cancer in humans is difficult to assess. As mentioned earlier, PAH are present in cigarette smoke, and a synergism is observed in the production of lung cancer in humans when a simultaneous exposure to asbestos and cigarette smoke occurs. Also mentioned earlier is the increased lung carcinogenesis observed among urban residents as compared to rural residents, which may be attributed to the higher levels of particles with associated PAH to which urban residents are exposed. However, both cigarette smoke and urban air pollution contain numerous compounds other than the PAH and some of these compounds also are carcinogenic as judged from animal testing. Clearly, it is not possible to study the effects of a single PAH, such as B[*a*]P, in association with particulates on human health because of ethical considerations. However, numerous animal and *in vitro* studies have been done in which the effects of particles and PAH were investigated. These experiments will be reviewed to assess the role of particulates and PAH in carcinogenesis.

B. Perspective

Many animal studies have revealed that administration of both particulates and PAH results in increased tumorigenesis as compared to administration of either particulates or PAH alone. The possible mechanisms by which particulates enhance PAH carcinogenesis have been evaluated using animal studies and *in vitro* techniques. In the animal studies, the PAH and particulates generally were administered by intratracheal instillation. The mechanism frequently proposed for the enhancement of PAH carcinogenesis by particulates based on these animal studies is that the particulate functions as a carrier to enhance the penetration of the PAH into the lungs (*133–135*). Alternatively, it has been proposed that the PAH are retained in the lungs longer by virtue of their association with particulates (*125, 149*). It also was recognized that it is important to consider not only the length of time the PAH remained in the lungs, but also the rate at which the PAH are eluted from the particulates into the surrounding tissue (*28, 132*). In fact, studies by Lakowicz and co-workers using *in vitro* techniques revealed that particulates may greatly enhance the rate at which PAH are taken up by biological membranes (*13, 88–91*). However, prior to discussing the phenomenon of particle-enhanced uptake, it is necessary to consider the nature of particulate–PAH associations.

B[*a*]P, a PAH that has been frequently used in carcinogenicity studies, has a solubility in water of 3 μg/liter (*32*). Consequently, at the B[*a*]P concentrations normally used for carcinogenicity testing, not all of the B[*a*]P is soluble in the aqueous medium used for instillation. In the absence of particulates, the majority of the B[*a*]P would exist as aggregates in aqueous suspension. Under these conditions, it is difficult to perform quantitative manipulations with the B[*a*]P because of its propensity to adsorb to surfaces with which it comes into contact.

By associating the B[*a*]P with a particulate, it is more effectively dispersed, thus increasing its effective dose.

However, the form of the B[*a*]P as it exists on the surface of the particulates also must be considered. That is, it could exist as adsorbed B[*a*]P molecules, as aggregates of B[*a*]P molecules associated with the particulates, or as aggregates of B[*a*]P not bound to the particulates. The bioavailability of the B[*a*]P in these different states most likely is different. The types of complexes that occur in a given preparation will depend on the method of preparation, the PAH, the type of particulate, and the size of the particulate.

The method most frequently used to prepare B[*a*]P and particulates for intratracheal instillation studies was a prolonged process of grinding the two materials together with a mortar and pestle (*135*). It was assumed that this procedure produced particulates with B[*a*]P associated with them. However, based on the method of preparation, the B[*a*]P most likely would exist as aggregates, some of which may have become attached to the particles. Another procedure used to prepare B[*a*]P and particulates for intratracheal instillation was a low temperature "precipitation" of B[*a*]P from acetone solution onto the surface of the particles (*68, 69*). Based on the description of the procedure, it would seem that this method also would produce aggregates of B[*a*]P bound to the particles. In the studies performed by Lakowicz and co-workers, the B[*a*]P was adsorbed to the surface of the particulates by a "heat adsorption" technique (*90*). Briefly, the particulates were added to a toluene solution of B[*a*]P, and the solvent was then removed at about 50°C under vacuum with constant agitation of the sample to ensure an even distribution of B[*a*]P on the particulates. It was with samples such as these that particle-enhanced membrane uptake of B[*a*]P was observed. These studies will be described, after which the mechanism of particle-enhanced PAH uptake will be related to other types of investigations used to study particulate–PAH interactions.

C. Particle-Enhanced Uptake of PAH into Membranes

Lakowicz and co-workers studied the effect of particulates on the uptake of PAH into model membranes (phospholipid bilayer vesicles) and rat liver microsomes (*88–91*). The particulates included four types of asbestos, hematite, silica, and carbon particles, among others. The properties of these particulates have been described in the literature (*91*). The uptake of adsorbed B[*a*]P from the particulates into rat liver microsomes and into phospholipid bilayer vesicles composed of dipalmitoylphosphatidylcholine (DPPC) was quantified and compared to that of B[*a*]P in the absence of particulates (*88, 89, 91*). At the concentrations of B[*a*]P used, B[*a*]P in the absence of particulates existed as aggregates, which were termed B[*a*]P microcrystals. In performing these experiments, a fluorescence spectroscopic technique was developed to monitor the transfer of B[*a*]P from the microcrystalline, or particle-adsorbed states, to the membranes. It was observed that, in general, particulates enhance the rate of B[*a*]P uptake into these membranes as compared to uptake from the microcrystalline state. The greatest enhancement of uptake was produced by asbestos particles. Release of B[*a*]P from carbon particles was not observed in these experiments. When particulates without adsorbed B[*a*]P were added to microcrystalline suspensions of B[*a*]P, the membrane uptake was not enhanced but rather resembled that from microcrystals alone. Additional experiments revealed that the mechanism of particle-enhanced uptake of B[*a*]P was a rate-limiting solubilization of B[*a*]P into the aqueous phase followed by a rapid uptake of solubilized B[*a*]P into the membranes.

Based on these experiments and on the proposed mechanism for particle-enhanced PAH uptake, one would predict that the equilibrium solubility of a PAH might determine whether or not its uptake into membranes would be enhanced by adsorption to particulates. Bevan *et al.* (*13*) studied the effects of particulates on the membrane uptake of PAH other than B[*a*]P. Specifically, they considered dibenzo[*a,h*]anthracene, benzo[*g,h,i*]perylene, 3-methylcholanthrene (3-MC), and dibenzo[*c,g*]carbazole. With dibenzo[*a,h*]anthracene, benzo[*g,h,i*]perylene, and 3-methylcholanthrene, particle-enhanced uptake into DPPC vesicles was observed. With dibenzo[*c,g*]carbazole, particle-enhanced uptake was not observed, but rather, the uptake rates were very rapid in both the presence and absence of particulates. These results are explained by considering the mechanism given above for particle-enhanced uptake and the aqueous solubilities of these PAH. Their solubilities are as follows: dibenzo[*a,h*]anthracene, 0.5 μg/liter at 27°C (*32*); benzo[*g,h,i*]perylene, 0.26 μg/liter at 25°C (*101*); 3-methylcholanthrene, 2.9 μg/liter at 25°C (*101*); and dibenzo[*c,g*]carbazole, 320 μg/liter at 25°C (*92*). Under the conditions of these *in vitro* experiments, $>60\%$ of the dibenzo[*c,g*]carbazole was solubilized in the aqueous phase in the presence of particulates prior to the addition of membrane vesicles, whereas $<1\%$ of the other PAH was solubilized. Because solubilized PAH are rapidly taken up by membranes, it is to be expected that particulates would not show further enhancement of dibenzo[*c,g*]carbazole uptake. However, for the other PAH used in this experiment most is associated with the particulates with little dissolved in the aqueous phase prior to addition of membranes. Thus, the particulates will influence the rate at which these PAH are transferred to the membranes.

As a result of particle-enhanced uptake, it is to be expected that PAH adsorbed to particulates will be delivered rapidly to cells. Consequently, the effective dose of the PAH is increased, and the incidence of carcinogenesis also would be expected to increase. Both *in vitro* and *in vivo* methods have been used to study the effects of particulates on PAH carcinogenesis. These experiments will now be described and related to the phenomenon of particle-enhanced membrane uptake of PAH.

D. *In Vitro* Methods

In the technique of tracheal grafts, the compound(s) to be tested are applied to tracheal explants, which are then implanted subcutaneously into syngeneic animals. In one experiment, the carcinogenicities of B[*a*]P and chromium carbonyl, applied singly and in combination, were tested in this system (*94*). Chromium carbonyl induced squamous cell carcinomas in 2 of 22 grafts, B[*a*]P induced similar tumors in 8 of 22 grafts, and the two together induced these tumors in 10 of 24 grafts. Clearly, the effect of these two agents was additive and not synergistic. In another application of this technique, hamster tracheal explants were exposed to 3-MC adsorbed to either hematite or crocidolite asbestos and then implanted into syngeneic hamsters (*26, 111*). Tumors developed when explants were exposed to 10 μg of 3-MC–hematite or 1.4 μg of 3-MC–crocidolite, though not upon exposure to lower doses. It was noted that PAH alone do not produce tumors in this system (*111*). Thus, adsorption of 3-MC to particulates increased the effectiveness of 3-MC in producing tumorigenesis, probably because greater amounts of the 3-MC would be taken up by the explants from the adsorbed state.

The effectiveness of chrysotile asbestos as a cocarcinogen also was evaluated using tracheal explants (*172*). Excised rat tracheas that were grafted subcutaneously into syngeneic animals were exposed to doses of dimethylbenzanthracene (DMBA) ranging from 12.5 to 100 μg. Chrysotile asbestos was subsequently introduced into the tracheas. DMBA produced tumorigenesis at the 100-μg dose level, but chrysotile did not enhance tumorigenesis. However, chrysotile did produce 15 and 23% incidences of tumors at 50- and 25-μg doses of DMBA, respectively, which are doses at which no tumorigenesis is seen in the absence of the asbestos. From these studies it was concluded that chrysotile asbestos can serve as a strong cocarcinogen with DMBA (*172*). The mechanism of cocarcinogenesis was not particle-enhanced PAH uptake in these studies because the DMBA was not adsorbed to the chrysotile, but rather was applied 4 weeks prior to the chrysotile. The enhancement of tumorigenesis in this study may result from the irritant properties of the particulate.

Both short- and long-term tracheal organ culture also have been applied to studies of PAH carcinogenesis (*82, 111*). In neither case have these studies directly compared PAH tumorigenesis in the absence and presence of particulates. However, they will be discussed to provide a brief description of another *in vitro* technique that is applicable to such studies. Kaufman *et al.* (*82*) studied the binding of B[*a*]P to DNA in short-term hamster tracheal cultures. Animals were exposed to hematite alone or to B[*a*]P–hematite mixtures (prepared by the grinding procedure) prior to preparation of the organ cultures. Tracheas from animals pretreated with hematite showed the same levels of B[*a*]P binding to DNA as those from untreated animals, whereas the binding in B[*a*]P–hematite-exposed animals was approximately fivefold higher. The effect of treatment with B[*a*]P alone was not considered. The increase in B[*a*]P binding to DNA after treatment with the B[*a*]P–hematite mixture may be the result of induction of the enzymes required for metabolism by the pretreatment with B[*a*]P–hematite. It is likely that pretreatment with B[*a*]P alone also would yield enhanced binding to DNA as compared to hematite-pretreated animals.

Warshawsky *et al.* (*176*) have applied another *in vitro* technique, the isolated perfused rabbit lung, to studies of the effect of particulates on B[*a*]P metabolism. In some cases, these experiments were performed by pretreating the animals with either B[*a*]P or particulates, but in all cases, the B[*a*]P metabolism was studied after excising the lungs. Three sets of experiments were performed. In the first set, B[*a*]P, B[*a*]P + hematite, or B[*a*]P + crude air particulate (CAP) was applied to the isolated perfused lungs obtained from animals that received no pretreatment. Both particulates were equally effective in slightly inhibiting the metabolism of B[*a*]P, and, in both cases, a change in the distribution of the metabolites was seen as compared to that observed when B[*a*]P was administered alone. In the second set of experiments, the animals were pretreated with B[*a*]P by intraperitoneal injection, and then B[*a*]P metabolism was studied in the isolated perfused lung. B[*a*]P pretreatment increased the metabolism of B[*a*]P applied to the isolated lungs as compared to the untreated animals, but the particulates again inhibited B[*a*]P metabolism. In the third set of experiments, the animals were pretreated with hematite or CAP by intratracheal instillation and then B[*a*]P metabolism was studied in the isolated perfused lungs. The particulate-treated animals showed an *increase* in the rate of B[*a*]P metabolism. To explain these data, it was proposed that the particulates affect B[*a*]P metabolism by two mechanisms (*176*). In one, the particulates must decrease the availability of B[*a*]P to the lung tissue, thereby slowing the metabolism. As a result, a decrease in B[*a*]P metabolism is observed when the B[*a*]P is administered in the presence of particulates as seen in the first two sets of experiments. The basis for the second mechanism comes from the third set of experiments in which particulate pretreatment increased the metabolism of B[*a*]P in the isolated perfused lungs. It is postulated that the particulates were in some way able to alter the activities of enzymes involved in the metabolism of B[*a*]P.

Particle-enhanced uptake of B[*a*]P would not occur in these experiments because the B[*a*]P was not adsorbed to the particulates prior to addition to the isolated perfused lungs. As noted by Warshawsky *et al.* (*176*), the observed effects could be due to differences in short- and long-term effects of particulates in the isolated perfused lung. It has been demonstrated by other investigators using microsomal systems that the activities of enzymes required for B[*a*]P metabolism are decreased by particulates (*79*). However, the longer exposures used by Warshawsky *et al.* in the pretreatment step have not been studied by other investigators and may, in fact, lead to increases in the enzyme activities.

E. Animal Studies

1. Development of the Intratracheal Instillation Technique

The technique that has been most widely used to expose experimental animals to particulates and PAH to study lung tumor formation is intratracheal instillation. The development of this technique will be reviewed briefly and specific examples will be given to illustrate the effect of particulates on the carcinogenicity of some PAH.

The intratracheal instillation technique was developed in an attempt to provide a means to administer potentially carcinogenic compounds to animals via a route that resembles the normal human route of exposure, namely, inhalation. In addition, it was desired that the method produce tumors similar to those observed in humans. The disadvantage of animal inhalation exposures are that elaborate exposure facilities are required and that the precise concentrations of the compounds in the lungs of the animals are not known. Other alternatives to inhalation exposures such as implantation of carcinogen-coated threads and pellets into the bronchi were successful in the production of tumors, but the means of administration were not similar to human exposures.

Early experiments using intratracheal instillation to investigate the lung carcinogenicity of PAH were unsuccessful. For example, Della Porta *et al.* (*35*) instilled DMBA in combination with tobacco smoke condensate into hamsters. Tumors were observed primarily in the larynx, trachea, and esophagus of the animals with only one animal developing an adenocarcinoma in the main bronchus. In similar experiments by these workers with B[*a*]P (*134*) and by other investigators with DMBA and B[*a*]P (*55*) lung cancers similar to those observed in humans also were not observed.

It had been observed by Pylev (*124, 125*) and Shabad (*149*) that intratracheal instillation of DMBA in rats resulted in high incidences of tumor formation if a particulate carrier was coinstilled with the DMBA. In their experiments they used a suspension of india ink powder. Lung tumors were observed in approximately 30% of the animals at the highest doses of DMBA employed (*125*), and the tumors resembled those found in humans. The use of particulates in conjunction with PAH to induce lung tumors in animals by intratracheal instillation was developed further by Saffiotti and co-workers (*135*). The principal experimental animal used by these investigators was the Syrian golden hamster. It had been observed in the studies of Pylev and Shabad that rats suffered from chronic inflammatory conditions that could contribute to the observed lung tumor induction (*124, 125, 149*). However, Syrian hamsters are much more resistant to these chronic inflammatory conditions and, in addition, exhibit a low incidence of spontaneous lung tumors (*135, 173*), though they are susceptible to the induction of lung tumors upon exposure to PAH (*35, 55*). Moreover, the tumors that develop in Syrian golden hamsters are histologically similar to tumors that occur in humans, with the majority being squamous cell carcinomas arising primarily in the tracheobronchial epithelium (*62, 63, 106, 123*).

The method developed by Saffiotti *et al.* (*135*) involved the intratracheal instillation of B[*a*]P which was attached to the surface of hematite particles by a prolonged grinding process. The dusts prepared in this manner generally contained 50% B[*a*]P and 50% hematite by weight, and they were suspended in saline for the instillation procedure. It was observed that once weekly instillations of 3 mg B[*a*]P + 3 mg hematite for 15 weeks resulted in tumor incidences of as high as 100% of the surviving animals. Instillation of hematite alone did not produce respiratory tumors (*135*). The predominant types of bronchogenic carcinomas observed in the hamsters were squamous cell, anaplastic, and adenocarcinomas, which are the 3 main types observed in humans (*135*). Based on these initial studies, the intratracheal instillation technique has been widely used. The application of this method to studies of tumorigenesis produced by various PAH will now be considered.

2. *Benzo[*a*]pyrene (B[*a*]P)*

The studies of Saffiotti *et al.* (*135*) were extended by many investigators to define further the conditions under which intratracheal instillation of B[*a*]P and hematite produced tumorigenesis in hamsters. For example, it was observed that repeated (5, 10, or 15 times) instillation of 3 mg B[*a*]P with 3 mg hematite was more effective in inducing tumors than were single instillations of either 37.5 mg B[*a*]P with 12.6 mg hematite or 5 mg B[*a*]P with 45 mg hematite (*136*). In addition, studies were performed to determine the dose–response relationship with respect to both B[*a*]P and hematite. It was found that tumorigenesis resulting from 30 weekly instillations of 2.0, 1.0, 0.5, and 0.25 mg B[*a*]P/dose with equal amounts of hematite was positively correlated with dose (*137*). However, the tumor induction was the same when a constant amount of B[*a*]P (3 mg) was used with varying amounts (3, 6, and 9 mg) of hematite (*146*). In this last set of experiments, the B[*a*]P–hematite mixtures were prepared in a 1:1 ratio, after which additional hematite was added to two portions of the mixture. Thus, the nature of the B[*a*]P–hematite association was the same at all hematite concentrations, which may have contributed to the observed results.

As mentioned above, the nature of the form of the PAH and its association with particulates may influence the carcinogenic potency of the PAH. For example, based on the work of Lakowicz and co-workers, it would be expected that

B[*a*]P would be taken up by the lung tissues more rapidly from the adsorbed state than from B[*a*]P aggregates. Animal experiments performed by Henry *et al.* (*68, 69*) are consistent with this hypothesis. In these experiments, the B[*a*]P–hematite samples were prepared by low-temperature precipitation from acetone and by grinding. The low-temperature precipitation technique produced large aggregates of B[*a*]P–hematite (94.1% > 15 μm), and, significantly, no particles of B[*a*]P alone were observed by microscopic analysis. The grinding procedure produced B[*a*]P–hematite suspensions of a smaller particle size (52.6% > 15 μm), but approximately 25% of the B[*a*]P existed as crystals not bound to hematite. Intratracheal instillation of the B[*a*]P–hematite in which all of the B[*a*]P was associated with the particulates resulted in earlier onset and greater incidence of tumorigenesis than when less of the B[*a*]P was bound to the hematite. In addition, it was observed by Henry *et al.* (*68, 69*) that simple mixtures of B[*a*]P and hematite yielded even fewer tumors, with the number being similar to that observed when B[*a*]P alone was instilled.

Studies also were performed to determine if the enhanced tumorigenesis observed upon simultaneous exposure to hematite and B[*a*]P resulted from an initiation–promotion relationship between the two agents (*147*). Large numbers of tumor-bearing animals (71%) were observed only when B[*a*]P in association with hematite was intratracheally instilled in the animals. Administration of the hematite either before or after the B[*a*]P produced tumors in only 13% of the animals, which was similar to that seen when B[*a*]P alone was administered. Thus, it was concluded that hematite does not serve as an initiator or promoter of B[*a*]P tumorigenesis and that physical association of the B[*a*]P and hematite is required to produce high incidence of tumors (*147*).

Numerous other particulates besides hematite have been used in conjunction with B[*a*]P to induce tumor formation in experimental animals. In a study that preceded much of the work with hematite, it was observed that intratracheal instillation of B[*a*]P with chrysotile asbestos resulted in enhanced tumorigenesis in hamsters as compared to exposure to B[*a*]P alone (*105*). However, it also was observed that the number of tumors produced by B[*a*]P was similar in the presence and absence of amosite asbestos (*105*). It was postulated that the different effects of chrysotile and amosite may result from the differences in the abilities of the types of asbestos to adsorb B[*a*]P (*60*). It was demonstrated in studies in rats that intratracheal instillation of chrysotile alone or B[*a*]P alone did not produce any precancerous or cancerous lesions in the respiratory tract, but that combined administration of chrysotile and B[*a*]P did yield numerous such lesions (*126, 154*). Interestingly, the number of lesions was greater when a mixture of B[*a*]P and chrysotile was used than when the B[*a*]P was adsorbed to the asbestos prior to administration. However, it should be noted that the dose of B[*a*]P administered in the mixture was approximately 10-fold greater than that administered in the adsorbed form. In addition, it is recognized that B[*a*]P adsorbs to asbestos so that some of the B[*a*]P most likely was adsorbed in the sample containing a mixture of B[*a*]P and chrysotile. Thus, it is difficult to evaluate differences in the extent of pathological lesions produced in these experiments when B[*a*]P was administered with chrysotile in the adsorbed and nonadsorbed states.

Particulates other than hematite and asbestos that have been used in combination with B[*a*]P in intratracheal instillation studies include talc, carbon, aluminum oxide, titanium dioxide, silica, magnesium oxide, lead oxide, arsenic trioxide, and nickel sulfide. With some of the particulates, a synergism in the induction of tumors is observed upon coinstillation with B[*a*]P. For example, Stenback *et al.* (*164*) studied titanium dioxide in combination with B[*a*]P and observed 52 respiratory lesions in 48 animals, which was similar to the number observed upon exposure to hematite and B[*a*]P. Moreover, a greater incidence of tumorigenesis was observed when silica and B[*a*]P were administered together than when B[*a*]P alone was used (*160*). In a study comparing magnesium oxide and B[*a*]P to hematite and B[*a*]P in respiratory tumor induction, it was observed that similar numbers of histologically similar tumors were produced with either particle, though hematite yielded tumors somewhat more frequently in the bronchi (*163*). Studies employing aluminum oxide and B[*a*]P do not show as great a synergism as those with hematite and B[*a*]P, though some enhancement of tumorigenesis was observed. For example, Stenback *et al.* (*164*) observed respiratory lesions in 2 of 48 animals exposed to B[*a*]P alone, 0 of 48 to aluminum oxide alone, and 6 of 48 to aluminum oxide plus B[*a*]P. Farrell and Davis (*40*) observed tumors in approximately 25% of the hamsters when aluminum oxide and B[*a*]P were administered together, whereas 50% developed tumors when hematite and B[*a*]P were instilled.

With some other particulates and B[*a*]P, enhancement of tumorigenesis also was observed though a direct comparison with hematite and B[*a*]P was not made. For example, Kobayashi and Okamoto (*85*) administered B[*a*]P and lead oxide intratracheally in hamsters and observed increased tumorigenesis when B[*a*]P and lead oxide were administered together as compared to either agent alone. In addition, intratracheal administration in rats of nickel sulfide or B[*a*]P alone resulted in fewer premalignant lesions than did administration of the two together (*80*). Also, intramuscular injection of nickel sulfide, B[*a*]P, or the two in combination in rats resulted in sarcoma production, but the time required for tumor production was shorter when the agents were injected together (*102*). Studies with talc and B[*a*]P are less definitive because these experiments were done without determining the tumorigenesis produced by B[*a*]P alone. However, intratracheal instillation of B[*a*]P in combination with talc in hamsters resulted in tumor formation in 33 of 45 animals with a total number of 123 respiratory lesions being produced, and instillation of talc alone produced no respiratory tumors (*158*).

In studies with carbon and B[*a*]P, a synergism in tumorigenesis was observed in some cases but not in others. Recall that in early experiments by Pylev (*124, 125*) and Shabad (*149*), india ink, which is a suspension of carbon particles, was effective in enhancing tumor formation in rats when administered with DMBA. In studies using hamsters, it was observed by Stenback *et al.* (*164*) that respiratory lesions were seen in 2 of 48 animals exposed to B[*a*]P alone, 0 of 48 animals exposed to carbon alone, and 13 of 48 animals exposed to a combination of B[*a*]P and carbon. In addition, other investigators observed that the response to B[*a*]P and carbon was similar to that to B[*a*]P and hematite in terms of the time of appearance and number of tumors observed (*40*). However, Davis *et al.* (*31*) intratracheally administered B[*a*]P alone, carbon alone, and B[*a*]P in combination with carbon. They observed squamous neoplasms in 1 of 16 animals in the carbon-exposed group, 24 of 48 animals in the B[*a*]P-exposed group, and 15 of 51 animals in the B[*a*]P/carbon-exposed group. It would appear from these data that carbon reduced tumor formation resulting from exposure to B[*a*]P, though the high incidence of tumors in the group exposed to B[*a*]P alone was unexpected and could not be explained by the investigators. In addition, in an investigation of the possible role of carbon as a promoter of B[*a*]P tumorigenesis, fewer respiratory tumors were observed with B[*a*]P and carbon than with B[*a*]P alone (*147*).

Intratracheal instillation of arsenicals in combination with B[*a*]P in rats also has been investigated (*78*). The arsenic-containing particulates included copper ore (3.95% arsenic), flue dust (10.6% arsenic), and arsenic trioxide. Administration of the particles alone yielded adenomas and adenocarcinomas. Administration of the particles in combination with B[*a*]P produced squamous cell carcinomas in numbers greater than those observed upon administration of B[*a*]P alone, but the numbers of surviving animals were not great enough to allow evaluation of the significance of these results.

As mentioned above, the particle–B[*a*]P system for inducing tumorigenesis was based on the idea that B[*a*]P alone was not carcinogenic upon intratracheal instillation because it could not readily penetrate into the lungs and be retained sufficiently long to be taken up into the lung tissues. Consequently, studies were performed to determine the rate at which B[*a*]P was cleared from the lungs when administered in conjunction with particulates. However, the routes by which the B[*a*]P was cleared were not considered, which complicates the interpretation of these experiments. That is, the B[*a*]P may be "cleared" from the lungs by three different pathways. These are (1) membrane uptake of B[*a*]P by lung tissue followed by absorption into the circulatory system, (2) excretion via the mucociliary escalator in the particle-bound form, and (3) excretion after phagocytosis by alveolar macrophages. Thus, it is difficult to interpret some of the studies to be cited in terms of the amount of B[*a*]P that was transferred to the lung tissues. In addition, the rate at which the particulates, with which the B[*a*]P was coinstilled, were cleared from the lungs usually was not considered. The studies of B[*a*]P retention will be discussed in light of these inadequacies.

Saffiotti *et al.* (*132–134*) studied the retention of B[*a*]P in the lungs of hamsters when administered in conjunction with hematite. From these studies several important observations were made. First, when a sample containing 3.4 mg each of B[*a*]P and hematite was intratracheally instilled into hamsters, 80–90% of the B[*a*]P was cleared from the lungs within 1 week. Moreover, the rate of B[*a*]P clearance was the same for a single instillation or for each of 10 successive instillations given at weekly intervals. Second, the rate of B[*a*]P clearance was faster for low concentrations of B[*a*]P–hematite (0.96 mg of each) than for higher concentrations (3.4 or 23.0 mg of each). Third, the rate of clearance of B[*a*]P was the same when administered with varying concentrations of hematite. Fourth, when the B[*a*]P was associated with the hematite by either the grinding procedure or the acetone precipitation technique, the rate of clearance of B[*a*]P was similar when the B[*a*]P–hematite aggregates were of similar sizes. Finally, the rate of B[*a*]P clearance increased as the particle size of the B[*a*]P–hematite aggregates decreased. Also, it was noted by Saffiotti (*132*) that most of the hematite was retained in the lungs over the course of these experiments.

In some cases, the retention data of Saffiotti *et al.* compare favorably with tumorigenesis studies. For example, as noted above, a positive dose–response effect was observed with increasing concentrations of B[*a*]P–hematite (*137*). This finding is consistent with the observation that lower doses of B[*a*]P are cleared from the lungs more rapidly than higher doses. In addition, similar numbers of tumors were observed when constant amounts of B[*a*]P were instilled with varying amounts of hematite (*146*), which is consistent with retention studies showing similar rates of clearance under these conditions. Finally, increased tumorigenesis was observed with larger aggregates of B[*a*]P–hematite (*68, 69*), which is consistent with the slower clearance of B[*a*]P when the B[*a*]P–hematite aggregate size is larger. However, two points should be made regarding these studies concerning tumorigenesis and B[*a*]P retention as a function of aggregate size. The first is that the retention data were shown only for the clearance of the last 1–8% of the administered B[*a*]P so it is not possible to compare the clearance rates for the large majority of the B[*a*]P. Second, as mentioned above, all of the B[*a*]P in the larger aggregates was particle associated (*68, 69*) so that the observed increase in tumorigenesis probably resulted from particle-enhanced membrane uptake of B[*a*]P.

Other investigators have not observed a correlation between particle size of B[*a*]P–hematite aggregates and B[*a*]P retention. For example, Henry *et al.* (*68*) studied the clearance of B[*a*]P from the lungs of hamsters when administered together with hematite in particle sizes of 0.1–5, 0.5–1, 2–5, 5–10, and 15–30 μm and observed no difference in B[*a*]P clearance rates as a function of particle size. In all cases, however, the clearance rates of B[*a*]P from the particle-associated state were slowed as compared to the clearance of B[*a*]P administered without particles. In addition, the clearance rates of B[*a*]P when present in mixtures with the particles, but not associated with the particles, were similar to those of B[*a*]P alone. These results are consistent with the tumorigenicity studies in which simple mixtures of B[*a*]P and hematite were observed to produce tumorigenesis similar to that of B[*a*]P alone (*68, 69*). In addition, recall that particulates did not enhance the membrane uptake of B[*a*]P when the two were present as a simple mixture (*88, 91*). Henry *et al.* (*69*) also studied the effect of the method used to combine the hematite and B[*a*]P on the clearance rates. The B[*a*]P was cleared from the lungs at similar rates both for particle–B[*a*]P mixtures prepared by low-temperature nucleation and for those prepared by grinding the hematite and B[*a*]P together. Again, the clearance rates for B[*a*]P when associated with particles were slower than those for mixtures of B[*a*]P and particles (*69*). In contrast, Feron *et al.* (*46*) did not observe a difference in B[*a*]P clearance rates when either B[*a*]P suspended in saline or B[*a*]P associated with hematite and suspended in saline was intratracheally instilled into hamsters. In that study, the B[*a*]P–particle preparations were done according to Saffiotti *et al.* (*135*) so the B[*a*]P presumably was associated with the particles. The authors hypothesized that the differences between their results and those of other investigators were due to differences in particle sizes. Although it was stated that large particles were present in their investigation (*46*), the sizes were not given so a more detailed comparison cannot be made.

The clearance of B[*a*]P from the lungs of experimental animals also has been studied using particles other than hematite. The effect of talc on B[*a*]P clearance was studied by intratracheally instilling in hamsters either 3 mg B[*a*]P, 3 mg B[*a*]P bound to 3 mg talc, or 3 mg B[*a*]P bound to 9 mg talc (*121*). The B[*a*]P was adsorbed to the talc by the nucleation procedure described above. Talc slowed the clearance of B[*a*]P such that, at 48 h after administration, 0% of the B[*a*]P remained in the lungs in the absence of talc whereas 20.6 and 24.1% of the B[*a*]P remained in the lungs when 3 and 9 mg, respectively, of talc were coinstilled. A comparison of the clearance rates of B[*a*]P when intratracheally instilled with carbon, aluminum oxide, or hematite was made using four size ranges of particles, these being 0.5–1, 2–5, 5–10, and 15–30 μm (*66*). The samples of B[*a*]P–hematite and B[*a*]P–aluminum oxide were prepared by low-temperature nucleation, and the B[*a*]P–carbon samples were prepared by heat adsorption. As noted earlier (*68*), no correlation was observed between hematite particle size and clearance rate and 95% of the administered B[*a*]P was cleared within 14–23 h (*66*). Similarly, no correlation was observed between aluminum oxide particle size and clearance rate though rates for clearance of 95% of the administered B[*a*]P ranged from 11 h to 14 days. A positive correlation was observed between B[*a*]P clearance and the sizes of the carbon particles and the times for clearance were considerably longer than with the other particles. The time for clearance of 95% of the B[*a*]P for the 0.5- to 1-μm particles was 3.9 weeks, for the 2- to 5-μm particles, 6.3 weeks, for the 5- to 10-μm particles, 71 weeks, and for the 15- to 30-μm particles, 72 weeks.

Pylev *et al.* (*127*) studied the clearance of B[*a*]P alone and in conjunction with carbon or crocidolite particles from the lungs of hamsters following intratracheal instillation. The clearance rates were essentially the same in all three cases when it is considered that the differences in clearance rates noted by Pylev and co-workers at 3 weeks after administration were representative of less than 1% of the administered dose. Because the method of preparation of the particle–B[*a*]P mixtures was not given, it is difficult to determine the reason for the similar clearance rates though the results could be explained if the B[*a*]P was not actually associated with the particles. Previously, Shabad *et al.* (*153*) also had investigated the retention of B[*a*]P in the presence of carbon particles. B[*a*]P in the presence and absence of carbon particles was suspended in either saline solution or Infusine, which contains casein, and instilled in the lungs of rats. Clearance was most rapid when B[*a*]P alone was suspended in saline and was slowest for B[*a*]P in the presence of carbon particles suspended in either saline or Infusine. When B[*a*]P alone was suspended in Infusine, the clearance rate was only slightly more rapid than that in the presence of carbon particles. This finding was attributed to an association of the B[*a*]P with the casein, which is cleared from the lungs somewhat more rapidly than are carbon particles (*153*). However, it was clear that association of the B[*a*]P with the carbon particles extended the time the B[*a*]P remained in the lungs.

Experiments also were performed in which the clearance rates of both B[*a*]P and carbon particles were considered (*27, 28, 116*). Carbon particles of two different size ranges, 0.5–1.0 and 15–30 μm, were either labeled with ^{103}Ru or coated with B[*a*]P. Samples of each were mixed and then intratracheally instilled in mice. The clearance rate of the carbon particles was monitored by ^{103}Ru radioactivity and that of B[*a*]P was measured spectrofluorometrically (*28*). The time for clearance of 50% of the administered B[*a*]P dose was as follows: B[*a*]P alone, 1.5 h; B[*a*]P–carbon (0.5–1.0 μm), 36 h; B[*a*]P–carbon (15–30 μm), 4–5 days. Clearance rates for the carbon particles were 7 days for the 0.5- to 1.0-μm particles and 4–5 days for the 15- to 30-μm particles. The similar clearance rates of the 15- to 30-μm carbon particles and the B[*a*]P adsorbed to these particles indicate that the B[*a*]P was not eluted from the particles in the lung but rather was cleared in the carbon-adsorbed state. That the B[*a*]P was cleared from the lungs when administered in association with 0.5- to 1.0-μm particles much more rapidly than were the particles indicates that B[*a*]P was eluted from the particles in the respiratory tract. These results explain a discrepancy observed in previous studies of the tumorigenic properties (*40*) and clearance rates (*66*) of B[*a*]P associated with carbon particles of several size ranges. That is, B[*a*]P–carbon particles of the smallest size range were the most tumorigenic though the B[*a*]P was cleared from the lungs most rapidly when associated with these particles. As demonstrated by Creasia *et al.* (*28*), the B[*a*]P is eluted from these particles in the lungs so that it is taken up by the surrounding tissue to produce tumorigenesis. Thus, the importance of quantifying B[*a*]P elution from the B[*a*]P–particulate samples is again emphasized.

The intratracheal instillation technique in which B[*a*]P and particulates were coinstilled was developed because early attempts to induce tumorigenesis upon instillation of B[*a*]P alone were unsuccessful. However, several investigators have subsequently succeeded in producing tumors in the lungs of hamsters upon instillation of B[*a*]P without particulates, thus demonstrating that the particulate is not indispensible. For example, Feron *et al.* (*45*) produced respiratory tract tumors in 93% of experimental animals when 1 mg of B[*a*]P was administered weekly for 52 weeks. A dose–response effect similar to that observed when B[*a*]P and hematite were coinstilled was observed, but the latent period of tumor formation was longer when the hematite was absent (*45*). Hilfrich *et al.* (*76*) also observed tumorigenesis in hamsters in which B[*a*]P alone was instilled. The number of tumors formed was similar to that observed by Feron *et al.* (*45*) though strict quantitative comparisons are not possible due to differences in experimental methods. Henry *et al.* (*67*) observed tumors in 26 of 65 experimental animals with a total of 69 respiratory lesions being observed. Their experimental technique differed from that of Feron *et al.* (*45*) in that eight weekly instillations of B[*a*]P averaging 13.9 mg per instillation were administered. Henry *et al.* (*67*) also studied the rate at which a single dose of intratracheally instilled B[*a*]P was cleared from the lungs and found that low doses of B[*a*]P were cleared much faster than high doses. These observations were extended by other investigators. The intratracheal instillation of two different sizes of B[*a*]P particles into hamsters demonstrated that instillation of the larger particles (64% greater than 10 μm) resulted in greater tumorigenesis than instillation of the smaller particles (98% less than 10 μm). Specifically, 5 of 48 animals developed respiratory tumors upon exposure to the smaller particles as compared to 31 of 48 animals exposed to the larger particles (*159*). Studies of the retention of the small and large particles in the lung revealed that the larger particles were retained for a longer period of time with 75% remaining in the lungs 6 h after instillation as compared to 17% of the smaller particles (*159*). Studies by Feron *et al.* (*47*) confirmed these observations. Again, it would be necessary to know the route by which the B[*a*]P was cleared to interpret these experiments further. That is, the smaller particles might be cleared from the lungs by dissolution in the surrounding tissue followed by absorption or by clearance on the mucociliary escalator followed by ingestion. It is reasonable to suppose that smaller particles would be more readily dissolved in lung tissue, but also that they would produce higher equilibrium concentrations of B[*a*]P in the tissue prior to absorption into the circulation. It would be expected that higher B[*a*]P concentrations would be more, not less, carcinogenic. Contrariwise, the larger particles, if dissolved more slowly, would result in the tissues being exposed to lower equilibrium concentrations of B[*a*]P, but over a longer period of time. This could be argued to be the reason for the greater carcinogenicity of the larger B[*a*]P particles. However, it should be realized that the size of larger particles instilled in these studies would ordinarily cause them to be deposited in the nasal region under inhalation exposure conditions. The lung probably was not designed to clear particles of this size.

3. Other PAH

The discussion up to now has dealt with B[*a*]P carcinogenesis. A few studies have been done using other PAH. As mentioned previously, in early studies by Pylev (*124, 125*) and Shabad (*149*) in which DMBA and india ink were intratracheally instilled in the lungs of rats, up to 30% of the animals developed tumors. In a subsequent study, the tumorigenicity of DMBA and hematite was noted in rats (*14*). In addition, the carcinogenic potency of DMBA and hematite in hamsters was demonstrated (*161*). After administration of relatively low amounts of DMBA (0.85 and 1.2 mg) in two groups of experimental animals, tumors developed in 11 of 28 and 15 of 46 of the animals, respectively. However, a comparison of the tumorigenesis induced by DMBA in the absence of particles was not considered.

Single instillations of 3-methylcholanthrene with chromic oxide, nickel oxide, aluminum oxide, or silicon dioxide were made in rats (*170*). Coinstillation of 3-MC and nickel oxide induced 12 hyperplastic (benign) and 5 neoplastic (invasive) tumors among 30 animals, 3-MC and chromic oxide, 4 hyperplastic and 3 neoplastic among 40 animals, 3-MC and silicon dioxide, 2 hyperplastic and 1 neoplastic among 15 animals, and 3-MC and aluminum oxide, 1 hyperplastic and 0 neoplastic among 15 animals.

Sellakumar and Shubik (*145*) intratracheally instilled benz[*a*]anthracene, benzo[*a*]fluoranthene, dibenz[*a,h*]anthracene, dibenz[*a,i*]pyrene, and pyrene in combination with hematite into hamsters. Only dibenz[*a,i*]pyrene was found to be tumorigenic, with 89% of the animals developing tumors at the highest dose level used (12 mg). It was noted that dibenz[*a,i*]pyrene also was carcinogenic in the absence of hematite, though data were not presented (*145*).

In an initial investigation of the carcinogenicity of 7*H*-dibenzo[*c,g*]carbazole (DBC), Sellakumar and Shubik (*144*) intratracheally instilled DBC together with hematite into hamsters. At both the lower (15 mg total) and higher (45 mg total) doses of DBC that were administered, tumors developed in 89% of the animals. A subsequent study was performed by Sellakumar *et al.* (*148*) in which the hematite was not used. Tumors developed in 72% of the animals receiving a total dose of 9 mg and 33% of the animals receiving 30 mg of DBC. The lower incidence of tumor-bearing animals at the higher administered dose may have resulted from the extreme toxicity of DBC such that many animals died early in the experiment. However, it appears from data at the lower doses that DBC is an effective carcinogen in the absence and presence of hematite particles.

Sufficient data are not available to interpret most of these experiments with PAH other than B[*a*]P in terms of the mechanisms involved in tumorigenesis. However, the data with DBC are consistent with the proposed mechanism of particle-enhanced uptake. As noted above, the aqueous solubility of DBC is 320 μg/liter (*92*). Because the PAH are taken up into membranes after solubilization, a PAH with high solubility will not show enhanced uptake when adsorbed to particles. Instead, the uptake is rapid in the absence and presence of particulates, and, consequently, tumorigenesis is high in the absence and presence of particulates.

F. Summary

Data from both *in vivo* and *in vitro* studies demonstrate that particulates increase the carcinogenicity of PAH. From some of the early studies in which particulates and PAH were coinstilled in hamsters, it was hypothesized that the particulates increased the penetration and retention of the PAH in the lungs and thereby increased tumorigenesis (*125, 133–135, 149*). More recent studies have demonstrated that particle-enhanced membrane uptake of PAH is consistent with the observed results and is the mechanism that most likely accounts for the increase in tumorigenesis upon exposure to both particulates and PAH (*88–91*).

In many previous studies, the nature of the particulate–PAH complex was not well defined. In cases where the samples were characterized, it appeared that much of the PAH was not adsorbed to the particles but rather existed as aggregates of PAH molecules, some of which may have been bound to the particles. A small fraction of the total PAH also may have been adsorbed to the particulates in these samples. The enhanced carcinogenicity of these samples may have resulted from the rapid uptake of the small fraction of adsorbed PAH.

Future experiments should be performed using particulates to which all of the PAH is *adsorbed.* Consequently, it probably will be possible to administer smaller amounts of the PAH as well as fewer particulates and thereby more closely approach exposure conditions similar to those of the human population. A comparison of the tumorigenicity of PAH in the particle-adsorbed state to that of PAH alone and of simple mixtures of PAH and particulates will reveal the extent of the synergism between particulates and PAH in carcinogenesis.

IV. N-NITROSO COMPOUNDS

N-Nitroso compounds are ubiquitous environmental pollutants, having been found in air, water, food, tobacco smoke, cosmetics, pesticides, and at a variety of industrial locations (*38a*). The first indication of the toxicity of the N-nitroso compounds was the observation by Freund (*51a*) of liver toxicity in two chemists working with dimethylnitrosamine (DMN). Subsequently, Magee and Barnes (*102a*) demonstrated that DMN is a liver carcinogen in rats.

In considering the effects of particulates on the carcinogenicity of the N-nitroso compounds, it is important to recognize that the compounds under consideration are extremely water soluble. Thus, the mechanism by which particulates alter N-nitroso compound carcinogenicity is likely to be different from that for PAH. In fact, in many of the studies to be discussed, the N-nitroso compounds were administered systemically and the particulates were administered by intratracheal instillation or inhalation. That is, the N-nitroso compounds were not bound to the particulates as were the PAH.

Most of the research regarding the effects of particulates on the carcinogenicity of N-nitroso compounds has been done with diethylnitrosamine (DEN) using the Syrian golden hamster as the experimental animal. It is important to make this observation because the sites at which tumorigenesis is produced with some of the N-nitroso compounds are species dependent (*107, 108*). Early studies with DEN revealed that intragastric, intratracheal, subcutaneous, intraperitoneal, intradermal, and topical administration of DEN alone in the hamster all resulted in similar localizations of tumors, the sites being the trachea, bronchi, anterior and posterior nasal cavity, and the liver (*71–74, 157*). Few, if any, peripheral lung tumors were observed upon administration of DEN.

Montesano *et al.* (*109*) investigated the effect of hematite particulates on DEN tumorigenesis. Hamsters treated with hematite by intratracheal instillation did not develop any tumors in the lower respiratory tract. Treatment with DEN by subcutaneous injection produced lower respiratory tract tumors in only 4% of the animals whereas injection of DEN followed by instillation of hematite resulted in lower respiratory tract tumors in 72% of the animals. Interestingly, when the order of DEN and hematite administrations was reversed, lower respiratory tract tumors were observed in only 4% of the hamsters. From these data, it was concluded that the mechanism for the hematite-enhanced carcinogenesis was linked to cellular reactions resulting from instillation of the particulate material (*109*).

Feron *et al.* (*44*) also investigated the effect of hematite on DEN carcinogenesis in hamsters. Both DEN and hematite, separately and in combination, were administered by intratracheal instillation. A threefold increase in lower respiratory tract tumors was observed in the animals treated with DEN and hematite as compared to those treated with DEN alone. It was proposed that the increase in tumors of the lower respiratory tract may have resulted from transplantation of tracheal papillomas into the lower respiratory tract during the experimental manipulations required for intratracheal instillation. To account for the increase when both DEN and hematite were administered, it was suggested that the particulates may facilitate transplantation (*44*). Studies by Stenback *et al.*

(*162*) supported the concept of tumor transplantation. Hamsters were exposed to DEN by subcutaneous injection and then to magnesium oxide, aluminum oxide, or carbon particles in 0.9% saline or to 0.9% saline without particles by intratracheal instillation. As expected, DEN alone produced tumors primarily in the upper respiratory tract and intratracheal instillation of the particulates resulted in a dramatic increase in tumors in the lower respiratory tract. However, instillation of the 0.9% saline produced an even greater number of tumors in the lower respiratory tract, suggesting that transplantation of tracheal papillomas had occurred (*162*). Similar results were obtained by Farrell and Davis (*41*) upon subcutaneous injection of DEN followed by intratracheal instillation of carbon, hematite, aluminum oxide, cobalt oxide, or nickel oxide suspended in a carrier, or of the particulate carrier alone.

Creasia and Nettesheim (*27*) extended these experiments by administering the DEN to hamsters subcutaneously while using inhalation exposure to the hematite particles to avoid the possible complication of transplantation. Exposure to DEN alone produced primarily upper respiratory tract tumors, whereas combined exposure to DEN and hematite resulted in a 2.7-fold increase in tumors in the lower respiratory tract. Thus, it was concluded that hematite exerts a mild cocarcinogenic effect in DEN carcinogenesis. Subsequent studies confirmed these results, and it also was observed that synthetic smog particulate is not a cocarcinogen in DEN carcinogenesis (*115*).

In contrast to the above results, one set of experiments indicated that mineral particulates did not enhance DEN carcinogenesis (*68*). DEN was administered subcutaneously to hamsters, after which carbon, hematite, aluminum oxide, cobalt oxide, or nickel oxide in size ranges of 0.5–1, 2–5, 5–10, and 15–30 μm was intratracheally instilled. It was stated that similar incidences of upper and lower respiratory tract tumors were observed in the absence and presence of the particulates though no attempt was made to explain why these results differed from those of other investigators (*68*).

In one study, the effect of chrysotile asbestos on DEN carcinogenesis was examined (*86*). DEN was administered to hamsters both orally and subcutaneously, and chrysotile was intratracheally instilled. Administration of DEN by either route produced tumors primarily in the upper respiratory tract. Addition of asbestos resulted in a much higher incidence of tumors in the lower respiratory tract when DEN was administered by either route. It was proposed that the observed increase in tumors upon instillation of chrysotile was not due to transplantation because instillation of the particulate carrier alone did not significantly increase the incidence of tumors in the lower respiratory tract (*86*).

The effect of chrysotile asbestos on *N*-methyl-*N*-nitrosourethane (MNU) carcinogenesis also has been studied (*83*). MNU and asbestos were injected intraperitoneally into mice. It was observed that lung *carcinomas* occurred more frequently and appeared more quickly upon exposure to MNU and chrysotile together as compared to either agent alone though the total number of pulmonary tumors was similar in the MNU and MNU + chrysotile groups (*83*).

V. PESTICIDES

Many pesticides become associated with particulate matter in soils and in sediments in aquatic systems. Consequently, most toxicological studies have been done with species that reside in these habitats. Studies have not been done to determine the effect of particulates on pesticide toxicity toward humans.

Generally, it is thought that adsorption of pesticides to particulates decreases the bioavailability of the pesticides and hence their bioactivity. These effects must depend upon numerous factors such as the nature of the pesticide, soil composition, pH, temperature, and humidity, all of which are involved in the adsorption process. These factors have been reviewed in detail (*10, 51, 178*). However, a brief description of the relationship of soil composition to pesticide adsorption is necessary to allow discussion of some recent studies regarding pesticide bioactivity.

The particulate matter in soil consists of a mixture of inorganic constituents such as sand, silt, clay, and metallic oxides and organic constituents such as humus, peat, muck, ashes, and carbon. The clay minerals are of two types, expanding (montmorillonite and vermiculite) and nonexpanding (kaolinite and illite). Pesticides may be absorbed on the external surface of nonexpanding clays and on the external and internal surfaces of the expanding clays. These clays possess cation-exchange capacity. The organic fraction of soil consists of a variety of organic materials including many humic materials that are believed to be responsible for the cation- and anion-exchange capacity of organic matter. These materials also possess hydrophilic and lipophilic sites. Iron and aluminum hydrous oxides in the soil are probably positively charged and thus may contribute to the adsorption of some anionic pesticides.

Studies with pesticides have revealed that cations and basic pesticides are adsorbed to both clays and organic matter. Acidic pesticides are weakly adsorbed to clay minerals though they are adsorbed in greater amounts by organic matter and metallic hydrous oxides. Finally, pesticides of low water solubility are adsorbed primarily to lipophilic organic matter. Thus, the adsorption of pesticides to soil particulates is dependent on the composition of the soil. When the soil composition is such that a particular pesticide is tightly adsorbed, the pesticide will not be bioavailable and hence will not exhibit biological activity.

Some recent studies illustrate the effects of factors that may alter adsorption capacity on the bioactivity of pesticides. The biological activity of fensulfothion [*O,O*-diethyl-*O*-[*p*-(methylsulfinyl)phenyl] phosphorothioate] as a function of cation content (NH_4^+, K^+, Ca^{2+}, Fe^{3+}, Al^{3+}) in plainfield sand was investigated (*61*). The insecticidal toxicity, as assessed in bioassays using first-stage crickets, decreased with increasing cation content, with the effectiveness of the cations in decreasing the activity being in the order trivalent > divalent > monovalent. These data were consistent with adsorption studies in which it was shown that fensulfothion adsorbed to montmorillonite in the presence of cations with the adsorption being greatest in the presence of trivalent cations and least in the presence of monovalent cations (*16*). In another study, the bioavailability of *S*-triazines adsorbed to montmorillonite was investigated (*70*). These widely used herbicides are basic, and the adsorption is pH dependent (*178*). The protonated forms of the *S*-triazines were adsorbed on the internal surface of the montmorillonite. It was observed that desorption was effected by a stronger base, NH_3, and that the desorbed form was bioactive in the *Chlamydomonas* bioassay (*70, 75*).

As mentioned earlier, lipophilic pesticides such as the chlorinated hydrocarbons also are adsorbed onto particulates. Studies have revealed that adsorbed chlorinated hydrocarbons may be bioavailable. For example, Aroclor 1254 that was adsorbed to silica was accumulated by juvenile Atlantic salmon whereas chlorinated paraffins were not (*188*). This finding was consistent with a previous result in which it was demonstrated that chlorinated paraffins were more strongly adsorbed to silica than was Aroclor 1254 (*187*). Similarly, in a study of the dynamics of DDT in a lentic environment, the accumulation of DDT in invertebrates and fish was dependent on adsorption and solubility properties of the DDT (*57*).

In the studies cited above, strict comparisons of the bioavailability or bioactivity in the absence and presence of particulates were not made. Thus, it is difficult to evaluate the effect of particulates on the toxicological properties of pesticides from these data.

Omann and Lakowicz (*118*) considered the effect of particulates on the uptake of several chlorinated hydrocarbon insecticides into model cell membranes composed of dipalmitoyl phosphatidylcholine. The transfer of the insecticides into the membranes was quantified by the fluorescence quenching of a carbazole-labeled phospholipid which was incorporated into the vesicles. The chlorinated hydrocarbons used were DDT, DDE, methoxychlor, lindane, stirofos, and three polychlorinated biphenyls. These insecticides were adsorbed onto porous glass, kaolinite, cellulose, or silica. From these experiments it was determined that membrane uptake of the chlorinated hydrocarbons was more rapid from the particle-adsorbed state than it was in the absence of particulates. Moreover, the uptake rate was faster when the chlorinated hydrocarbons were dispersed over a larger surface area. The rate of transfer from particle to membrane correlated with the aqueous solubilities of the insecticides. A mechanism similar to that proposed for the particle-enhanced uptake of PAH (*88, 91*) was invoked to explain the insecticide transfer rates. That is, the rate-limiting step is the solubilization of the compound in the aqueous phase after which rapid uptake into the membranes occurs (*118*). Thus, it is apparent that adsorption of pesticides to particulates does not necessarily reduce the bioavailability of the pesticides. Rather, dispersal of pesticides on particulates may increase their rate of transfer into lipoidal phases.

VI. GASES AND VAPORS

Particulates in the form of liquid and solid aerosols may alter the toxicity of inhaled gases and vapors. Possible mechanisms by which aerosols could alter these toxicities include effects on the penetration of gases and vapors into the respiratory tract, chemical conversion of the inspired compounds by the aerosols to more toxic forms, and dilution of the gases and vapors by the aerosols.

In early studies by LaBelle *et al.* (*87*), liquid aerosols of triethylene glycol, ethylene glycol, mineral oil, and glycerin and solid aerosols of sodium chloride, dicalite, celite, attapulgus clay, and Santocel CF (silica gel) were administered via inhalation to mice in combination with formaldehyde, acrolein, or nitric acid fumes. The toxicity of formaldehyde, acrolein, and nitric acid fumes in the absence and presence of the aerosols was evaluated from the survival times of the mice. The addition of the aerosols to formaldehyde resulted, in general, in shorter survival times as compared to formaldehyde alone, whereas the aerosols generally increased the survival times of animals exposed to nitric acid fumes. Addition of aerosols to acrolein produced either variable effects or no effect. The differences in toxicity observed among the vapors and among the particles were attributed to relative differences in penetration of the vapors and particles into the lungs (*87*), though the simple dilution effects of the aerosols must not be ignored.

Subsequent studies of the effects of particulates on gases and vapors have been done primarily with SO_2, with a few being done with NO_2 and HCl. Recent reviews related to the occurrence and effects of these gases have appeared (*21, 22, 98*). High concentrations of SO_2 alone are required to produce deleterious effects on lung function. In fact, at the concentrations generally encountered in the environment, SO_2 would be considered a minor respiratory irritant. This probably results from the high solubility of SO_2 in tissue fluids such that most SO_2 normally is removed in the upper airways with less than 1% reaching the alveoli (*5, 166*).

Conflicting reports have appeared concerning the effects of particulates on SO_2 toxicity. One series of experiments in which particulates enhanced the toxicity of SO_2 was reported by Amdur and Underhill (*6*). These investigators exposed guinea pigs to various aerosols and SO_2 and monitored lung function by measuring pulmonary flow resistance. When aerosols that were soluble in the airways were used (NaCl, KCl, $MnCl_2$, NH_4SCN, $FeSO_4$, and Na_3VO_4), a potentiation of the effect of SO_2 on flow resistance was observed even though the aerosols alone showed no effect. However, administration of insoluble aerosols (carbon, manganese dioxide, hematite, open hearth dust, fly ash, and triphenyl phosphate) in combination with SO_2 did not alter the change in pulmonary flow resistance observed upon exposure to SO_2 alone. It was postulated that the SO_2 was chemically converted to sulfuric acid in the airway in the presence of the soluble aerosols because the SO_2 would be dissolved in the liquid droplets under conditions in which oxidation was known to occur (*6*). The increase in pulmonary flow resistance was then attributed to the H_2SO_4, which is a stronger lung irritant than SO_2 (*5*). Objections have been raised to this hypothesis. Little of the SO_2 would have been associated with the particles prior to inhalation because they were dry, so association was postulated to occur after formation of liquid droplets in the respiratory tract. Even though the formation of droplets may occur within milliseconds (*119*), the SO_2 also is quite rapidly absorbed in the upper airways. Because of the much greater surface area of the nasal mucosa as compared to the aerosols, most of the SO_2 would probably be absorbed in the tissues before it became associated with the droplets. In addition, the oxidation of the SO_2 in the droplets proceeds too slowly to allow the observed changes to be due to H_2SO_4 (*166*).

The role of relative humidity in the synergism between SO_2 and soluble NaCl aerosols in altering lung function also has been assessed. McJilton *et al.* (*104*) performed experiments in which the SO_2 and NaCl were mixed at low and high humidities for 8–10 min prior to exposing guinea pigs to the mixture. Under the conditions employed, it was known that NaCl droplets would form only at high relative humidity. In these experiments, a dramatic change in pulmonary flow resistance was seen only when the animals were exposed to both SO_2 and NaCl at high relative humidity (*104*). Sulfuric acid was not detected in the samples, but bisulfite, which also is reported to be a respiratory irritant (*2*), was found.

Lung clearance of viable and nonviable bacteria (*Escherichia coli*) also has been used to monitor the effect of SO_2 and particulates on lung function. Guinea pigs were exposed to either coal dust or manganese dioxide particles and SO_2, separately and in combination, and the clearance rates were determined (*130, 131*). A significant decrease in lung clearance rate was observed only when the particulates and SO_2 were administered simultaneously.

Numerous other studies show no synergism between SO_2 and particulates in altering lung function. For example, exposure of humans to SO_2 alone or SO_2 and NaCl aerosol provoked similar changes in lung mechanical parameters (*18, 49, 155*). Likewise, the changes in lung compliance and pulmonary flow resistance were similar in cats exposed to either SO_2 or SO_2 + NaCl (*25*). Synergistic effects also were absent in monkeys and guinea pigs exposed to mixtures of SO_2, H_2SO_4 mist, and fly ash (*3*). The adsorption of SO_2 to carbon particles was studied and it was determined that some of the adsorbed SO_2 was rapidly converted to H_2SO_4 (*30*). However, a synergistic effect on the ciliary movement in rabbit trachea was not observed upon exposure to carbon with adsorbed SO_2 and H_2SO_4 as compared to exposure to carbon alone or SO_2 alone. Other investigators have studied the adsorption of SO_2 to insoluble particulates and observed oxidation of the SO_2 to sulfates (*19, 65, 117*), though studies of the health effects of these samples were not performed.

A few additional studies have been done to determine the effect of particulates and SO_2 on lung function. In the studies to be cited, the effect of SO_2 alone was not monitored so the compounding effect of particulates cannot be evaluated. Fraser *et al.* (*50*) exposed rats to 1 ppm of SO_2 and 1 mg/m^3 of a graphite dust and observed no change in the ciliary activity or number of cells that contained dust. Fenters *et al.* (*42*) evaluated the effect of sulfuric acid mist and carbon particles on the immunologic state of mice and observed alterations of several immunological parameters to include a decreased resistance to respiratory infection. Finally, in recent studies the effect of SO_2 and zinc oxide particles on respiratory function and lung morphology and biochemistry in guinea pigs has been evaluated (*23, 93*). Decreases in lung volumes, diffusing capacity for carbon monoxide, and alveolar volume were noted in exposed groups. These were interpreted as being due to a reaction in the periphery of the lung that resulted from penetration of SO_2 or other sulfur species into peripheral regions because of association with the zinc oxide particles (*93*). In addition, pulmonary edema and increased DNA synthesis were noted in the SO_2–zinc oxide-exposed animals (*23*).

A limited number of studies have been performed to determine the effect of particulates on NO_2 toxicity. Boren (*15*) exposed mice via inhalation to carbon, NO_2, and the two in combination. Carbon alone produced no abnormalities in the lungs. Single exposures to high levels (250 ppm) of NO_2 produced pulmonary edema though repetitive exposures to 25 ppm NO_2 produced no edema or lesions of any type. However, exposure to carbon with adsorbed NO_2 resulted in focal destructive pulmonary lesions, presumably because carbon could act as a carrier to produce high local concentrations of NO_2 (*15*). Furiosi *et al.* (52) exposed monkeys and rats to NO_2 in the absence and presence of NaCl aerosol. The NaCl did not alter the morphological changes seen upon exposure to NO_2 alone, and it was hypothesized that this resulted from the lack of adsorption of NO_2 to the aerosol (*52*). More detailed analysis of factors that may be involved in the effects of particulates on NO_2 toxicity, as has been done with SO_2, has not been performed.

An effect of particulates on the toxicity of HCl gas also has been noted. A serious problem in fire fighting is the release of noxious fumes, and among the most serious is HCl gas which is released during polyvinyl chloride combustion. The HCl gas is adsorbed onto soot particles which also are produced in the combustion process (*165*). Consequently, the HCl, which normally would cause minor irritation in the upper airways, is carried into the alveoli in which it exerts a more profound toxicity. Episodes of toxicity, including one fatality, among fire fighters have been reported which are attributed to HCl resulting from combustion of polyvinyl chloride (*37*).

VII. METAL COMPOUNDS

It has been recognized for many years that metal compounds may be toxic and/or carcinogenic (*99, 103, 167, 168*). Many of these metals are emitted into the atmosphere in concentrations greatly in excess of their natural crustal abundance as a result of coal and petroleum combustion and industrial processes (*96, 97*). Characterization of the fly ash emitted from coal-fired power plants indicated that the metals were most concentrated on particles in the respirable size range (*33, 112, 114*) which magnifies the hazard to human health.

The effect of particulates on the toxicity of metal compounds has been considered in only a few studies. An *in vitro* approach that has been applied is that of cellular toxicity (*177*). Aranyi *et al.* (*8*) evaluated the cytotoxicity to rabbit alveolar macrophages (RAM) of size-fractionated fly ash with surface enrichment of PbO, NiO, or MnO_2. The PbO-coated particles were the most cytotoxic, the NiO- and MnO_2-coated particles had intermediate cytotoxicity, and the untreated fly ash particles were the least cytotoxic. Because leachate from the particles was not cytotoxic, it was concluded that the toxicity was due not to solubilization of the metals from the particles but rather to a surface interaction of particles and macrophages (*8*). A reason for the enhancement of cytotoxicity with the metal oxide-coated particles was not forwarded. Garrett *et al.* (*53*) studied the cytotoxicity of fly ash particles with and without NiO, PbO, and CdO adsorbed to the surfaces. In these experiments, both RAM and Chinese hamster ovary (CHO) cells were used. The cytotoxicity of all of the metal oxide-coated particles was greater than that of the untreated particles in both assay systems. Neither PbO nor NiO dissociated from the particles, again leading to the suggestion that the cytotoxicity resulted from interactions between the cells and the particles. However, a significant portion of the CdO was solubilized into the medium though the cytotoxicity of the CdO-coated fly ash was greater than would have been predicted based on the amount of CdO solubilized. The enhanced toxicity was attributed to the association between the CdO and the fly ash (*53*).

Wehner *et al.* (*184*) studied the effects of Ni-enriched fly ash when administered to hamsters via inhalation. Low toxicity was observed and no differences were seen between the fly ash and the Ni-enriched fly ash. The lack of differences was attributed by the authors to a low solubility of the Ni compounds in the body fluids. The disparity between the results of Wehner *et al.* (*184*) and Aranyi *et al.* (*8*) and Garrett *et al.* (*53*) regarding the effects of fly ash on the cytotoxicity of Ni compounds may result from the different parameters being used to monitor toxicity.

VIII. GENERAL SUMMARY

Human epidemiological studies indicate a synergism between asbestos inhalation and cigarette smoking in the induction of bronchogenic carcinoma. Sufficient data are not available to determine if a similar synergism exists between cigarette smoking and inhalation of other particulates in the induction of bronchogenic carcinoma or between cigarette smoking and inhalation of particulates in the production of other disease states.

In vivo animal studies, in general, show increased carcinogenicity of PAH when administered in conjunction with particulates. It appears from these studies that the PAH must be associated with the particulates for a synergism to be observed. Moreover, *in vitro* studies demonstrated that the mechanism by which particulates enhance PAH carcinogenesis is a particle-enhanced membrane uptake of PAH. As a result of this phenomenon, the PAH may be delivered to cells more rapidly. Evidence also exists for a synergism between injected DEN and inhaled or intratracheally instilled particulates in the induction of tumorigenesis. The mechanism for the synergism must be different from that for particulates and PAH and may involve the irritant properties of the particulates.

Most data concerning pesticides indicate that adsorption of pesticides to soil particulates decreases the bioactivity of the pesticides. However, some *in vitro* studies demonstrate that particle-enhanced membrane uptake of pesticides also occurs and therefore should not be ignored when considering pesticide toxicity.

Some data indicate that particulates enhance the toxicity of gases and vapors, probably by enhancing the penetration of these compounds into the lower airways. However, other studies show no effect of particulates on the toxicity of gases and vapors, and more studies are required to determine the mechanisms by which particulates interact with gases and vapors. Similarly, not all data concerning the effect of particulates on metal toxicity are in agreement. That is, *in vitro* studies indicate an enhancement of the toxicity of metal oxides adsorbed to fly ash particles, but *in vivo* studies do not show the enhancement.

IX. PROSPECTUS

From the available literature, it is clear that some particulates alter the toxicological properties of certain classes of chemicals. The mechanisms by which particulates alter toxicities are not completely elucidated, though for water-insoluble PAH it appears that particulates may greatly enhance the bioavailability of the compounds. However, it is clear from the particle–PAH studies that it is necessary to consider the nature of the interaction between the toxicant and the particulate. Thus, in future studies the particulates must be well characterized and the association of the chemical compounds with the particulates must be understood. In addition, the particulates and chemical compounds that should be investigated in greater detail are those that are pollutants in the environment, such as fly ash particles with adsorbed PAH and metal oxides.

Some newer analytical techniques have been applied to the study of surfaces. For example, photoelectron spectroscopy (ESCA), Auger electron spectroscopy (AES), and secondary ion mass spectrometry (SIMS) have been applied to studies of the elemental composition of the surfaces of fly ash and diesel particulates (*52a, 69a, 98a*). In addition, photoacoustic spectroscopy (PAS) has been used to study the surface chemistry of PAH adsorbed to particulates (*18a*). By extending techniques such as these to further studies of the nature of chemical compounds adsorbed on particulates, it will be possible to understand better the mechanisms by which particulates may alter the bioavailability and bioactivity of adsorbed toxicants.

References

1. Akin, F. J., Snook, M. E., Severson, R. W., Chamberlain, W. J., and Walters, D. B. (1976). *J. Natl. Cancer Inst.* **57,** 191–195.
2. Alarie, Y., Wakisaka, I., and Oka, S. (1973). *Environ. Physiol. Biochem.* **3,** 182–184.
3. Alarie, Y. C., Krumm, A. A., Busey, W. M., Ulrich, C. E., and Kantz, R. J., II. (1975). *Arch. Environ. Health* **30,** 254–262.
4. Albagli, A., Oja, H., and Dubois, L. (1974). *Environ. Lett.* **6,** 241–251.
5. Amdur, M. O. (1969). *J. Air Pollut. Control Assoc.* **19,** 638–644.
6. Amdur, M. O., and Underhill, D. (1968). *Arch. Environ. Health* **16,** 460–468.
7. Andelman, J. B., and Snodgrass, J. E. (1974). *CRC Crit. Rev. Environ. Control* **4,** 69–83.
8. Aranyi, C., Miller, F. J., Andres, S., Ehrlich, R., Fenters, J., Gardner, D. E., and Waters, M. D. (1979). *Environ. Res.* **20,** 14–23.
9. Archer, V. E., Wagoner, J. K., and Lundin, F. E., Jr. (1973). *J. Occup. Med.* **15,** 204–211.
10. Bailey, G. W., and White, J. L. (1964). *J. Agric. Food Chem.* **12,** 324–332.
11. Baum, E. J. (1978). *In* "Polycyclic Hydrocarbons and Cancer" (H. V. Gelboin and P. O. P. Ts'O, eds.), Vol. 1, pp. 45–70. Academic Press, New York.
12. Berenblum, I. (1944). *Arch. Pathol.* **38,** 233–244.
13. Bevan, D. R., Riemer, S. C., and Lakowicz, J. R. (1981). *J. Toxicol. Environ. Health* **8,** 241–250.
14. Blair, W. H., Otero, N., and Rao, H. (1973). *Proc. Am. Assoc. Cancer Res.* **14,** 125.
15. Boren, H. G. (1964). *Arch. Environ. Health* **8,** 127–132.
16. Bowman, B. T. (1973). *Soil Sci. Soc. Am. Proc.* **37,** 200–207.
17. Broddin, G., Van Vaeck, L., and VanCauwenberghe, K. (1977). *Atmos. Environ.* **11,** 1061–1064.
18. Burton, G. G., Corn, M., Gee, J. B. L., Vasallo, C., and Thomas, A. P. (1969). *Arch. Environ. Health* **18,** 681–692.

18a. Cabaniss, G. E., and Linton, R. W. (1981). *In* "Chemical Analysis and Biological Fate: Polynuclear Aromatic Hydrocarbons" (M. Cooke and A. J. Dennis, eds.), pp. 277–286. Battelle, Columbus, Ohio.
19. Cofer, W. R., III, Schryer, D. R., and Rogowski, R. S. (1980). *Atmos. Environ.* **14,** 571–575.
20. Cohen, B. H. (1978). *Lancet* **8098,** 1024–1027.
21. Committee on Medical and Biological Effects of Environmental Pollutants, National Research Council (1977). "Nitrogen Oxides." Nat. Acad. Sci., Washington, D.C.
22. Committee on Sulfur Oxides, National Research Council (1978). "Sulfur Oxides." Nat. Acad. Sci., Washington, D.C.
23. Conner, M. W., Rogers, A. E., and Amdur, M. O. (1982). *Toxicol. Appl. Pharmacol.* **66,** 434–442.
24. Conney, A. H. (1982). *Cancer Res.* **42,** 4875–4917.
25. Corn, M., Kotsko, N., Stanton, D., Bell, W., and Thomas, A. P. (1972). *Arch. Environ. Health* **24,** 248–256.
26. Craighead, J. E., and Mossman, B. T. (1979). *Prog. Exp. Tumor Res.* **24,** 48–60.
27. Creasia, D. A., and Nettesheim, P. (1974). *In* "Experimental Lung Cancer" (E. Karbe and J. F. Park, eds.), pp. 234–245. Springer-Verlag, Berlin and New York.
28. Creasia, D. A., Poggenburg, J. K., Jr., and Nettesheim, P. (1976). *J. Toxicol. Environ. Health* **1,** 967–975.
29. Crittenden, B. D., and Long, R. (1976). *In* "Polynuclear Aromatic Hydrocarbons: Chemistry, Metabolism, and Carcinogenesis" (R. I. Freudenthal and P. W. Jones, eds.), pp. 209–223. Raven, New York.
30. Dalhamn, T., and Strandberg, L. (1963). *Int. J. Air Water Pollut.* **7,** 517–529.
31. Davis, B. R., Whitehead, J. K., Gill, M. E., Lee, P. N., Butterworth, A. D., and Roe, F. J. R. (1975). *Br. J. Cancer* **31,** 443–452.
32. Davis, W. W., Krahl, M. E., and Clowes, G. H. A. (1942). *J. Am. Chem. Soc.* **64,** 101–107.
33. Davison, R. L., Natusch, D. F. S., Wallace, J. R., and Evans, C. A., Jr. (1974). *Environ. Sci. Technol.* **8,** 1107–1113.
34. Deelman, H. T. (1927). *Proc. R. Soc. Med.* **20,** 19–20.
35. Della Porta, G., Kolb, L., and Shubik, P. (1958). *Cancer Res.* **18,** 592–597.
36. Doll, R. (1978). *Environ. Health Perspect.* **22,** 23–31.
37. Dyer, R. F., and Esch, V. H. (1976). *J. Am. Med. Assoc.* **235,** 393–397.
38. Egan, B., Waxweiler, R. J., Blade, L., Wolfe, J., and Wagoner, J. K. (1979). *J. Environ. Pathol. Toxicol.* **2,** 259–272.
38a. Ember, L. R. (1980). *Chem. Eng. News* **58,** 20–26.
39. Falk, H. L. (1970). *In* "Inhalation Carcinogenesis" (M. G. Hanna, Jr., P. Nettesheim, and J. R. Gilbert, eds.), pp. 13–26. U.S. Atomic Energy Commission, Oak Ridge, Tennessee.
40. Farrell, R. L., and Davis, G. W. (1974). *In* "Experimental Lung Cancer" (E. Karbe and J. F. Park, eds.), pp. 186–198. Springer-Verlag, Berlin and New York.
41. Farrell, R. L., and Davis, G. W. (1974). *In* "Experimental Lung Cancer" (E. Karbe and J. F. Park, eds.), pp. 219–233. Springer-Verlag, Berlin and New York.
42. Fenters, J. D., Bradof, J. N., Aranyi, C., Ketels, K., Ehrlich, R., and Gardner, D. E. (1979). *Environ. Res.* **19,** 244–257.
43. Ferin, J., and Leach, L. J. (1976). *Environ. Res.* **12,** 250–254.
44. Feron, V. J., Emmelot, P., and Vossenaar, T. (1972). *Eur. J. Cancer* **8,** 445–449.
45. Feron, V. J., DeJong, D., and Emmelot, P. (1973). *Eur. J. Cancer* **9,** 387–390.
46. Feron, V. J., DeJong, D., and Rijk, M. A. H. (1976). *Zbl. Bakt. Hyg., I. Abt. Orig. B* **163,** 441–447.
47. Feron, V. J., VanDenHeuvel, P. D., Koeter, H. B. W. M., and Beems, R. B. (1980). *Int. J. Cancer* **25,** 301–307.

48. Filipy, R. E., Stuart, B. O., Palmer, R. F., Ragan, H. A., and Hackett, P. L. (1974). *In* "Experimental Lung Cancer—Carcinogenesis and Bioassays" (E. Karbe and J. F. Park, eds.), pp. 403–410. Springer-Verlag, Berlin and New York.
49. Frank, N. R., Amdur, M. O., and Whittenberger, J. L. (1964). *Int. J. Air Water Pollut.* **8,** 125–133.
50. Fraser, D. A., Battigelli, M. C., and Cole, H. M. (1968). *J. Air Pollut. Control Assoc.* **18,** 821–823.
51. Freed, V. H., Chiou, C. T., and Haque, R. (1977). *Environ. Health Perspec.* **20,** 55–70.
51a. Freund, H. A. (1937). *Ann. Intern. Med.* **10,** 1144–1155.
52. Furiosi, N. J., Crane, S. C., and Freeman, G. (1973). *Arch. Environ. Health* **27,** 405–408.
52a. Gardella, J. A., Jr., and Hercules, D. M. (1979). *Int. J. Environ. Anal. Chem.* **7,** 121–136.
53. Garrett, N. E., Campbell, J. A., Stack, H. F., Waters, M. D., and Lewtas, J. (1981). *Environ. Res.* **24,** 345–365.
54. Gibson, E. S., Martin, R. H., and Lockington, J. N. (1977). *J. Occup. Med.* **19,** 807–812.
55. Gross, P., Tolker, E., Babyak, M. A., and Kaschak, M. (1965). *Arch. Environ. Health* **11,** 59–65.
56. Guerin, M. R. (1978). *In* "Polycyclic Hydrocarbons and Cancer" (H. V. Gelboin and P. O. P. Ts'O, eds.), Vol. 1, pp. 3–42. Academic Press, New York.
57. Hamelink, J. L., Waybrant, R. C., and Ball, R. C. (1971). *Trans. Am. Fish. Soc.* **100,** 207–214.
58. Hammond, E. C., and Selikoff, I. J. (1973). *In* "Biological Effects of Asbestos" (P. Bogovski, V. Timbrell, J. C. Gibson, and J. C. Wagner, eds.), pp. 312–317. Int. Agency Res. Cancer, Lyon, France.
59. Harington, J. S., and Roe, F. J. C. (1965). *Ann. N.Y. Acad. Sci.* **132,** 439–450.
60. Harington, J. S., and Smith, M. (1964). *Arch. Environ. Health* **8,** 453–458.
61. Harris, C. R., and Bowman, B. T. (1976). *Soil Sci. Soc. Am. J.* **40,** 385–389.
62. Harris, C. C., Sporn, M. B., Kaufman, D. G., Smith, J. M., Baker, M. S., and Saffiotti, U. (1971). *Cancer Res.* **31,** 1977–1989.
63. Harris, C. C., Kaufman, D. G., Sporn, M. B., and Saffiotti, U. (1973). *Cancer Chemother. Rep., Part 3* **4,** 43–54.
64. Hatch, T. F., and Gross, P. (1964). "Pulmonary Deposition and Retention of Inhaled Aerosols." Academic Press, New York.
65. Haury, G., Jordan, S., and Hofmann, C. (1978). *Atmos. Environ.* **12,** 281–287.
66. Henry, M. C., and Kaufman, D. G. (1973). *J. Natl. Cancer Inst.* **51,** 1961–1964.
67. Henry, M. C., Port, C. D., Bates, R. R., and Kaufman, D. G. (1973). *Cancer Res.* **33,** 1585–1592.
68. Henry, M. C., Port, C. D., and Kaufman, D. G. (1974). *In* "Experimental Lung Cancer" (E. Karbe and J. F. Park, eds.), pp. 173–185. Springer-Verlag, Berlin and New York.
69. Henry, M. C., Port, C. D., and Kaufman, D. G. (1975). *Cancer Res.* **35,** 207–217.
69a. Hercules, D. M. (1978). *Anal. Chem.* **50,** 734A–744A.
70. Hermosin, M. C., Cornejo, J., White, J. L., and Hess, F. D. (1982). *J. Agric. Food Chem.* **30,** 728–733.
71. Herrold, K. M. (1964). *Arch. Pathol.* **78,** 189–195.
72. Herrold, K. M. (1964). *Br. J. Cancer* **18,** 763–767.
73. Herrold, K. M. (1964). *Cancer* **17,** 114–121.
74. Herrold, K. M., and Dunham, L. J. (1963). *Cancer Res.* **23,** 773–777.
75. Hess, F. D. (1980). *Weed Sci.* **28,** 515–520.
76. Hilfrich, J., Bresch, H., Misfeld, J., and Mohr, U. (1973). *Zbl. Bakt. Hyg., I. Abt. Orig. B* **158,** 59–61.
77. Hoffman, D., Adams, J. D., Brunnemann, K. D., and Hecht, S. S. (1979). *Cancer Res.* **39,** 2505–2509.

78. Ishinishi, N., Kodama, Y., Nobutomo, K., and Hisanaga, A. (1977). *Environ. Health Perspect.* **19,** 191–196.
79. Kandaswami, C., and O'Brien, P. J. (1980). *Biochem. Biophys. Res. Commun.* **97,** 794–801.
80. Kasprzak, K. S., Marchow, L., and Breborowicz, J. (1973). *Res. Commun. Chem. Pathol. Pharmacol.* **6,** 237–245.
81. Katz, M., and Pierce, R. C. (1976). *In* "Polynuclear Aromatic Hydrocarbons: Chemistry, Metabolism, and Carcinogenesis" (R. I. Freudenthal and P. W. Jones, eds.), pp. 413–429. Raven, New York.
82. Kaufman, D. G., Genta, V. M., and Harris, C. C. (1974). *In* "Experimental Lung Cancer" (E. Karbe and J. F. Park, eds.), pp. 564–574. Springer-Verlag, Berlin and New York.
83. Kawai, T. (1979). *Acta Pathol. Jpn.* **29,** 421–433.
84. Keith, C. H., and Tesh, P. G. (1965). *Tobacco Sci.* **9,** 61–64.
85. Kobayashi, N., and Okamoto, T. (1974). *J. Natl. Cancer Inst.* **52,** 1605–1610.
86. Kung-Vosamae, A., and Vinkmann, F. (1980). *In* "Biological Effects of Mineral Fibers" (J. C. Wagner, ed.), pp. 305–310. Int. Agency Res. Cancer, Lyon, France.
87. LaBelle, C. W., Long, J. E., and Christofano, E. E. (1955). *Arch. Ind. Health* **11,** 297–304.
88. Lakowicz, J. R., and Bevan, D. R. (1979). *Biochemistry* **18,** 5170–5176.
89. Lakowicz, J. R., and Bevan, D. R. (1980). *Chem.-Biol. Interact.* **29,** 129–138.
90. Lakowicz, J. R., and Hylden, J. L. (1978). *Nature (London)* **275,** 446–448.
91. Lakowicz, J. R., Bevan, D. R., and Riemer, S. C. (1980). *Biochim. Biophys. Acta* **629,** 243–258.
92. Lakowicz, J. R., Bevan, D. R., and Riemer, S. C. (1980). Unpublished observation.
93. Lam, H. F., Peisch, R., and Amdur, M. O. (1982). *Toxicol. Appl. Pharmacol.* **66,** 427–433.
94. Lane, B. P., and Mass, M. J. (1977). *Cancer Res.* **37,** 1476–1479.
95. LeBouffant, L., Martin, J. C., Daniel, H., Henin, J. P., and Normand, C. (1980). *J. Natl. Cancer Inst.* **64,** 273–284.
96. Lee, R. E., Jr., Goranson, S. S., Enrione, R. E., and Morgan, G. B. (1972). *Environ. Sci. Technol.* **6,** 1025–1030.
97. Lee, R. E., Jr., and von Lehmden, D. J. (1973). *J. Air Pollut. Control Assoc.* **23,** 853–857.
98. Lee, S. D. (1980). "Nitrogen Oxides and Their Effects on Health." Ann Arbor Sci. Publ., Ann Arbor, Michigan.
98a. Linton, R. W., Williams, P., Evans, C. A., Jr., and Natusch, D. F. S. (1977). *Anal. Chem.* **49,** 1514–1521.
99. Louria, D. B., Joselow, M. M., and Browder, A. A. (1972). *Ann. Intern. Med.* **76,** 307–319.
100. Lundin, F. E., Jr., Lloyd, J. W., Smith, E. M., Archer, V. E., and Holaday, D. A. (1969). *Health Phys.* **16,** 571–578.
101. MacKay, D., and Shiu, W. Y. (1977). *J. Chem. Eng. Data* **22,** 399–402.
102. Maenza, R. M., Pradhan, A. M., and Sunderman, F. W., Jr. (1971). *Cancer Res.* **31,** 2067–2071.
102a. Magee, P. N., and Barnes, J. M. (1956). *Br. J. Cancer* **10,** 114–122.
103. Mastromatteo, E. (1967). *J. Occup. Med.* **9,** 127–136.
104. McJilton, C., Frank, R., and Charlson, R. (1973). *Science* **182,** 503–504.
105. Miller, L., Smith, W. E., and Berliner, S. W. (1965). *Ann. N.Y. Acad. Sci.* **132,** 489–500.
106. Mohr, U. (1979). *Prog. Exp. Tumor Res.* **24,** 245–252.
107. Montesano, R. (1970). *Tumori* **56,** 335–344.
108. Montesano, R., and Saffiotti, U. (1968). *Cancer Res.* **28,** 2197–2210.
109. Montesano, R., Saffiotti, U., and Shubik, P. (1969). *In* "Inhalation Carcinogenesis" (M. G. Hanna, Jr., P. Nettesheim, and J. R. Gilbert, eds.), pp. 353–368. U.S. Atomic Energy Commission, Oak Ridge, Tennessee.
110. Mossman, B. T., and Craighead, J. E. (1974). *In* "Experimental Lung Cancer" (E. Karbe and J. F. Park, eds.), pp. 514–520. Springer-Verlag, Berlin and New York.

111. Mossman, B. T., and Craighead, J. E. (1979). *Prog. Exp. Tumor Res.* **24,** 37–47.
112. Natusch, D. F. S. (1978). *Environ. Health Persp.* **22,** 79–90.
113. Natusch, D. F. S., and Tomkins, B. A. (1978). *In* "Carcinogenesis, Vol. 3: Polynuclear Aromatic Hydrocarbons" (P. W. Jones and R. I. Freudenthal, eds.), pp. 145–153. Raven, New York.
114. Natusch, D. F. S., Wallace, J. R., and Evans, C. A., Jr. (1974). *Science* **183,** 202–204.
115. Nettesheim, P., Creasia, D. A., and Mitchell, T. J. (1975). *J. Natl. Cancer Inst.* **55,** 159–169.
116. Nettesheim, P., Topping, D. C., and Jamasbi, R. (1981). *Annu. Rev. Pharmacol. Toxicol.* **21,** 133–163.
117. Novakov, T., Chang, S. G., and Harker, A. B. (1974). *Science* **186,** 259–261.
118. Omann, G. M., and Lakowicz, J. R. (1981). *Pest. Biochem. Physiol.* **16,** 231–248.
119. Orr, C., Jr., Hurd, F. K., and Corbett, W. J. (1958). *J. Colloid Sci.* **13,** 472–482.
120. Palmer, W. G., and Scott, W. D. (1981). *Am. Ind. Hyg. Assoc. J.* **42,** 329–340.
121. Pelfrene, A. F. (1976). *Am. Ind. Hyg. Assoc. J.* **37,** 706–710.
122. Perera, F. (1981). *Environ. Health Perspect.* **42,** 163–185.
123. Port, C. D., Henry, M. C., Kaufman, D. G., Harris, C. C., and Ketels, K. V. (1973). *Cancer Res.* **33,** 2498–2506.
124. Pylev, L. N. (1961). *Bull. Exp. Biol. Med.* **52,** 1316–1319.
125. Pylev, L. N. (1962). *Acta Unio Int. Cancrum* **19,** 688–691.
126. Pylev, L. N., and Shabad, L. M. (1973). *In* "Biological Effects of Asbestos" (P. Bogovski, J. C. Gilson, V. Timbrell, and J. C. Wagner, eds.), pp. 99–105. Int. Agency Res. Cancer, Lyon, France.
127. Pylev, L. N., Roe, F. J. C., and Warwick, G. P. (1969). *Br. J. Cancer* **23,** 103–115.
128. Raeburn, C., and Spencer, H. (1957). *Br. J. Tuberc. Dis. Chest* **51,** 237–245.
129. Ripstein, C. B., Spain, D. M., and Bluth, I. (1968). *J. Thorac. Cardiovasc. Surg.* **56,** 362–368.
130. Rylander, R. (1969). *Arch. Environ. Health* **18,** 551–555.
131. Rylander, R., Ohrstrom, M., Hellstrom, P. A., and Bergstrom, R. (1971). *In* "Inhaled Particles III" (W. H. Walton, ed.), Vol. 1, pp. 535–540. Gresham, Surrey, England.
132. Saffiotti, U. (1970). *In* "Inhalation Carcinogenesis" (M. G. Hanna, Jr., P. Nettesheim, and J. R. Gilbert, eds.), pp. 27–51. U.S. Atomic Energy Commission, Oak Ridge, Tennessee.
133. Saffiotti, U., Borg, S. A., Grote, M. I., and Karp, D. B. (1964). *Chicago Med. Sch. Q.* **24,** 10–17.
134. Saffiotti, U., Cefis, F., Kolb, L. H., and Shubik, P. (1965). *J. Air Pollut. Control Assoc.* **15,** 23–25.
135. Saffiotti, U., Cefis, F., and Kolb, L. H. (1968). *Cancer Res.* **28,** 104–124.
136. Saffiotti, U., Montesano, R., Sellakumar, A. R., Cefis, F., and Kaufman, D. G. (1972). *Cancer Res.* **32,** 1073–1081.
137. Saffiotti, U., Montesano, R., Sellakumar, A. R., and Kaufman, D. G. (1972). *J. Natl. Cancer Inst.* **49,** 1199–1204.
138. Samet, J. M., Epler, G. R., Gaensler, E. A., and Rosner, B. (1979). *Am. Rev. Respir. Dis.* **120,** 75–82.
139. Sawicki, E. (1976). *In* "Environmental Pollution and Carcinogenic Risks" (C. Rosenfeld and W. Davis, eds.), pp. 337–354. INSERM, Paris.
140. Schmeltz, I., and Hoffmann, D. (1976). *In* "Polynuclear Aromatic Hydrocarbons: Chemistry, Metabolism, and Carcinogenesis" (R. I. Freudenthal and P. W. Jones, eds.), pp. 225–239. Raven, New York.
141. Selikoff, I. J., Churg, J., and Hammond, E. C. (1964). *J. Am. Med. Assoc.* **188,** 142–146.
142. Selikoff, I. J., Hammond, E. C., and Churg, J. (1968). *J. Am. Med. Assoc.* **204,** 104–110.
143. Selikoff, I. J., Seidman, H., and Hammond, E. C. (1980). *J. Natl. Cancer Inst.* **65,** 507–513.
144. Sellakumar, A., and Shubik, P. (1972). *J. Natl. Cancer Inst.* **48,** 1641–1646.

145. Sellakumar, A., and Shubik, P. (1974). *J. Natl. Cancer Inst.* **53,** 1713–1719.
146. Sellakumar, A. R., Montesano, R., Saffiotti, U., and Kaufman, D. G. (1973). *J. Natl. Cancer Inst.* **50,** 507–510.
147. Sellakumar, A., Stenback, F., and Rowland, J. (1976). *Eur. J. Cancer* **12,** 313–319.
148. Sellakumar, A., Stenback, F., Rowland, J., and Shubik, P. (1977). *J. Toxicol. Environ. Health* **3,** 935–939.
149. Shabad, L. M. (1962). *J. Natl. Cancer Inst.* **28,** 1305–1332.
150. Shabad, L. M. (1967). *Cancer Res.* **27,** 1132–1137.
151. Shabad, L. M. (1977). *In* "Air Pollution and Cancer in Man" (U. Mohr, D. Schmahl, and L. Tomatis, eds.), pp. 61–67. Int. Agency Res. Cancer, Lyon, France.
152. Shabad, L. M. (1980). *J. Natl. Cancer Inst.* **64,** 405–410.
153. Shabad, L. M., Pylev, L. N., and Kolesnichenko, T. S. (1964). *J. Natl. Cancer Inst.* **33,** 135–142.
154. Shabad, L. M., Pylev, L. N., Krivosheeva, L. V., Kulagina, T. F., and Nemenko, B. A. (1974). *J. Natl. Cancer Inst.* **52,** 1175–1187.
155. Snell, R. E., and Luchsinger, P. C. (1969). *Arch. Environ. Health* **18,** 693–698.
156. Stedman, R. L. (1968). *Chem. Rev.* **68,** 153–207.
157. Stenback, F. (1974). *In* "Experimental Lung Cancer" (E. Karbe and J. F. Park, eds.), pp. 161–172. Springer-Verlag, Berlin and New York.
158. Stenback, F., and Rowland, J. (1978). *Scand. J. Respir. Dis.* **59,** 130–140.
159. Stenback, F., and Rowland, J. (1978). *Eur. J. Cancer* **14,** 321–326.
160. Stenback, F., and Rowland, J. (1979). *Oncology* **26,** 63–71.
161. Stenback, F., and Sellakumar, A. (1974). *Eur. J. Cancer* **10,** 483–486.
162. Stenback, F. G., Ferrero, A., and Shubik, P. (1973). *Cancer Res.* **33,** 2209–2214.
163. Stenback, F., Sellakumar, A., and Shubik, P. (1975). *J. Natl. Cancer Inst.* **54,** 861–867.
164. Stenback, F., Rowland, J., and Sellakumar, A. (1976). *Oncology* **33,** 29–34.
165. Stone, J. P., Hazlett, R. N., Johnson, J. E., and Carhart, H. W. (1973). *J. Fire Flammability* **4,** 42–51.
166. Subcommittee on Airborne Particles, National Research Council (1979). "Airborne Particles." Univ. Park Press, Baltimore.
167. Sunderman, F. W., Jr. (1976). *Prevent. Med.* **5,** 279–294.
168. Sunderman, F. W., Jr. (1979). *Biol. Trace Element Res.* **1,** 63–86.
169. Thomas, J. F., Mukai, M., and Tebbens, B. D. (1968). *Environ. Sci. Technol.* **2,** 33–39.
170. Toda, M. (1962). *Bull. Tokyo Med. Dent. Univ.* **9,** 440–441.
171. Tola, S., Koskela, R. S., Hernberg, S., and Jarvinen, E. (1979). *J. Occup. Med.* **21,** 753–760.
172. Topping, D. C., and Nettesheim, P. (1980). *J. Natl. Cancer Inst.* **65,** 627–630.
173. Toth, B., Tomatis, L., and Shubik, P. (1961). *Cancer Res.* **21,** 1537–1541.
174. U.S. Dept. of Health, Education, and Welfare (1979). "Smoking and Health—A Report of the Surgeon General." U.S. Govt. Printing Office, Washington, D.C.
175. Wagoner, J. K., Archer, V. E., Lundin, F. E., Jr., Holaday, D. A., and Lloyd, J. W. (1965). *N. Engl. J. Med.* **273,** 181–188.
176. Warshawsky, D., Niemeier, R. W., and Bingham, E. (1978). *In* "Carcinogenesis, Vol. 3: Polynuclear Aromatic Hydrocarbons" (P. W. Jones and R. I. Freudenthal, eds.), pp. 347–360. Raven, New York.
177. Waters, M. D., Huisingh, J. L., and Garrett, N. E. (1979). *In* "Application of Short-Term Bioassays in the Fractionation and Analysis of Complex Environmental Mixtures" (M. D. Waters, S. Nesnow, J. L. Huisingh, S. S. Sandhu, and L. Claxton, eds.), pp. 126–167. Plenum, New York.
178. Weber, J. B., and Weed, S. B. (1974). *In* "Pesticides in Soil and Water" (W. D. Guenzi, ed.), pp. 223–256. Soil Sci. Soc. Am., Madison.
179. Wehner, A. P., Busch, R. H., and Olson, R. J. (1974). *In* "Experimental Lung Cancer—

Carcinogenesis and Bioassays'' (E. Karbe and J. F. Parke, eds.), pp. 360–368. Springer-Verlag, Berlin and New York.

180. Wehner, A. P., Busch, R. H., Olson, R. J., and Craig, D. K. (1975). *Am. Ind. Hyg. Assoc. J.* **36,** 801–810.
181. Wehner, A. P., Busch, R. H., Olson, R. J., and Craig, D. K. (1975). *Environ. Res.* **10,** 368–383.
182. Wehner, A. P., Busch, R. H., Olson, R. J., and Craig, D. K. (1977). *Am. Ind. Hyg. Assoc. J.* **38,** 338–346.
183. Wehner, A. P., Stuart, B. O., and Sanders, C. L. (1979). *Prog. Exp. Tumor Res.* **24,** 177–198.
184. Wehner, A. P., Moss, O. R., Milliman, E. M., Dagle, G. E., and Shirmer, R. E. (1979). *Environ. Res.* **19,** 355–370.
185. Weiss, W. (1971). *Am. Rev. Respir. Dis.* **104,** 223–227.
186. Wynder, E. L., and Hoffman, D. (1968). *Science* **162,** 862–871.
187. Zitko, V. (1973). *J. Chromatogr.* **81,** 152–157.
188. Zitko, V. (1974). *Bull. Environ. Contam. Toxicol.* **12,** 406–412.

Aquatic Animal Neoplasia as an Indicator for Carcinogenic Hazards to Man

John J. Black

Department of Experimental Biology
Roswell Park Memorial Institute
Buffalo, New York

I. Introduction 181
II. Environmental Pollution 183
III. Environmental Carcinogenesis 184
  A. Cancer as an Environmental Disease 184
  B. Oncogenes and Viruses 185
  C. Initiation, Promotion, and Metabolism 187
IV. Diagnostic Considerations 189
  A. Definition of Neoplasia 189
  B. Mimicry of Neoplasia 190
  C. Diagnosis of Neoplasia in Aquatic Animals 193
V. Assessment of Exposure to Chemical Agents 195
VI. Epizootiologic Studies of Spontaneous Neoplasms in Aquatic Animals 196
  A. Fish 196
  B. Mollusks 199
  C. Detailed Case Histories for Which There Is Good Evidence of a Pollutant Etiology 202
VII. Linkage of Environmental Pollution to Aquatic Animal Neoplasia ... 218
  A. Indirect Experimental Evidence 218
  B. Direct Evidence 218
VIII. Conclusion 225
  References 226

I. INTRODUCTION

Water and, therefore, ultimately aquatic environments are simultaneously important sources of essential nourishment for both the body and spirit of man.

HAZARD ASSESSMENT OF CHEMICALS:
Current Developments, Vol. 3

ISBN 0-12-312403-4

Unfortunately, either by design or by chance, water is used as a major medium of disposal for agricultural, industrial, and urban wastes. The potential hazards created in this relationship include not only damage to delicately balanced ecosystems and their associated biota, but also potential risks to human health.

In this regard, there have been numerous reports of spontaneous neoplasms occurring in aquatic animals and these have been extensively reviewed (*39, 41, 113, 114, 141, 156, 194*). Overall, data indicate that most kinds of neoplasms found in aquatic animals today were observed also in the past. There are suggestions, however, that in some locations some kinds of neoplasms presently occur with greater frequencies than in the past (*66*). Although it is not known who first advanced the hypothesis that aquatic animals could serve as "sentinel animals" to warn man of the presence of noxious substances in the aquatic environment, the idea parallels and may in part have sprung from the example of early miners who employed caged canaries to warn them of the asphyxiation hazards of firedamp in the coal mines. From a less intuitive viewpoint, Tomatis (*182*) has reviewed evidence showing predictive correlations between a carcinogenic response in laboratory animals and the tumorigenic potential of chemicals to humans. It should be noted that no attempt has been made here to catalog all aquatic animal neoplasms that may have relevance to environmental carcinogens. For a concise tabular summary of epizootic neoplasms occurring in feral marine and estuarine species the reader is referred to Sinderman *et al.* (*163*).

Although fish neoplasia has received the greatest emphasis as an indicator system for early detection of waterborne carcinogens, this should not discourage interest in the study and utility of other aquatic animals as indicators of carcinogenic hazards to man. There has long been speculation that neoplasia may eventually be shown to occur in all metazoan species. What is offered in this article is not a hard and fast choice of which organism(s) to employ, but rather some general discussions and guidelines regarding relevance of feral aquatic animal neoplasms to carcinogenic pollutants. From this vantage point, it should be remembered that discovery of the carcinogenic action of aflatoxins can be attributed largely to the observation of epizootics of liver neoplasia in an aquatic animal, the rainbow trout (*Salmo gairdneri*). Historical aspects of the trout hepatoma problem have been discussed by Wales (*187a*). The liver neoplasia was detected in hatchery-reared trout (*155*) shortly after the wide-scale introduction of synthetic pelleted diets which replaced those composed of fresh or frozen fish and ground meats. This circumstantial evidence led Halver (*63*) to fractionate the suspected carcinogenic diets into lipid, protein, and carbohydrate moieties. The lipid fraction was highly carcinogenic, as it contained the lipophilic mold metabolites that were detected as fluorescent bands on thin-layer chromatography. This is now history, but it exemplifies the premise that aquatic animals are sensitive to carcinogenic stimuli and can serve as indicators of carcinogenic hazards to man.

II. ENVIRONMENTAL POLLUTION

Superimposed patterns of waste disposal and water reuse may pose direct risks to humans through the introduction of toxic and carcinogenic chemicals into drinking water supplies. An additional area of concern is the bioaccumulation of chemical pollutants by fish and marine invertebrates that are important staples in the human food chain. Kraybill (*95, 96*) has previously reviewed the subject of carcinogens in the aquatic environment. Polycyclic aromatic hydrocarbons (PAH), a family of hydrocarbons with numerous carcinogenic members, are well-documented aquatic pollutants and the subject of a recent book compiled by Neff (*132*).

A compilation of aquatic pollutants, including many that have been identified as contaminants in fishery products, would be lengthy and would contain numerous chemicals with demonstrated or suspected carcinogenic potential, and is beyond the scope of this article. Garrison (*55*) has pointed out that in a fairly comprehensive literature search (5500 entries) 1296 different organic compounds have been identified from water. Among these, 51 were scheduled for some kind of regulatory action. Several of these 51 compounds have been detected from water in only one or two instances. Perhaps more importantly, several compounds of seemingly important toxicological significance that had been detected 10 or more times as water pollutants were not on the list of 51 compounds. Recent studies comparing similarities in qualitative and quantitative aspects of surface water pollution with ground water pollution in the state of New Jersey (*137*) provide sobering evidence that all components of the hydrosphere may eventually be contaminated with toxic chemicals.

Carcinogenic pollution of drinking water has been implicated as a possible causative factor for some human cancers (*31, 137a*). Although the potential hazards of waterborne carcinogens may not be limited to ingestion of chemical contaminants in drinking water, the risks posed to human health by other mechanisms may be less well appreciated. Clearly, contaminated fishery products are capable of delivering considerable fluxes of persistent chemicals to the human population. An extreme example is illustrated in the case of the Great Lakes, which, as a result of successful fisheries management efforts, presently support major sport fisheries for trout and salmon. The magnitude of these fisheries is impressive, with recent catch statistics for the state of Michigan alone indicating an annual harvest of approximately 2.9 million salmonids (*119*). This represents in the range of 30 million pounds, not including an additional sizable catch of lake trout. Persistent chemicals, especially chlorinated hydrocarbons, present in the water in minute concentrations have accumulated to high concentrations in these fish. Accumulation of these chemicals occurs rapidly by direct uptake from water via the gills, as well as through food chain mechanisms, and appears to depend more on physical–chemical properties (e.g., octanol:water partition coefficient) and aspects of metabolism of the chemical than on ecological considerations (*63a*). The problem is particularly acute in salmonid species because they concentrate high levels of these chemicals in their lipid-rich flesh. The relative importance of the "fish vs drinking water" in this situation can be appreciated by considering that one would have to consume water (contaminated with PCB at 3×10^{-9} g/liter) for longer than 1000 years in order to ingest the amount of PCB that is contained in one 500 g portion of Great Lakes fish (contaminated at 5×10^{-3} g/kg).

III. ENVIRONMENTAL CARCINOGENESIS

A. Cancer as an Environmental Disease

Historically, the first recognition of cancer as an environmentally induced disease must be credited to an English physician, Dr. Percival Pott (*147*), who noted the occurrence of scrotal cancer in chimney sweeps. Although Pott's observations were published in 1775, the carcinogenic effect of coal tar was not proven experimentally until 1915 (*197*). Another 15 years would pass before the PAH 3,4-benzo[*a*]pyrene, an active carcinogenic agent, would be successfully isolated from coal tar (*35*). In the past, the assemblage of diseases that is recognized as human cancer was thought of as spontaneous, but, as a result of evidence accumulated from epidemiology and the laboratory, cancer is presently considered mainly an environmental disease (*46, 64, 76, 121, 122, 192*). This does not mean that pollutant chemicals per se are the cause of all, or even most, human cancer. Doll and Pitot (*44*) have suggested that only 2–5% of all human cancers in the United States are directly attributable to total effects of air, water, and food pollution. However, their estimates have isolated specific factors exclusive of possible synergistic effects between multiple-agent exposures, i.e., pollution plus occupational plus tobacco, etc. Synergistic effects of carcinogens have been documented in complex exposures both in the laboratory (*33, 62, 180*) and in the human population (*23, 154, 159*). In addition, the estimate of 2–5% is suggested as an average, and one would suspect, in view of the uneven geographic distribution of cancer mortality (*111*), that in some local situations the relative importance of carcinogenic pollution may be somewhat greater.

The development of neoplasia involves a complex interaction between environmental factors and factors endogenous to the host. Many aspects are not understood and no single unifying theory incorporates all known mechanisms of carcinogenesis. The pathogenesis of neoplasia is viewed as a multistep, multifactored process requiring a long period of latency. The molecular basis of oncogenic transformation remains the subject of intensive investigation. Although details of current theory and experimental investigations relative to mechanisms of carcinogenesis are beyond the scope of this article, the following broad outlines may help the nonspecialist gain an appreciation for the complexity of the problem.

B. Oncogenes and Viruses

Heubner and Todaro (*81*) postulated that most vertebrates carry cancer genes, or oncogenes, as repressed genetic information, derived possibly from C-type RNA viruses. Oncogenes are now considered to be a small but diverse group of cellular genes (of nonviral origin) found in all animal cells. Normally the gene is inactive and is transcribed in only a few protein copies per cell. But when activated the gene is transcribed at a higher frequency leading to neoplasia. Activation of the gene is thought to occur through rearrangements or transposition in specific short sequences of DNA (*30, 134*). Many potent chemical carcinogens produce chromosome abnormalities and sister-chromatid exchanges (*80, 166*). These exchanges may represent morphological signs associated with the necessary transposition(s) that lead to the required activation or enhanced activity of oncogenes and the formation of the incipient neoplasm.

Some studies have indicated activation of a virus (*32, 83, 184*) or enhancement of virally induced neoplasia (*65*) by chemical carcinogens. But the story appears to have come "full circle" by demonstration *in vitro* that DNA from chemically transformed cells can cause phenotypic changes characteristic of neoplasia when experimentally introduced into nontransformed cells (*161*). The recent confirmation of human and animal oncogenes and the elucidation of their gene product(s) are exciting in that the promise of unraveling the molecular basis of neoplasia may occur in the foreseeable future. It is intriguing that human bladder cancer, one of the cancers recognized as having a chemical etiology via worker exposure to aromatic amines in the workplace (*120, 190*) and more recently suspected of having a linkage to halomethanes in drinking water (*31*), was among the first human neoplasms from which an oncogene was isolated and characterized (*59*).

Oncogenic viruses are recognized as causative agents of neoplasia in diverse groups of animals from warm-blooded vertebrates to poikilothermic vertebrates and even invertebrates, including aquatic animal species. In some aquatic animals, such as the leopard frog (*Rana pipiens*), a viral agent has been characterized as the specific etiologic agent leading to the development of renal adenocarcinoma (*53, 103, 117, 118*) and was the first oncogenic herpesvirus

TABLE I

Neoplasms for Which There Is Evidence of an Etiologic Role for Virus

Phylum/class	Species/neoplasm	Location	Evidence for viral agent	References
Mollusca/Pelecypoda	Soft-shell clams (*Mya arenaria*)/			
	Germinoma	Searsport, Maine	Intranuclear inclusions	Harshbarger (*70*)
	Hematopoietic	New England	Virus transmission	Brown (*26*); Oprandy (*136*)
Chordata/amphibia	Leopard frog (*Rana pipiens*)/ renal adenocarcinoma	North America	Virus transmission, EM, herpes-type virus	Fawcett (*53*); Lunger (*103*)
Teleost/Osteichthyes	Chum salmon (*Oncorhynchus keta*)/ epithelial neoplasms at several sites	Japan	Virus transmission, herpes-type virus	T. Kimura (*89*)
	Northern pike and muskellunge (*Esox lucius* and *Esox masquinongy*)/ lymphosarcoma	Europe and North America	Epizootiology, cell-free transmission, reverse transcriptase	Ljungberg (*101*); Mulcahy (*130, 131*); Papas (*138*); Sonstegard (*171*)
	European eel (*Anguilla anguilla*)/ oral papillomas	Elbe River, Germany	Viruses demonstrated by EM	Schwanz-Pfitzner (*158*); McAllister (*115*)
	White sucker (*Catostomus commersonii*)/ oral papillomas	Great Lakes	C-type particles by EM and reverse transcriptase	Sonstegard (*170, 173*)
	Walleye pike (*Stizostedion vitreum*)/ dermal fibroma, epithelial hyperplasia	Lake Oneida, New York	Viruses demonstrated by EM	Walker (*189*)

identified. In other animals, viruses may be integrated into the host's genome, and the presence of the virus may be indicated only by successful cell-free transmission of the disease or by the presence of viral gene products such as unique viral enzymes (*138, 173*). In other cases, viruses of a particular morphological type appear presumptively to be associated with a particular neoplasm on the basis of particles visualized by electron microscopy (*115, 158, 189*). In yet other instances, the role of a virus may be as yet unrecognized. Isolation of new oncogenic viruses and confirmation of suspected viral agents of neoplasms in aquatic animals can be illustrated only by the methods of the experimental virologist. A listing of neoplastic diseases found in aquatic animals for which there is evidence of a viral etiology is shown in Table I.

C. Initiation, Promotion, and Metabolism

1. Initiating Agents

Current concepts of chemical carcinogenesis state that at least two somewhat independent processes, or stages, may be distinguished. The first process, termed initiation, can occur upon a single exposure to a carcinogen and is essentially irreversible. The second stage, promotion, requires multiple exposures and is reversible, at least in the early stages. The discovery of two-stage carcinogenesis is credited to Berenblum and Shubik (*9*), who discovered that croton oil, although inactive as a carcinogenic agent by itself, resulted in a high incidence of neoplasms (skin papillomas) when repeatedly applied to the back of a mouse that had previously received a single subcarcinogenic dose of a PAH. Although the concepts of initiation and promotion were applied initially as operational definitions to describe two different stages of experimental skin carcinogenesis in laboratory mice, subsequent studies have identified distinct properties of agents recognized as initiators and promoters. For example, it is thought that the primary target of the initiators is the cellular DNA. Thus, initiators can be chemical, physical, or biological agents that impair or alter the integrity of the DNA molecule. Although covalent binding to DNA is a feature of some chemical carcinogens (or more often of a metabolite), effects on the DNA may take the form of distortion, breakage, rearrangement, or interference with DNA repair. Though these attributes do not completely prove a role for DNA as the critical macromolecule upon which the initiation action of carcinogenesis hinges, they are consistent with the irreversible aspect of initiation, i.e., DNA, the hereditary molecule, "remembers."

2. Promoting Agents and Effects

The promoting agent may differ from the initiator in that it appears to act by altering gene expression. The promoting agent tetradecanoylphorbol acetate (TPA), the active agent in croton oil, has been the most intensively studied model promoter compound. Available evidence suggests that the effects of TPA involve interference with or modulation of cellular differentiation, which is triggered by binding of the TPA molecule to a receptor in the cell membrane (*191*) and which perhaps leads to failure of the cell-to-cell communication necessary to maintain the normal nonproliferative state (*185a*). Although many biochemical effects such as changes in cell surface charges, cell adhesion properties, membrane phospholipids, Na^+, K^+-ATPase, and ion transport effects have been documented (*191*), it is not yet clear how these changes relate to the stimulation of cell division and the ultimate outgrowth of a neoplasm. One attractive hypothesis is that the promoter TPA may bind to a cell surface receptor that normally functions as an (as yet unidentified) hormonal receptor. This is supported by the fact that some hormones play a major role in cancer growth and development, and the hormonal dependency of some cancers is exploited in their treatment (*78*). An alternative mechanism of action postulated for promoters involves the production of free oxygen radicals which secondarily attack DNA and otherwise affect chromosomal structure (*110*). Promoting effects have been observed in cell culture systems (*129*) and in liver carcinogenesis, including fish liver (*50, 165*) and mouse skin (*9, 20*), although it is not clear that a common biochemical mechanism is involved in all of these various forms of experimental carcinogenesis promotion.

3. *Metabolism of Carcinogens*

An important aspect of the host's reaction upon exposure to an exogenous chemical agent is the manner in which a chemical is metabolized. Most chemicals require metabolic conversion of an inactive parent compound to an active metabolite in order to manifest their carcinogenic activity (*122, 192*), although some chemicals can act directly. In general, most chemical carcinogens are metabolized via a two-component system of enzymatic reactions. The first group of enzymes (phase 1 enzymes) introduces a polar handle, e.g., —OH or —NH group, onto the parent compound as part of an apparent detoxification pathway for the conversion of xenobiotic chemicals from a relatively nonpolar to a more polar structure as part of a biochemical pathway leading to urinary excretion. This is carried out by enzymes of the general group known as microsomal mixed-function oxidases. Unfortunately, in this process some chemicals are converted to highly reactive intermediates (*121*) such as epoxides and/or free radicals that are believed to alter the integrity of some critical macromolecule and thus lead to carcinogenesis. The second phase of xenobiotic metabolism (phase 2 enzymes) acts mainly to attach a glucuronide or sulfate group to the polar metabolites formed by the phase 1 enzymes. These conjugation reactions represent a true detoxification mechanism in preparation for removal of the newly modified chemical by the kidney. In some mammalian species, these glucuronide conjugated forms may be released from the glucuronide moiety to do further damage by the hydrolytic action of acidic urine or through the action of β-glucuronidase produced by bacterial flora in the gut (*121, 192*).

IV. DIAGNOSTIC CONSIDERATIONS

A. Definition of Neoplasia

The terms "tumor" and "cancer" have been commonly used in the popular communication media, often interchangeably. However, the terms have significantly different meanings. The term tumor has both a primary archaic definition, i.e., simply stated, a swelling, and a secondary modern usage, i.e., a neoplasm. In contrast, the term cancer denotes a malignant neoplasm, i.e., a neoplasm that has the capacity to metastasize or spread to distant sites. Whereas all cancers are neoplasms, all neoplasms are not cancers. Within the group of proliferative diseases that is recognized as neoplasia, there is a spectrum of growth potentials ranging from those with totally benign behavior and exhibiting little significant tendency to metastasize to those that grow rapidly and have seemingly unlimited ability to invade the host tissues. While the terms benign and malignant originally were applied in the clinical setting and referred to the prognosis for a patient undergoing treatment, the terms have acquired meaning relating to certain frequently associated morphological and histological criteria. Although distinguishing between benign neoplasia and malignant neoplasia is one of the most important tasks of the clinical pathologist, the academic pathologist studying aquatic animals may have only casual interest in whether a neoplasm is malignant cancer or a relatively noninvasive growth. The primary distinction for the academic pathologist is whether a particular lesion is neoplastic or nonneoplastic in origin.

The most commonly accepted definition of a neoplasm is that offered by Willis (*195*): "an abnormal mass of tissue, the growth of which exceeds and is uncoordinated with that of normal tissue and persists in the same excessive manner after cessation of the stimuli which evoked the change." An alternative definition simply states that "a neoplasm is a relatively autonomous growth of tissue" (*148*). However widely accepted, a definition alone is insufficient to enable the diagnosis of neoplasia in any species. That is to say, a diagnosis of neoplasia is ultimately a judgment made on the part of a trained specialist. The judgement entails a thorough knowledge of the structure of the normal tissue as well as knowledge of and the ability to recognize and associate various morphological patterns with the counterparts of a particular cell, tissue, or organ. According to Dawe (*39*), "it is an empiric art—science depending heavily on past experience." A precise diagnosis relative to the tissue of origin and classification of a particular neoplasm is sometimes difficult. There are many conditions in lower animals that mimic neoplasia and sometimes can deceive even the experienced pathologist. In such cases, numerous tissue sections must be examined and a thorough review of the slides and case history made before a final diagnosis can be made.

Although there is no single criterion that can be used to arrive at a diagnosis, the principal feature of all neoplasms is cellular proliferation. In this respect, neoplasia can be thought of as a form of abnormal hyperplasia that involves an inheritable change in the neoplastic cell. An increase in cellularity or hyperplasia is frequently an antecedent of the development of neoplasia (*20*). It is even probable that many hyperplastic conditions may represent an early stage of neoplasia wherein certain molecular and biochemical changes leading to development of a cellular defect have already occurred but are impossible to ascertain from their appearance. As expressed by Stewart (*176*), "no immutable histologic criteria exist for the certain determination of the early cellular proliferations of cancers as purely hyperplastic changes." Although the precise molecular basis of these changes is not yet completely understood, both genetic and epigenetic factors appear to be important (*192*).

B. Mimicry of Neoplasia

The distinction between neoplasia and hyperplasia may be subtle and difficult to interpret. A classic example in cold-blooded vertebrates may be thyroid hyperplasia, common in many aquarium and some feral fish species. The introduction of Pacific salmon into the Great Lakes has resulted in marine-adapted (high iodine) animals placed into a low-iodine environment, and epizootics of thyroid hyperplasia (goiter) have been observed in Great Lakes coho salmon, especially in Lakes Erie and Ontario (*18, 174*). Although the low iodine content of these lakes appears to be an underlying factor, careful evaluations of the goiters throughout the lakes suggest that possible antithyroid agent(s) may be responsible for the large interlake differences that have been observed (*128a, 174*).

Examination of tissue sections from these hyperplastic thyroid growths reveals

Fig. 1. Nonneoplastic lesions that mimic neoplasia. (a) Thyroid hyperplasia in a coho salmon (*Oncorhynchus kisutch*) from Lake Erie. A prominent focus of thyroid cells in a microfollicular pattern occurs on the right, with scattered thyroid cells infiltrating fibrous connective tissue on the left. Bar = 50 μm. (b) Granulomatous response in kidney of freshwater drum (*Aplodinotus grunniens*). Micrograph shows granulomas (arrows) in various stages of formation surrounding and encapsulating protozoan parasites. H&E. Bar = 75 μm. (c) Lymphocystis disease in a walleye pike (*Stizostedion vitreum*). Micrograph shows numerous hypertrophic virus-infected cells with hyaline capsules adjacent to dermal scales (S). H&E. Bar = 75 μm. (d) X cell lesion in Atlantic cod (*Gadus morrhua*). An extensive field of pale-staining degenerating X cells (organisms) surrounds lamellar structure of pseudobranch (P). Inset (10×) shows X cells surrounding a thin-walled blood vessel (V). Bar = 125 μm. With kind permission, Registry for Tumors in Lower Animals (RTLA), Dr. Clyde J. Dawe (contributor).

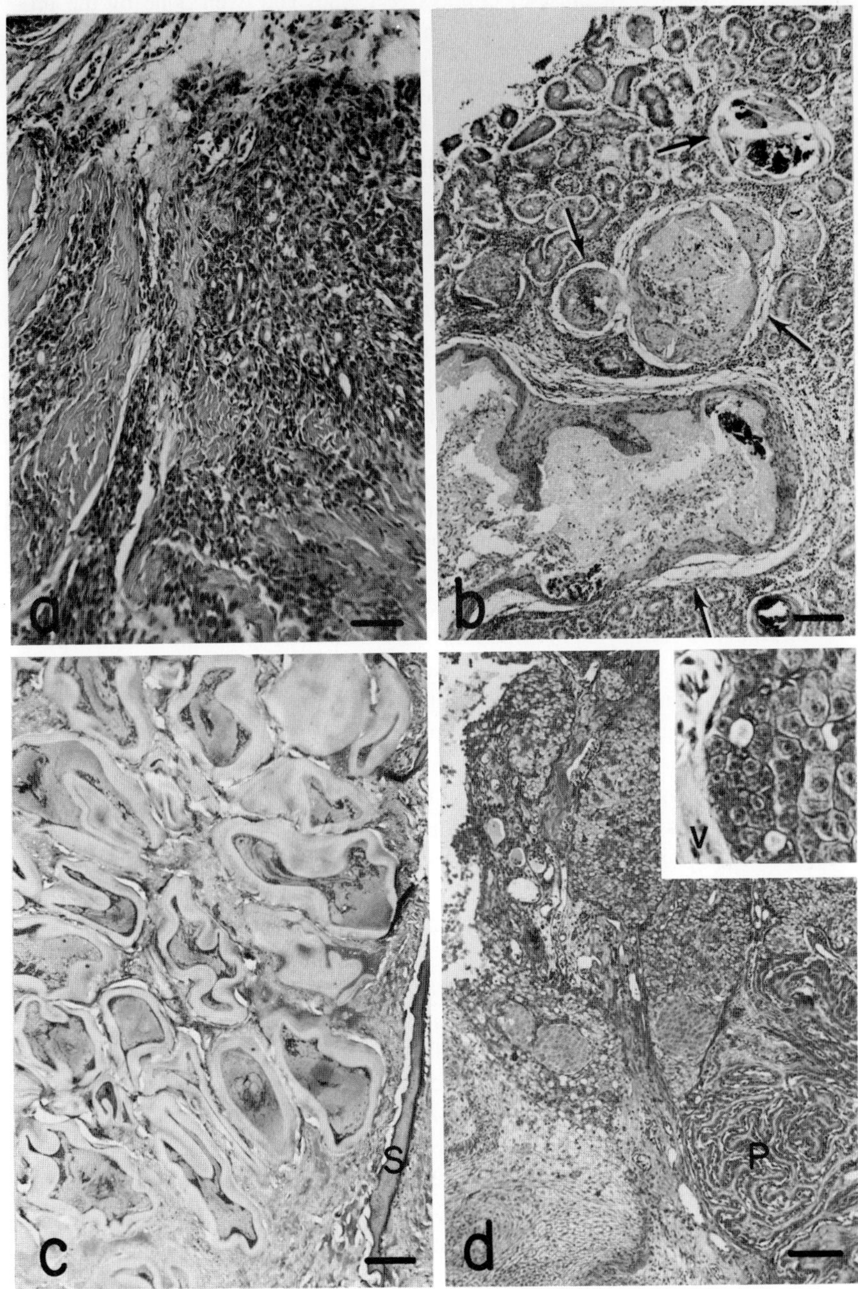
a
b
c
S
d
V
P

extensive infiltration of muscle, connective tissue, and even bone by the proliferating thyroid cells (Fig. 1a). This infiltrative behavior is considered one of the hallmarks of a malignant neoplasm. However, the teleost thyroid is not encapsulated. Therefore, when stimulated to a hyperplastic condition, the thyroid cells easily spread out into neighboring structures. This difference in the comparative anatomy of the fish vs mammalian thyroid contributed in part to early classic studies by Gaylord (*56*) in which thyroid hyperplasia in trout was described as thyroid carcinoma. Marine and Lenhart (*107–109*) were able to show regression of these lesions in response to iodine administration. To make matters even more confusing, it is possible that ectopic thyroid, which occasionally occurs in fish, can be mistaken for metastases (*156*). There is also the possibility that, in occasional specimens, carcinoma is superimposed upon hyperplasia (*18*) and if so the dividing line between the two conditions may be very difficult to ascertain. A recent case history of metastasizing thyroid adenocarcinomas in a captive population of kelp bass (*Paralabrax clathratus*) has been reported (*19*). It is of interest that these lesions were suspected to have a chemical etiology perhaps in response to an agent(s) contained in an aquarium paint.

Inflammatory and infectious conditions may result in exuberant growths of tissue that mimic neoplasia. For example, granulomatous infections in fish due to acid-fast bacteria frequently produce neoplastic-appearing growths, especially proliferations of fibrocytes in which the only clue to differentiate the growth as having an infectious etiology is an occasional small granuloma (Fig. 1b). Another infectious disease, lymphocystis, is a viral condition that sometimes produces quite large growths which affect both freshwater and marine fish species. In this disease, the tumorlike growths are seen to be composed of hypertrophic virus-infected cells (Fig. 1c). Although hyperplasia may be present, the growth of these masses is due mainly to cellular hypertrophy not cell proliferation.

Certain skin lesions that occur in pleuronectid fishes (flounder and sole) and lesions of the parabranchial (pseudobranch) body in members of the Gadidae (codfish) were widely described as neoplasms by numerous investigators (*177–179, 193*). It was later discerned that both of these proliferative lesions contained a morphologically similar cell (Fig. 1d), the so-called X cell (*1, 22*). In the pseudobranch lesions, some investigations focused on the possibility that the X cells originated from acidophilic cells of the pseudobranch and that functionally they were producing endocrine or vasoactive substances (*97, 98*). However, chromatographic analysis of tumor tissue extracts was negative for the vasoactive amines epinephrine, norepinephrine, dopamine, and serotonin, indicating that these lesions did not originate from neural crest tissues. Alpers *et al.* (*1*) also discussed this possibility, but in addition suggested that many of the features of the X cells differed from the host's cells and some resembled unicellular organisms. Peters *et al.* (*144*) suggested that X cells in flatfishes are degenerating cells, similar in some respects to lymphocystis cells. It was pointed out that X-cell papillomatosis and lymphocystis seem to be mutually exclusive in distribution and the suggestion was made that both diseases may result from infection by a closely related virus. Dawe, in the same volume, reviewed the history of the X-cell problem and presented rather convincing evidence that the X cells in pseudobranch lesions of cod exhibited synchronous division stages in nuclei of multinucleated cells, suggestive of a plasmodialike protistan possibly of the family Hartmannellidae (*40*). In addition, the X cells differed from fish cells in numerous other ways, including autofluorescence, much lower DNA content, and the presence of a glycocalyxlike coat on the surface of the X-cell membrane (*40*). This interesting problematic lesion clearly points out the potential pitfalls and diligent investigations that are sometimes required to arrive at a correct diagnosis. The reader is referred to Harshbarger (*69*) for further discussions and descriptions of pseudoneoplastic conditions that affect fish.

C. Diagnosis of Neoplasia in Aquatic Animals

The Registry for Tumors in Lower Animals (RTLA), established in 1966 at the Smithsonian Institution, Washington, D.C., 20560, serves as a diagnostic center and specimen repository for neoplasms and related disorders in invertebrate and cold-blooded vertebrate animals. Established in a joint venture by the National Cancer Institute and the Smithsonian Institution, the Registry serves also as a research center and clearinghouse for information on neoplasia in lower animals (*66*). One of the premises that led to the establishment of this facility is the focus of this article, namely that lower animals may serve as an early detection system for carcinogenic stimuli in the environment.

As there are few opportunities to acquire formal training in pathology of a diverse range of aquatic animals, most investigators have attempted to extrapolate objective criteria applicable to neoplasia in higher vertebrates to the diagnosis of tumors in cold-blooded animals. These criteria include changes in the normal tissue architecture and cellular morphology. Changes in morphology include loss of the normal differentiated cell and tissue structures (anaplasia) and increased cell-to-cell variation (pleomorphism). These alterations may range from a neoplasm in which the cells rather faithfully reproduce the morphological characteristics of their tissue of origin to a neoplasm in which nearly all resemblances to the original cell/tissue type have been lost. The tumor cells in this latter type of neoplasia are usually extremely variable in size and shape. Along with these changes in tissue architecture, the neoplastic cells tend to exhibit increased nuclear to cytoplasmic ratios and increased nucleolar to nuclear ratios. Mitotic activity may be present in increased frequency and, in some cases, mitoses may be abnormal. The neoplastic cells usually exhibit an increased affinity for basic dyes (basophilia) such as hemotoxylin. The increased basophilia results in accumulation of a blue dye that, when visualized microscopically in fixed and stained tissue sections, is also one of the hallmarks used by the pathologist in deciding the status of a lesion. The changes in staining reflect alterations in nucleic acid distribution within the neoplastic cell, i.e., increase in free ribosomes in cytoplasm, increased DNA content of the nucleus, and increased nuclear to cytoplasmic ratios. The reader is referred to Mawdesley-Thomas (*113*) for a classification scheme for neoplasia in fish.

The combination of basophilia, anaplasia, and pleomorphism in mammalian neoplasms is frequently correlated with malignancy, i.e., a tendency to invade or metastasize leading to death of the host. In cold-blooded animals, distant metastases are seldom observed, although local invasion and spread to adjoining structures are fairly common. The precise reasons for this are not clear but one reason that has been advanced is that in the harsh natural environment the animal dies before metastases develop. Another possible explanation for the infrequent observation of metastases is that environmental conditions may favor slow growth rates. In this situation, immune surveillance may limit the survival of the few cells that do immigrate into the bloodstream. Temperature-dependent growth including regression of neoplasms has been observed in several poikilothermic vertebrate neoplasms including papillomas in European eels (*143*), renal tumors in leopard frogs (*118*), and lymphosarcoma in northern pike and muskellunge (*171*). In the case of the northern pike, the regressive growth of lymphosarcomas is apparently linked to critical temperature optima required for a viral reverse transcriptase (*138*). The growth characteristics of the animals under study should be borne in mind when interpreting a histologic specimen. What may be a relatively slowly developing neoplasm in a rodent could be the equivalent of a rapidly developing neoplasm in a fish.

Although expertise in diagnostic tumor pathology is acquired only through years of training and experience, the aquatic scientist equipped with a basic foundation of training in histology should not hesitate to ''get his feet wet'' in the study of neoplastic diseases in aquatic animals. The gross signs of a neoplasm are often readily detectable providing the observer has a thorough knowledge of the anatomy and morphology of the animal suspected of harboring a neoplasm. In general, most neoplasms grossly appear as nodules or masses of tissue that have no counterpart in the normal animal. Other signs include distortions and/or discolorations of the involved tissue parts. It should be noted that the gross aspects of a number of nonneoplastic diseases such as parasitic pseudotumors (e.g., myxosporidian cysts) and granulomatous infections are easily mistaken for neoplasms. It is important, therefore, that a diagnosis of neoplasia be based on microscopic examination of the diseased tissue.

However, occasionally a particular neoplasm may have a characteristic gross morphology and occur with sufficient frequency in a population that representative specimens of the diseased animals can be collected and submitted to the tumor pathologist for a definitive diagnosis. In this way helpful leads to interesting problems sometimes develop. This idea was the basis for a recent workshop and manual to aid field biologists in the recognition and monitoring of several more or less common site- and species-specific types of neoplasia found in Great Lakes fish (*175*).

The status of preservation of the tissues for microscopic examination is an important aspect of the histologic and diagnostic study of neoplasms. In general, most tissue-processing methods applicable to mammalian tissues are suitable for aquatic animals, although a number of specific tissue fixative, dehydration, embedding, and staining procedures have been used for cold-blooded animals such as fish (*29, 175*) and mollusks (*126*). Two important factors are the procurement of adequate amounts of the lesion together with some adjoining normal tissues, and whenever possible the tissues should be from freshly expired animals since tissues begin to break down after the animal dies. If an animal cannot be fixed immediately the tissues should be kept cool or chilled until fixation is possible. The reader is referred to Sonstegard and Leatherland (*175*) or Davenport (*37*) for discussions of techniques and aids to accomplish the fixation procedure. The importance of good tissue preservation can be appreciated if one considers that even the most knowledgeable pathologist cannot render an accurate diagnosis if the tissue processed for microscopic examination is badly autolyzed.

V. ASSESSMENT OF EXPOSURE TO CHEMICAL AGENTS

The study of neoplasms in wild animal populations suffers from many inherent difficulties, not the least of which is the difficulty in obtaining comparable representative samples. Because the incidence of neoplasms frequently varies as a function of the age, size, and, sometimes, sex of the animal, these potential variables must be taken into account in the development of sampling strategies and in the interpretation of results. It is important to consider what is known of life history, feeding and migrating habits, intermixing of populations, and other ecological factors when interpreting data relative to aquatic animal neoplasms. Fisheries and aquatic biologists routinely deal with this type of problem and are often a source of helpful information. Because studies of neoplasia in aquatic animals cross the boundaries of many scientific fields, e.g., pathology, epizootiology, oncology, ecology, and chemistry, a multidisciplinary team approach is highly recommended.

Since development of neoplasia usually requires a long latent period, the long time interval between exposure to carcinogenic agent(s) and the development of a neoplasm may make it very difficult to assess the role of environmental exposures, particularly for situations in which highly mobile species encountering multiple-agent exposures are involved. This may be further complicated by differences in environment and habitat requirements of different life history stages of some aquatic animals. At present, very little is known of differential susceptibility to carcinogenesis in aquatic animal populations as a function of life history stages. That fish are susceptible to carcinogenic effects resulting from embryonic exposures was shown in experiments by Wales *et al.* (*188*) in which aqueous exposures of fertilized rainbow trout eggs to aflatoxin B_1 at a concentration of 0.5 ppm (μg/ml) resulted in a high incidence of neoplasms among fish examined 9–12 months after exposure. This work has been extended by Hendricks *et al.* (*73*) to include the induction of liver and renal tumors (nephroblastomas) in response to similar exposures to dimethylnitrosamine (DMN) and *N*-methyl-*N*′-nitro-*N*-nitrosoguanidine (MNNG). Japanese scientists demonstrated the development of adenomas of the air bladder and gastrointestinal tract as well as the production of gonadal tumors in rainbow trout (in addition to liver and renal tumors) through embryonic exposures to MNNG (*87*).

From the standpoint of a chemical etiology, it is important to characterize or monitor the environmental distribution of carcinogenic pollutants. Analytical studies of bottom sediments are generally preferable because many chemical pollutants are sequestered by sediments (*84, 102*). A recently developed technique used successfully for monitoring PAH in water employs artificial substrates made from polypropylene cloth (*15*). This technique may be useful where sediments are difficult to obtain.

A problem that arises when dealing with a highly mobile or migratory species is that it is often impossible to assess the degree to which an individual specimen may have been exposed to chemical agent(s) under investigation. A technique that may have utility in this situation is measurement and correlation of chemicals in tissues and fluids as markers for exposure in populations suspected of harboring chemically induced neoplasms. Recent studies have indicated that parent PAHs such as naphthalene and benzo[*a*]pyrene and their metabolite complexes can be measured in bile and tissues (*13, 34, 45, 58, 94, 187*). Several investigators have also utilized biological assays for detection of chemical mutagens/carcinogens in environmental samples, tissues, and bile (*15, 71, 90, 91, 93, 139, 140, 142*).

VI. EPIZOOTIOLOGIC STUDIES OF SPONTANEOUS NEOPLASMS IN AQUATIC ANIMALS

A. Fish

In studies of native bottom-feeding fish species from Deep Creek Lake, Maryland, the presence of bile duct neoplasms (cholangiomas) in 3 of 12 common white suckers (*Catostomus commersonii*) and the presence of hepatocellular nodules suspected as a possible minimal deviation neoplasm in a brown bullhead (*Ictalurus nebulosus*) led Dawe *et al.* (*42*) to suggest that bottom-feeding fish species may prove to be useful indicators of environmental carcinogens. This idea has been advanced by several studies that have utilized epizootiology, i.e., tumor-rate comparisons between exposed and control groups. Studies by Brown *et al.* (*24*) documented the frequency of tumors occurring in 16 species of fish from the Fox River watershed (Illinois) and compared these incidences to data for the same species from a nonpolluted Canadian watershed. This study documented an overall fourfold increase in tumor incidence among the fish from the Fox River system.

Investigations by Sonstegard (*172*) have used a similar approach to demonstrate an association between environmental pollution and neoplasia in aquatic animals. Sonstegard described highly variable incidences of oral papillomas in common white suckers from Lake Ontario. The papillomas increased dramatically in frequency near the Oakville–Burlington area of the lake, and then decreased along a gradient as one moved away from these sites. In addition, the tumor sizes increased in the Oakville–Burlington region and were seen to shift anatomically from a more or less random body distribution to an exclusive location on the lips. Sonstegard's original work on this form of neoplasia in the white sucker has subsequently been reproduced in independent monitoring studies (Fig. 2) conducted by the Great Lakes Fisheries Research Branch, Fisheries and Oceans, Canada (*3*).

As the white sucker is primarily a detritus feeder that uses its fleshy lips to feed on materials sifted from the bottom sediments, a logical mechanism of exposure to carcinogenic materials sequestered in bottom sediments was postulated. By observing the behavior of these fish in the laboratory, Sonstegard noted that during both feeding and resting the upper lip of the fish has more direct contact with bottom substrates and thus might be more prone to abrasion and/or contact with chemically contaminated sediments (*172*). This observation was consistent with field observations that papillomas occurred more frequently on the upper lip than the lower lip (61.4% upper vs 38.6% lower). In the white croaker (*Genyonemus lineatus*) from the coastal waters of southern California, mechanical abrasion of the lips was thought to be involved in the etiology of oral papillomas, based on elevated tumor frequencies among fish collected in areas with coarse bottom materials (*144a*). In the white suckers, the presence of a C-type particle visualized by electron microscopy, together with reverse transcriptase activity detected in fractions prepared from frozen papilloma tissue, strongly suggested to Sonstegard that an RNA virus that may be activated by chemical pollutants is involved in the etiology of this neoplasm.

In addition to oral papillomas in the white sucker, Sonstegard also studied gonadal neoplasms in carp × goldfish hybrids from the Great Lakes (*172*). Although hybrids captured from relatively nonpolluted sites on the Lakes exhib-

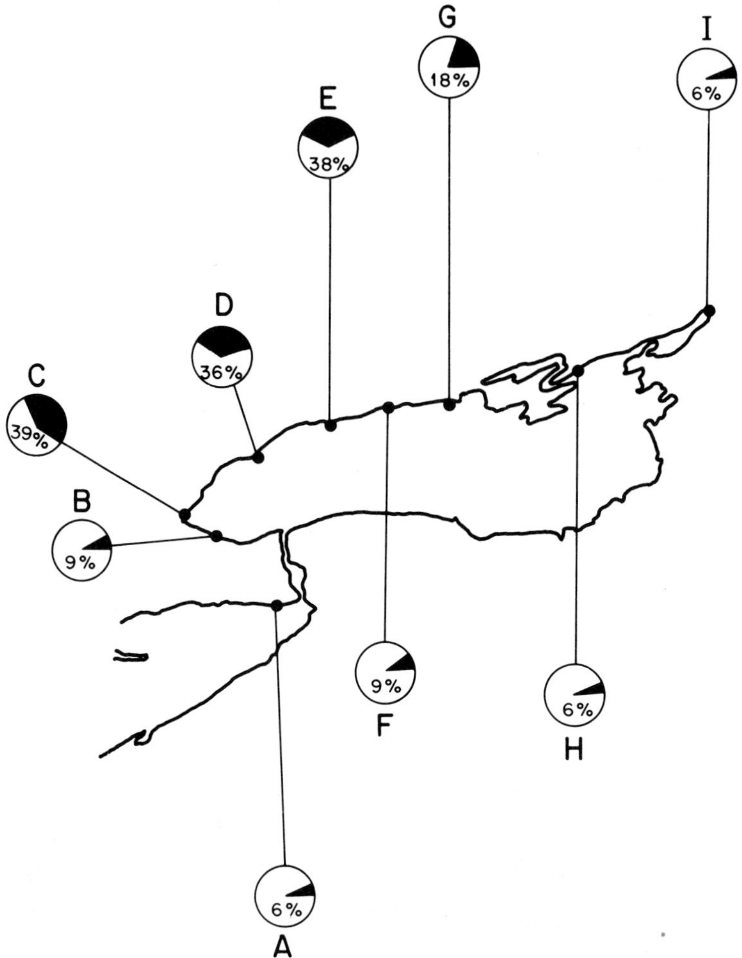

Fig. 2. Incidence of oral papillomas in white suckers (*Catostomus commersonii*) at various locations in Canadian waters of Lake Ontario. A, Fort Erie; B, Hamilton; C, Burlington; D, Oakville; E, Toronto; F, Oshawa; G, Colborne; H, Kingston; and I, Brockville. With kind permission, Vic Cairns, Great Lakes Fisheries Research Branch, Fisheries and Oceans, Canada, Burlington, Ontario.

ited a high frequency of these growths (16 of 26 specimens), an interesting and thought-provoking observation was made by comparing the present day incidence of these gonadal masses to that found in museum specimens. Among a series of museum hybrid collections including specimens held at the Royal Ontario Museum, the University of Michigan, Ohio State University, and the Smithsonian Institution not one single tumor was found. Particularly relevant was a moderately large collection at the University of Michigan. This collection of 38 hybrids, captured in 1952 near the mouth of the Rouge River at Detroit, Michigan, was entirely free of the gonadal neoplasms whereas collections at the same site 25 years later revealed tumor frequencies of 100% among selected subgroups of older males. The results of these studies suggest that most hybrid populations in the Great Lakes now exhibit elevated frequencies of gonadal neoplasms and, therefore, a now ubiquitous environmental agent may be implicated.

B. Mollusks

If one were to design an ideal aquatic animal for monitoring waterborne carcinogens in terms of an *in vivo* neoplastic response, the organisms would most likely be (1) sedentary, in order to eliminate some of the difficulties encountered in assessing exposure in a highly mobile animal species; (2) relatively long lived, in order to properly integrate episodic exposures to carcinogenic pollutants and to allow for a latent period of tumor induction; (3) sensitive to oncogenic agents and yet relatively tolerant of noncarcinogenic pollutants; and (4) intimately associated with bottom sediments, either through substrate habitat preference, a detritus food chain, or both. The animals that come closest to satisfying all these criteria are the filter-feeding mollusks.

A fairly large number of cryptogenic proliferative lesions from bivalve mollusks have been described (*68, 141*). Most reports of molluscan neoplasms have come from the marine environment, an understandable fact considering the economic importance of marine shellfish as a food source. Papillomas from mantle tissue (Fig. 3a) described from Australian rock oysters (*Crassostrea commercialis*) were originally observed during inspection of an oyster packing facility on the Hawksbury River, New South Wales, Australia (*196*). This area was not subject to major industrial and urban effluents, although the river does drain agricultural lands, support moderate pleasure boating, and receive effluent from a paper pulp mill.

Another type of proliferative lesion described by Yevich and Barszcz (*198*) as a gonadal neoplasm in soft-shelled clams (*Mya arenaria*) appeared to be consistently associated with an area in Long Cove, Searsport, Maine impacted by a spill of No. 2 fuel oil and JP-5 fuel. Soft-shelled clams from another site (Harpswell Neck, Maine) that was subjected to a spill of JP-4 fuel also exhibited neoplasia, but some of these were considered to be of hematopoietic origin (*198*). Farley (*51, 52*) described hematopoietic disease(s) in oysters (*Crassostrea virginica, Crassostrea gigas,* and *Ostra lurida*) and mussels (*Mytilus edulis*). Farley observed (*52*) that remarkable similarities existed between undifferentiated germ cells in gonadal tubules and the atypical hemocytes (Fig. 3b). He also developed a procedure for obtaining samples of circulating blood cells, thus enabling a fairly simple and rapid screening diagnosis of the leukemialike disor-

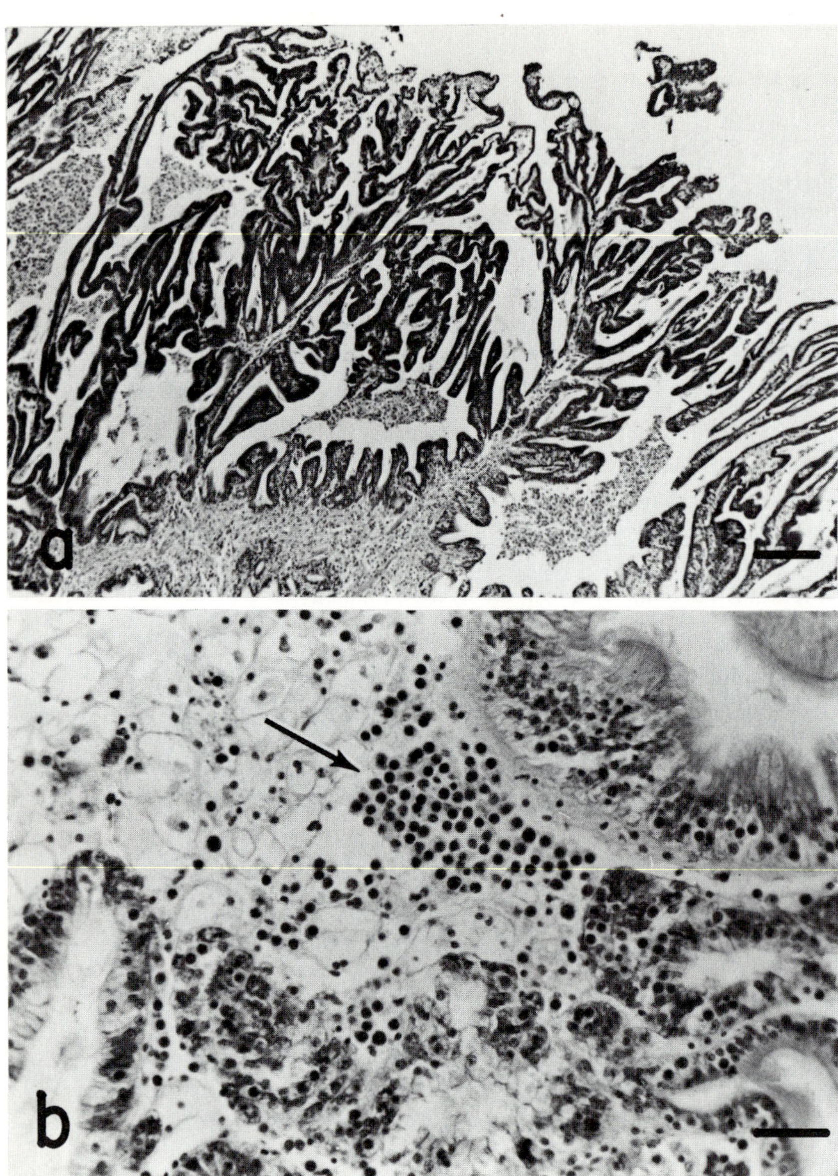

Fig. 3. Two types of molluscan neoplasia. (a) Papilloma of the mantle in an Australian rock oyster (*Crassostrea commercialis*). H&E. Bar = 75 μm. With kind permission, RTLA, Dr. Peter H. Wolf, New South Wales State Fisheries, Scientific Section, Sydney, Australia (contributor). (b) Hematopoietic neoplasia in an oyster (*Crassostrea virginica*). A focus of neoplastic cells (arrow) is located adjacent to the gut with neoplastic cells scattered throughout vesicular tissue. H&E. Bar = 20 μm. With kind permission, RTLA, C. Austin Farley, Northeast Fisheries Center, National Oceanic and Atmospheric Administration, Oxford, Maryland (contributor).

der (*28*). However, in view of the possible similarity between germinal cells and the atypical hemocytes, histopathology is recommended to confirm the origin of the disease.

Other studies of hematopoietic neoplasia in Virginia oysters conducted at the Virginia Institute of Marine Science (*54*) demonstrated that among many experimentally bred laboratory groups of oysters two groups of the inbred animals were much more susceptible to development of the disease when transferred to the natural environment. It was not clear that these experiments may have reflected increased genetic susceptibility to the hematopoietic neoplasia contracted through exogenous agents or as a result of vertical transmission of a virus.

Studies by Brown (*26*) and Cooper *et al.* (*36*) of hematopoietic neoplasms in mollusks (*M. arenaria*) have indicated primarily an infectious etiology based upon epizootiology and evidence of transmission and progression of the disease through a series of experimental flowing seawater trays. Some aspects of the study by Brown (*26*) were suggestive of a chemical influence upon the disease. For example, in experimental trays with clean sediments, 21% of the clams developed neoplasia. On a 1:1 mixture of clean and oil-contaminated sediment, 68% developed neoplasia, and on oil-polluted sediment alone, only 17% developed the disease. These data are suggestive of influence by the oil pollution, however, clams in tanks without any sediment had a 72% incidence of the disease. These somewhat ambiguous results are difficult to assess, but the decreased incidence observed among those clams exposed to oil-polluted sand only is suggestive of a toxic response, i.e., metabolic shutdown, decreased siphoning, etc. In subsequent studies, virus particles described as 120-nm β-type enveloped particles with an eccentric nucleoid were isolated from infected clam tissues. When purified virus was injected into healthy clams, these animals developed the disease within 2 months. Reisolated virus from these clams caused the disease in a second group of healthy animals (*136*). Efforts to isolate a reverse transcriptase on sucrose gradients from diseased clam tissues indicated a band that purified at 1.17–1.18 g/cm^3, but significant levels of reverse transcriptase enzymatic activity were not readily demonstrable (*173*).

To date, there are relatively few laboratory studies that support the supposition that neoplasms in feral marine bivalves are caused by exposures to waterborne carcinogens. Results from experiments by Khudoley and Syrenko (*85*) using a freshwater mollusk (*Unio pictorium*) are consistent with the hypothesis that mollusks develop a neoplastic response upon exposure to chemical agents classified as carcinogens on the basis of their activity in animals. These studies were particularly exciting since excessive proliferation of hemocytes or hemocytelike cells was observed. Another study cited by Khudoley (*85*) indicated that neoplasia had been produced in mollusks by administration of PAH. Mix and colleagues at Oregon State University have conducted numerous studies of hematopoietic neoplasia in mussels, *M. edulis* (*123–126*), including several that have focused on the question of tissue and body burdens of PAH in mussels in relation to the prevalence of the disease at several sites (*127, 128*). Among four sites the whole-body concentrations of benzo[*a*]pyrene, measured by UV spectroscopy, averaged 2.5, 31.66, 32.54, and 0.54 mg/kg of wet tissue. The prevalence of the neoplasia among clams in these experiments was elevated only in the two sites in which the highest benzo[*a*]pyrene concentrations were measured (*128*). In these two sites, approximately 10% of the mussels in the population exhibited the disease, whereas the other two sites were disease free. Brown *et al.* (*27*) previously thought that hematopoietic neoplasia in *M. arenaria* was also correlated to tissue PAH levels.

C. Detailed Case Histories for Which There Is Good Evidence of a Pollutant Etiology

1. Neoplasms in Black River Bullheads

The Black River enters Lake Erie near the city of Lorain, Ohio. The lower portion of the river has been dredged for navigation and, at that point, has a wide estuarine character as a result of very low gradients and low current velocities. Near the head of the estuarine portion of the river a coking facility discharges wastewater. Sediments collected near this area have elevated concentrations of PAH ranging from 5 to 400 μg/g wet weight for individual hydrocarbon species. Benzo[*a*]pyrene and benz[*a*]anthracene were reported present at 43 and 51 μg/g, respectively (*6*). The poorly flushed estuarine character of the Black River environment greatly facilitates the build-up and retention of these contaminants in the sediments.

This Great Lakes estuary was selected by Baumann (*6*) for an investigation of hepatic neoplasia in brown bullheads (*I. nebulosus*) as a possible indicator system for assessment of the impact of industrial pollutants, especially PAH. The liver was selected as the target organ for neoplasia, in part based upon several reports of elevated frequencies of hepatic neoplasms in fish, including the brown bullhead. Elevated frequencies of hepatomas (12.2%, $N = 283$ fish) had been

Fig. 4. Neoplasms of brown bullheads (*Ictalurus nebulosus*) from the Black River, Lorain County, Ohio. (a) Multiple cream-colored nodules visible within the liver substance are grossly visible signs of neoplasms. (b) Bile duct neoplasia (cholangioma vs cholangiocarcinoma). H&E. Bar = 50 μm. (c) Moderately well-differentiated hepatocellular neoplasm. Neoplastic nodule (above) is seen compressing normal liver tissue below. H&E. Bar = 50 μm. (d) Epidermal neoplasms. Ventral location is suggestive of an etiologic role through contact with heavily contaminated bottom sediments. A high percentage of these neoplasms was low-grade invasive carcinomas. With kind permission, RTLA and Dr. Paul C. Baumann, Columbia National Fisheries Laboratory, United States Fish and Wildlife Service.

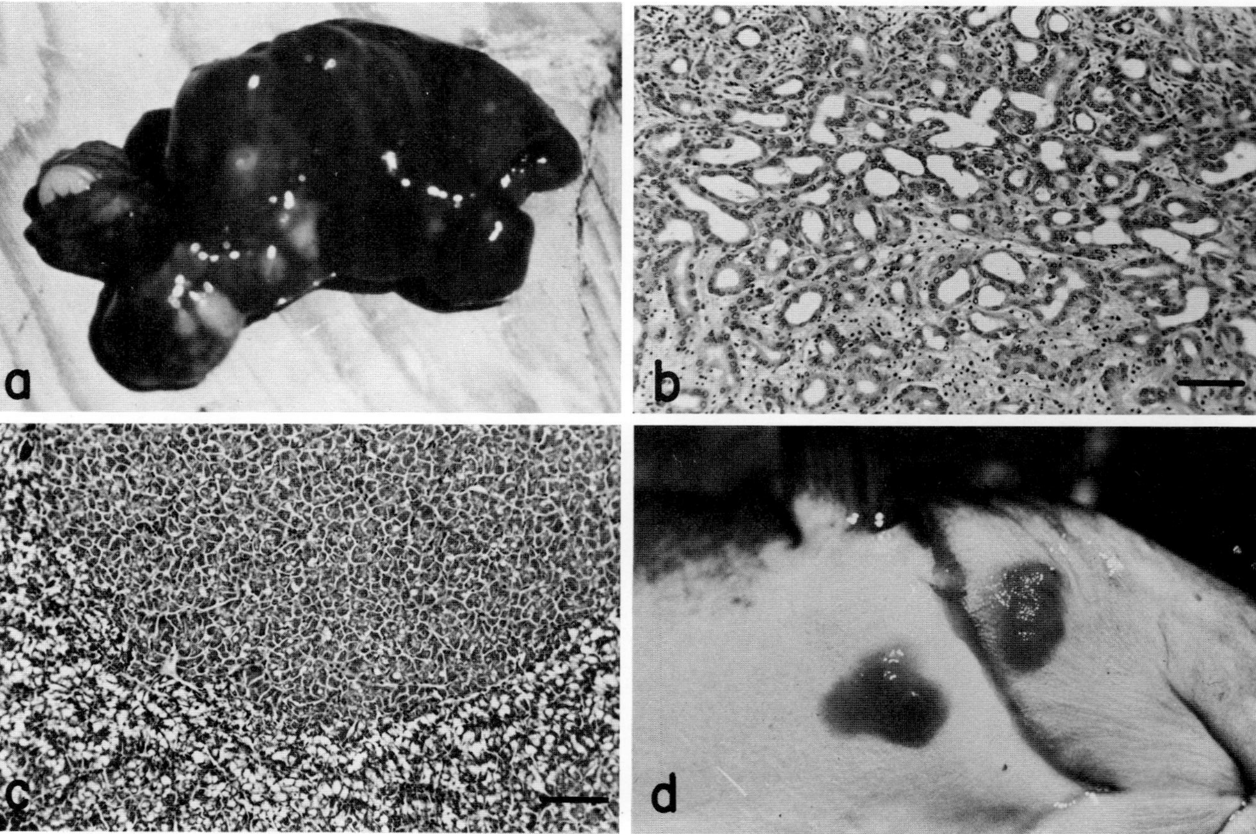
a
b
c
d

previously reported in brown bullheads from the Fox River in the study by Brown *et al.* (*24*), although no descriptions of the pathology were given.

Gross examination of the Black River bullheads revealed that, generally, the livers were prominently enlarged and frequently darker in color than the normal reddish-tan. Although most of the livers had a smooth capsular surface, frequent specimens exhibited numerous small (2–4 mm), pale to cream-colored foci, and/or occasional larger (0.5–1.5 cm), slightly raised nodules (Fig. 4a). Some livers also exhibited indistinct, pale blotchy areas visible within the liver substance. Microscopic examination of affected livers revealed a range of degenerative changes, including both focal and zonal necrosis and fatty vacuolation. In addition to these apparent toxic responses, two types of neoplasms were found: proliferations of the bile duct (cholangiomas) and hepatocellular neoplasms (hepatocellular carcinoma, well-differentiated type), as shown in Fig. 4b and c. During the open water seasons of 1980 and 1981, a total of 275 Black River fish of age ≥ 3 years were examined. Based on the gross pathology, liver neoplasms were present in an estimated 28% of the population. In addition to liver neoplasms, Baumann documented epidermal and oral neoplasms in bullheads from the Black River. These occurred frequently near the lips, head, and barbels (Fig. 4d), and occasionally elsewhere. Although most of these were epidermal papillomas, among a series of 12 specimens submitted to the Registry for Tumors in Lower Animals, 3 were diagnosed as carcinomas (*67*). In a control population from a relatively nonpolluted lake (Buckeye Lake, Ohio), no liver or skin tumors were present among 80 bullheads, also ≥ 3 years of age (Fig. 5).

This recent study by Baumann represents a well-documented investigation integrating appropriate measurements of environmental pollution with assessment of environmental pathology in a logical target species, the brown bullhead. An important feature of the investigation was the inclusion of analytical measurements of PAH in sediments as well as in the edible muscle tissues of the fish themselves. PAH compounds were present only in traces (≤ 200 ng total PAH/g tissue) in bullheads from the control environment, whereas bullheads from the Black River had significantly elevated levels (≥ 14,088 ng total PAH/g tissue). An interesting aspect of the study is that reported values for PAH in the tumor-bearing fish appeared to be slightly lower than the levels found in nontumorous members of the population. The differences reported may reflect normal variation associated with the type of analytical measurement, but a tantalizing possibility exists that the lower PAH concentrations in the tumorous fish reflect more active metabolism of these compounds by the tumorous animals' liver tissue.

In view of the liver's central role in detoxification and activation of xenobiotic chemicals, together with a recent report describing induction of hepatocellular neoplasms (and lymphosarcomas) in a species of laboratory-reared aquarium fish (*Poeciliopsis* sp.) in response to 7,12-dimethylbenz[*a*]anthracene (*157*), it is

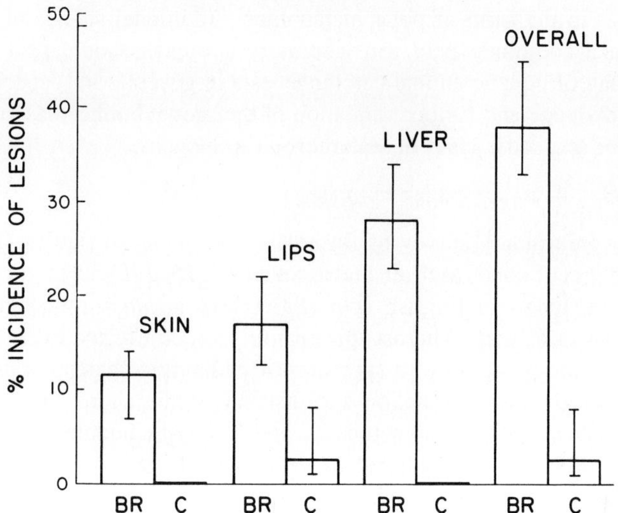

Fig. 5. Frequency of occurrence of three types of neoplasms (2-year average) in two populations of brown bullheads (*Ictalurus nebulosus*). BR, Black River and C, control site, Buckeye Lake, Ohio. Vertical bars represent approximate 95% confidence intervals. With kind permission, Dr. Paul C. Baumann, Columbia National Fisheries Research Laboratory, United States Fish and Wildlife Service.

likely that PAH exposure represents the primary etiologic factor that produced the liver pathology observed in the Black River bullheads. Previous attempts to induce fish neoplasms with this compound failed (*146*), although the two experiments varied in their routes of administration. It should be borne in mind that although the aqueous exposures employed in the laboratory experiments with *Poeciliopsis* sp. were high (5 ppm with 3–4 exposures of 5–20 h) tumor formation was fairly rapid and a high percentage of the animals was affected, i.e., 22 of 46 fish surviving approximately 8 months. In the "real world" polluted Black River ecosystem, fish are exposed more or less continuously via three different routes: aqueous exposure with uptake via gills and skin, direct contact with polluted sediments, and the dietary route through ingestion of detritus and contaminated food chain organisms. This latter route of exposure may have considerable significance to the Black River bullheads. In this river, the bullheads forage heavily on crayfish (*12*). The ability of crustaceans, including crayfish, to accumulate PAH in their tissues is well documented (*16, 132, 167, 168*). Studies have shown that the lobster hepatopancreas sequesters very high levels of PAH (*168*) and, perhaps even more importantly, PAH metabolites. A study by Lee *et al.* (*100*) indicated rapid uptake of radiolabeled PAH compounds by juvenile crabs (*Callinectes sapidus*), with up to 50% of the total radioactivity accumulated by the hepatopancreas. Of the PAH accumulated by this organ, approximately one-half was in the form of polar metabolites. Additional studies of the Black River using a combined field and laboratory approach should lead to greater understanding of aquatic animal carcinogenesis in general, and, specifically, to detailed knowledge and further validation of the brown bullhead as an indicator organism for a specific class of waterborne carcinogens.

2. *Hepatomas in Puget Sound Flatfishes*

During a preliminary survey of fin erosion occurring in flatfish (Pleuronectidae) from Puget Sound, McCain and associates (*116, 145*) noted grossly abnormal-appearing livers in English sole (*Parophrys vetulus*) collected from the Duwamish River estuary. Microscopic examination of affected livers from these fish indicated the presence of a spectrum of pathologic changes, ranging from mildly increased fatty vacuolation and disarray of the chord structure and sinusoids, to several types of preneoplastic and neoplastic nodules (Fig. 6). These findings, in part, led to a detailed investigation of the role of chemical contamination in abnormalities in fish and invertebrates from Puget Sound. The investigations were conducted under the Micro Ecosystems Analysis (MESA) Puget Sound Project (*105, 106*) using a large number of investigators representing multiple disciplines.

In brief, the approach in these investigations was based on statistical correlations of various aquatic animal pathologies, including neoplasms and several types of preneoplastic lesions, to site-specific similarities in organic contaminants (i.e., cluster analysis). In establishing the site groupings, approximately 50 locations were analyzed for chemical contamination of sediments, and these were grouped on the basis of similarities in qualitative and quantitative pollution loadings. Four major groups of chemical pollutants were utilized in establishing the site similarity groupings: metals, selected metals plus polychlorinated biphenyls (PCBs), chlorinated organics exclusive of PCBs, and PAH. Analytical values for sediment concentrations of 27 target PAH compounds, reported as the

---

Fig. 6. Hepatic neoplasia in English sole (*Parophrys vetulus*) from Puget Sound, Seattle, Washington. (a) Typical structure of normal liver. Central vein, C; pancreatic tissue, P. H&E. Bar = 75 μm. (b) Liver showing enlarged hyperchromatic nuclei and increased hepatocellular diameters characteristic of the degenerative lesion, megalocytic hepatosis. Interspersed throughout the liver parenchyma are islands of regeneration characterized by small, basophilic hepatocytes. H&E. Bar = 50 μm. (c) Mixed hepatocellular–cholangiocellular carcinoma. The tubular elements of this neoplasm are composed of biliary epithelium, with the remaining neoplastic cells composed of hepatocytes. H&E. Bar = 50 μm. (d) Hepatocellular adenoma with central carcinomatous focus. The highly basophilic cells composing the adenoma exhibit well-organized mural architecture. The carcinomatous focus is composed of enlarged anaplastic hepatocytes in an abnormal architecture. Normal hepatic parenchyma is visible at the far left. H&E. Bar = 75 μm. With kind permission, Mark S. Meyers, Environmental Conservation Division, Northwest and Alaska Fisheries Center, Seattle, Washington.

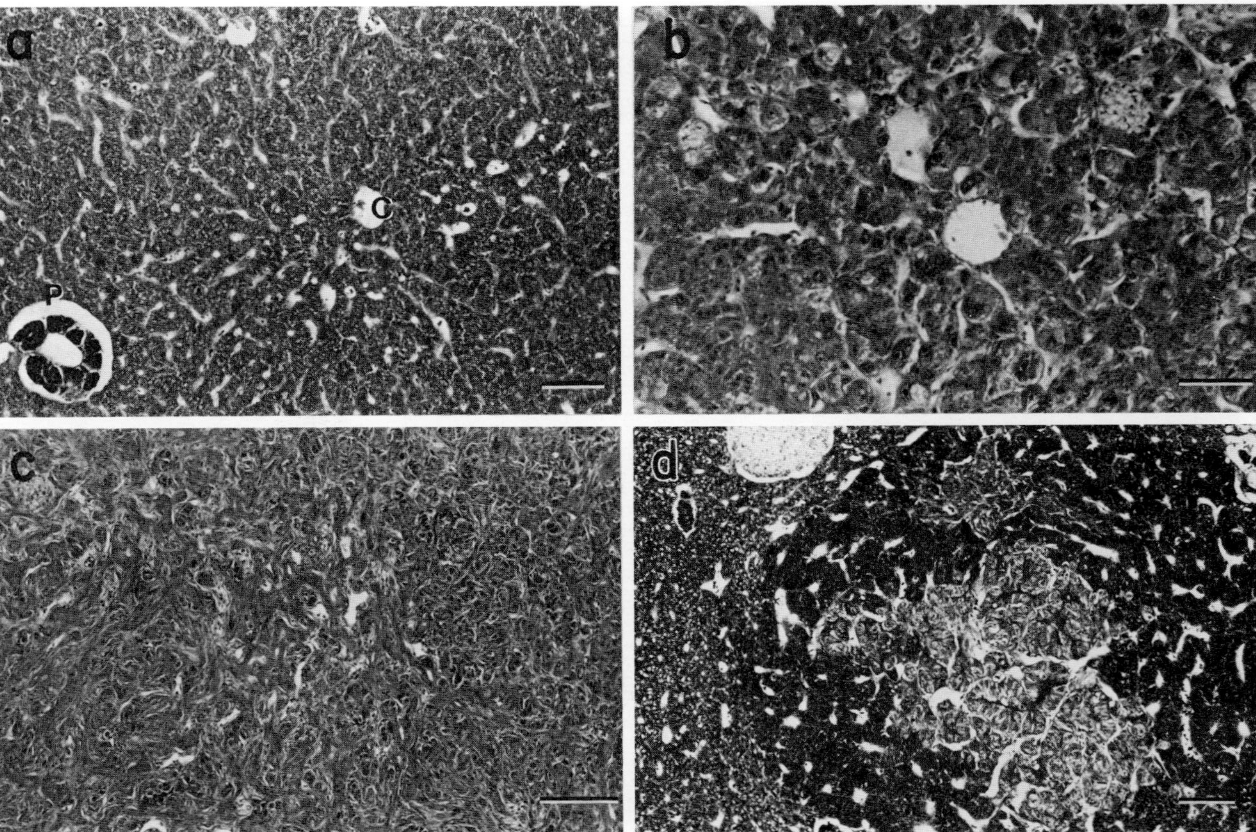
a
P
c
b
c
d

sum of individual concentrations, varied by as much as 3000% between the most heavily polluted sites and the cleanest locations. English sole and rock sole (*Lepidopsetta bilineata*) were selected as target fish species for histopathological investigation, in large part because they are the most widely distributed bottom-feeding fish species in the Sound and because liver neoplasms had been discovered in English sole in the preliminary study (*116, 145*).

In evaluating histopathology, tissue slides were examined without prior knowledge of the location of collection. Pathologists examining the slides were provided only with information relative to length, weight, sex, and a description of gross pathology. This blind procedure mitigated biased interpretation of the pathology. Upon tabulating the results, a statistical method was used to test a null hypothesis that the prevalence of lesions in sole collected from each group location did not differ from the average prevalence of that lesion among all sole collected, i.e., the average prevalence for all cluster stations. In this analysis, sole with hepatic lesions, including both neoplastic and preneoplastic and specific degenerative/necrotic changes (hepatic changes not associated with microbial agents), were "consistently most prevalent" at two locations: the Duwamish Waterway and the Sitcum-Hylebos Waterway (Fig. 7). Relationships among the four groups of chemical pollutants within and between the cluster group sites were complex, however, the types of chemicals that were consistently present at moderate to high levels in the stations comprising these two cluster groups were aromatics and toxic metals. Although these results do not identify the specific etiologic pollutant(s) that caused these pathologic lesions in fish livers, they provide convincing, systematic, objective evidence of a role for environmental chemical pollutants as the cause of liver neoplasms and related lesions in two bottom-feeding fish species, the English sole and rock sole.

3. *Sewage-Associated Neoplasia in Tiger Salamanders*

The studies of Dr. Francis Rose have centered on the study of several types of neoplastic lesions in tiger salamanders (*Ambystoma tigrinum*). Rose has described the results of these investigations in considerable detail (*149–153*), but only the relevant outlines of his work will be presented here.

In northwest Texas and eastern New Mexico is a large area in which the primary source of surface water are the small depressions known as playas. These are heavily used by wildlife, including many amphibian species. A population of neotenic (nonmetamorphosing larval stage) tiger salamanders inhabiting a playa located on the Reese Air Force Base at Hurlwood, Texas has been studied by Rose since 1969. Rose, an ecologist who was interested in the neoteny phenomenon, selected the playa on the Air Force base to do an in-depth ecological study. He chose this particular site because it had a large population of salamanders and the site was more or less unavailable to the general public. In 1970, while examining samples of the population, he noticed a single abnormal

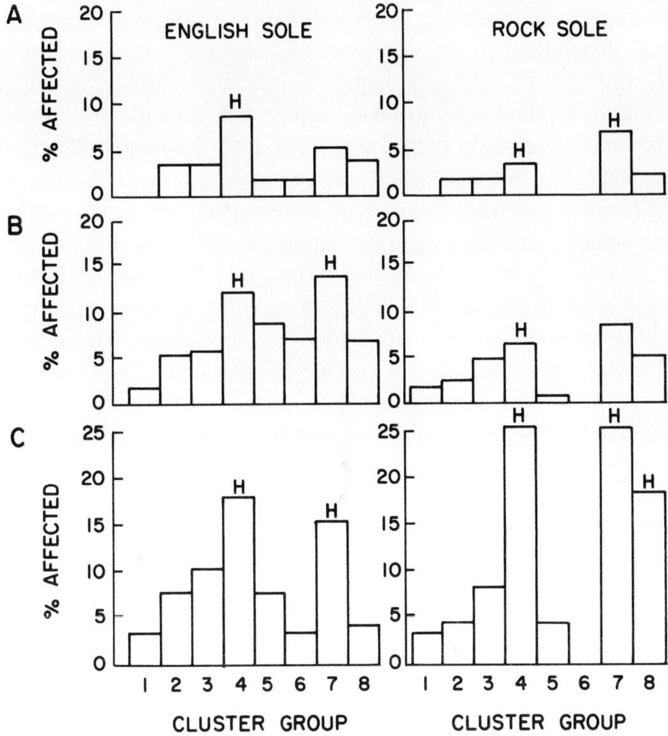

Fig. 7. Prevalence of (A) neoplasms, (B) preneoplastic lesions, and (C) degenerative lesions in English sole (*Parophrys vetulus*) and rock sole (*Lepidopsetta bilineata*) at various Puget Sound locations, grouped according to similarity in sediment pollution composition (cluster groups). H, significantly higher. Cluster groups 4 and 7 include the polluted Duwamish and Sitcum–Hylebos areas, respectively. Adapted from Malins *et al.* (*106*).

specimen with a roundish growth (a fibroma) on the tail (Fig. 8a). Although Rose was not conducting a study of, nor was he trained in the study of, tumors, he had a keen interest in the playas and their associated biota. The above-mentioned salamander was the 275th animal examined among a total of 2430 salamanders. Having handled thousands of normal salamanders, Rose immediately recognized the growth on the tail as something unusual. Fortunately, the growth was detected early in the 1970 sampling and, as a result, Rose decided to examine all the animals for the presence of tissue abnormalities. Based on these observations, only 1 of the 2430 animals collected in 1970 exhibited any obvious external growths. In the following year (1971), approximately 25% of the salamanders from the Air Force base were observed to be affected with various kinds of external abnormalities (most of these were papillomas). Representative examples were sent to the RTLA, where eight kinds of tissue abnormalities were diagnosed. Neoplasms included epidermal papillomas, melanocytomas (and/or melanomas, depending on the degree of invasiveness), dermal fibromas, and dermal myxofibromas (Fig. 8a, b, and d). Nonneoplastic lesions included epidermal hyperplasia, epidermal inclusion cysts (Fig. 8c), cephalic cysts, and abdominal ascites. A 10-year study of the neoplasia, from 1970 to 1980, has recently been published (*152*). After an initial sudden increase in the incidence of lesions, the overall incidence of tissue abnormalities remained fairly constant until 1974, when the incidence increased again to approximately 40%, and reached a peak incidence of 50% 1 year later, in 1975. Among 19,802 salamanders examined from ostensibly clean playas (non-sewage-receiving), no neoplasms were found. Among 1900 individuals examined from six other sewage playas, only 2 had external lesions and these were low-grade hyperplasias. Thus, sewage per se does not seem to be the causative factor.

In addition to epizootiologic observations, several experimental findings point to an environmental agent unique to the contaminated air base playa. By caging newly hatched larval salamanders, age dependence of the salamander papillomas was demonstrated. In this experiment, a few salamanders began to develop hyperplasia by the fifth month and papillomas by the twelfth month. By the twenty-sixth month, approximately 7% of the animals exhibited these neoplasms. Because early-stage lesions are detectable in fairly young animals, it seems unlikely that age differences between the salamanders on the air base and the other playas are responsible for the failure to observe lesions in salamanders from either the clean or the other sewage-contaminated ponds. A fascinating observation was that when larvae bearing papillomas were removed from their contaminated environment and placed into clean water the papillomas regressed. This characteristic is typical of chemically induced neoplasms seen in experimental mouse skin carcinogenesis, i.e., at the papilloma stage, skin neoplasms often will regress when the application of the chemical is discontinued (*20*).

The pollution history of the Reese Air Force Base playa has several interesting features. Aside from the presence of chlorinated sewage effluent and possible surface contamination from aircraft exhaust condensate, two suspicious events/agents stand out. In 1959, the lake was poisoned with toxaphene in an attempt at fisheries management, and as part of a mosquito control program the lake received annual sprayings of diesel oil (no mention of pesticides in the diesel fuel). Toxaphene, a complex mixture of chlorinated camphenes, has been shown to be carcinogenic in mice (*185*). Presently, the EPA is taking action to cancel the registration of this pesticide based mainly on the potential long-term ecological hazards and bioaccumulation potential (*185*).

Diesel oil sprayings as a potential source of chemical carcinogenesis have recently been evaluated (*150*). The amount of diesel oil dispersed to this small, 30-acre pond through 1976 was estimated by Rose to range from 3360 to 8400 liters per year, based on a rate of application of approximately 60 liters/acre and

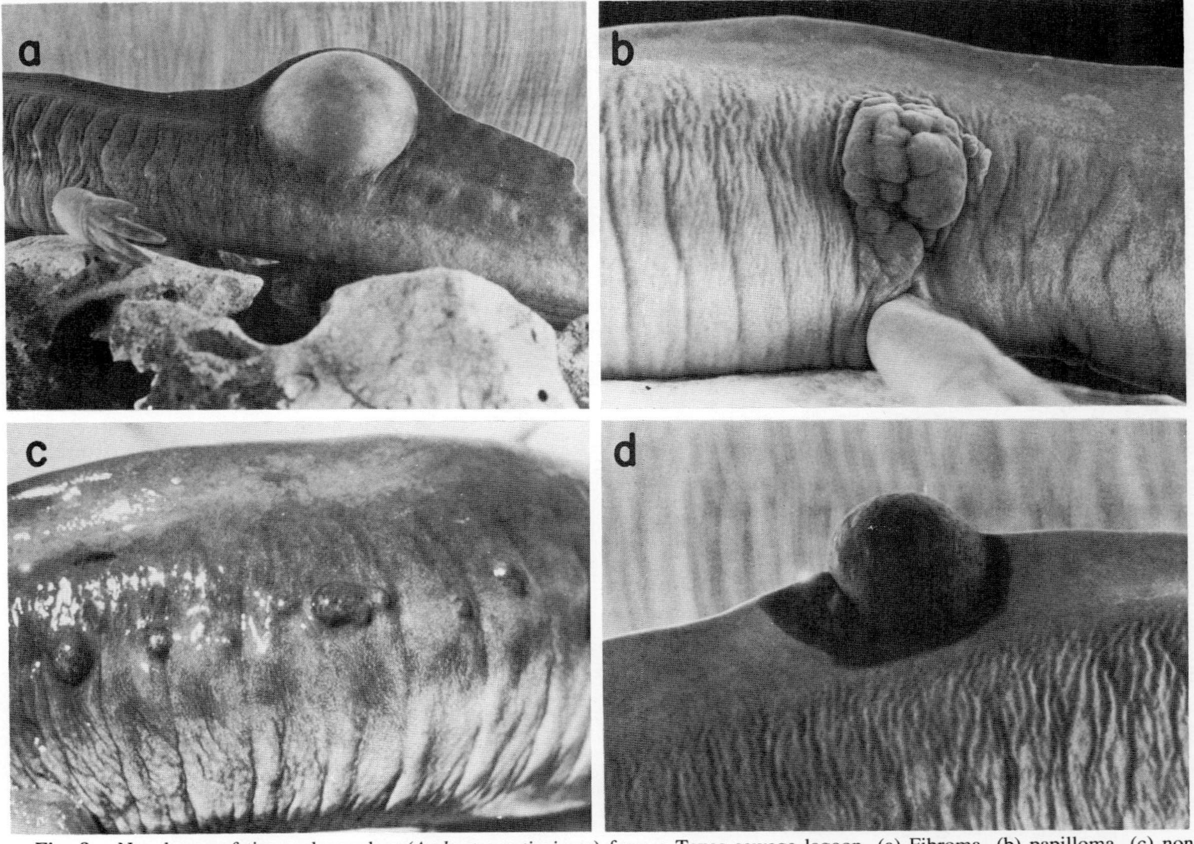

Fig. 8. Neoplasms of tiger salamanders (*Ambystoma tigrinum*) from a Texas sewage lagoon. (a) Fibroma, (b) papilloma, (c) non-neoplastic epidermal inclusion cysts, and (d) melanophoroma vs melanocarcinoma. With kind permission, Dr. Francis Rose, Department of Biological Sciences, Texas Tech University, Lubbock, Texas.

sprayings from two to five times per year. In his evaluation, there was no information as to the number of years this practice was carried out. However, this base has been in operation at various degrees of activity since 1941. Fuel oils and other petroleum products are well-known sources of PAH (*25*) and various oils are sources of occupational cancers in man (*82*). It also may be relevant that oils sometimes exhibit greater carcinogenic effects than would be expected from their PAH content alone (*82*). Asphalt, also a source of PAH compounds, was also dumped into the lake and then removed around 1969–1970.

From a historical perspective it appeared that the salamander's lesions developed well after the 1959 application of toxaphene and, therefore, the most plausible agents would have appeared to be PAH and/or heterocyclic compounds associated with the fuel oil dispersions. Perylene has previously been determined to be present in the sediments at 300 ppm. However, recent analyses of surface sediments (*12*) indicated only trace amounts of PAH, e.g., benzanthracene, 33.6; chrysene, 35.7; benzo[*k*]fluoranthrene, 26.4; benzo[*a*]pyrene, 31.4; and dibenz[*a,h*]anthracene, 13.6 ng/g of sediment, dry weight.

Unfortunately, the Reese Air Force Base salamander population appears to be on the edge of extinction, with recruitment of new individuals to the population near zero. The true identity of the causative agent(s) may never be known. However, there may still be important clues "locked up" in the playa's sediments. Core sampling techniques and estimates of sedimentation rates might enable sediment of the 1968–1969 stratum to be analyzed by the relatively sensitive methods of high-pressure liquid chromatography and/or computerized gas chromatography/mass spectrometry. This time period in the sediments would coincide with the known lag time for the development of papillomas, as demonstrated by Rose's field induction experiments. These data could then be compared to sediments from strata just prior to this point in time. If a diesel oil chromatographic "fingerprint" was undetected prior to about 1968, or present in increased amounts at about the expected time frame, this would augment the available evidence implicating fuel oils as the source of the carcinogenic insult.

4. *Liver Neoplasia in Hudson River Tomcod*

In 1977–1978, in conjunction with studies of the impact of power plants on the fisheries of the Hudson River, a high percentage of enlarged and abnormal-appearing livers was detected in Atlantic tomcod (*Microgadus tomcod*). Some of these livers had obvious dark-colored, nodular growths. Based upon gross signs, approximately 25% of the population appeared to be affected with liver neoplasia. The livers ranged in appearance from normal reddish-tan with a smooth capsular surface, to those with grossly abnormal, nodular, dark red to purple lesions that varied greatly in size. The nodules ranged from approximately 3.0 mm in size to some that covered nearly the entire liver. The frequency of

TABLE II

Gross Pathology in Tomcod Livers[a]

Group	Description	Number of fish[b]	Percentage
Normal		84	31.8
A	Diffuse hemorrhagic areas	114 (1)	43.2
B	1–3 mm light gray, pustulelike lesions on liver surface	61 (1)	23.1
C	Variably sized, dark red to purple nodules/lesions	5 (2)	1.9
	Total fish	264	

[a] Based on data from Smith *et al.* (*169*).
[b] Numbers in parentheses indicate number of fish examined microscopically.

occurrence, morphology, and histology of these lesions have been described by Smith *et al.* (*169*). The present discussions are based on these authors' account.

Based upon the severity of the gross pathology, livers were placed into one of three groups (Table II). Microscopic examination of four livers (one from Group A, one from Group B, and two from Group C) demonstrated the presence of hepatocellular neoplasms including carcinoma (Fig. 9) in the three livers from Groups B and C. The single liver examined from Group A had excessive vacuolation, focal subcapsular congestion, and mild hemorrhage, but no obvious neoplasms. Clearly the small number of livers studied microscopically does not rule out the possibility of early-stage hepatocellular neoplasms in the Group A livers. It is probable that the incidence of liver neoplasia in these fish is significantly higher than the reported 25%.

A major source of PCBs to the river has been traced to the General Electric (G.E.) capacitor manufacturing facilities on the upper part of the river at Hudson Falls and at Fort Edwards, New York (*75, 79*). It has been estimated that PCBs are being transported at a rate that will deliver between 32 and 54 metric tons to the estuary over the next 20 years (*79*). When the sheer magnitude of the resource is considered, PCB contamination of the Hudson exceeds that of any other aquatic environment. PCBs have been measured in Hudson River tomcod (*4*), including tumorous and normal specimens, by Klauda *et al.* (*92*). Any correlation between tissue concentration and severity of gross pathology was inapparent. Unfortunately, a relatively small number of specimens were analyzed and data were not measured on a lipid weight basis. Because PCB accumulation is frequently correlated with the lipid content of the fish, any relation between liver pathology and tissue burdens of PCBs would easily have been obscured by variations in lipid content of the fish. PCBs have exhibited carcinogenicity, producing both hepatocellular and cholangiocellular neoplasms in

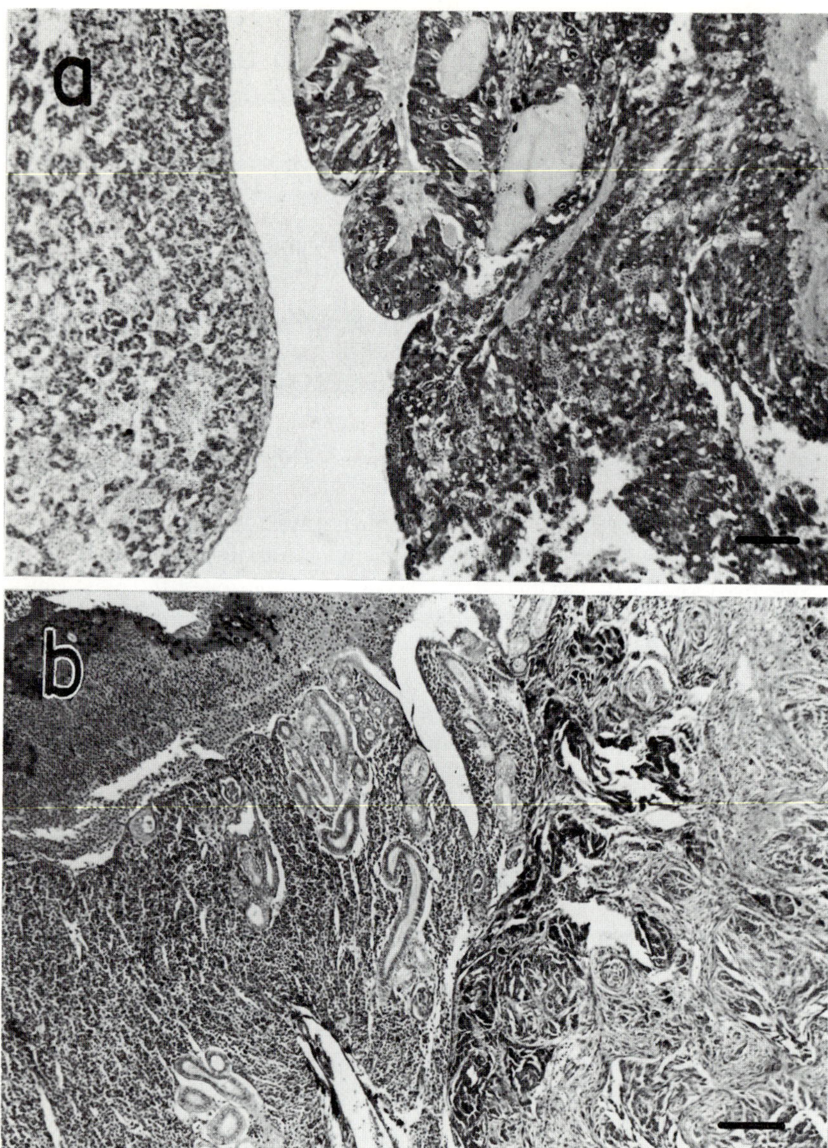

Fig. 9. Hepatocellular carcinoma in Atlantic tomcod (*Microgadus tomcod*). (a) Neoplasm (right) is composed of basophilic hepatocytes exhibiting considerable pleomorphism. Relatively normal liver is shown at left. H&E. Bar = 75 μm. (b) Metastases to kidney. Poorly differentiated hepatocytes (right) can be seen invading kidney substance (left). H&E. Bar = 75 μm. With kind permission, C. E. Smith, Fish Cultural Development Center, United States Fish and Wildlife Service, Bozeman, Montana.

laboratory rodents when fed at approximately 100 ppm for 24 months (*86*). These chlorinated hydrocarbons appear to be an obvious candidate for the etiologic agent that caused the tumors in the Hudson River tomcod. However, it should be noted that PCBs have not yet been shown to be carcinogenic in fish. When fed at 100 ppm for a period of 1 year to Shasta strain rainbow trout, no neoplasms developed. However, when maternal trout were fed a diet of 200 ppm, eggs from these fish exhibited an enhanced carcinogenic response (approximately 30% increase in tumorous fish) to aflatoxin B_1 (*74*). Alternatively, another well-documented chlorinated hydrocarbon pollutant, DDT, has produced liver neoplasia in trout including both hepatocellular carcinomas and cholangiomas/cholangiocarcinomas (*72*). This is notable, for previous tests with DDT produced mostly a toxic response at dietary concentrations as low as 70 ppm in the diet (*5*). Cholangiomas have also been observed in some of the Hudson River tomcod livers (*169*).

Falkmer (*48, 49*) correlated decreases in the observed frequencies of hepatocellular carcinomas occurring in hagfish (*Myxine glutinosa*) from the Gullmar Fjord, Sweden, to the legislative restriction of the use of PCBs in Sweden in 1971–1972. In a 5-year period during which 28,000 hagfish were examined for neoplasms, hepatomas decreased in frequency from 5.8% of the population in 1972 to 0.6% in 1974–1976. In 1000 hagfish caught from the open sea (12 km from the fjord) hepatomas were also observed but at a lower frequency. In this control group, the prevalence also declined and was observed to decrease from 2.8% in 1972 to 0.9% in 1974.

In the case of the Hudson, the discharge of PCBs from the major source (G.E. transformer plant) has recently been reduced from an average daily discharge rate of 13.6 kg/day (as "unspecified chlorinated hydrocarbons") in 1972 to less than 1 g/day as of March 1977 (*75*). This recent reduction in the release of PCBs to the Hudson should eventually be accompanied by decreases in the prevalence of the cod neoplasms if PCBs are the offending etiologic agent. Clearly, further studies of this problem should be undertaken.

5. *Neoplasms from a Lake Polluted with Copper Mining Wastes*

Torch Lake is a large, deep lake (surface area, 1077 ha; maximum depth, 35 m) located near the middle of the Keweenaw Peninsula, Houghton County, northern Michigan. This part of the state has been extensively mined, especially for copper. Since the 1800s, copper ore extracted from mines in the area has been crushed at mills located along the shore. Crushed ore tailings dumped in the area were subsequently reprocessed and the very fine residual materials were dumped into the lake. Eventually, over 20% of the entire lake volume was filled in by these extremely fine mining waste particulates. In addition, the lake receives some sewage effluent and it has received high-salinity water from mine "de-watering" operations. Presently, sewage effluent from the small communities along the lake is spray-irrigated onto old mine tailings in an effort to foster vegetation growth. Copper concentrations range from 1700 to 2400 mg/liter (dry weight basis) in the upper sediment layer, which consists of a light brown ooze covering the tailings to a depth of several centimeters. Copper concentrations in the water average 30–45 μg/liter, which is approximately 10 times higher than ambient levels recorded for Lake Superior.

Abnormalities in fish from this lake and from waters of the Keweenaw Canal system were noted in the early 1970s by Tomljanovich (*183*). Although Tomljanovich noted dermal fibromas, lipogranulomas, and lesions diagnosed at that time as lipofibromas (visceral cavity masses), liver neoplasms were either not present or simply not noticed at the time. Recent examination of sauger (*Stizostedion canadense*) and walleyes (*Stizostedion vitreum*) from Torch Lake has revealed high frequencies of hepatocellular neoplasms in both species (*14*). On the basis of obvious nodules (Fig. 10a), liver neoplasms appeared to be present in every sauger examined (i.e., 100%, $N = 23$), whereas among 11 walleyes, only 3 fish exhibited obvious tumors (*14*). Microscopically, the livers were all similar in their histology, with neoplasms present as single to multiple nodules composed of well-differentiated hepatocytes exhibiting increased basophilia and mild pleomorphism. In addition to the hepatocellular neoplasms, both walleye and sauger exhibited dermal neoplasms (ossifying fibromas) and large perivisceral masses of uncertain histopathologic classification. These growths took the form of large, rubbery masses that were frequently attached to the spleen (Fig. 10b). Based upon evidence of secretion and possible origin from elements of the spleen capsule, a tentative diagnosis of mesothelioma has been tendered. The presence of occasional granulomas in these perispleenic masses also raises the possibility that these are of a reactive nature rather than truly neoplasmic.

The presence of copper mining waste implies a causal role for some agent or associated aspect of the mine waste. Although no evidence presently indicates that copper is intrinsically carcinogenic, in view of the high concentrations and chronic lifetime exposures to very old fish (average age of sauger and walleye in 1979 was 11 and 6 years, respectively), copper could conceivably be the offending agent. Conversely, copper ores may contain carcinogenic metals such as selenium and arsenic (*57*). Another possibility is that the carcinogenic agent may be associated with the mining particulate composition. Mining-related agents such as asbestos, silica dust, and hematite have been implicated in human health effects, including cancer (*21, 38, 160, 162*). An alternative hypothesis might involve a *de novo* synthetic interaction between metallic elements and organic compounds present in sewage, e.g., metal-catalyzed formation of nitrosamine(s). Although the tumors in Torch Lake sauger and walleyes appear circumstantially to be related to the copper mining waste, further study will be required to determine the precise role of these materials in the fish neoplasia.

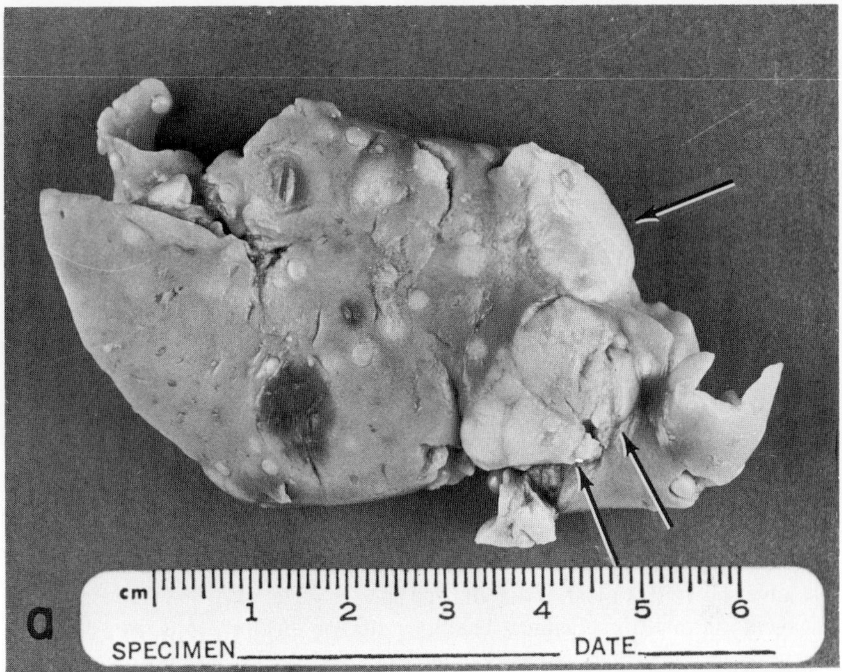

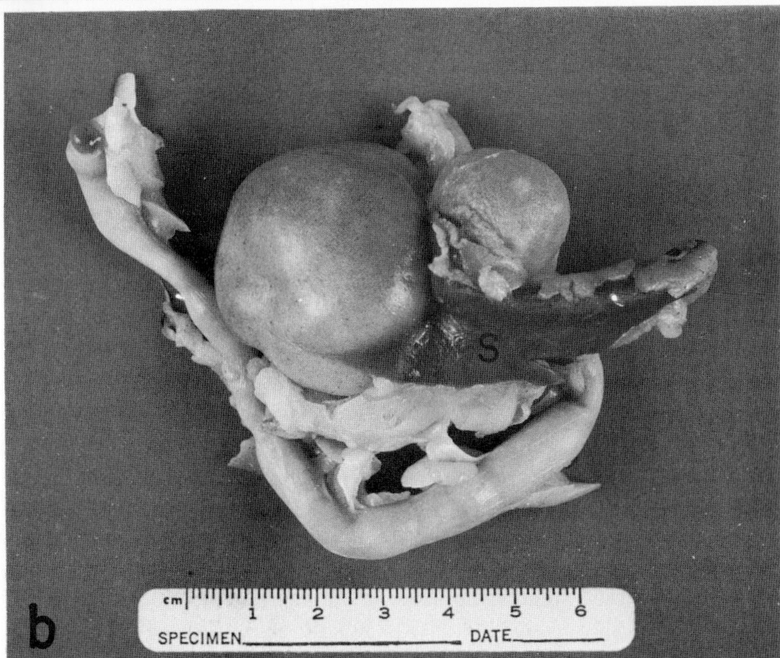

Fig. 10. Gross aspect of neoplasms in sauger (*Stizostedion canadense*) from Torch Lake, Houghton County, Michigan. (a) Hepatocellular neoplasia (arrows). Smaller nodules are parasitic cysts. (b) Large mass of uncertain histologic classification, originating from the spleen (S).

VII. LINKAGE OF ENVIRONMENTAL POLLUTION TO AQUATIC ANIMAL NEOPLASIA

A. Indirect Experimental Evidence

The previously described investigations support circumstantially the contention that aquatic animals can be useful as early warning indicators of waterborne carcinogenic hazards. However, there has been little direct experimental evidence to indicate which animal neoplasias truly represent a response to environmental pollutants. That pollutants are causative agents of some kinds of tumors in wild fish populations has been supported indirectly, largely by laboratory experiments in which neoplasms have been produced in a limited number of organs in rainbow trout (*5, 63, 72, 73, 164, 165, 188*) or by induction of hepatic neoplasms in small aquarium species (*112, 146*). The subject of chemical carcinogenesis in fish has recently been reviewed by Hendricks (*72*). The usefulness of feral aquatic species in detecting carcinogenic events is also indicated by data that have shown that most, if not all, fish species metabolize and activate chemical agents through biochemical reactions similar to thosc associated with oncogenesis in higher vertebrates (*77, 99, 104, 135, 181*). Various other aquatic animals including frogs (*135*), salamanders (*2*), marine mollusks (*7*), crustacea (*8*), and even aquatic insects (*47*) have some functional ability to metabolize carcinogens via mixed-function oxidase-type reactions. In general, data from these *in vitro* studies seem to suggest that the effectiveness of this ability decreases as one moves down the phylogenetic tree. However, data gathered *in vitro* may not accurately reflect the capacities of the intact organism. For example, methods for measurement of aryl hydrocarbon hydroxylase (AHH) in aquatic animals are usually modifications of the mammalian AHH assay, and the requirements of the analogous nonmammalian enzymes may be considerably different. In addition, proteases released from tissues such as the hepatopancreas would probably digest the enzyme and this might explain the failure in some experiments to measure significant activity.

B. Direct Evidence

1. General

Although classical carcinogenesis assays utilizing single compound exposures in laboratory rodents are of significant value in assessment of the oncogenic potential of a chemical, this type of experiment possibly will not identify specific neoplasias in free-living populations of aquatic animals as resulting from exposure to chemical pollutants. In attempting to demonstrate cause and effect relationships in wild populations, detection and quantification of chemicals with demonstrated oncogenic potential are of some value, particularly where associations and appraisal of environmental exposures can be made. However valuable, the epizootiological approach, when used alone, can only suggest which aquatic animal neoplasms are relevant to environmental exposures. Furthermore, a single etiologic compound may not always be identifiable. In "real world" polluted ecosystems, whole populations of animals may be continuously or episodically exposed to chemicals as complex mixtures. Assessment of the role of an individual compound in the mixture may be difficult due to synergistic and/or other modifying effects that may occur in multiple agent exposures.

2. *Experimental Induction of Hyperplasia and Papillomas in Bullheads*

As a first approximation to establishment of a cause and effect relationship between organic and chemical pollution and neoplasia in an indigenous fish species, Black (*10, 11*) chose the brown bullhead as a test animal. The brown bullhead (*I. nebulosus*) is a common bottom-dwelling/feeding freshwater fish species belonging to the catfish family (Ictaluridae). It is subject to various idiopathic neoplasms including hepatic neoplasms (hepatocellular carcinomas and cholangiomas), epidermal (and oral) papillomas, epitheliomas, melanocytomas, and their more aggressive carcinomatous counterparts (*67*). Epidermal neoplasms in brown bullheads sometimes occur in anatomic sites near the mouth and/or the ventral aspect of the fish, suggesting that carcinogens encountered through contact with contaminated bottom sediments may be involved. To explore the possible validity of this hypothesis, feral brown bullheads collected from a nonpolluted pond and acclimated to laboratory conditions were directly exposed to an organic solvent extract of polluted river sediment. As brown bullheads have a fairly simple scaleless skin, the exposure route of choice was direct topical application of the river sediment extract (RSE). The sediment was obtained from a heavily industrialized section of the Buffalo River at Buffalo, New York. Previous investigations of chemical pollution have indicated that both sediments and biota from the river are contaminated with mutagenic substances (*17*) including aromatic amines (*43, 133*) and PAH (*13*). The sediments used to prepare the RSE consisted of a fine silt that was dark in color and exhibited a strong petroleumlike odor. Dried sediments contained approximately 7% ignitable organic matter and yielded up to 0.7 g of organic extractable material per 100 g of dried sediment (Soxhlet extracted for 12 h using hexane:acetone, 1:1 v/v). The extract was made up to contain 5% organic residue (w/v). Except for a slight relative enrichment of higher molecular weight components in the RSE, both the sediment and the RSE contained qualitatively similar mixtures of PAH that were present at generally two to five times greater concentration in the RSE than in the dried sediments (Table III).

The extract (RSE) was applied once a week to an area between the occiput and the lip with control fish receiving solvent only. Fish were placed in a recovery

TABLE III

Concentrations of PAH in Sediment and Sediment Extract

Compound	Sediment (μg/g)	Sediment extract (μg/ml)
Phenanthrene	25.20	32.97
Anthracene	7.77	5.96
Fluoranthene	29.00	76.86
2-Methylanthracene	2.24	4.19
Benzanthracene	7.32	16.02
Chrysene	4.30	14.04
Perylene	8.73	29.36
Benzo[*k*]fluoranthene	2.78	9.03
Benzo[*a*]pyrene	4.45	15.35
Dibenz[*a,h*]anthracene	0.99	3.30
Benzo[*ghi*]perylene	1.52	16.30

tank prior to their return to the holding tanks. This procedure was used to minimize transfer of unabsorbed RSE to the holding tanks. Fish exposed to the RSE underwent a series of localized skin reactions. Initially these consisted of a mild coarsening and blanching of the normal coloration over the painted area. These mild symptoms persisted with little change for approximately 12 months. Gradually, the blanching gave way to a darkening. Light microscopic examination of skin sections taken from animals during this stage revealed no obvious differences between experimental and control fish. Between 12 and 18 months, progressive changes including increases in skin thickness (hyperplasia) and darkening of the skin were noted. In addition, several fish developed single or multiple skin papillomas (*10, 11*), as shown in Fig. 11. At 24 months, papillomas and hyperplasia were present in 8 of 22 surviving fish from an original group of 69 animals placed at risk. The skin of the solvent-treated controls remained normal-appearing throughout the experiment. The survival rates in the control and experimental groups were identical. Sensitivity to liver carcinogens in the sediments was also indicated by focal proliferations and nodules of several morphologic types (*12a*), including clear cell, basophilic, and eosinophilic foci

Fig. 11. Epidermal hyperplasia and neoplasia in brown bullheads (*Ictalurus nebulosus*) in response to repeated skin paintings with an organic solvent extract of polluted river sediment (RSE). (a) Close-up view of fish head showing a small papilloma (arrow) and irregular thickening (hyperplasia) of the skin over the exposed area. (b) Multiple skin papillomas induced by RSE. (c) Microscopic view of the epidermis from a control animal. Normal epidermis (E) shows relatively uniform thickness, dermal papillae (D), and numerous alarm cells (A). H&E. Bar = 75 μm. (d) RSE-induced papilloma showing papillary formations of grossly hyperplastic epithelium separated by intervening strands of connective tissue. H&E. Bar = 75 μm.

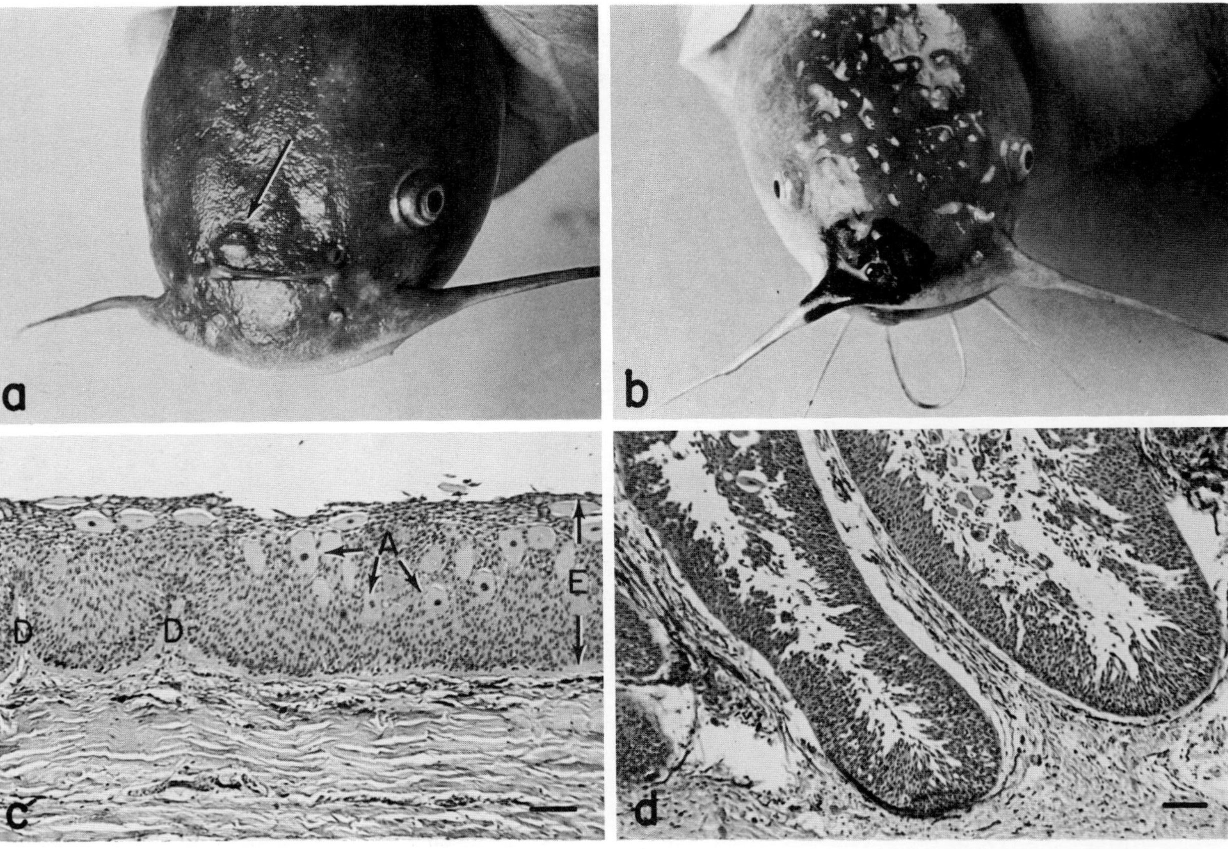
a
b
A
E
D
D
c
d

and nodules (see Fig. 12a and b) similar to those described as type A hepatocellular neoplasms in rodents (*175a*).

Similar research was conducted by Grizzle and Melius (*60*) to explore the cause of oral papillomas in black bullheads (*Ictalurus melas*) living in the final oxidation pond of the Tuskegee, Alabama sewage treatment plant. The water in the pond contained chlorinated effluent from the sewage treatment plant. Up to 75% of the bullheads in the pond exhibited oral papillomas that localized to the fornices of the mouth (*61, 181*). Acidic organic solvent extractions contained a substance(s) that was mutagenic in the Ames test in both strain TA98 and TA100 in the presence of an S-9 arochlor-induced rat liver homogenate fraction (*60*). The chemical identity of the mutagen is not yet known. To date, attempts to induce the neoplasm with cell-free tumor homogenates have produced negative results. Nonneoplastic bullheads obtained from a clean farm pond (no tumors) were placed in the oxidation pond in cages. After 168 days, 55 surviving fish exhibited a type of focal hyperplasia in the fornices of the mouth, but no obvious papillomas were observed. The cause of the papillomas and the relationship of the focal hyperplastic lesions to the papillomas are as yet unresolved.

The bullhead is a relatively hardy fish that adapts rapidly to the laboratory. Wild fish will often feed readily on an artificial pelleted diet (trout pellets) within 1 to 2 days of entering the laboratory environment. Although larger animals have a tendency to become aggressive and mortality can occur due to fighting and stress-induced bacterial infections, on the whole bullheads appear to be an adaptable and useful animal for experimental carcinogenesis. They have one advantage over trout: as warm-water fishes, they do not require regulated cold-water temperatures. Development of complementary *in vivo* and *in vitro* carcinogenesis techniques, such as embryo exposure or cell culture methods, awaits the hand and eye of the interested investigator.

3. *Experimental Induction of Pigmented Lesions in Croaker*

Kimura and colleagues (*88*) have explored the etiology of spontaneous, pigmented neoplasms (chromatophoromas) in the nibe (*Nibea mitsukurii*), a species of marine croaker (family Sciaenidae) inhabiting the coastal waters of Japan. A survey of the chromatophoroma incidence in the nibe populations from 25 stations along the coastline of the islands of Honshu, Shikoku, and Kyushu indicated highly variable frequencies of the neoplasia, with clusterings in the vicinity of the mouth of the Kumano River. Interestingly, a histologically similar epizootic neoplasm occurs in freshwater drum (*Aplodinotus grunniens*), also a member of the family Sciaenidae, from eastern Lake Erie and from the Niagara River in the vicinity of Buffalo, New York and Fort Erie, Ontario (*11*). A viral etiology for the neoplasms occurring in the Japanese fish was eliminated from consideration on the basis of negative results from electron microscopy, *in vitro* iodo-2′-deoxyuridine (IUdR) uptake, and assays for reverse transcriptase activity. Al-

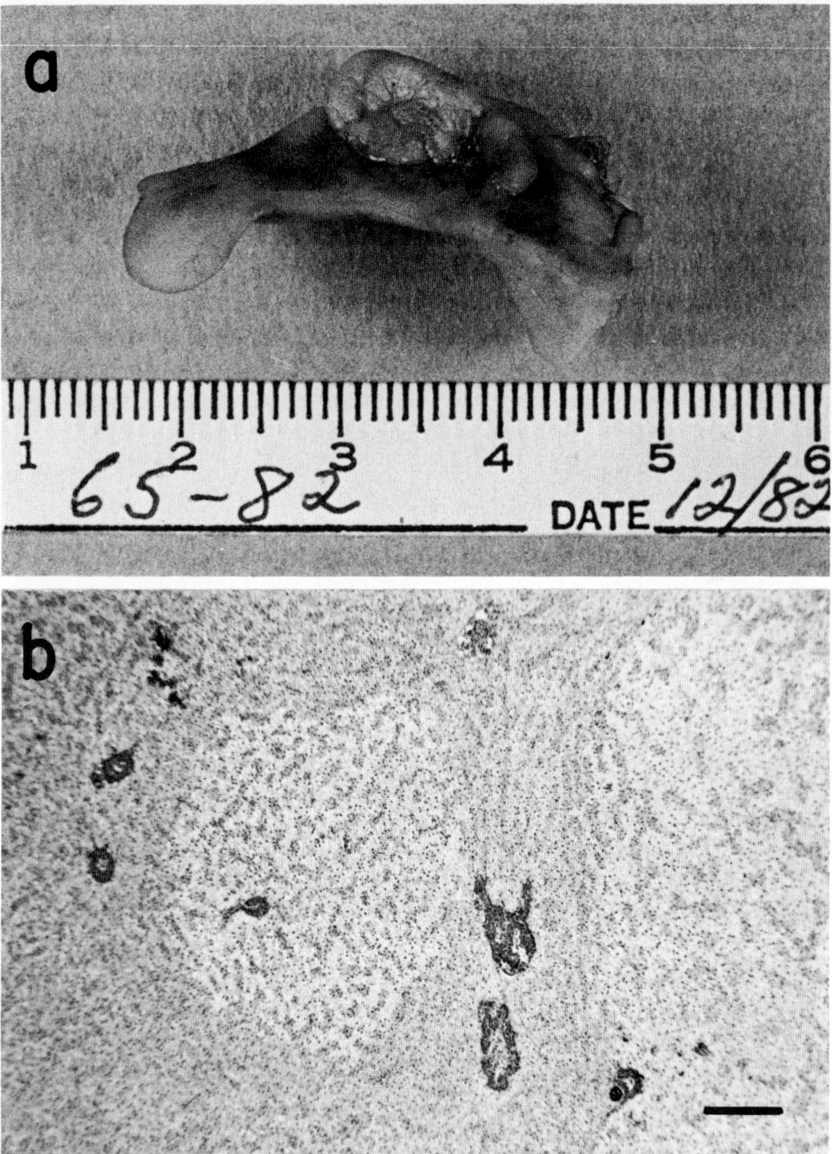

Fig. 12. Liver neoplasms in laboratory-adapted brown bullheads exposed to polluted river sediment extract. (a) Large solitary liver nodule. Microscopically, nodule was composed of well-differentiated eosinophilic hepatocytes with occasional areas composed of more basophilic hepatocytes. Nodule was noncapsulated and compression of surrounding liver parenchyma was evident. (b) Small, clear cell focus and large neoplastic nodule, also of the clear cell type. Note distortion and compression of normal liver structures by expanding nodule. Morphologically similar nodules in laboratory rodents are considered developmental stages in chemically induced hepatocellular carcinoma. H&E. Bar = 125 μm.

though the region is generally nonpolluted, several paper mills are present. Organic solvent extracts from river effluents and sediments and from livers of the nibe caught from the station exhibiting the highest frequency of nibe chromatophoroma development exhibited low but statistically significant mutagenic effects in two types of *in vitro* mutagenicity assays (*90, 91*). Subsequent *in vivo* testing of injected extracts of materials concentrated from industrial effluents and sediments collected near the industry (a paper mill) indicated enhanced development of pulmonary tumors by these substances in ICR mice.

To test the possibility that chromatophoroma development in the nibe may be caused by chemicals, the inducibility of these neoplasms was studied in captive populations of nibe (*88*). Fifty fish averaging 13 cm in length (5 months of age) were given single subcutaneous injections of 7,12-dimethylbenz[*a*]anthracene (DMBA) in olive oil or a single oral dose of *N*-methyl-*N'*-nitro-*N*-nitrosoguanidine (MNNG) approximating 100 mg/kg of body weight. A sham control group was administered a subcutaneous injection of olive oil and a single oral dose of saline. Approximately 300 other nibe were maintained as untreated controls. After 7 months the incidence of small melanotic lesions, thought to be precursory to chromatophoroma development, was 73% in the DMBA group, 59% in the MNNG group, and 23 and 13% in the sham and untreated groups, respectively. Subsequently, differences between the experimental and sham controls became smaller due to increases of the pigmented lesions in the controls. The experiment was repeated with similar results.

Furanase [6-hydroxymethyl-2[2-(5-nitro-2-furyl)vinyl]pyridine], a water-soluble antibacterial agent widely used in aquaculture, had been employed in the experiments as a prophylactic agent. To clarify a possible role of the furanase treatments in the production of the relatively high incidence of pigmented lesions in the control groups, additional lots of 5-month-old nibe (approximately 50 fish/group) were exposed twice daily for a period of 1 h to seawater containing 2, 1, 0.5, and 0 ppm of the drug for 7 days. Because the groups, including the 0-ppm group, had their water supply shut off during the experiment and sham treatments, a nontreated, negative control group was added, i.e., no cut-off in water supply and no drug given. Most of the fish in the 2-ppm group died within 40 days, apparently due to toxic effects of the drug. When the surviving groups were examined at 20 months, a dose-dependent response to the drug treatment, both in terms of the percentage of fish with lesions and the number of pigmented nevi per fish, was observed. Pigmented nevi were present in 38% of the fish receiving the 1 ppm treatment, in 29% of the 0.5-ppm treated fish, and in 6 and 0%, respectively, of the sham and untreated fish. The average number of lesions per fish went from 11.5 in the 1-ppm group to 4.0 in the 0.5-ppm group. The sham and untreated groups had an average of 0.8 and 0.0 nevi per fish, respectively. Kimura and colleagues interpreted these results to indicate that nibe had a high sensitivity to develop the pigmented nevi in response to chemicals, and these nevi were thought to be precursory to chromatophoroma development. In subsequent experiments, furanase demonstrated mutagenicity and carcinogenicity. When furanase was injected subcutaneously into ICR mice, it produced high incidences of pulmonary neoplasms in both sexes and hepatocellular neoplasms in male mice (*88*). When given by intubation, the drug produced forestomach papillomas in addition to pulmonary neoplasms. Currently, furanase is widely used as an antibacterial agent in veterinary medicine. It is possible that the nibe fish has fortuitously warned us of a human carcinogenic hazard as a result of serendipity in Kimura's experimentation.

VIII. CONCLUSION

Summarized data and detailed case histories of neoplasia in wild, free-living populations of aquatic animals have been presented and discussed. In aggregate, data strongly support the usefulness of aquatic animals as indicator organisms for waterborne carcinogens and thus in the early detection of a potential hazard for the human population.

There is considerable interest in the field of aquatic animal carcinogenesis in both the environmental and laboratory settings. A recent symposium on the "Use of Small Fish Species in Carcinogenicity Testing" (*186*) will undoubtedly stimulate further interest and lead to refinements in both *in vitro* and *in vivo* systems utilizing fish as models of carcinogenesis. Gaps in knowledge remain, especially in areas concerning neoplasia in wild populations. The difficulties in this problem cannot be surmounted by use of field epizootiological methods alone. Research is needed to demonstrate linkage of epizootic neoplasia in feral aquatic animals to causal factors and/or agents. Innovative new approaches utilizing both *in vitro* and *in vivo* systems and employing wild but laboratory-adapted target species (or surrogate species) may be helpful in establishing specific linkages.

The inherent difficulties imposed by a lack of unequivocal objective criteria upon which to base a diagnosis of neoplasia will remain an open question for aquatic pathologists and is one of many challenges faced in the study of aquatic animal neoplasia.

Acknowledgment

The author wishes to thank Dr. John Harshbarger, Registry for Tumors in Lower Animals, Smithsonian Institution, for helpful suggestions.

References

1. Alpers, C. E., McCain, B. B., Myers, M., Wellings, S. R., Poore, M., Bagshaw, J., and Dawe, C. J. (1977). *J. Natl. Cancer Inst.* **59,** 377–398.
2. Anderson, R. S., Doos, J. E., and Rose, F. L. (1982). *Cancer Lett.* **16,** 33–41.
3. "Annual Review, Great Lakes Fisheries Research Branch, Canada" (1981). Unpublished rep., Fisheries and Oceans, Canada, Burlington, Ontario.
4. Armstrong, R. W., and Sloan, R. J. (1980). "Trends in Levels of Several Known Chemical Contaminants in Fish from New York State Waters," Tech. Rep. 80–2. New York State Dept. of Environ. Conserv., Albany, New York.
5. Ashley, L. M. (1970). *In* "A Symposium on Diseases of Fishes and Shellfishes" (S. F. Snieszko, ed.), pp. 366–379. Am. Fish. Soc., Washington, D.C.
6. Baumann, P. C., Smith, W. D., and Ribick, M. (1982). *In* "Polynuclear Aromatic Hydrocarbons: Physical and Biological Fate" (M. Cooke, A. J. Dennis, and G. L. Fisher, eds.), pp. 93–102. Battelle, Columbus, Ohio.
7. Bend, J. R., James, M. O., and Dansette, P. M. (1977). *Ann. N.Y. Acad. Sci.* **298,** 505–521.
8. Bend, J. R., James, M. O., Little, P. J., and Foureman, G. L. (1981). *In* "Phyletic Approaches to Cancer" (C. J. Dawe, J. C. Harshbarger, S. Kondo, T. Sugimura, and S. Takayama, eds.), pp. 179–194. Japan Sci. Soc., Tokyo.
9. Berenblum, I., and Shubik, P. (1947). *Br. J. Cancer* **1,** 379–382.
10. Black, J. J. (1983). *In* "Polynuclear Aromatic Hydrocarbons: Formation, Metabolism and Measurement" (M. Cooke and A. J. Dennis, eds.), pp. 99–111. Battelle, Columbus, Ohio.
11. Black, J. J. (1983). *J. Great Lakes Res.* **9,** 326–334.
12. Black, J. J. (1983). Personal observation.

12a. Black, J. J. (1983). *Mar. Environ. Res.* (in press).

13. Black, J. J., Dymerski, P. P., and Zapisek, W. F. (1981). *In* "Aquatic Toxicology and Hazard Assessment" (D. R. Branson and K. L. Dickson, eds.), pp. 215–225. Am. Soc. Test. Mater., Philadelphia.
14. Black, J. J., Evans, E. D., Harshbarger, J. C., and Zeigel, R. F. (1982). *J. Natl. Cancer Inst.* **69,** 915–926.
15. Black, J. J., Hart, T. F., Jr., and Black, P. J. (1982). *Environ. Sci. Technol.* **16,** 247–250.
16. Black, J. J., Hart, T. F., Jr., and Evans, E. (1980). *In* "Polynuclear Aromatic Hydrocarbons: Chemical Analysis and Biological Fate" (M. Cooke and A. J. Dennis, eds.), pp. 343–356. Battelle, Columbus, Ohio.
17. Black, J. J., Holmes, M., Dymerski, P. P., and Zapisek, W. F. (1980). *In* "Hydrocarbons and Halogenated Hydrocarbons in the Aquatic Environment" (B. K. Afghan and D. MacKay, eds.), pp. 559–566. Plenum, New York.
18. Black, J. J., and Simpson, C. L. (1974). *J. Natl. Cancer Inst.* **53,** 725–729.
19. Blasiola, G. C., Turnier, J. C., and Hurst, E. E. (1981). *J. Natl. Cancer Inst.* **66,** 51–59.
20. Boutwell, R. K. (1964). *Prog. Exp. Tumor Res.* **4,** 207–250.
21. Boyd, J. T., Doll, R., Faulds, J. S., and Leiper, J. (1970). *Br. J. Ind. Med.* **27,** 97–105.
22. Brooks, R. E., McArn, G. E., and Wellings, S. R. (1969). *J. Natl. Cancer Inst.* **43,** 97–100.
23. Bross, I. D. J., and Coombs, J. (1976). *Oncology* **33,** 136–139.
24. Brown, E. R., Hazdra, J. J., Keith, L., Greenspan, I., and Kwapinski, J. B. G. (1973). *Cancer Res.* **33,** 189–198.
25. Brown, R. A., and Weiss, F. T. (1978). "Fate and Effects of Polynuclear Aromatic Hydrocarbons in the Aquatic Environment, " Publ. No. 4297. Am. Petroleum Inst., Washington, D.C.
26. Brown, R. S. (1980). *Rapp. P—V. Reun., Cons. Int. Explor. Mer.* **179,** 125–128.
27. Brown, R. S., Wolke, R. E., Brown, C. W., and Saila, S. B. (1979). *In* "Animals as Monitors of Environmental Pollutants," pp. 41–51. Natl. Acad. Sci., Washington, D.C.

28. Brown, R. S., Wolke, R. E., Saila, S. B., and Brown, C. W. (1977). *Ann. N.Y. Acad. Sci.* **298,** 522–534.
29. Bucke, D. (1972). *Symp. Zool. Soc. London* **30,** 153–190.
30. Cairns, J. (1981). *Nature (London)* **289,** 353–357.
31. Cantor, K. P., Hoover, R., Mason, T. J., and McCabe, L. J. (1978). *J. Natl. Cancer Inst.* **61,** 979–985.
32. Casto, B. C., and DiPaolo, J. A. (1973). *Prog. Med. Virol.* **16,** 1–47.
33. Chang, M. J., Singh, N. P., and Hart, R. W. (1982). *Proc. Am. Assoc. Cancer Res.* **23,** 199.
34. Collier, T. K., Krahn, M. M., and Malins, D. C. (1980). *Environ. Res.* **23,** 35–41.
35. Cook, J. W., Hewett, C. L., and Hieger, I. (1933). *J. Chem. Soc.*, Pt. 1, 395–405.
36. Cooper, K. R., Brown, R. S., and Chang, P. W. (1982). *J. Invertebr. Pathol.* **38,** 149–157.
37. Davenport, H. A. (1960). "Histological and Histochemical Techniques." Saunders, Philadelphia.
38. David, A., Hurych, J., Effenbergerova, J., Holusa, R., and Simcek, J. (1981). *Environ. Res.* **24,** 140–151.
39. Dawe, C. J. (1973). *In* "Cancer Medicine" (J. Holland and E. Frei, eds.), pp. 193–240. Lea & Febiger, Philadelphia.
40. Dawe, C. J. (1981). *In* "Phyletic Approaches to Cancer" (C. J. Dawe, J. C. Harshbarger, S. Kondo, T. Sugimura, and S. Takayama, eds.), pp. 19–49. Japan Sci. Soc., Tokyo.
41. Dawe, C. J., and Harshbarger, J. C. (1975). *In* "Pathology of Fishes" (W. E. Ribelin and G. Migaki, eds.), pp. 871–894. Univ. of Wisconsin Press, Madison.
42. Dawe, C. J., Stanton, M. F., and Schwartz, F. J. (1964). *Cancer Res.* **24,** 1194–1201.
43. Diachenko, G. W. (1979). *Environ. Sci. Technol.* **13,** 329–333.
44. Doll, R., and Peto, R. (1981). *J. Natl. Cancer Inst.* **66,** 1193–1308.
45. Dunn, B. P. (1979). *In* "Polynuclear Aromatic Hydrocarbons: Chemistry and Biological Effects" (A. Bjorseth and A. J. Dennis, eds.), pp. 367–378. Battelle, Columbus, Ohio.
46. Epstein, S. S. (1974). *Cancer Res.* **34,** 2425–2435.
47. Estenik, J. F., and Collins, W. J. (1979). *In* "Pesticide and Xenobiotic Metabolism in Aquatic Organisms" (M. A. Q. Khan, J. J. Lech, and J. J. Menn, eds.), pp. 349–370. Am. Chem. Soc., Washington, D.C.
48. Falkmer, S., Emdin, S. O., Ostberg, Y., Mattisson, A., Johansson-Sjobech, M.-L., and Fange, R. (1976). *Prog. Exp. Tumor Res.* **20,** 217–250.
49. Falkmer, S., Marklund, S., Mattisson, P. E., and Rappe, C. (1977). *Ann. N.Y. Acad. Sci.* **298,** 342–355.
50. Farber, E., and Cameron, R. (1980). *Adv. Cancer Res.* **31,** 125–226.
51. Farley, C. A. (1969). *J. Natl. Cancer Inst.* **43,** 509–516.
52. Farley, C. A. (1969). *Natl. Cancer Inst. Monogr.* **31,** 541–555.
53. Fawcett, D. W. (1956). *J. Cell Sci.* **2,** 725–741.
54. Frierman, E. M., and Andrews, J. D. (1976). *J. Natl. Cancer Inst.* **56,** 319–324.
55. Garrison, A. W. (1977). *Ann. N.Y. Acad. Sci.* **298,** 2–19.
56. Gaylord, H. R., and Marsh, M. C. (1914). "Carcinoma of the Thyroid in Salmonid Fishes." U.S. Govt. Printing Office, Washington, D.C.
57. Germani, M. S., Small, M., and Zoller, W. H. (1981). *Environ. Sci. Technol.* **15,** 299–305.
58. Gmur, D. J., and Varanasi, U. (1982). *Carcinogenesis* **12,** 1397–1403.
59. Goldfarb, M., Shimizu, K., Perucho, M., and Wigler, M. (1982). *Nature (London)* **296,** 404–409.
60. Grizzle, J. M., and Melius, P. (1984). "Causes of Papillomas on Fish Living in Chlorinated Sewage Effluent," Proj. Rep. U.S. Environ. Prot. Agency, Gulf Breeze, Florida (in press).
61. Grizzle, J. M., Schwedler, T. E., and Scott, A. L. (1981). *J. Fish Dis.* **4,** 345–351.
62. Gurkalo, V. K., and Volfson, N. I. (1982). *Arch. Geschwulstforsch.* **52,** 259–265.

63. Halver, J. E. (1967). *In* "Trout Hepatoma Research Conference Papers" (J. E. Halver and I. A. Mitchell, eds.), pp. 78–102, Res. Rep. 70. Bureau Fisheries and Wildlife, Washington, D.C.
63a. Hamelink, J. L., and Spacie, A. (1977). *Annu. Rev. Pharmacol. Toxicol.* **17,** 167–177.
64. Hammond, E. C. (1975). *Cancer* **35,** 652–654.
65. Haran-Ghera, N. (1966). *Int. J. Cancer* **1,** 81–87.
66. Harshbarger, J. C. (1977). *Ann. N.Y. Acad. Sci.* **298,** 280–289.
67. Harshbarger, J. C. (1981). "Activities Report Registry of Tumors in Lower Animals, 1981 Supplement." Random House (Smithsonian Inst. Press), New York.
68. Harshbarger, J. C. (1982). *In* "Advances in Comparative Leukemia Research 1981" (D. S. Yohn and J. R. Blakeslee, eds.), pp. 39–46. Elsevier North-Holland, Amsterdam.
69. Harshbarger, J. C. (1984). *Natl. Cancer Inst. Monogr.* (in press).
70. Harshbarger, J. C., Otto, S. V., and Chen Chang, S. (1979). *Haliotis* **8,** 243–248.
71. Hart, R. W., Hays, S., Brash, D., Daniel, F. B., Davis, M. T., and Lewis, N. J. (1977). *Ann. N.Y. Acad. Sci.* **298,** 141–158.
72. Hendricks, J. D. (1982). *In* "Aquatic Toxicology" (L. J. Weber, ed.), Vol. 1, pp. 149–211. Raven, New York.
73. Hendricks, J. D., Scanlan, R. A., Williams, J. L., Sinnhuber, R. O., and Grieco, M. P. (1980). *J. Natl. Cancer Inst.* **64,** 1511–1519.
74. Hendricks, J. D., Stott, W. T., Putnam, T. P., and Sinnhuber, R. O. (1980). *In* "Aquatic Toxicology and Hazard Assessment" (D. R. Branson and K. L. Dickson, eds.), pp. 203–214. Am. Soc. Test. Mater., Philadelphia.
75. Hetling, L. J., Tofflemire, T. J., Horn, E. G., Thomas, R., and Mt. Pleasant, R. (1979). *Ann. N.Y. Acad. Sci.* **320,** 630–650.
76. Higginson, J. (1969). *Proc. Can. Cancer Conf.* **8,** 40–75.
77. Hinton, D. E., Klaunig, J. E., Jack, R. M., Lipsky, M. M., and Trump, B. F. (1981). *In* "Aquatic Toxicology and Hazard Assessment" (D. R. Branson and K. L. Dickson, eds.), pp. 226–246. Am. Soc. Test. Mater., Philadelphia.
78. Iacobelli, S., King, R. J. B., Lindner, H. R., and Lippman, M. E., eds. (1980). "Hormones and Cancer, " Vol. 14. Raven, New York.
79. Horn, E. G., Hetling, L. J., and Tofflemire, T. J. (1979). *Ann. N.Y. Acad. Sci.* **320,** 591–609.
80. Huang, C. C., and Furukawa, M. (1978). *Exp. Cell Res.* **111,** 458–461.
81. Heubner, R. J., and Todaro, G. J. (1969). *Proc. Natl. Acad. Sci. U.S.A.* **64,** 1087–1094.
82. IARC (1973). "IARC Monograph: Evaluation of Carcinogenic Risk of the Chemical to Man," Vol. 3. Int. Agency Res. Cancer, Lyon, France.
83. Igel, H. J., Heubner, R. J., Turner, H. C., Kotin, O., and Falk, H. L. (1969). *Science* **166,** 1624–1626.
84. Jungclaus, G. A., Lopez-Avila, V., and Hites, R. A. (1978). *Environ. Sci. Technol.* **12,** 88–96.
85. Khudoley, V. V., and Syrenko, O. A. (1978). *Cancer Lett.* **4,** 349–354.
86. Kimbrough, R. D., Squire, R. A., Linder, R. E., Strandberg, J. D., Montali, R. J., and Burse, V. W. (1975). *J. Natl. Cancer Inst.* **55,** 1453–1459.
87. Kimura, I., Kitaori, H., Yoshizaki, K., Tayama, K., Ito, M., and Yamada, S. (1981). *In* "Phyletic Approaches to Cancer" (C. J. Dawe, J. C. Harshbarger, S. Kondo, T. Sugimura, and S. Takayama, eds.), pp. 241–252. Japan Sci. Soc., Tokyo.
88. Kimura, I., Taniguchi, Y., Kumai, H., Nakamura, M., Tomita, I., Kinae, N., Kitaori, H., Yoshizaki, K., Ito, M., Yamada, S., Kubota, S., Miyazaki, T., Miyake, T., and Nagayo, T. (1984). *Natl. Cancer Inst. Monogr.* (in press).
89. Kimura, T., Yoshimizu, M., and Tanaka, M. (1981). *In* "Phyletic Approaches to Cancer" (C. J. Dawe, J. C. Harshbarger, S. Kondo, T. Sugimura, and S. Takayama, eds.), pp. 59–68. Japan Sci. Soc., Tokyo.

90. Kinae, N., Hashizume, T., Makita, T., Tomita, I., Kimura, I., and Kanamori, H. (1981). *Water Res.* **15,** 17–24.
91. Kinae, N., Hashizume, T., Makita, T., Tomita, I., Kimura, I., and Kanamori, H. (1981). *Water Res.* **15,** 25–30.
92. Klauda, R. J., Peck, T. H., and Rice, G. K. (1981). *Bull. Environ. Contam. Toxicol.* **27,** 829–835.
93. Kligerman, M. D., Bloom, S. E., and Howell, W. M. (1975). *Mutat. Res.* **31,** 225–233.
94. Krahn, M. M., Schnell, J. V., Uyeda, M. Y., and MacLeod, W. D., Jr. (1981). *Anal. Biochem.* **113,** 27–33.
95. Kraybill, H. F. (1976). *Prog. Exp. Tumor Res.* **20,** 3–34.
96. Kraybill, H. F. (1977). *Ann. N.Y. Acad. Sci.* **298,** 80–89.
97. Lange, E. (1973). *Comp. Biochem. Physiol.* **45A,** 477–481.
98. Lange, E., and Johannessen, J. V. (1977). *Lab. Invest.* **37,** 96–104.
99. Lech, J. J., Vodinik, M. J., and Elcombe, C. R. (1982). *In* "Aquatic Toxicology" (L. J. Weber, ed.), Vol. 1, pp. 107–148. Raven, New York.
100. Lee, R. F., Ryan, C., and Neuhauser, M. L. (1976). *Mar. Biol.* **37,** 363–370.
101. Ljungberg, O. (1976). *Prog. Exp. Tumor Res.* **20,** 156–165.
102. Lopez-Avila, V., and Hites, R. A. (1980). *Environ. Sci. Technol.* **14,** 1382–1390.
103. Lunger, P. O. (1964). *J. Virol.* **28,** 318–324.
104. Malins, D. C. (1977). *Ann. N.Y. Acad. Sci.* **298,** 482–496.
105. Malins, D. C., McCain, B. B., Brown, D. W., Sparks, A. K., and Hodgins, H. O. (1980). "NOAA Technical Memorandum OMPA-2, Chemical Contaminants and Biological Abnormalities in Central and Southern Puget Sound," MESA Rep. Micro EcoSystem Analysis, Boulder, Colorado.
106. Malins, D. C., McCain, B. B., Brown, D. W., Sparks, A. K., Hodgins, H. O., and Chan, S.-L. (1984). "NOAA Technical Memorandum, Chemical Contaminants and Abnormalities in Fish and Invertebrates from Puget Sound," MESA Rep. Micro EcoSystem Analysis, Boulder, Colorado (in press).
107. Marine, D., and Lenhart, C. H. (1910). *J. Exp. Med.* **12,** 311–337.
108. Marine, D., and Lenhart, C. H. (1911). *J. Exp. Med.* **13,** 455–475.
109. Marine, D., and Lenhart, C. H. (1914). *J. Exp. Med.* **19,** 70–88.
110. Marx, J. L. (1983). *Science* **219,** 158–159.
111. Mason, T. J., and MacKay, F. W. (1975). "U.S. Cancer Mortality by County, 1950–1969," DHEW Publ. No. (NIH) 74–615. Natl. Cancer Inst., Bethesda, Maryland.
112. Matsushima, T., and Sugimura, T. (1976). *Prog. Exp. Tumor Res.* **20,** 367–379.
113. Mawdesley-Thomas, L. E. (1972). *Symp. Zool. Soc. London* **30,** 191–283.
114. Mawdesley-Thomas, L. E. (1975). *In* "Pathology of Fishes" (W. E. Ribelin and G. Migaki, eds.), pp. 805–870. Univ. of Wisconsin Press, Madison.
115. McAllister, P. E., Nagabayash, T., and Wolf, K. (1977). *Ann. N.Y. Acad. Sci.* **298,** 233–244.
116. McCain, B. B., Pierce, K. V., Wellings, S. R., and Miller, B. S. (1977). *Bull. Environ. Contam. Toxicol.* **18,** 1–2.
117. McKinnell, R. G. (1973). *Am. J. Zool.* **13,** 97–114.
118. McKinnell, R. G. (1981). *In* "Phyletic Approaches to Cancer" (C. J. Dawe, J. C. Harshbarger, S. Kondo, T. Sugimura, and S. Takayama, eds.), pp. 101–110. Japan Sci. Soc., Tokyo.
119. Michigan Dept. of Nat. Resources (1983). Unpublished rep., Lansing, Michigan.
120. Melick, W. F., Naryka, J. J., and Kelley, R. E. (1971). *J. Urol.* **106,** 220–226.
121. Miller, E. C. (1978). *Cancer Res.* **38,** 1479–1496.
122. Miller, J. A. (1970). *Cancer Res.* **30,** 559–576.
123. Mix, M. C. (1975). *J. Invertebr. Pathol.* **26,** 289–298.
124. Mix, M. C. (1976). *Prog. Exp. Tumor Res.* **20,** 275–282.

125. Mix, M. C., Hawkes, J. W., and Sparks, A. K. (1979). *J. Invertebr. Pathol.* **34,** 41–56.
126. Mix, M. C., Pribble, J., Riley, R. T., and Tomasovic, S. P. (1977). *Ann. N.Y. Acad. Sci.* **298,** 356–373.
127. Mix, M. C., and Schaffer, R. L. (1979). *Bull. Environ. Contam. Toxicol.* **23,** 677–684.
128. Mix, M. C., Trenholm, S. R., and King, K. I. (1979). *In* "Animals as Monitors of Environmental Pollutants, " pp. 52–62. Natl. Acad. Sci., Washington, D.C.
128a. Moccia, R. D., Leatherland, J. F., and Sonstegard, R. A. (1977). *Science* **198,** 425–426.
129. Mondal, S., Brankow, D. W., and Heidelberger, C. (1976). *Cancer Res.* **36,** 2254–2260.
130. Mulcahy, M. F. (1976). *Prog. Exp. Tumor Res.* **20,** 129–140.
131. Mulcahy, M. F., and O'Leary, A. (1970). *Experientia* **26,** 891.
132. Neff, J. M. (1979). "Polycyclic Aromatic Hydrocarbons in the Aquatic Environment—Sources, Fates and Biological Effects." Applied Science, Essex, England.
133. Nelson, C. R., and Hites, R. A. (1980). *Environ. Sci. Technol.* **14,** 1147–1149.
134. O'Conner, T. E. (1984). "Oncogenes and Retroviruses: Evaluation of Basic Findings and Clinical Potential." Liss, New York (in press).
135. Omura, T., Marmura, S., and Noshiro, M. (1981). *In* "Phyletic Approaches to Cancer" (C. J. Dawe, J. C. Harshbarger, S. Kondo, T. Sugimura, and S. Takayama, eds.), pp. 195–202. Japan Sci. Soc., Tokyo.
136. Oprandy, J. J., Chang, P. W., Pronovost, A. D., Cooper, K. R., Brown, R. S., and Yates, V. J. (1981). *J. Invertebr. Pathol.* **38,** 45–51.
137. Page, W. G. (1981). *Environ. Sci. Technol.* **15,** 1475–1481.
137a. Page, T., Harris, R. H., and Epstein, S. S. (1976). *Science* **193,** 55–57.
138. Papas, T. S., Pry, T. W., Schafer, M. P., and Sonstegard, R. A. (1977). *Cancer Res.* **37,** 3214–3217.
139. Parry, J. M., Kadhim, M., Barnes, W., and Danford, N. (1981). *In* "Phyletic Approaches to Cancer" (C. J. Dawe, J. C. Harshbarger, S. Kondo, T. Sugimura, and S. Takayama, eds.), pp. 141–166. Japan Sci. Soc., Tokyo.
140. Parry, J. M., Tweats, D. J., and Al-Mossawi, M. A. J. (1976). *Nature (London)* **264,** 538–540.
141. Pauley, G. B. (1969). *Natl. Cancer Inst. Monogr.* **31,** 509–539.
142. Payne, J. F., Martins, I., and Rahimtula, A. (1978). *Science* **200,** 329–330.
143. Peters, G., and Peters, N. (1977). *Ann. N.Y. Acad. Sci.* **298,** 245–260.
144. Peters, N., Stich, H. F., and Kranz, H. (1981). *In* "Phyletic Approaches to Cancer" (C. J. Dawe, J. C. Harshbarger, S. Kondo, T. Sugimura, and S. Takayama, eds.), pp. 111–121. Japan Sci. Soc., Tokyo.
144a. Phillips, M. L., Warner, N. E., and Puffer, H. W. (1976). *Prog. Exp. Tumor Res.* **20,** 108–112.
145. Pierce, K. V., McCain, B. B., and Wellings, S. R. (1978). *J. Natl. Cancer Inst.* **60,** 1445–1453.
146. Pliss, G. B., and Khudoley, V. V. (1975). *J. Natl. Cancer Inst.* **55,** 129–136.
147. Pott, P. (1963). Printed in *Natl. Cancer Inst. Monogr.* **10,** 7–13.
148. LaVia, M. F., and Hill, R. B., Jr., eds. (1971). "Principles of Pathobiology." Oxford Univ. Press, London and New York.
149. Rose, F. L. (1976). *Prog. Exp. Tumor Res.* **20,** 251–262.
150. Rose, F. L. (1977). *Science* **198,** 1280.
151. Rose, F. L. (1977). *Ann. N.Y. Acad. Sci.* **298,** 270–279.
152. Rose, F. L. (1981). *In* "Phyletic Approaches to Cancer" (C. J. Dawe, J. C. Harshbarger, S. Kondo, T. Sugimura, and S. Takayama, eds.), pp. 91–100. Japan Sci. Soc., Tokyo.
153. Rose, F. L., and Harshbarger, J. C. (1977). *Science* **196,** 315–317.
154. Rothman, K., and Keller, A. (1972). *J. Chronic Dis.* **25,** 711–716.
155. Rucker, R. R., Yasutake, W. T., and Wolf, H. (1961). *Prog. Fish-Cult.* **23,** 3–7.

156. Schlumberger, H. G., and Lucke, B. (1948). *Cancer Res.* **8,** 657–754.
157. Schultz, M. E., and Schultz, R. J. (1982). *Environ. Res.* **27,** 337–351.
158. Schwanz-Pfitzner, I. (1976). *Prog. Exp. Tumor Res.* **20,** 101–107.
159. Selikoff, I. J., Hammond, E. C., and Churg, J. (1968). *J. Am. Med. Assoc.* **204,** 104–112.
160. Selikoff, I., Churg, J., and Hammond, E. C. (1965). *N. Engl. J. Med.* **272,** 560–565.
161. Shih, C., Shilo, B.-Z., Goldfarb, M. P., Dannenberg, A., and Weinburg, R. A. (1979). *Proc. Natl. Acad. Sci. U.S.A.* **76,** 5714–5718.
162. Shin, M. L., and Firminger, H. I. (1973). *Am. J. Pathol.* **70,** 291–313.
163. Sinderman, C. J., Bang, F. B., Christensen, N. O., Dethlefsen, V., Harshbarger, J. C., Mitchell, J. R., and Mulcahy, M. F. (1980). *Rapp. P.—V. Reun., Cons. Int. Explor. Mer.* **179,** 135–151.
164. Sinnhuber, R. O., Hendricks, J. D., Wales, J. H., and Putnam, G. B. (1977). *Ann. N.Y. Acad. Sci.* **298,** 389–407.
165. Sinnhuber, R. O., Lee, D. J., Wales, J. H., Landers, M. K., and Keyl, A. C. (1974). *J. Natl. Cancer Inst.* **53,** 1285–1288.
166. Sirianni, S., and Huang, C. C. (1978). *Proc. Soc. Exp. Biol. Med.* **158,** 269–274.
167. Sirota, G. R., and Uthe, J. F. (1981). *In* "Polynuclear Aromatic Hydrocarbons: Chemical Analysis and Biological Fate" (M. Cooke and A. J. Dennis, eds.), 329–341. Battelle, Columbus, Ohio.
168. Sirota, G. R. (1983). *In* "Polynuclear Aromatic Hydrocarbons: Formation, Metabolism and Measurement" (M. Cooke and A. J. Dennis, eds.), pp. 1123–1136. Batelle, Columbus, Ohio.
169. Smith, C. E., Peck, T. H., Klauda, R. J., and McLaren, J. B. (1979). *J. Fish Dis.* **2,** 313–319.
170. Sonstegard, R. A. (1975). *In* "Virological and Epizootiological Studies of Fish Neoplasms in Polluted and Non-polluted Waters of the Great Lakes" (M. S. Mahdy and R. J. Dutka, eds.). Unpublished rep., Environment Canada.
171. Sonstegard, R. A. (1976). *Prog. Exp. Tumor Res.* **20,** 141–155.
172. Sonstegard, R. A. (1977). *Ann. N.Y. Acad. Sci.* **298,** 261–269.
173. Sonstegard, R. A. (1980). *In* "Advances in Comparative Leukemia Research 1979" (D. S. Yohn, B. A. Lapin, and J. R. Blakeslee, eds.), pp. 227. Elsevier North-Holland, Amsterdam.
174. Sonstegard, R. A., and Leatherland, J. F. (1976). *Cancer Res.* **36,** 4467–4475.
175. Sonstegard, R. A., and Leatherland, J. F. (1983). "Handbook for Identification of Tumors in Great Lakes Fishes." Tech. Rep., Canadian Fish. and Aquatic Sci., Ottawa (in press).
175a. Squire, R. A., and Levitt, M. H. (1975). *Cancer Res.* **35,** 3214–3233.
176. Stewart, H. L. (1977). *Ann. N.Y. Acad. Sci.* **298,** 305–315.
177. Stich, H. F., and Acton, A. B. (1976). *Prog. Exp. Tumor Res.* **20,** 44–54.
178. Stich, H. F., Acton, A. B., and Forrester, C. R. (1976). *J. Fish. Res. Board Can.* **33,** 1993–2001.
179. Stich, H. F., Acton, A. B., Dunn, B. P., Oishi, J., Yamazaki, F., Harada, T., Peters, G., and Peters, N. (1977). *Int. J. Cancer* **20,** 780–791.
180. Stinson, S. F., Reznik, G., and Levitt, M. H. (1982). *Proc. Am. Assoc. Cancer Res.* **23,** 225.
181. Tan, B., Melius, P., and Grizzle, J. (1981). *In* "Polynuclear Aromatic Hydrocarbons: Chemical Analysis and Biological Fate" (M. Cooke and A. J. Dennis, eds.), pp. 377–386. Battelle, Columbus, Ohio.
182. Tomatis, L. (1979). *Annu. Rev. Pharmacol. Toxicol.* **19,** 511–530.
183. Tomljanovich, D. A. (1974). "Growth Phenomena and Abnormalties of the Sauger, *Stizostedion canadense* (Smith), of the Keweenaw Waterway." M. S. Thesis, Michigan Technological Institute, Houghton, Michigan.
184. Toth, B. (1963). *Proc. Soc. Exp. Biol. Med.* **112,** 873–875.
185. "Toxaphene Federal Register Notice" (1982). *Fed. Regist.* **47**(229), 53784–53793.

185a. Trosco, J. E., Yotti, L. P., Warren, S. T., Tsushimoto, G., and Chang, C. (1982). *In* "Carcinogenesis" (E. Hecker, N. E. Fusenig, W. Kunz, F. Marks, and H. W. Thielmann, eds.), Vol. 7, pp. 565–585. Raven, New York.
186. "Use of Small Fish Species in Carcinogenicity Testing" (1983). *Natl. Cancer Inst. Monogr.* (in press).
187. Vassilaros, D. L., Stoker, P. W., Booth, G. M., and Lee, M. L. (1982). *Anal. Chem.* **54,** 106–112.
187a. Wales, J. H. (1970). *In* "A Symposium on Diseases of Fishes and Shellfishes" (S. F. Snieszko, ed.), pp. 351–365. Am. Fish. Soc., Washington, D.C.
188. Wales, J. H., Sinnhuber, R. O., Hendricks, J. D., Nixon, J. E., and Eisele, T. A. (1978). *J. Natl. Cancer Inst.* **60,** 1133–1139.
189. Walker, R. (1969). *Natl. Cancer Inst. Monogr.* **31,** 195–207.
190. Walpole, A. L., and Williams, M. H. C. (1958). *Br. Med. Bull.* **14,** 141–145.
191. Weinstein, I. B., Mufson, R. A., Lee, L.-S., Fisher, P. B., Laskin, J., Horowitz, A. D., and Ivanovic, V. (1980). *In* "Carcinogenesis: Fundamental Mechanisms and Environmental Effects" (B. Pullman, P. O. P. Ts's, and H. Gelboin, eds.), pp. 543–563. Reidel Publ., Amsterdam.
192. Weisberger, J. H., and Williams, G. M. (1975). *In* "Cancer—A Comprehensive Treatise" (F. F. Becker, ed.), Vol. 1, pp. 185–234. Plenum, New York.
193. Wellings, S. R., McCain, B. B., and Miller, B. S. (1976). *Prog. Exp. Tumor Res.* **20,** 55–74.
194. Wellings, S. R. (1969). *Natl. Cancer Inst. Monogr.* **31,** 59–128.
195. Willis, R. A. (1976). "Pathology of Tumours, " 4th Ed. Butterworth, London.
196. Wolf, P. H. (1969). *Natl. Cancer Inst. Monogr.* **31,** 563–573.
197. Yamagiwa, K., and Ichikawa, K. (1915). *Mit. Med. Fak. Tokio* **15,** 295–344.
198. Yevich, P. P., and Barszcz, C. A. (1977). *Ann. N.Y. Acad. Sci.* **298,** 409–426.

Behavioral Effects of Industrial Chemicals on Aquatic Animals

G. F. Westlake

Environment Canada
Dartmouth, Nova Scotia, Canada

I. Introduction ... 233
II. Variability in Behavior of Aquatic Animals ... 235
III. Preference or Avoidance of Industrial Chemicals by Aquatic Animals ... 237
IV. Stress Behavior for Biological Monitoring ... 241
V. Effects of Industrial Chemicals on Other Types of Animal Behavior ... 243
VI. Conclusions ... 244
References ... 244

I. INTRODUCTION

Behavioral changes in aquatic animals induced by the liquid wastes of industry are as much a threat to survival of their populations as are the outright lethal effects of these effluents. Sublethal effects including behavioral changes occur at concentrations that are commonly found in nature yet have been less thoroughly studied. During the last 15 to 20 years, efforts to improve environmental quality have required pragmatic approaches employing tests for acute lethality as the primary criterion of control. However, it is the sublethal effects that are most important in nature. In fact, fish kills, which are the most obvious environmental analog of tests for acute lethality, are relatively rare events caused by animals being overwhelmed by sudden changes in conditions. Kleerekoper (*96*) points out that acutely lethal levels are most likely restricted to areas near the point of discharge so that most animals are exposed only to sublethal levels. More commonly, a gradual deterioration in water quality causes adjustments in populations by sublethal mechanisms. This article deals with behavioral effects, which constitute an important group of the sublethal consequences of industrial wastewater discharge.

HAZARD ASSESSMENT OF CHEMICALS:
Current Developments, Vol. 3

ISBN 0-12-312403-4

Most people have an intuitive understanding of the importance of the effects of industrial effluents on the behavior of aquatic animals. If, for example, a migratory fish can take the left side of a river when the right side is contaminated by a toxic effluent, then its chances for survival increase. Less obvious are possible effects of chemicals or effluents on predator/prey relationships, on ability to swim against a current, or on responses to environmental stimuli. All of these and many other related effects, though difficult to assess, can greatly affect survival.

Since behavior is the expression of so many bodily functions, it stands to reason that it is readily affected by changes in environmental conditions. It is furthermore clear from the available literature, some of which will be discussed below, that any form of behavior may be affected.

The complexity of behavior makes it particularly difficult to study. The reported literature, which by necessity consists primarily of accounts of behavior in tubes, troughs, and tanks, is difficult to interpret in terms of the distribution of populations of animals in nature. Comparisons between responses in large vs small tanks, between responses to shallow vs steep gradients, and between the activity of animals in the field vs in the lab, in which results can appear to be in completely opposite directions, all point to the fact that changes in behavior produced by substances are complex and not yet well understood.

The variability of the behavioral response has lead Scherer (*144*) to suggest that uniform tests or protocols be developed for the study of the effects of industrial chemicals on behavior. On the one hand, the variety of methods that are being applied to the problem presents us with opportunities to discover the underlying mechanisms of behavioral effects. Most discoveries come from new approaches to problems or from chance observations, not from routine repetition of established techniques. On the other hand, it is important that enough replication take place so that the true result of these studies can be distinguished from high natural variation in behavior and from differences between labs. It is one goal of this article to give an impression for how variable both behavior and the approaches for studying it are.

In order to present some concepts that I feel are important, examples are taken primarily from my experience in sensory physiology and behavior of fish and the design and testing of automated biological monitoring devices. The areas in which deficiencies are most likely to occur involve a growing invertebrate literature and effects of chemicals on feeding and other related forms of behavior. An attempt has been made to include these subjects, but the reader is warned that proper discussion of these effects may be found elsewhere. See, for example, reviews by Olla *et al.* (*128, 130*), Olla and Studholme (*129*), Sutterlin (*161*), and Sprague (*154*).

The surroundings of aquatic animals are drastically different from our own. The greater density of the medium makes for entirely different locomotor mechanisms. Due to the relatively poor light transmission, it is also a medium in which the chemical senses predominate. Much of the study of behavior of these animals, therefore, involves study of their preference or avoidance of chemicals.

It is of some value in this discussion to distinguish between observed behavioral responses to chemicals that are due to olfactory or other chemical senses and those that are premortal stress responses. It may not always be possible to determine which is taking place, however, generally, stress occurs at orders-of-magnitude-higher concentrations.

Because of a need to develop automatic monitoring techniques for complex industrial effluents, premortal behavioral signs of stress in animals exposed to discharges or sensitive water supplies have been proposed as viable mechanisms for pollution detection. Indeed, several such devices have been developed and successfully tested.

II. VARIABILITY IN BEHAVIOR OF AQUATIC ANIMALS

Anyone who has tried to study animal behavior will agree that these responses are highly variable. Even under constant conditions one can expect large differences among individuals as well as temporal variations in response related to diurnal periodicity and exploratory behavior.

The need to compensate for this extreme variability by increasing sample sizes necessitates the use of automatic recording and computer analysis. It is necessary only to add up the time required for a small behavioral experiment to see that manual methods have severe practical limitations. For example, a series of trials to test one compound at five concentrations with a modest 10 animals for each and an observation period of 2 h each (i.e., 1 h control and 1 h exposure) would require a total of 100 h of observation and probably 100 separate days due to the need for acclimation periods and due to fatigue of the observer. Then double or even triple this time for analysis and we are faced with a 2- or 3-month experiment, all for a single simple exposure series.

At one time the "unfriendliness" of computers and their great cost put consideration of automatic methods out of reach of all but the most fortunate or persistent of researchers. This barrier is no longer present. Small and inexpensive computers have capabilities that are far greater than the large machines that were used in the lab only 10 years ago. Equipment is now available that is friendlier and cheaper, and one no longer needs to be an electronics engineer or a programmer to use it. A small home computer, for example, can be interfaced to four photocells or four thermistors using only one resistor each and just a few wires (*13*). This quick interfacing job produces a system that is capable of watching a fish, and is much more stable than one designed totally from hardware because it can automatically adjust to ambient conditions. Although this is an inefficient use of a computer, it can be a viable use for computers with considerable periods of inactivity.

Variability in behavior is most apparent from studies in which it has been possible to automatically record animals continuously for long periods of time. For example, Kleerekoper *et al.* (*97*) found that goldfish change their center of activity significantly on successive periods varying from 5 to 40 min. It is clear that, with such an underlying pattern of activity, a record based on 10 or 15 min or even 1 h is easily misinterpreted.

In addition to the apparently random variability of behavior, there are variations that can be accounted for by differences between individuals. These are frequently large, necessitating the use of a control period on the same individual for comparisons. Left-handed (or perhaps more accurately, left-finned) fish and right-handed (right-finned) fish have been observed by Kleerekoper (*94*).

The activities of most animals vary with time of day. Cycles in behavior may be diurnal or related to other natural phenomena such as tides. Winter flounder, for example, are normally most active at low tide (*185*). These periodicities must be considered when studying behavioral effects, either by observing only at a specific time of day or by recording over a period of at least 24 h.

Animals exposed to new surroundings usually undergo a period of increased activity that is thought to be exploratory behavior (*95, 97*). This may last some time and may be reinitiated by some small change in the animal's surroundings. In fact, a change in overall levels of activity frequently observed after introduction of a chemical may be largely related to this phenomenon. Most researchers have eliminated most of this effect by waiting for some time after placing an animal in an observation chamber before starting to record.

The fundamental responses may not be behavioral at all but instead related to some other aspect of physiology that indirectly induces changes in behavior. Jones (*88*), for example, suggested that responses he had found in sticklebacks to lead nitrate may have been due to osmotic pressure from the solutions tested. Changes in sensory or locomotor capabilities in particular have profound effects on behavioral expression. Jacobson and Boyland (*86*) found that the seawater-soluble fraction of kerosene interfered with attraction of the snail *Narassius obsoletus* to food extracts. Hara *et al.* (*72*) and Hara and Thompson (*71*) observed that perfusing toxicants into the nares of fish could alter electrical responses in the olfactory bulb. Hidaka (*76*) measured depressed activity in carp palatal chemoreceptors in response to various metals. Gardner and LaRoche (*59*) found histological changes in the lateral line and olfactory receptors of estuarine fish after exposure to copper. Blocking of receptor mechanisms may produce fundamental changes in the behavior of fish (*162*). Dodson and Mayfield (*49*) found that exposure of rainbow trout to the herbicides diquat and Simazine affected their responses to a moving striped background.

The variability itself may change in response to chemicals. The behavior of animals during control periods is often highly irregular, with exposure to chemicals resulting in much more uniform activity (*181*). It may be hard to tell whether this is due to low activity levels frequently found during control periods or whether it is a true response. A response involving changes in variability is difficult to deal with statistically.

As with all toxicity, it is easy to misinterpret a behavioral response because the chemistry is acting unexpectedly. Ishio (*84*) presented evidence that the apparent attraction of fish to ammonia and copper was actually due to the indirect effect these materials had on the concentration of carbonic acid. Granett *et al.* (*62*) discovered that the carrier they were using to expose salmon to the larvicide Dimilin-G1 was causing the observed avoidance. Such things as adsorption of the toxicant to surfaces are important as is reaction with other constituents or changes in speciation. A fish exposed to a pH gradient, for example, may also be exposed to gradients of other materials indirectly altered by pH. pH can affect such things as the dissociation of ammonia, the solubility and speciation of metals, and many other aspects of the chemical environment that an animal can respond to. In saltwater tests, the response to a freshwater effluent can be attributed primarily to the changes in salinity.

III. PREFERENCE OR AVOIDANCE OF INDUSTRIAL CHEMICALS BY AQUATIC ANIMALS

The olfactory capabilities of most aquatic animals are extraordinary and it is reasonable to expect that the imposition of complex wastewaters on the environment since the industrial revolution has had dramatic effects on the behavior of these animals. There is no reason to expect, however, that these few hundred years is enough time for significant evolutionary adaptation to the chemicals. A lethal material, therefore, may be repulsive, indifferent, or even attractive to aquatic animals. Sprague and Drury (*156*), for example, found that although trout avoided alkylbenzene sulfonate (ABS) detergent, chlorine, and zinc, they appeared to be indifferent to phenol even at lethal concentrations. Jones (*88*) found that sticklebacks were attracted to lethal concentrations of ammonia.

There has been a variety of apparatuses developed to study the responses of animals to chemical gradients. The nature of the gradient tank may greatly affect the response. Since several thorough reviews already exist regarding these gradient tanks (*32, 39, 57, 102, 116, 144*), they will be described only briefly here. These tanks have consisted of "Y" mazes, double Y mazes, "rosette" mazes, troughs, or tubes with central drains. Figure 1 shows a number of them. While these apparatuses vary in size, none is large enough to completely simulate field conditions. Höglund's "fluvarium" (*78*) has a particularly interesting approach for creating gradients that involves the use of various shaped plates in front of

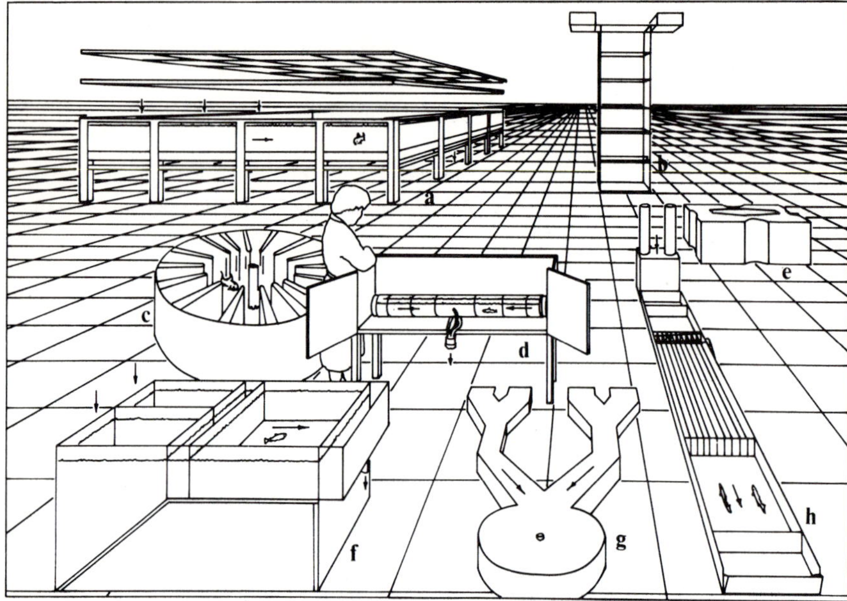

Fig. 1. Representative preference/avoidance apparatuses used with aquatic animals and their approximate overall dimensions. (a) Kleerekoper (*94*), 500 × 500 × 50 cm; (b) Birtwell (*22*), 50 × 50 × 600 cm; (c) Kleerekoper (*94*), 200 cm diameter; (d) Sprague (*152*), 115 × 15 cm diameter; (e) Reynolds (*139*), 120 × 120 × 46 cm; (f) Westlake and Lubinski (*180*), 50 × 100 × 50 cm; (g) Hansen (*65*), 112 cm long × 7 cm deep; (h) Höglund (*78*), 250 × 32 × 12 cm. Arrows indicate direction of flow.

two inlets at either side of a trough, so that a variety of gradients can be created by the same apparatus. Reynolds (*139*) designed another unique device he called an "ichthiotron," which is capable of testing two variables simultaneously.

This complexity in tank design is further complicated by the fact that flow patterns are difficult to create reproducibly, particularly at the low flow rates that are necessary in most labs. Limited resources in terms of water supply or chemical or transportation costs of effluents frequently restrict the researcher to small chambers and low flows. Unexpected circulation patterns are a constant concern under these conditions. If inlet and outlet flows are not critically balanced, the entire tank contents might circulate and some parts of the observation area may be going in the opposite direction to what is designed. One way of getting around these problems, for industrial effluents at least, is to move the lab to the source of the effluent where larger volumes are available. Another method of overcoming the flow problem that has been used in most of the preference/avoidance research is to create flows in opposite directions in a tube or trough with inlets at either end and outlets in the center (*148, 152*). The complication with this approach is that the test confounds exposure to a gradient of chemicals with exposure to a change in direction of flows.

Another major difficulty with small tanks in the laboratory is that the response may be totally different, in fact in the opposite direction, from the response in larger tanks or in the field. Kleerekoper *et al.* (*99*) and Timms *et al.* (*168*) found attraction to copper by goldfish in a large tank (5 × 5 × 0.5 m) with a shallow gradient, yet, under the same conditions in the same lab but in a smaller tank, Westlake *et al.* (*183*) observed that the fish avoided copper. Laughlin *et al.* (*103*) in a combined lab and field study with blue crabs collected numerous animals under conditions that were strongly avoided in the lab. Very few field studies on avoidance of industrial chemicals have yet been done. Saunders and Sprague (*142*) observed changes in salmon migration after the start-up of a heavy metal mine in northern New Brunswick. Birtwell (*22*) studied changes in movements of caged salmon in the vicinity of a pulp mill. Kelso (*93*) followed the movements of fishes in relation to a pulp mill effluent both by sonic tagging and by acoustic echo analysis. Elson *et al.* (*51*) did similar work for salmon in the Miramichi River in New Brunswick. There are also a number of chance observations (e.g., *141*) that may have relevance to avoidance.

Just because an animal responds to a chemical while all other stimuli are held constant, as in a lab test, does not mean that the same animal would respond to that chemical when some stronger stimulus was present. For example, the migratory urge of salmon is so strong that they will try even to exhaustion to get past some physical barrier such as a dam. Would they not suffer the unpleasant odor while passing some obnoxious effluent plume as long as it did not totally mask the odor of their spawning stream? Kleerekoper *et al.* (*98*) found that in the presence of a 0.4°C temperature increase, goldfish no longer avoided but were attracted to copper. Hill (*77*), studying the reactions of a cave fish to both light and low oxygen, found that the fish more strongly avoided light than low oxygen. Although animals may be capable of concentrating on the inputs from one sensory modality, they seldom do. They respond to all in a coordinated way and in some circumstances may even substitute one modality for another.

The actual mechanism of response to a gradient may be so obscure that it is difficult to interpret. The response does not have to be up or down the gradient but may instead be primarily kinetic. Kinetic responses to environmental stimuli do not at first appear to be directed toward the source of stimulus yet can cause net movement along a gradient (*57*). Suppose, for example, that an animal were to react to a chemical by increasing the speed of locomotion. It would then tend to be quiescent in dilute areas and active in concentrated areas, resulting in eventual movement away from the source. Even taxes or oriented reactions are rarely absolute since it takes only a slight bias toward or away from the source to result in significant population distribution shifts.

The statistics used for this work is tricky because many effects are spacially confounded and one move made by the animal is serially connected to the next. That is, the animal's current location and activity within the observation chamber have a great influence on what the animal will do next.

Researchers then are faced with the dilemma that, with such great variability of many types, heavy reliance must be placed on statistical inference rather than on graphical interpretation or intuitive evaluation of data. However, it may be impossible to completely randomize an experiment because the animal cannot be reused without the problem of separating learned from naive responses, and it cannot be assigned randomly to various areas of the observation apparatus since it is normally free to move between them. For example, if for the purpose of statistics one considers parts of an observation or choice apparatus as treatments in addition to the exposure concentrations, then a completely randomized design is impossible because one would have to assign fish randomly to one or other parts of the chamber but they are free to move. Some researchers have gotten around this problem by not permitting the animal to move between parts of the box but to choose between alternatives (e.g., *68*). This makes the data able to fit a chi square analysis with a theoretical distribution of equal numbers in each of the possible choices.It may, however, be somewhat unnatural for an animal to respond in this manner. For example, fish migration is thought by some to be related to the relationship of rheotaxis to chemical stimuli. As long as the odor of the home stream is present, the fish has positive rheotaxis. This means that the initial choice of the fish might be random but the final result comes from falling back to remake the choice. This design makes it difficult to evaluate such behavior.

Another possibility is to accept the fact that a completely randomized design is difficult and use a more complex one. For example, Westlake *et al.* (*184*) used a split-plots analysis of variance, split on areas of the tank. Significant changes in behavior related to a gradient in the tank could then be tested using conservative degrees of freedom since the covariance matrix associated with the areas did not, and could not, satisfy the assumption of compound symmetry (*188*). Perhaps the best approach is more complex yet, and involves the study of the series of moves made by the fish (*97*). By study of the mechanism and pattern of movement it should be possible to determine when the behavior has changed significantly. The price paid is that statistics is pressed to the limits of presently available theory and a point is reached at which the mathematics must be developed hand in hand with the behavioral research. For the moment, we may have to be satisfied with one of the more standard analyses.

Many researchers using various of these related methods have found significant behavioral responses to industrial chemicals and effluents. Table I lists many of these studies.

The chemical senses of fish surpass the capabilities of our most sensitive instruments in many respects. Wisby (*189*), for example, determined that salmon could be trained to detect concentrations of morpholine as low as 0.000001 ppm.

TABLE I

Preference/Avoidance and Related Studies on Aquatic Animals with Various Chemicals, Effluents, and Perturbations

Chemical/effluent/perturbation	References
Zinc	*14, 15, 24, 87, 88, 115, 142, 152, 153, 155, 160, 167, 179*
Copper	*23, 48, 53, 54, 59, 69, 98, 99, 101, 107, 113, 142, 152, 155, 160, 167, 168, 183*
Chlorine	*25, 33–38, 40, 41, 52, 61, 81, 123, 145, 157*
Phenolic compounds	*16, 73, 84, 89, 91, 107, 140, 156, 159, 160*
Detergents	*12, 56, 156, 163, 166, 187*
Ammonia	*88, 110, 111, 146, 148*
pH	*20, 80, 103, 147, 148, 178*
Dissolved oxygen	*1, 20, 77, 79, 80, 90, 147, 148, 158, 159, 171, 190*
Herbicides and pesticides	*3, 16, 27, 47, 49, 50, 53–55, 58, 62, 65–68, 100, 120, 127, 134, 137, 138, 143, 173, 174*
Petroleum hydrocarbons and oil refinery effluents	*16, 19, 86, 132, 133, 184*
Pulp and paper mill effluents	*6, 26, 42, 51, 63, 79, 82, 92, 93, 105, 118, 119, 156, 181, 186*
Other	*2, 43, 104, 111, 117, 167, 169, 179*

It should be possible to use this sensitivity for practical applications, such as low-level detection and biological monitoring. There is a great need, for example, as the world begins to look more and more at recycling water as a viable source of supply, particularly in arid regions, for a method of monitoring this water on a continuous basis. Techniques for training aquatic animals to detect chemical changes in their environment have been developed (see the review by Höglund, *78*) and should be adaptable to a great number of circumstances. A particularly interesting method of training, in which animals were provided with the ability to regulate the temperature of their surroundings by their own movements, was described by Neill *et al.* (*126*). The same approach could probably be used with chemical changes. It is conceivable that in the future fish may be trained to detect problems in our drinking water before we receive it.

Another important area for study involves the determination of the most active components of complex industrial effluents causing avoidance or preference reactions. One of the few efforts along this line was an attempt by Wildish *et al.* (*186*) to identify active components of a pulp mill effluent.

IV. STRESS BEHAVIOR FOR BIOLOGICAL MONITORING

Often, at concentrations closer to the lethal concentration, stress behavior is observed. This stress behavior has been used in a practical way to design devices

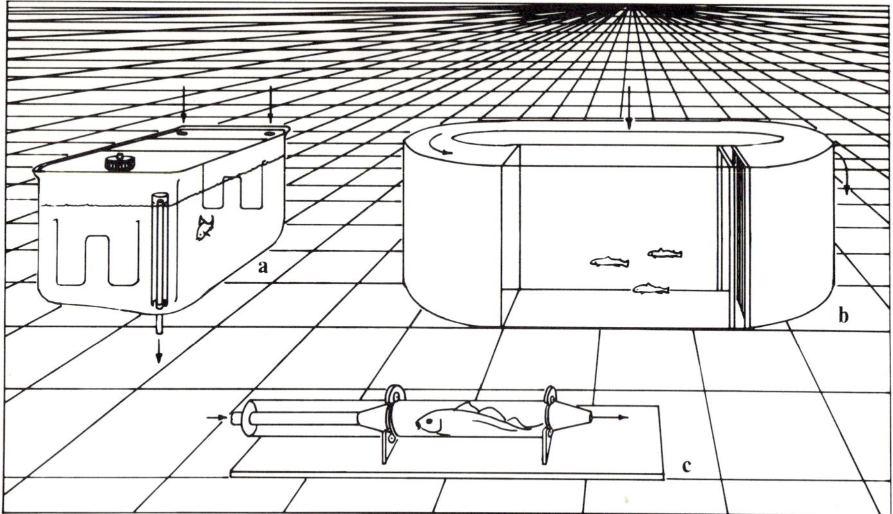

Fig. 2. Representative biological monitoring devices with approximate overall dimensions. (a) Westlake and van der Schalie (*182*), 60 × 35 × 35 cm; (b) Besch *et al.* (*17*), 120 × 60 × 40 cm; (c) Lindahl (*108*), 30 × 5 cm diameter (size only of tube containing fish). Arrows indicate direction of flows.

to biologically monitor effluents continuously (Fig. 2). The advantage of this is that, environmentally, significant changes can be detected soon enough to take corrective action even without understanding them in detail.

Waller and Cairns (*172*) describe a monitoring device consisting of a number of aquaria each of which is equipped with three photocells. By means of electronic counters, these photocells, illuminated by light beams traversing the length of the tanks, permitted continuous records to be made of the fish activity. Even though these monitors were tested with prepared zinc solutions, the authors speculated that a similar system could be used to monitor industrial discharges. This idea was pursued in the same laboratories and extended to include monitoring of respiratory movements of the fish by means of a pair of electrodes attached to the ends of the tank (*28, 30, 149*). Similar monitoring devices have been described by Morgan (*124*), Morgan and Kuhn (*125*), and Spoor *et al.* (*150*). These monitors were tested under industrial conditions and shown to be viable (*64, 170, 182*). Lindahl and Schwanbom (*108, 109*) and Lindahl *et al.* (*106*) describe an apparatus capable of continuous monitoring that utilizes a rotating tube which measures the ability of fish to maintain an upright position. The fish were exposed to increasing speeds of rotation while being exposed to various toxic materials until they were unable to stay upright. Besch *et al.* (*17, 18*) and Peols (*135*) employed a similar monitor using the ability of groups of fish to maintain themselves against a strong current.

Frequently, a concern is raised regarding the sensitivity of these biological monitors for detecting industrial upsets. However, in practical terms, the biomonitor can be used to detect not only sublethal effects but also acute lethality. During upsets it is more common for the effluent to go all the way over to acute lethality than to stay at conditions just below. In fact, a viable monitor might be produced that concentrates entirely on automatically detecting sudden mortalities in animals exposed continuously to an effluent.

The relevance of these biological monitors to the animals living in the receiving water is of only minor importance. The biological monitor is an instrument, as is a pH meter, that has been found to change when things go wrong in an effluent. The effects of that change on the environment have to be evaluated separately. There is sufficient practical advantage in knowing that such a change has occurred in time to take corrective measures.

As with preference/avoidance studies, the statistics used in biological monitoring is difficult. An initial calibration period for each new animal introduced to the system is normally required, during which the normal limits of behavior for the individual are established (*29, 30*).

V. EFFECTS OF INDUSTRIAL CHEMICALS ON OTHER TYPES OF ANIMAL BEHAVIOR

Changes in several other types of behavior have been observed in aquatic animals after or during exposure to various chemicals. A good example of the range of behavioral effects that are possible may be found in a paper by Bull and McInerney (*27*). They found changes in various agonistic behaviors as well as in feeding and comfort behaviors in coho salmon exposed to the insecticide fenitrothion.

Feeding behavior and chemosensory responses to food or food extracts have been the subjects of several studies. Drummond *et al.* (*48*) found both feeding behavior and locomotor activity altered in brook trout after exposure to copper. Manley and Davenport (*114*) encountered total and partial value closure in various mollusks in response to copper concentrations slightly higher than normally encountered in the marine environment. A number of studies in the literature involve the behavioral effects of petroleum hydrocarbons. These materials have been shown to cause changes in feeding behavior of lobsters (*7–10*), mud snails (*8, 83*), and crabs (*131, 165*).

Related to feeding effects, modifications to predator/prey relationships have been observed in response to various materials. Predator/prey changes have been found in decapods after exposure to chromium (*31*), in salmon after exposure to DDT (*74*) and mirex (*164*), and in fish after exposure to ammonia (*191*).

A number of researchers have studied changes in fish learning after exposure to insecticides (*4, 5, 75, 85, 121, 122, 175*). Many other observations have been made on the effects of chemicals on behavior: on schooling behavior and activity of goldfish exposed to DDT by Weis and Weis (*177*), on righting responses of sea urchins under the influence of crude oil by Axiak and Saliba (*11*), on detachment of limpets after exposure to crude oil by Dicks (*44, 45*), and on the settlement of barnacles around an oil refinery outfall by Dicks (*45*). The early development of behavior in juvenile Atlantic salmon was found to be delayed by exposure to DDT (*46*).

VI. CONCLUSIONS

It is clear from the literature that a great variety of behavioral responses can occur to industrial chemicals. However, although many studies have shown that the behavior of aquatic animals was significantly affected, few addressed at any length the nature of the response or its relevance to field behavior. It would contribute much to the understanding of behavioral effects of chemicals if the more distinct responses were studied further. I am convinced that clues to the mechanism of these responses can be found by testing one experimental design beside another. It is also essential that more field verification be attempted. Information on how animals react to real effluent plumes is very scanty.

While complicated biological monitoring devices have been effectively demonstrated, it may make more sense in practical terms to simplify the equipment so that it detects only whether or not the animals are alive. The lowered costs and decreased numbers of test animals required may far outweigh the loss in sensitivity.

There is currently a revolution going on in television equipment and, just as with home computers, prices are dropping and capabilities and availability are increasing. It appears that video disc recording and computer control of video equipment are at most a few years away. The implications of these developments on behavioral research are staggering. First, very large storage may be readily available with the flexibility to mix video signals with data. Second, frame by frame analysis may soon be within reach of the researcher with a modest budget. Recent developments in software are equally important. It may be possible, for example, with the use of artificial intelligence, to train a computer to look for patterns in animal behavior in an unstructured way.

References

1. Aardt, W. J., and Frey, B. J. (1979). *S. Afr. J. Zool.* **14,** 202–207.
2. Amiard, J. C. (1976). *Mar. Biol.* **34,** 239–245.

3. Anderson, J. M. (1971). *Proc. R. Soc. London, Ser. B* **177,** 307–320.
4. Anderson, J. M., and Prins, H. B. (1970). *J. Fish. Res. Board Can.* **27,** 331–334.
5. Anderson, J. M., and Peterson, M. R. (1969). *Science* **164,** 440–441.
6. Anonymous (1979). Can. Cooperative Pollut. Abatement Res., Prog. Summ. 8, Sept. 1978. Water Pollut. Control Div., Environ. Can., Ottawa.
7. Atema, J., Karnofsky, E. B., and Oleszko-Szuts, S. (1979). *In* "Advances in Marine Environmental Research" (F. Jacoff, ed.), EPA Rep. No. EPA-600/9-79-035, pp. 122–134. U.S. Environ. Prot. Agency, Cincinnati, Ohio.
8. Atema, J. (1976). *In* "Estuarine Processes, Volume 1, Uses, Stresses, and Adaptation to the Estuary" (M. Wiley, ed.), pp. 302–312. Academic Press, New York.
9. Atema, J., and Stein, L. S. (1974). *Environ. Pollut.* **6,** 77–86.
10. Atema, J., Leavitt, D. F., Barshaw, D. E., and Cuomo, M. C. (1982). *Can. J. Fish. Aquat. Sci.* **39,** 675–690.
11. Axiak, V., and Saliba, L. J. (1981). *Mar. Pollut. Bull.* **12,** 14–19.
12. Bardach, J. E., Fujiya, M., and Holl, A. (1965). *Science* **148,** 1605–1607.
13. Barden, W., Jr. (1981). *Byte* **6,** 134–160.
14. Bengtsson, B. E. (1974). *Water Res.* **8,** 829–833.
15. Bengtsson, B. E. (1975). *Bull. Environ. Contam. Toxicol.* **12,** 654–658.
16. Bergquist, B. L., and Bovee, E. C. (1976). *Acta Protozool.* **15,** 471–483.
17. Besch, W. K., Kemball, A., Meyer-Waarden, K., and Scharf, B. (1977). *ASTM Spec. Tech. Publ.* **607,** 56–74.
18. Besch, W. K., Loseries, H. G., Meyer-Waarden, K., and Schmitz, W. (1974). *Arch. Hydrobiol.* **74,** 551–565.
19. Bigford, T. E. (1977). *Mar. Biol.* **43,** 137–148.
20. Bishai, H. M. (1962). *J. Cons., Cons. Int. Explor. Mer* **27,** 167–180.
21. Bishai, H. M. (1962). *J. Cons., Cons. Int. Explor. Mer* **27,** 181–191.
22. Birtwell, I. K. (1977). *Econ. Tech. Rev. Rep. EPS (Can., Environ. Prot. Serv. EPS-5-AR-77-1,* 69–86.
23. Black, J. A., and Birge, W. J. (1980). U.S. Dept. Inter. Res. Rep. No. 123 (A-077-Ky). Univ. of Kentucky Water Resources Res. Inst., Lexington.
24. Bloom, H. D., Perlmutter, A., and Seeley, R. J. (1978). *Environ. Pollut.* **17,** 127–131.
25. Bogardus, R. B., Teppen, T. C., Boies, D. B., and Harvath, F. J. (1978). *Water Chlorination Environ. Impact Health Eff., Proc. Conf. 1975, 1977* **2,** 123–133.
26. Brett, J. R., and MacKinnon, D. (1954). *J. Fish. Res. Board Can.* **11,** 310–318.
27. Bull, C. J., and McInerney, J. E. (1974). *J. Fish. Res. Board Can.* **31,** 1867–1872.
28. Cairns, J., Jr., Dickson, K. L., and Westlake, G. F. (1977). "Biological Monitoring of Water and Effluent Quality, " *ASTM Spec. Tech. Publ.* **607.**
29. Cairns, J., Jr., Dickson, K. L., and Westlake, G. F. (1975). Advances in water pollution research. *In* "Proceedings of the 7th International Conference on Water Pollution Control Research, Paris, Sept. 1974" (S. H. Jenkins, ed.). Pergamon, Oxford.
30. Cairns, J., Jr., Hall, J., Morgan, E. L., Sparks, R. E., Waller, W. T., and Westlake, G. F. (1973). *Trace Subst. Environ. Health* **7,** 43–55.
31. Chaisemartin, C. (1979). *C. R. Seances Soc. Biol. Ses. Fil.* **173,** 613–619.
32. Cherry, D. S., and Cairns, J., Jr. (1982). *Water Res.* **16,** 263–301.
33. Cherry, D. S., Hoehn, R. C., Waldo, S. S., Willis, D. H., Cairns, J., Jr., and Dickson, K. L. (1977). *Water Resour. Bull.* **13,** 1047–1055.
34. Cherry, D. S., Larrick, S. R., Gaittina, J. D., Dickson, K. L., and Cairns, J., Jr. (1979). *Environ. Int.* **2,** 1–6.
35. Cherry, D. S., Larrick, S. R., Cairns, J., Jr., and Dickson, K. L. (1977). *Trace Subst. Environ. Health* **11,** 413–418.

36. Cherry, D. S., Larrick, S. R., Giattina, J. D., Cairns, J., Jr., and van Hassel, J. (1982). *Can. J. Fish. Aquat. Sci.* **39,** 162–173.
37. Cherry, D. S., Larrick, S. R., Dickson, K. L., Hoehn, R. C., and Cairns, J., Jr. (1977). *J. Fish. Res. Board Can.* **34,** 1365–1372.
38. Cherry, D. S., Larrick, S. R., Giattina, J. D., Dickson, K. L., and Cairns, J., Jr. (1979). *Environ. Int.* **2,** 85–90.
39. Coutant, C. C. (1977). *J. Fish. Res. Board Can.* **34,** 739–745.
40. Cripe, C. R. (1979). *J. Fish. Res. Board Can.* **36,** 11–16.
41. Dandy, J. W. T. (1972). *Can. J. Zool.* **50,** 405.
42. Davis, J. C. (1976). *J. Fish. Res. Board Can.* **33,** 2031–2035.
43. DeCoursey, P. J., and Vernberg, W. B. (1972). *Oikos* **23,** 241–247.
44. Dicks, B. (1973). *Environ. Pollut.* **5,** 219–229.
45. Dicks, B. (1976). *Mar. Ecol. Oil Pollut. (Proc. Meet.), 1975,* 303–319.
46. Dill, P. A., and Saunders, R. C. (1974). *J. Fish. Res. Board Can.* **31,** 1936–1938.
47. Domanik, A. M., and Zar, T. H. (1978). *Arch. Environ. Contam. Toxicol* **7,** 207–220.
48. Drummond, R. A., Spoor, W. R., and Olson, G. F. (1973). *J. Fish. Res. Board Can.* **30,** 698–701.
49. Dodson, J. J., and Mayfield, C. I. (1979). *Environ. Pollut.* **18,** 147–157.
50. Ellgaard, E. G., Ochsner, J. C., and Cox, J. K. (1977). *Can. J. Zool.* **55,** 1077–1081.
51. Elson, P. F., Lauzier, L. M., and Zitko, V. (1972). *In* "Marine Pollution and Sea Life," FAO Bull., pp. 325–330. Fishing News Ltd., London.
52. Fava, J. A., Jr., and Tsai, C. F. (1978). *Comp. Biochem. Physiol. C* **60,** 123–128.
53. Folmar, L. C. (1976). *Bull. Environ. Contam. Toxicol.* **15,** 509–514.
54. Folmar, L. C. (1978). *Bull. Environ. Contam. Toxicol.* **19,** 312–318.
55. Folmar, L. C. (1979). *Arch. Environ. Contam. Toxicol.* **8,** 269–278.
56. Foster, N. R., Scheier, A., and Cairns, J. (1966). *Trans. Am. Fish. Soc.* **95,** 109–110.
57. Fry, F. E. J. (1958). *Proc., Indo-Pac. Fish Counc.* **3,** 37–42.
58. Gardner, D. R. (1973). *Pestic. Biochem. Physiol.* **2,** 437.
59. Gardner, D. R., and LaRoche, G. (1973). *J. Fish. Res. Board Can.* **30,** 363–368.
61. Ginn, T. C., and O'Connor, J. M. (1978). *Estuarine Coastal Mar. Sci.* **6,** 459–469.
62. Granett, J., Morang, S., and Hatch, R. (1978). *Bull. Environ. Contam. Toxicol.* **19,** 462–464.
63. Greer, G. L., and Kosakoski, G. J. (1978). *Tech. Rep.—Fish. Mar. Serv. (Can.)* **831.**
64. Gruber, D., Cairns, J., Jr., and Hendricks, R. C. (1981). *J. Water Pollut. Control Fed.* **53,** 505–511.
64a. Hadjinicoloau, J., and Spraggs, L. D. (1981). *Econ. Tech. Rep. Fisheries Aquatic Sci.* **1151,** 68–82.
65. Hansen, D. J. (1969). *Trans. Am. Fish. Soc.* **98,** 426–429.
66. Hansen, D. J., Schimmel, S. C., and Matthews, E. (1974). *Bull. Environ. Contam. Toxicol.* **12,** 253–256.
67. Hansen, D. J., Schimmel, S. C., and Keltner, J. M., Jr. (1973). *Bull. Environ. Contam. Toxicol.* **9,** 129.
68. Hansen, D. J., Matthews, E., Nall, S. L., and Dumas, D. P. (1972). *Bull. Environ. Contam. Toxicol.* **8,** 46–51.
69. Hara, T. J. (1972). *J. Fish. Res. Board Can.* **29,** 1351–1355.
71. Hara, T. J., and Thompson, B. E. (1978). *Water Res.* **12,** 893–897.
72. Hara, T. J., Law, Y. M. C., and MacDonald, S. (1976). *J. Fish. Res. Board Can.* **33,** 1568–1573.
73. Hasler, R. D., and Wisby, W. J. (1949). *Trans. Am. Fish. Soc.* **79,** 64–70.
74. Hatfield, C. T., and Anderson, J. M. (1972). *J. Fish. Res. Board Can.* **29,** 27–29.
75. Hatfield, C. T., and Johansen, P. H. (1972). *J. Fish. Res. Board Can.* **29,** 315–321.

76. Hidaka, I. (1970). *Jpn. J. Physiol.* **20,** 599–609.
77. Hill, L. G. (1968). *Trans. Am. Fish. Soc.* **97,** 448–454.
78. Höglund, L. B. (1951). *Oikos* **3,** 247–267.
79. Höglund, L. B. (1961). *Inst. Freshwater Res. Drottningholm Rep.* **43,** 1–147.
80. Höglund, L. B., and Harding, J. (1969). *Inst. Freshwater Res. Drottningholm Rep.* **49,** 76–119.
81. Hose, J. E., and Stoffel, R. J. (1980). *Bull. Environ. Contam. Toxicol.* **25,** 929–935.
82. Howard, T. E. (1975). *J. Fish. Res. Board Can.* **32,** 789–793.
83. Hyland, J. L., and Miller, D. C. (1979). *In* "Proceedings of the 1979 Oil Spill Conference," API Publ. No. 4308, pp. 603–607. Am. Petroleum Inst., Washington, D.C.
84. Ishio, S. (1965). *In* "Proceedings of the 2nd International Conference on Water Pollution Research, Tokyo, 1964," Vol. 1, pp. 19–33. Pergamon, Oxford.
85. Jackson, D. A., Anderson, J. M., and Gardner, D. R. (1970). *Can. J. Zool.* **48,** 577–580.
86. Jacobson, S. M., and Boylan, D. B. (1973). *Nature (London)* **241,** 213–215.
87. Jones, J. R. E. (1947). *J. Exp. Biol.* **24,** 110–122.
88. Jones, J. R. E. (1948). *J. Exp. Biol.* **25,** 22–34.
89. Jones, J. R. E. (1951). *J. Exp. Biol.* **28,** 261–270.
90. Jones, J. R. E. (1952). *J. Exp. Biol.* **29,** 403–415.
91. Jones, J. R. E. (1964). "Fish and River Pollution." Butterworth, London.
92. Jones, B. F., Warren, C. E., Bond, C. E., and Doudoroff, P. (1956). *Sewage Ind. Wastes* **28,** 1403–1413.
93. Kelso, J. R. M. (1977). *J. Fish. Res. Board Can.* **34,** 879–885.
94. Kleerekoper, H. (1967). *Am. Zool.* **7,** 385–395.
95. Kleerekoper, H., Timms, A. M., Westlake, G. F., Davy, F. B., Malar, T., and Anderson, V. M. (1970). *Anim. Behav.* **18,** 317–330.
96. Kleerekoper, H. (1976). *J. Fish. Res. Board Can.* **33,** 2036–2039.
97. Kleerekoper, H., Matis, J., Gensler, P., and Matis, J. (1974). *Anim. Behav.* **22,** 124–132.
98. Kleerekoper, H., Waxman, J. B., and Matis, J. (1973). *J. Fish. Res. Board Can.* **30,** 725–728.
99. Kleerekoper, H., Westlake, G. F., Matis, J. H., and Gensler, P. J. (1972). *J. Fish. Res. Board Can.* **29,** 45–54.
100. Kynard, B. (1974). *Trans. Am. Fish. Soc.* **102,** 557.
101. Lang, W. H., Lawrence, S., and Miller, D. C. (1979). *In* "Advances in Marine Environmental Research" (F. Jacoff, ed.), EPA Rep. No. EPA-600/9-79-035, pp. 273–289. U.S. Environ. Prot. Agency, Cincinnati, Ohio.
102. Larrick, S. R., Dickson, K. L., Cherry, D. S., and Cairns, J., Jr. (1978). *Hydrobiologia* **61,** 257–265.
103. Laughlin, R. R., Cripe, C. R., and Livingston, R. J. (1978). *Trans. Am. Fish. Soc.* **107,** 78–86.
104. Lawrence, M., and Scherer, E. (1974). *Tech. Rep.—Fish. Mar. Serv. (Can.)* **502.**
105. Lewis, F. G., and Livingston, R. J. (1977). *J. Fish. Res. Board Can.* **34,** 568–570.
106. Lindahl, P. E., Olofsson, S., and Schwanbom, E. (1977). *ASTM Spec. Tech. Publ.* **607,** 75–84.
107. Lindahl, P. E., and Marcstrom, A. (1958). *J. Fish. Res. Board Can.* **15,** 685–694.
108. Lindahl, P. E., and Schwanbom, E. (1971). *Oikos* **22,** 210–214.
109. Lindahl, P. E., and Schwanbom, E. (1971). *Oikos* **22,** 354–357.
110. Lubinski, K. S., Cairns, J., Jr., and Dickson, K. L. (1978). *Trace Subst. Environ. Health* **12,** 508–514.
111. Lubinski, K. S., Dickson, K. L., and Cairns, J., Jr. (1980). *ASTM Spec. Tech. Publ.* **707,** 320–340.

112. Maciorowski, A. F., Benfield, E. F., and Cairns, J., Jr. (1980). *Hydrobiologia* **74,** 105–112.
113. Maciorowski, H. D., Clarke, R. McV., and Scherer, E. (1977). *Econ. Tech. Rev. Rep. EPS (Can., Environ. Prot. Serv. EPS-5AR-77-***1,** 49–58.
114. Manley, A. R., and Davenport, J. (1979). *Bull. Environ. Contam. Toxicol.* **22,** 739–744.
115. Manton, M., Karr, A., and Shrenfeld, D. W. (1972). *Biol. Bull.* **143,** 184–195.
116. McCauley, R. W. (1977). *J. Fish. Res. Board Can.* **34,** 749–752.
117. McGreer, E. R. (1979). *Mar. Pollut. Bull.* **10,** 259–262.
118. McLeese, D. W. (1970). *J. Fish. Res. Board Can.* **27,** 731–736.
119. McLeese, D. W. (1973). *J. Fish Res. Board Can.* **30,** 279–282.
120. McMahon, T. E., and Kynard, B. E. (1979). *Southwestern Naturalist* **24,** 87–92.
121. McNichol, P. G., and Mackay, W. C. (1975). *J. Fish. Res. Board Can.* **32,** 661–665.
122. McNichol, P. G., and Mackay, W. C. (1975). *J. Fish. Res. Board Can.* **32,** 785–788.
123. Middaugh, D. P., Crane, A. M., and Couch, J. R. (1977). *Water Res.* **11,** 1089–1096.
124. Morgan, W. S. G. (1977). *ASTM Spec. Tech. Publ.* **607,** 38–55.
125. Morgan, W. S. G., and Kuhn, P. C. (1974). *Water Res.* **8,** 67–77.
126. Neill, W. H., Magnuson, J. J., and Chipman, G. (1972). *Science* **176,** 1443–1445.
127. Ogilvie, D. M., and Anderson, J. M. (1965). *J. Fish. Res. Board Can.* **22,** 503–512.
128. Olla, B. L., Forward, R., Kittredge, J., Livingston, R. J., Mcleese, D. W., Miller, D. C., Vernberg, K., Wilson, K., Atema, J., and Wells, P. G. (1980). *Rapp. P.—V. Reun., Cons. Int. Explor. Mer* **179,** 174–181.
129. Olla, B. L., and Studholme, A. L. (1975). *In* "Second Joint U.S./U.S.S.R. Symposium on the Comprehensive Analysis of the Environment," pp. 25–31. U.S. Environ. Prot. Agency, Washington, D.C.
130. Olla, B. L., Pearson, W. H., and Studholme, A. L. (1983). *In* "Proceedings of ICES Workshop on the Problems of Monitoring Biological Effects of Pollution in the Sea." Unpublished Rep., Int. Council Explor. Sea.
131. Pearson, W. H., and Olla, B. L. (1979). *Estuaries* **2,** 63–64.
132. Percy, J. A. (1977). *Environ. Pollut.* **13,** 1–10.
133. Percy, J. A., and Mullin, T. C. (1977). *Mar. Pollut. Bull.* **8,** 35–40.
134. Peterson, R. H. (1974). *J. Fish. Res. Board Can.* **31,** 1757–1762.
135. Poels, C. L. M. (1977). *ASTM Spec. Tech. Publ.* **607,** 85–95.
136. Price, R. B., and Ache, B. W. (1977). *Comp. Biochem. Physiol.* **57,** 249–253.
137. Rand, G. M., and Barthalmus, G. T. (1980). *ASTM Spec. Tech. Rep.* **707,** 341–353.
138. Rand, G., Kleerekoper, H., and Matis, J. (1975). *J. Fish Biol.* **7,** 497–504.
139. Reynolds, W. W. (1977). *J. Fish. Res. Board Can.* **34,** 300–304.
140. Saarikoski, J., and Kaila, K. (1977). *Bull. Environ. Contam. Toxicol.* **17,** 40–48.
141. Saunders, J. W. (1969). *J. Fish. Res. Board Can.* **26,** 695–699.
142. Saunders, R. L., and Sprague, J. B. (1967). *Water Res.* **1,** 419–432.
143. Scherer, E. (1975). *Bull. Environ. Contam. Toxicol.* **13,** 492–496.
144. Scherer, E. (1977). *Econ. Tech. Rev. Rep. EPS (Can. Environ. Prot. Serv.) EPS-5AR-77-1,* 33–40.
145. Schumacher, P. D., and Ney, J. J. (1980). *Water Res.* **14,** 651–655.
146. Shelford, V. E. (1917). *Bull—Ill. Nat. Hist. Surv.* **11,** 380–412.
147. Shelford, V. E., and Powers, E. B. (1915). *Biol. Bull.* **28,** 315.
148. Shelford, V. E., and Allee, W. C. (1913). *J. Exp. Zool.* **14,** 207–266.
149. Sparks, R. E., Cairns, J., and Heath, A. G. (1972). *Water Res.* **6,** 895–911.
150. Spoor, W. A., Neiheisel, T. W., and Drummond, R. A. (1971). *Trans. Am. Fish. Soc.* **100,** 22–28.
152. Sprague, J. B. (1964). *J. Water Pollut. Control Fed.* **36,** 990–1004.
153. Sprague, J. B. (1968). *Water Res.* **2,** 367–372.
154. Sprague, J. B. (1971). *Water Res.* **5,** 245–266.

155. Sprague, J. B., Elson, P. F., and Saunders, R. L. (1965). *Int. J. Air Water Pollut.* **9,** 531–543.
156. Sprague, J. B., and Drury, D. E. (1969). *Adv. Water Pollut. Res., Proc. Int. Conf., 4th,* 169–179.
157. Stober, Q. J., Dinnel, P. A., Hurlburt, E. F., and DiJulio, D. H. (1980). *Water Res.* **14,** 347–354.
158. Stott, B. (1976). *In* "Lectures Presented at the Fourth FAO SIDA Training Course on Aquatic Pollution in Relation to Protection of Living Resources. Bioassays and Toxicity Testing. Lysekill, Sweden, Oct. 1975, " pp. 79–86. FAO, Rome.
159. Stott, B. (1979). *J. Fish Biol.* **14,** 135–146.
160. Summerfelt, R. C., and Lewis, W. M. (1967). *J. Water Pollut. Control Fed.* **39,** 2030–2038.
161. Sutterlin, A. M. (1974). *Chem. Senses & Flavour* **1,** 167–178.
162. Sutterlin, A. M., and Sutterlin, N. (1971). *J. Fish. Res. Board Can.* **28,** 565–571.
163. Sutterlin, A., Sutterlin, N., and Rand, S. (1971). *Fisheries Research Board of Canada Tech. Rep. No. 207.* Fisheries Res. Bd. Can., St. Andrews, New Brunswick.
164. Targatz, M. E. (1976). *Trans. Am. Fish. Soc.* **105,** 546–549.
165. Takahashi, F. T., and Kittredge, J. S. (1973). *In* "Microbial Degradation of Oil Pollutants" (D. G. Ahearn and S. P. Meyers, eds.), pp. 259–264, Louisiana State Univ. Publ. No. LSU-SG-73-01. Louisiana State Univ. Press, Baton Rouge.
166. Tatsukawa, B., and Hidaka, H. (1978). *J. Agric. Chem. Soc. Jpn. (Nippon Nogei Kagaku Kaishi)* **52,** 263–270.
167. MacInnes, J. R., and Thurberg, F. P. (1973). *Mar. Pollut. Bull.* **4,** 185–186.
168. Timms, A. M., Kleerekoper, H., and Matis, J. (1972). *Water Resour. Res.* **8,** 1574–1580.
169. Updegraff, K. F., and Sykora, J. L. (1976). *Environ. Sci. Technol.* **10,** 51–54.
170. van der Schalie, W. H., Dickson, K. L., Westlake, G. F., and Cairns, J., Jr. (1979). *Environ. Manage.* **3,** 217–235.
171. van Sommers, P. (1962). *Science* **137,** 678–679.
172. Waller, W. T., and Cairns, J., Jr. (1972). *Water Res.* **6,** 257–269.
173. Ward, D. V., and Busch, D. R. (1976). *Oikos* **27,** 331–335.
174. Ward, D. V., Howes, B. L., and Ludwig, D. F. (1976). *Mar. Biol.* **35,** 119–126.
175. Warner, R. E., Peterson, K. K., and Borgmann, L. (1966). *J. Appl. Ecol. (Suppl.)* **3,** 223–247.
177. Weis, P., and Weis, J. S. (1974). *Environ. Res.* **7,** 68–74.
178. Wells, M. M. (1915). *Biol. Bull.* **29,** 221–257.
179. Wentsel, R., McIntosh, A., McCafferty, W. P., Atchison, G., and Anderson, V. (1977). *Hydrobiologia* **55,** 171–175.
180. Westlake, G. F., and Lubinski, K. S. (1976). *In* "Proceedings of the 1976 National Conference on Control of Hazardous Material Spills, April 1976. New Orleans, " pp. 64–69.
181. Westlake, G. F., Bourdreau, C. A., and Guilcher, M. (1983). *Econ. Tech. Rev. Rep. EPS (Can., Environ. Prot. Serv.) ESP-5-AR-83-3,* 1–60.
182. Westlake, G. F., and van der Schalie, W. H. (1977). *ASTM Spec. Tech. Publ.* **607,** 30–37.
183. Westlake, G. F., Kleerekoper, H., and Matis, J. (1974). *Water Resour. Res.* **10,** 103–105.
184. Westlake, G. F., Sprague, J. B., and Brown, I. T. (1983). *Aquatic Toxicology* **4,** 235–245.
185. Wildish, D. J. (1974). *Water Res.* **8,** 579–583.
186. Wildish, D. J., Akagi, H., and Poole, N. J. (1977). *Bull. Environ. Contam. Toxicol.* **18,** 521–525.
187. Wilson, R. S. (1973). *In* "The Early Life History of Fish" (J. H. S. Bloxter, ed.), pp. 587–602. Dunstaffnage Mar. Res. Lab., Scottish Marine Biol. Assoc., Oban, Scotland.
188. Winer, B. J. (1962). "Statistical Principles in Experimental Design." McGraw-Hill, New York.

189. Wisby, W. J. (1952). Ph.D. Thesis, Univ. of Wisconsin, Madison.
190. Whitmore, C. M., Warren, C. E., and Doudoroff, P. (1960). *Trans. Am. Fish. Soc.* **89,** 17–26.
191. Woltering, D. M., Hedtke, J. L., and Weber, L. J. (1978). *Trans. Am. Fish. Soc.* **107,** 500–504.

Stratospheric Ozone Modification by Man's Influence

A. J. Owens,* A. Yokozeki,† and J. M. Steed†,[1]

**Engineering Department and †Petrochemicals Department*
E. I. du Pont de Nemours and Co, Inc.
Wilmington, Delaware

I.	Introduction	252
II.	Chemistry and Physics of Ozone	255
	A. Production and Loss of Ozone	255
	B. Chemistry of Odd Nitrogen	258
	C. Chemistry of Odd Hydrogen	259
	D. Chemistry of Odd Chlorine	261
	E. Coupling of NO_x/HO_x/ClO_x Families	263
	F. Other Chemistry	264
	G. Transport of Chemical Species in the Atmosphere	265
	H. Radiation	267
III.	Modeling the Ambient Atmosphere	269
	A. Model Structure	269
	B. Model Applications	273
	C. Long-Lived Chemical Species	275
	D. Active Radicals and Sink Gases	280
	E. Temporary Sink Species	286
	F. Measurement Programs to Further Test Models	288
IV.	Model Calculations to Characterize Potential Changes	289
	A. Understanding Perturbations	289
	B. Odd Nitrogen Perturbations	291
	C. Halogen Perturbations	294
	D. Other Chemical Perturbations	300
	E. Greenhouse Effects	302
V.	A Comprehensive Approach—Multiple Perturbations	306
	A. Multiple-Perturbation Scenarios	306
	B. Steady-State Multiple Scenario Calculations	308
	C. Time-Dependent Multiple Scenario Calculations	315
	D. Limits of Model "Predictions"	318

[1]Present address: Petrochemicals Department, Corpus Christi Plant, E.I. du Pont de Nemours and Co., Inc., Ingleside, Texas.

HAZARD ASSESSMENT OF CHEMICALS:
Current Developments, Vol. 3

ISBN 0-12-312403-4

VI. Ozone Measurements 320
A. Total Ozone 320
B. Vertical Ozone Concentration Profile 325
VII. Prospects for Future Research 328
References 330

I. INTRODUCTION

As a focal point of environmental concerns over atmospheric pollution during the past decade or longer, ozone has presented atmospheric scientists with an intriguing paradox. Despite its toxicity as a constituent of smog pollution in the lower atmosphere (troposphere), its presence at higher altitudes in the stratosphere is acknowledged to be important as a filter for ultraviolet light. This article addresses the research associated with the potential influence of human activities on the amount of ozone in the stratospheric "ozone layer."

In 1840, the similarity between the odors accompanying electrical discharges in air and decomposition of water by "voltaic current" led Schonbein (*94*) to conclude that the common source was a gas which he called ozone (from the Greek for odor). The respiratory irritation accompanying this odor now underlies concerns over its presence at ground level as a component of photochemical smog.

The more esoteric problem of maintaining the continual presence of ozone in the stratosphere (i.e., between 12 and 50 km altitude) can be traced to Hartley's (*46*) conclusion that the ultraviolet cutoff in solar radiation can be attributed to absorption by ozone existing in the upper atmosphere at concentrations well above those common at ground level. The sensitivity of life forms to high levels of ultraviolet radiation led scientists to conclude that the emergence of life forms from the oceans onto land occurred after the ozone layer developed in the early atmosphere.

In the same paper, Hartley further provided a key stimulus to the study of ozone by noting that ground level concentration appeared to vary with wind direction, indicating that ozone measurements may be a valuable tool in meteorological studies. The absorption by ozone of ultraviolet light in turn led to methods for determination of the total column of ozone above the earth, with much of the early work being done by Fabry, Buisson, and Gotz. Technical advances enabled Dobson in the 1920s to design and build the first version of a spectrometer that has become a standard instrument for measuring ozone (*32*).

Dobson's interest had been aroused by noting that the deceleration rate of meteors in the 30- to 50-km-altitude range implied denser air and hence higher temperatures at those altitudes than expected based on extrapolation from below.

He concluded that UV absorption by ozone might be responsible for the heating, and set out to investigate the role of ozone in atmospheric dynamics. That foresight led to an initial network of 5 instruments, broadened to over 50 by 1958, and eventually to well over 100 before their numbers declined recently. These instruments were used to define the geographical variations in the ozone concentration. From this work, we now know that the total ozone column amount varies seasonally by a factor of greater than 2 near the polar regions, decreasing to about a 10% variation at the equator. Annual average column amounts are as much as 35% higher at the poles than at the equator. Furthermore, the distribution is not constant, and daily fluctuations of up to 10% are common. Throughout the series of measurements, there has been no evidence of significant long-term changes (see ref. *51* for further discussion).

Other work by Dobson and others was aimed at determining the vertical distribution of ozone. Hartley (*46*) had shown that peak concentrations were well above ground level, but only with the regular application of the Umkehr (inversion) technique, developed first by Gotz and Dobson (*44*), has the distribution become well known. The ozone "layer" is actually a relatively smooth distribution over altitude, with a peak in concentration of 6×10^{12} molecules/cm^3 between 20 and 25 km at middle and low latitudes. At higher latitudes, the peak concentration is larger, but occurs generally below 20 km.

As the meteorological interest in ozone was growing, Chapman (*20*) developed and published the first photochemical theory attempting to explain the presence of the ozone layer. This work underlies most of the recent research on stratospheric ozone. Chapman's theory included only oxygen species (O, O_2, and O_3) and sunlight. Hunt (*52*), Dutsch (*35*), Hessvedt (*47*), and Crutzen (*23*), among others, recognized the importance of hydrogen oxides (HO_x) in the late 1960s. This was followed still more recently by Johnston's (*53*) inclusion of nitrogen oxides (NO_x) and finally with consideration of chlorine oxides (ClO_x) by Stolarski and Cicerone (*98*) and by Wofsy and McElroy (*114*).

Only with these more complete theories did scientists begin to recognize the possibility of inadvertent modification of ozone concentrations by man. Chemists recognized that radical species from each of the hydrogen, nitrogen, and chlorine families participated in reactions of the form:

$$X + O_3 \rightarrow XO + O_2 \quad (1)$$

$$XO + O \rightarrow X + O_2 \quad (2)$$

$$\text{Net:} \quad O + O_3 \rightarrow 2O_2 \quad (3)$$

By mimicking Chapman's oxygen-only reaction [Eq. (3)], which destroys the so-called odd oxygen species, these cycles help to control the dynamic balance between production and destruction of ozone in a catalytic fashion, regenerating the active species. If anthropogenic releases of appropriate source gases were to increase the average concentration of the key hydrogen, nitrogen, or chlorine radicals, then perhaps the average amount of ozone present could be reduced.

In 1971, Johnston pointed out that projected fleets of supersonic transport (SST) aircraft then under consideration would directly release in their exhaust both H_2O (the source of active odd hydrogen radicals OH and HO_2) and NO (which along with NO_2 comprise the active odd nitrogen radicals), in quantities large enough to add significantly to the natural sources. Theorists formulated computer models to simulate the chemistry and transport involved, and calculated the effects of possible anthropogenic perturbations. An almost explosive expansion of stratospheric research followed quickly, including balloon-borne measurements in the stratosphere, more sophisticated modeling, and numerous laboratory studies of photolytic and chemical reaction rates. The scientific discussions of potential ozone reduction led to reviews of the subject by the Inadvertent Modification of the Stratosphere (IMOS) task force, the Climatic Impact Assessment Program (CIAP), and the National Academy of Sciences.

At about the time the proposed SST was rejected, primarily on economic grounds, Molina and Rowland (*69*) postulated a growing source for stratospheric chlorine through emission at ground level of chlorofluorocarbons (CFCs). The high stability of these gases, one of the keys to their utility, enables them to survive slow transport from the ground to the stratosphere. There, ultraviolet photolysis releases chlorine atoms, which, with ClO, comprise the active chlorine radicals. In contrast to the SST exhausts, CFCs were already being released, and the slow nature of their transport to the stratosphere implied that any effects might be delayed by up to several decades. The pace of research intensified, and government support was supplemented by the chlorofluorocarbon industry through the Chemical Manufacturers Association (CMA) Fluorocarbon Program Panel (FPP). The FFP had been organized in 1973 to investigate the long-term atmospheric fate of these stable compounds.

While researchers worked to better understand the potential reduction in ozone, officials in some countries, including the United States, initiated regulations to eliminate nonessential use of CFCs as aerosol propellants. International groups like the United Nations Environment Program's Coordinating Committee on the Ozone Layer were formed to consider the global aspects of the issue and to coordinate research efforts. Industry vigorously pursued the so far unsuccessful search for alternatives to CFCs in their various important uses. The atmospheric research, however, was uncovering still more complications within the chemistry and physics of the stratosphere. Many of the studies since 1976 have demonstrated connections among the various atmospheric chemical "families" which make quantitative assessment of a single perturbation such as CFCs increasingly difficult.

Measurements of nitrous oxide (N_2O, the major natural source gas for odd nitrogen radicals) demonstrated a trend toward increasing concentrations, and consequently toward a possible reduction in ozone (*109*). It became apparent that the true effect of CFCs would depend, in part, on future concentrations of N_2O. The documented increase in carbon dioxide (CO_2) concentrations (*56*) could also be important if, as theorized, it reduces temperatures in the stratosphere, thereby altering the rates of chemical processes and increasing the amount of ozone. During 1982, methane (CH_4) was added to the list of potential perturbants as a possible source of increased ozone (*10, 39, 81*).

The growing complexity of the problem has stimulated renewed efforts to return to the record of ozone measurements by Dobson instruments to look for any indication of a developing long-term trend. Such evidence could provide a potential verification of computer models used to forecast changes and provide an early warning of ongoing changes. Meanwhile, the most recent theoretical results indicate that the net effect of all ongoing perturbations may be little or no change in the total amount of ozone, but rather a partial redistribution from the upper stratosphere to somewhat lower altitudes. Such a result, if true, would eliminate concerns over increased ground-level ultraviolet exposure. However, the known connection between ozone and atmospheric dynamics raises questions about possible impacts on climate. At present, the understanding of this area is poor, although, at most, only minor effects have been suggested. Thus, research has, in a sense, come full circle to the questions posed by Hartley concerning the relationship between ozone and weather/climate experienced at the earth's surface.

Despite impressive advances in understanding of the stratosphere, especially during the past decade, the story is by no means complete. The remainder of this article will review the current understanding of man's potential impact on ozone, beginning in Section II with a discussion of the underlying chemical and physical processes. The expression of these processes in mathematical models of the ambient atmosphere and the comparison of the models with atmospheric measurements are covered in Section III. Sections IV and V review extrapolations of the model calculations into the future for scenarios involving hypothetical individual perturbations and more realistic, but also more complex, combined perturbations. A review of the available analyses of ozone measurements in Section VI is followed by some comments on the future directions of research in Section VII.

II. CHEMISTRY AND PHYSICS OF OZONE

A. Production and Loss of Ozone

Photodissociation of molecular oxygen by solar radiation at wavelengths below 240 nm liberates oxygen atoms, most of which recombine with oxygen molecules to form ozone.

$$O_2 + h\nu \rightarrow O + O \tag{4}$$

$$O + O_2 \xrightarrow{M} O_3 \tag{5}$$

Photodissociation of ozone establishes a very rapid equilibrium between O and O_3, such that loss of either is equivalent to ozone destruction.

$$O_3 + h\nu \rightarrow O_2 + O(^1D) \qquad (\lambda < 310 \text{ nm}) \tag{6a}$$

$$O_3 + h\nu \rightarrow O_2 + O(^3P) \qquad (\lambda < 1140 \text{ nm}) \tag{6b}$$

The continual production of oxygen atoms and ozone (together called "odd oxygen") is regulated in the atmosphere by several loss mechanisms. The simplest of these are the recombinations of odd oxygen.

$$O + O_3 \rightarrow 2O_2 \tag{7}$$

$$O + O \xrightarrow{M} O_2 \tag{8}$$

Other natural loss processes involve active radical species, and proceed via catalytic cycles. The dominant cycles follow one of two pathways:

$$\begin{array}{ll} & X + O_3 \rightarrow XO + O_2 \\ & XO + O \rightarrow X + O_2 \\ \text{net:} & O + O_3 \rightarrow 2O_2 \qquad (X = OH, Cl, NO) \end{array} \tag{Cycle 1}$$

or

$$\begin{array}{ll} & X + O_3 \rightarrow XO + O_2 \\ & XO + O_3 \rightarrow X + 2O_2 \\ \text{net:} & O_3 + O_3 \rightarrow 3O_2 \qquad (X = OH) \end{array} \tag{Cycle 2}$$

Other examples, as discussed later, consist of more than two steps in the catalytic cycles (*27*).

On a global scale, ozone is produced mainly in the low latitude stratosphere and transported from this major production region to higher latitudes. Ozone formed in the lower stratosphere also moves downward to the troposphere. This redistribution of O_3 with season, latitude, and altitude is governed by atmospheric transport phenomena, which are discussed in Section II,G.

The catalytic ozone destruction cycles are of greatest concern, since a small amount of X can potentially remove a great deal of ozone before being removed from the cycle by other processes. Concentrations of some source gases for the X species are known to be increasing with time. Therefore, understanding the chemical processes, sources, and sinks of X species is one of the most important tasks in understanding the possible modification of stratospheric ozone. The most important X species (NO, OH, and Cl) are conventionally categorized by the three family groups into which they fall, although they are not completely isolated (see Section II, E). The NO_x, or odd nitrogen group, is commonly taken to include NO, NO_2, NO_3, N_2O_5, HNO_3, HO_2NO_2, and frequently $ClONO_2$. Odd chlorine, or ClO_x, includes Cl, ClO, HCl, HOCl, and $ClONO_2$. Finally, odd hydrogen, HO_x, can be usefully defined as H, OH, HO_2, and H_2O_2.

The relative contributions of the various catalytic cycles to the removal of odd oxygen depend upon the budget and partitioning of HO_x, NO_x, and ClO_x radicals, which are governed by upward flux of the source species for the radicals, reaction rate constants (dependent on pressure and temperature), transport of chemical species, couplings among the three family groups, etc. It is, however, instructive to show a simplified picture of the relative importance of the HO_x, NO_x, and ClO_x catalytic effects as a function of altitude based on the rate-limiting step of each major ozone-removing reaction cycle in the present atmosphere. Such an example is given in Table I, and includes the direct loss from odd oxygen [$O_x = O(^3P)$, $O(^1D)$, and O_3]. Around the ozone concentration peak (20–30 km), ozone production is thought to be controlled mainly by the NO_x catalytic reactions (~ 70%) and by odd oxygen recombination (~ 10%). However, changes in the assumed chemical reaction rates for related reactions tend to alter the relative importance of various cycles. In 1979, for instance, HO_x chemistry was thought to be considerably more influential in controlling ozone than it is today.

TABLE I

Relative Contributions of the Major Rate-Limiting Chemical Reactions for the Catalytic Destruction of Ozone[a,b]

Altitude (km)	$k(O)(O_3)$ (%)	$k(NO_2)(O)$ (%)	$k(ClO)(O)$ (%)	$k(HO_2)(O)$ (%)	$k(HO_2)(O_3)$ (%)
20	11	70	1	1	26
25	12	78	5	1	8
30	10	69	8	2	3
35	11	68	13	4	1
40	18	53	16	10	0
45	29	24	10	31	0
50	25	7	4	52	0

[a] From Hudson *et al.*, 1982 (*51*), based on present day model atmosphere.
[b] Percentages are relative to the photolytic production of odd oxygen and do not include transport sources or sinks.

B. Chemistry of Odd Nitrogen

The important photochemical interactions among the major odd-nitrogen species are shown schematically in Fig. 1. Nitrous oxide (N_2O) is the main source gas for stratospheric NO_x. It is emitted from various sources such as fossil fuel combustion, biomass burning, oceans, soils, and nitrogen fertilizer application. The dominant source is thought to be microbiological activities (oceans and soils). The atmospheric N_2O concentration is increasing slowly. During the period from 1976 to 1980, a rate of increase of 0.2%/year was observed (*111*). Extrapolated from those data, the concentration of N_2O by the year 2000 is projected to be 5–7% above the present value (i.e., 315–320 ppb). Because of its chemical inertness in the troposphere, N_2O penetrates into the stratosphere, from which it is removed by photolysis:

$$N_2O + h\nu \rightarrow N_2 + O(^1D) \tag{9}$$

and by the reaction

$$N_2O + O(^1D) \rightarrow NO + NO \tag{10}$$

Approximately 3% of the global flux of N_2O is converted into NO by the latter reaction [Eq. (10)], which constitutes the major source for stratospheric NO_x. Other sources of NO_x are direct injection to the stratosphere from high-flying aircraft (jet engine exhaust) and formation there by nuclear bomb tests, lightning, and cosmic ray ionization.

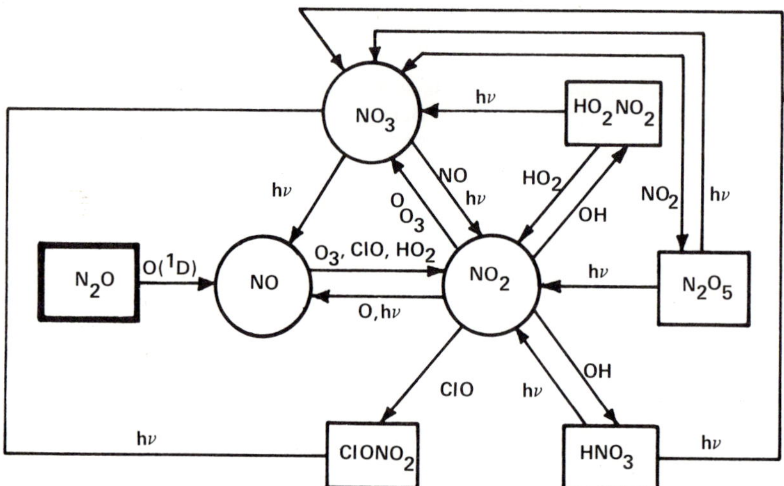

Fig. 1. Schematic representation of stratospheric NO_x chemistry. Long-lived source species are in heavy boxes, reservoir species in boxes, and free radicals in circles. The species next to the arrows participate in the major reaction pathways.

There are two catalytic cycles that remove odd oxygen:

$$NO + O_3 \rightarrow NO_2 + O_2$$
$$NO_2 + O \rightarrow NO + O_2$$
$$\text{net:}\quad O + O_3 \rightarrow 2O_2 \qquad \text{(Cycle 1)}$$

$$NO + O_3 \rightarrow NO_2 + O_2$$
$$NO_2 + O_3 \rightarrow NO_3 + O_2$$
$$NO_3 + h\nu \rightarrow NO + O_2$$
$$\text{net:}\quad O_3 + O_3 \rightarrow 3O_2 \qquad \text{(Cycle 2)}$$

Cycle 1 is far more important than Cycle 2, and is rate limited by the reaction $NO_2 + O \rightarrow NO + O_2$. As illustrated in Fig. 1, the two cycles are not independent reaction paths, but are regulated through interactions with other chemical species, with HO_x and ClO_x radicals playing important roles. These couplings among NO_x, HO_x, and ClO_x families will be discussed in Section II,E.

The removal of NO_x from the stratosphere is accomplished by sink species, HO_2NO_2 and HNO_3, and to some extent NO_2, all of which rain out after downward transport to the troposphere.

C. Chemistry of Odd Hydrogen

The HO_x and related hydrogen family species are shown in Fig. 2 with their major reaction paths in the stratosphere. The dominant source of HO_x radicals throughout the stratosphere is the reaction of $O(^1D)$ with H_2O.

$$H_2O + O(^1D) \rightarrow OH + OH \qquad (11)$$

The H_2O mixing ratio in the stratosphere is ~ 5 ppm (parts per million) and the major source of stratospheric H_2O is the exchange of air masses between the troposphere and the stratosphere. A significant amount of H_2O in the upper stratosphere is also produced through hydrocarbon (mainly CH_4) oxidation.

HO_x radicals affect odd oxygen via four major catalytic cyles.

$$O + OH \rightarrow O_2 + H$$
$$H + O_2 \rightarrow HO_2$$
$$O + HO_2 \rightarrow OH + O_2$$
$$\text{net:}\quad O + O \rightarrow O_2 \qquad \text{(Cycle 1)}$$

$$O + OH \rightarrow O_2 + H$$
$$H + O_3 \rightarrow OH + O_2$$
$$\text{net:}\quad O + O_3 \rightarrow 2O_2 \qquad \text{(Cycle 2)}$$

$$OH + O_3 \rightarrow HO_2 + O_2$$
$$HO_2 + O \rightarrow OH + O_2$$
$$\text{net:}\quad O + O_3 \rightarrow 2O_2 \qquad \text{(Cycle 3)}$$

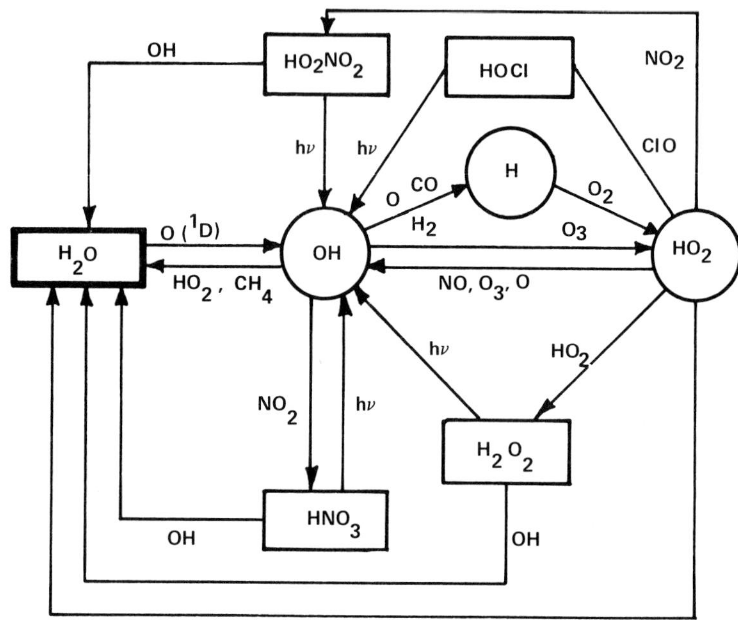

Fig. 2. Schematic representation of stratospheric HO_x chemistry. Symbols are as in Fig. 1.

$$OH + O_3 \rightarrow HO_2 + O_2$$
$$HO_2 + O_3 \rightarrow OH + 2O_2$$
$$\text{net:} \quad O_3 + O_3 \rightarrow 3O_2 \qquad \text{(Cycle 4)}$$

Above 40 km, HO_x Cycles 1 and 2 are of major importance to the destruction of odd oxygen, whereas below 40 km, Cycle 3 becomes more important as the O to O_3 ratio and the H atom concentration decrease. Below 30 km, Cycle 4 dominates.

The major removal mechanisms for HO_x radicals in the stratosphere are

$$OH + HO_2 \rightarrow H_2O + O_2 \qquad (12)$$

$$OH + HNO_3 \rightarrow H_2O + NO_3 \qquad (13)$$

$$OH + HO_2NO_2 \rightarrow H_2O + NO_2 + O_2 \qquad (14)$$

Between 40 and 50 km, reaction (12) is the major route of loss of HO_x. In the region 25–40 km, reactions (12), (13), and (14) control the HO_x budget; below 25 km, HO_x is essentially controlled by (13) and (14). It should be noted that the reactions between OH and nitric/pernitric acids [Eqs. (13) and (14)] are the rate-limiting steps of the HO_x recombination cycles. That is, these reactions produce NO_2-catalyzed HO_x removal:

$$\begin{array}{llll} & HO_2 + NO_2 & \xrightarrow{M} HO_2NO_2 & \\ & OH + HO_2NO_2 & \rightarrow H_2O & + O_2 + NO_2 \\ \text{net:} & OH + HO_2 & \rightarrow H_2O & + O_2 \end{array}$$

$$\begin{array}{llll} & OH + NO_2 & \xrightarrow{M} HNO_3 & \\ & OH + HNO_3 & \rightarrow H_2O & + NO_3 \\ & NO_3 + h\nu & \rightarrow NO_2 & + O \\ \text{net:} & OH + OH & \rightarrow H_2O & + O \end{array}$$

D. Chemistry of Odd Chlorine

The important chlorine chemistry in the stratosphere is shown schematically in Fig. 3. The major Cl_x source in the stratosphere is the photolysis of various chlorofluorocarbons (CFCs) and chlorocarbons: $CFCl_3$ (CFC-11), CF_2Cl_2 (CFC-12), CH_3Cl, CCl_4, and CH_3CCl_3. All are believed to be anthropogenic in origin except for CH_3Cl, a natural constituent with the ocean as its main source. The major uses of CFCs are for refrigeration, air-conditioning, plastic foams, food freezants, cleaning agents, and aerosal propellants.

In the contemporary atmosphere, the principal sources of stratospheric chlorine are presently thought to be CH_3Cl, CCl_4, CH_3CCl_3, CFC-11, and CFC-12. Table II shows their globally averaged tropospheric abundances, their rates of increase in early 1980, and their lifetimes in the atmosphere. The relatively short lifetimes of CH_3Cl and CH_3CCl_3 are due to the reaction with OH radicals in the troposphere. CFC-11, CFC-12, and CCl_4 are very unreactive in the troposphere and eventually are destroyed via photolysis to produce free chlorine atoms in the stratosphere. The rate of increase of CFC concentration in the stratosphere has

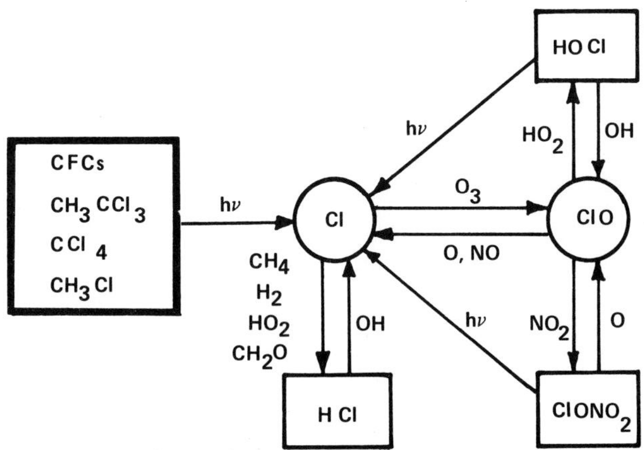

Fig. 3. Schematic representation of stratospheric ClO_x chemistry. Symbols are as in Fig. 1.

TABLE II

Major Chlorine Source Gases in the Atmosphere[a]

	CH_3Cl	CF_2Cl_2	$CFCl_3$	CH_3CCl_3	CCl_4
Tropospheric abundance (pptv, 1981)	620	285	168	123	118
Rate of increase per year (%)	0	6.0	5.7	8.7	1.8
Atmospheric lifetime (years)	1	120	67	10	50

[a] Data from refs. *26, 51, 84, 95.*

decreased in recent years; during the mid-1970s, the rate was about 12%/year, and in January 1980 about 6%/year. The rate of ground-level release declined about 11% between 1974 and 1982 (*20a*).

There are three major catalytic cycles involving chlorine.

$$\begin{array}{llll} Cl & + O_3 & \rightarrow ClO & + O_2 \\ ClO & + O & \rightarrow Cl & + O_2 \\ \text{net:}\ O & + O_3 & \rightarrow 2O_2 & \end{array} \quad \text{(Cycle 1)}$$

$$\begin{array}{llll} Cl & + O_3 & \rightarrow ClO & + O_2 \\ ClO & + HO_2 & \rightarrow HOCl & + O_2 \\ OH & + O_3 & \rightarrow HO_2 & + O_2 \\ HOCl & + h\nu & \rightarrow HO & + Cl \\ \text{net:}\ O_3 & + O_3 & \rightarrow 3O_2 & \end{array} \quad \text{(Cycle 2)}$$

$$\begin{array}{llll} Cl & + O_3 & \rightarrow ClO & + O_2 \\ ClO & + NO_2 & \rightarrow ClONO_2 & \\ NO & + O_3 & \rightarrow NO_2 & + O_2 \\ ClONO_2 & + h\nu & \rightarrow Cl & + NO_3 \\ NO_3 & + h\nu & \rightarrow NO & + O_2 \\ \text{net:}\ O_3 & + O_3 & \rightarrow 3O_2 & \end{array} \quad \text{(Cycle 3)}$$

At altitudes above ~ 25 km, Cycle 1 is dominant, with the contribution from Cycle 2 approaching 25% at 25 km and diminishing to less than 1% at 40 km. Below 25 km, the three cycles are competitive. However, in this lower part of the stratosphere, the reactions coupling the NO_x, HO_x, and ClO_x families become increasingly important, as discussed in Section II,E, and the concentration of the rate-limiting radical, ClO, is extremely sensitive to the OH radical concentration. The calculated contribution of these catalytic cycles to ozone destruction in the lower stratosphere is minimal according to the current set of kinetic data, and the chlorine-induced destruction of ozone in current atmospheric models is confined to altitudes above 30 km.

The major removal of ClO_x radicals from the stratosphere is due to downward transport of the sink gas, HCl, which rains out in the troposphere.

E. Coupling of $NO_x/HO_x/ClO_x$ Families

The three families NO_x, HO_x, and ClO_x, although convenient for an initial overview of stratospheric chemistry, are not independent of each other. The many reactions coupling the families with each other cause significant changes in the anticipated ozone destruction efficiency of each catalytic cycle. The following examples illustrate some of the more important couplings.

$$ClO + HO_2 \rightarrow HOCl + O_2 \quad (15)$$

$$ClO + NO_2 \xrightarrow{M} ClONO_2 \quad (16)$$

$$ClO + NO \rightarrow Cl + NO_2 \quad (17)$$

In reactions (15) and (16), the rate-limiting species in the major catalytic cycles (ClO, NO_2, and HO_2 radicals) are temporarily trapped as the less reactive species HOCl and $ClONO_2$. These are eventually photolyzed, producing Cl, OH, and NO radicals, but this formation and destruction process temporarily removes the radicals from the respective catalytic cycles. Thus, HOCl and $ClONO_2$ are properly called "temporary" sinks or "reservoir" species. N_2O_5, HO_2NO_2, and H_2O_2 also act as temporary sinks of HO_2 and NO_2 radicals. The photochemical lifetime of these temporary sinks in the daytime is generally a few hours.

The third reaction [Eq. (17)] can be called a "switching reaction" in that it transfers the odd oxygen atom from the ClO_x group to the NO_x group. This type of reaction is important in determining the relative contribution of ClO_x and NO_x to odd oxygen removal.

The important coupling between ClO and OH in the lower stratosphere has already been mentioned. Below ~ 25 km, NO and NO_2 partitioning is mainly controlled by the following reactions (*75*):

$$NO_2 + OH \xrightarrow{M} HNO_3 \quad (18)$$

$$HNO_3 + h\nu \rightarrow NO_2 + OH \quad (19)$$

$$NO_2 + h\nu \rightarrow NO + O \quad (20)$$

$$NO + O_3 \rightarrow NO_2 + O_2 \quad (21)$$

In a steady-state approximation

$$[NO_2] \propto 1/[OH] \quad (22)$$

$$[NO] \propto 1/[OH] \quad (23)$$

That is, both nitrogen dioxide and nitric oxide concentrations vary inversely as the concentration of OH. On the other hand, ClO_x radicals are mainly controlled by the series of reactions below.

$$HCl + OH \rightarrow H_2O + Cl \qquad (24)$$

$$Cl + O_3 \rightarrow ClO + O_2 \qquad (25)$$

$$Cl + (CH_4, H_2, H_2CO) \rightarrow HCl + (CH_3, H, HCO) \qquad (26)$$

$$ClO + NO \rightarrow NO_2 + Cl \qquad (27)$$

Assuming steady-state conditions, [ClO] will be proportional to [OH]/[NO], and this relationship along with Eq. (23) shows that [ClO] increases approximately as $[ClO] \propto [OH]^2$.

The important point here is that the OH budget sensitively affects the ClO abundance in the lower stratosphere. Thus, the rates for the OH sink reactions (12), (13), and (14) significantly influence NO_x- and ClO_x-catalyzed ozone destruction. The recent changes in these reaction rates have significantly modified our view of the relative impact of various perturbations (e.g., N_2O, CFCs) on ozone.

The coupling discussed here is predominantly important in the lower stratosphere, and considerably complicates chemistry there. Many of the reactions involved have been relatively recent additions to the list of those included in atmospheric models. While in the earlier efforts to evaluate potential changes to ozone it was thought to be sufficient to isolate a given family group and consider its impact on ozone loss rates, the recent assessments have been much broader. Clearly, the impact of an increase in stratospheric NO_x is quantitatively dependent upon the total amount of ClO_x present, and vice versa. The effectiveness of each as an ozone loss mechanism, in turn, depends strongly on the HO_x budget. The impact of this expanded view becomes most apparent in a review of the model calculations, and is the subject of Section IV.

F. Other Chemistry

The photochemical reactions mentioned above (including those referred to in Figs. 1, 2, and 3) are by no means an exhaustive list of the chemical reactions that are being used in current model calculations. About 50 substances and 150 reactions are included in a realistic stratospheric photochemical scheme. The complete list of recent photochemical reactions and their rate constants is given in a JPL publication [JPL 82–57, (*55*)] and in the recent review by WMO/NASA (*118*). Heterogeneous chemistry (except rainout of soluble species and dry deposition of ozone) and ion reactions have not so far been incorporated into most models, given the general consensus that they have little impact on ozone production or loss. It should be mentioned, however, that stratospheric aerosols consisting of H_2SO_4, H_2O, HNO_3, and other stratospheric gases have been detected (*8*). The nucleation mechanism of stratospheric aerosols and the possible connections between ions and aerosols are poorly understood. In general, though, the small relative concentrations of particles and ions seem to preclude significant interactions with gas phase neutral species.

There has been speculation that some important sodium reactions might exist in the stratosphere (*92*). The photochemical reactions with Na, NaO_2, NaOH, and NaCl could influence Cl_x partitioning if the rates of key reactions were fast enough. Most of the reaction rates, however, have not yet been measured.

Chemistry of the other halogens (I, Br, and F) parallels that of chlorine. Iodine and bromine are unimportant in their impact on stratospheric chemistry due to extremely low abundances. The atmospheric budget of fluorine-containing compounds is as large as that of Cl_x. However, the extremely strong chemical bonding of fluorocarbons, and the rapid conversion of released fluorine to the photolytically stable compound HF, virtually prohibit the release of free fluorine radicals in the stratosphere. For instance, the reaction $HF + OH \rightarrow H_2O + F$ does not occur under stratospheric conditions because of its high endothermicity, in contrast with the analogous Cl reaction [Eq. (24)].

G. Transport of Chemical Species in the Atmosphere

The atmosphere can be visualized as a complex, three-dimensional chemical reactor with incomplete mixing. Many of the important trace gases have their source near the earth's surface (for example, the biospheric releases of the source gases H_2, CO, and CH_4). The source gases are transported by the (three-dimensional) wind patterns of the atmosphere, and they are gradually mixed with other air masses (e.g., in the stratosphere) having lower concentrations of these species. Chemical reactions take place along the trajectory of each air mass.

The fundamental equation governing the time evolution of the mixing ratio (x) of a gas species in a given packet of air is Liouville's equation

$$dx/dt = P - L \qquad (28)$$

in which P and L represent the net production and loss rates in the gas volume. Here, the time derivative must be evaluated along the trajectory of the air packet. This is the "Lagrangian" formulation of the problem, in which the coordinate system moves with the selected particles. In this reference frame, only microscopic mixing, based on very localized turbulence or molecular diffusion, can occur.

For computational convenience, it is conventional to adopt an "Eulerian" view, in which the coordinates are fixed to the earth and air packets move through it. Then the transport equation becomes

$$\partial x/\partial t + \mathbf{v} \cdot \nabla x = P - L \qquad (29)$$

where $\mathbf{v}$ represents the instantaneous wind velocity. In principle, this equation gives a complete description of atmospheric transport on a macroscopic scale, ignoring microturbulence. However, it requires a three-dimensional model and very fine resolution to capture the variations of the wind field. Although a few three-dimensional models exist (see *51*, Chapter 2), both their resolution and their description of chemistry are very limited. A finite resolution results in "subgrid-scale diffusion," and sometimes additional parameterized diffusion processes are added to represent transport processes on scales too small to be modeled.

There is a hierarchy of mixing times evident in the atmosphere. As indicated by studies of long-lived tracers, including nuclear particles from bomb tests, tropospheric mixing takes place relatively rapidly in longitude (a few weeks) but much more slowly in latitude (~ 14 months). Vertical mixing between the surface and the top of the troposphere at 10–15 km takes about 4 months. The relatively short times are due to the nature of the troposphere, which is a region of turbulent mixing caused by a rapid decrease in temperature with altitude (~ 6.5 K/km) leading to convective instabilities. Above the tropopause (10–15 km), however, the temperature increases with altitude. This permanent inversion layer, the stratosphere, is stable and not as well mixed as the troposphere. The transport of species from the tropopause to the mid-stratosphere takes several years.

As discussed above, it is prohibitively expensive to model the chemistry of the stratosphere in three dimensions. A two-dimensional approximation can be obtained by integrating Eq. (29) over longitude. The integration or averaging process introduces terms involving covariances between zonal velocity and mixing ratio deviations from the means. The effects of these covariance terms are assumed to be characterizable via a diffusive eddy transport tensor (*88*). The resulting transport equation (30) is expressed here in coordinates: $y = r_e \sin(\lambda)$ and $z = -\ln(p/p_0)$, where r_e is the radius of the earth, λ = latitude, p is the local atmospheric pressure, and p_0 is the pressure at the earth's surface.

$$\begin{aligned}\partial x/\partial t + v_y\, \partial x/\partial y + v_z\, \partial x/\partial z &= P - L \\ &+ (\partial/\partial y)(K_{yz}\, \partial x/\partial z + K_{yy}\, \partial x/\partial y) \\ &+ (\partial/\partial z)(K_{zy}\, \partial x/\partial y + K_{zz}\, \partial x/\partial z) \qquad (30)\end{aligned}$$

The eddy diffusion coefficients (K_{ij}) are typically "derived" by fitting observations of long-lived trace species, including potential temperature, radioactive decay products from nuclear tests, and certain chemical species (N_2O, CH_4, O_3). The appropriate wind field (**v**) was until recently thought to be the mean observed (Eulerian) transport (e.g., *62*). More recently, however, Lagrangian-mean concepts (*7, 34, 49*) have been advanced, suggesting that a more accurate picture is obtained in Eq. (30) if an approximation to the diabatic circulation is used and the K tensor is assumed to be symmetric. The Lagrangian-mean circulation has a strong direct Hadley cell (see Fig. 4), giving rising motions in the tropics and receding motions at high latitudes in the lower stratosphere. The ability of such a

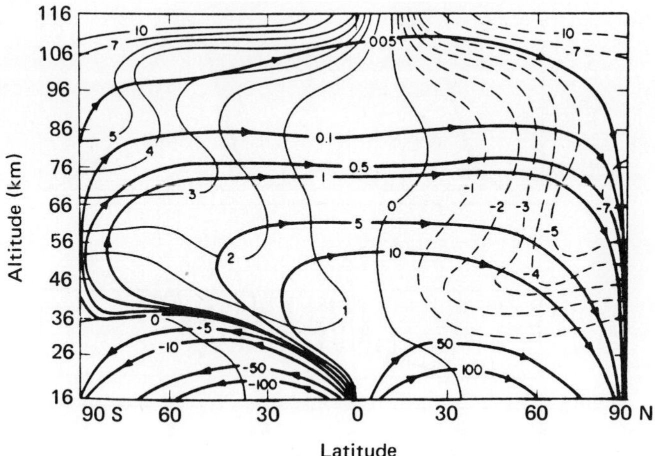

Fig. 4. Calculated meridional stream function (heavy solid lines, in kg/m/sec) and the distribution of net diabatic heating (light lines, in K/day) for northern hemisphere winter solstice. From Garcia and Solomon, 1983 (*42*), with permission.

Lagrangian-mean transport description to better characterize the observed distributions of major stratospheric trace species has recently been demonstrated by Miller *et al.* (*68*) and Solomon and Garcia (*96*).

A further level of approximation involves integrating Eq. (30) over latitude to obtain a one-dimensional (altitude) model. One-dimensional atmospheric models were until relatively recently the only ones with sufficiently complete chemistry to simulate potential man-made perturbations to stratospheric chemistry. These models, however, neglect potentially important latitudinal covariances between transport and photochemistry. For example, latitudinal covariances between species concentrations and destruction mechanisms (i.e., by the OH radical or photolysis) can be important in determining the atmospheric removal rates of species like CH_3CCl_3 (*30*) and CFCs (*80*). They can also be important in determining the local response of atmospheric temperatures to changes in the global CO_2 concentration (*76*). For example, the polar regions are expected to show much larger temperature changes than are equatorial regions.

H. Radiation

The earth's atmosphere and surface are heated by ultraviolet and visible radiation from the sun, whose effective radiative temperature is ~ 5800 K, and cooled by emitting thermal infrared radiation to space. Figure 5 illustrates the opacity of the atmosphere to electromagnetic radiation in the relevant wavelength region. The atmosphere is opaque to radiation with wavelength shorter than about 300

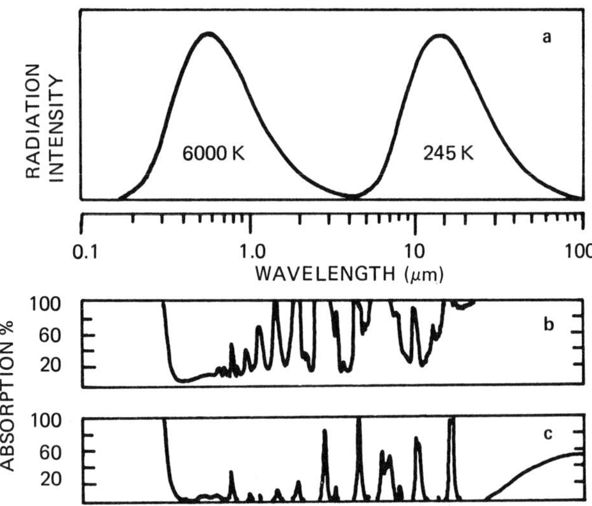

Fig. 5. (a) Black-body curves for the sun (6000 K) and midtroposphere (245 K). (b) Absorption spectrum of atmospheric gases above the surface. (c) Absorption spectrum of atmospheric gases above 11 km altitude. From Goody, 1964 (*43a*), with permission.

nm due to absorption by O_2 (< 200 nm) and O_3 (200–300 nm). This ultraviolet absorption gives the dominant local solar heating above the tropopause, and the absorption by ozone is directly responsible for maintaining the stratospheric temperature inversion. Other species that absorb a significant fraction of the sun's visible and ultraviolet radiation are H_2O, CO_2, and NO_2.

Solar radiation is also scattered by molecules in the lower atmosphere. This scattering, adequately characterized by Rayleigh's scattering phase function, is important because it increases the amount of ultraviolet light available to photolyze important trace chemical species. The importance of Rayleigh scattering to atmospheric photochemistry models was first suggested only relatively recently (*63*).

A warm body emits radiation with wavelength characteristic of its temperature. For the atmosphere (T = 200–300 K), the thermal radiation falls in the infrared portion of the spectrum (~ 5–20 μm). In this wavelength region, the three major atmospheric absorbing gases are H_2O, CO_2, and O_3. As shown in Fig. 6, water vapor absorbs efficiently at both short and long infrared wavelengths. The H_2O absorption takes place primarily in the troposphere, in which the water vapor mixing ratio is thousands of times larger than it is in the stratosphere. This mixing ratio difference is thought to be due to the "cold trap," i.e., the temperature minimum at the tropopause which freezes water and keeps most of it from reaching the stratosphere (*36a*).

In the atmospheric "window" region near 10 μm, where infrared radiation is

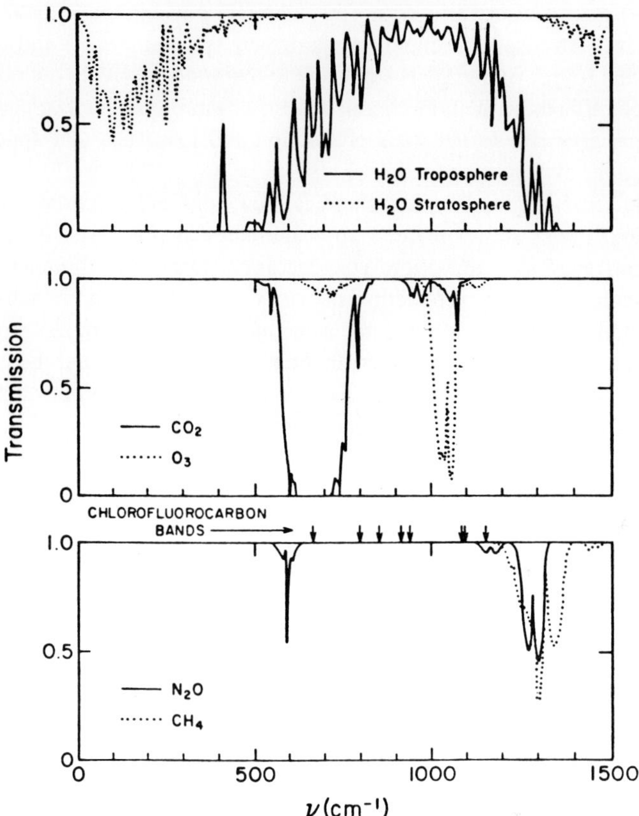

Fig. 6. Transmission of thermal radiation by atmospheric gases. Arrows indicate locations of CFC bands. From Lacis *et al.*, 1981 (*58*), with permission.

not absorbed by water, CO_2 and O_3 absorb significantly. Other gases with strong absorption bands near the 10 μm window (e.g., CH_4, N_2O, fluorocarbons) can help trap infrared radiation near the earth's surface and contribute to global warming. This so-called greenhouse effect will be discussed in Section IV,E.

III. MODELING THE AMBIENT ATMOSPHERE

A. Model Structure

Because the chemistry of minor trace gases in the troposphere and stratosphere is quite complicated and highly nonlinear, numerical models of atmospheric chemistry are necessarily large and complex. Typical models include on the order of 30 chemical species with 150 chemical and photochemical reactions relating them. Atmospheric chemistry is driven by solar visible and ultraviolet light, and this source of energy is sufficient to maintain relatively high concentrations of free radicals with short chemical destruction lifetimes. The photochemical reactions give rise to the local production and loss rates that appear as the right-hand side of Eqs. (29) and (30).

Basically, three types of photochemical reactions are important in the troposphere and lower stratosphere: simple bimolecular reactions, three-body recombination reactions, and photolytic decompositions. The bimolecular reactions are included in the production and loss terms for each chemical species using the standard rate expression for a second-order reaction: the product of a rate constant (usually specified as a function of temperature) and the concentrations of the two reactants. In some cases, rate constant expressions also include experimentally observed pressure dependences.

Three-body reactions are included essentially with a standard third-order rate expression, with the third-body collision effectiveness based on a weighted average of the rates for O_2 and N_2. The range of pressures encountered (<1 to 10^3 mbar) requires explicit consideration of low and high pressure limiting rate constants. For both two- and three-body reactions, knowledge of pressure, temperature, and species concentrations permits calculation of production and loss contributions for the species involved in a reaction at a given location in the atmosphere. For photolytic reactions, the required information includes initial concentration of the reactant, its photoabsorption cross section as a function of frequency, quantum yields for particular products, and frequency-dependent solar flux as a function of altitude.

The amount of solar radiation in a small wavelength interval reaching a given altitude in a nonscattering atmosphere is

$$F(\tau) = F_0 \exp(-\tau) \tag{31}$$

where F_0 is the flux (photons/cm^2/sec) at the top of the atmosphere and τ is the optical depth at that altitude. The optical depth is calculated using Eq. (32) to account for absorption at higher altitudes.

$$\tau = \sum_i \int_z^\infty [A_i]\sigma_i dz' \tag{32}$$

The integral is the product of the altitude-dependent concentration, $[A_i]$, and the absorption coefficient σ_i, and the sum is over all species i that absorb in the wavelength region of interest. In practice, only the species O_2, O_3, and NO_2 absorb visible and/or ultraviolet light sufficiently that their contributions must be included in the sum of Eq. (32). Models often use a multiple regression scheme to derive the effective photolysis rates for all species based on several precalculated "base" atmospheres with varying concentrations of O_3 and NO_2.

In the earth's atmosphere, scattering of light by air molecules is also important, particularly below 30 km. The rate equation then must contain an additional multiplicative factor to account for scattering. Several methods have been developed to include multiple scattering in the calculation of photolytic reaction rates (*63, 101, 121*).

As discussed in Section II,G, the size of the atmospheric models (i.e., the length of the computer program) depends primarily on the dimensionality used. A one-dimensional (1-D) model, with altitude as the only coordinate, typically has about 20 vertical grid points. Reduction of the partial differential mass balance equation to a set of coupled ordinary differential equations is usually done using the method of lines on a fixed, finite grid (*37, 104*). For typical one-dimensional models, there are approximately (30 species) $\times$ (20 vertical grid points) = 600 variables. The time scales of these variables cover many orders of magnitude, from seconds to centuries. Such systems are mathematically designated as "stiff."

The approximately 600 coupled, stiff ordinary differential equations have generally been solved by two methods. "Family grouping" reduces the stiffness problem by directly calculating only the relatively longer lived family groups (e.g., $NO_y = NO + NO_2 + HONO_3 + HO_2NO_2 + 2 \times N_2O_5 + NO_3$). It also reduces the total number of equations to be solved prognostically, with the breakdown of the families into individual species concentrations being done diagnostically off-line. This method requires care and insight in the construction of the grouping scheme and in the subsequent partitioning of the resulting family group into individual species concentrations. It can handle the important cross-linking species (e.g., HO_2NO_2 = HO_x and NO_x, $ClONO_2$ = NO_x and ClO_x) and coupling reactions (e.g., $ClO + NO \rightarrow Cl + NO_2$) only approximately, usually through successive iterations.

A more straightforward method is to solve the 600 equations directly, using a stiff equation solver (*19, 66*). The stiff methods (*43*) use a large Jacobean matrix (600 $\times$ 600), so they are implemented on computers with large core storage.

Two-dimensional (2-D) models introduce a latitudinal coordinate as an additional variable, expanding the size of the computational and data base problem by a factor of 10–20. The du Pont 2-D model, for example, solves approximately 7000 simultaneous coupled ordinary differential equations. The 7000 $\times$ 7000 Jacobean matrix cannot fit in the memory of any existing computer, but the matrix is block banded and special techniques have been developed that allow the model to run on a class VI computer. The simulation of a single year's chemistry costs about $100.

A three-dimensional (3-D) model with self-consistently determined transport and complete chemistry would involve an increase in complexity and size by about two orders of magnitude. Not only are additional grid points added in longitude, but the calculation of the atmospheric circulation adds an additional complication. Although there are special-purpose 3-D models that include summary chemistry (*25*) and parameterized approaches are currently being developed, it is unlikely that a full three-dimensional treatment of atmospheric chemistry will be feasible with currently available computers.

Boundary conditions must also be specified in order to solve the set of differential equations. Most atmospheric chemistry models have a lower boundary at the ground or the tropopause (~ 12 km) and an upper boundary above the stratopause (~ 50 km). At the lower boundary, fixed mixing ratios are ordinarily specified for species with lifetimes longer than a few years (e.g., N_2O = 300 ppbv, CH_4 = 1.6 ppmv, CCl_4 = 125 pptv). Short-lived species (e.g., O, OH) are assumed to have a zero flux at the ground. Ozone is usually treated with a fixed dry deposition velocity. Finally, soluble gases are "rained out" of the troposphere with a fixed time constant (5–10 days).

At the "top" of the model, the boundary condition for most species is zero flux. Occasionally, gases with either well-characterized outflow rates (e.g., H_2) or a mesospheric source (e.g., NO) are given fixed fluxes at the upper boundary. For calculating solar flux at the top boundary, a layer of O_2 and O_3 with fixed mixing ratio above the top grid point is usually assumed.

Given this entire set of differential equations and boundary conditions, the chemistry–transport models are used to solve for the concentration of each chemical species at each grid point in the model atmosphere. The solution may be either a time-dependent integration of the differential equations to demonstrate evolution of the atmosphere under a given set of circumstances, or a direct calculation of the dynamic equilibrium reached after many years of constant input fluxes. The latter solution is referred to as steady state.

A second important class of atmospheric models consists of those used to calculate the thermal structure and dynamics of the atmosphere. Although several complex models (general circulation models or GCMs) of tropospheric dynamics and climate are available, few such models have much detail in the stratosphere. A summary of existing two- and three-dimensional dynamic models of the stratosphere are given in WMO/NASA (*118*), Chapter 3. For brevity here, we will not discuss these models.

The climatic analog of the one-dimensional stratospheric chemistry model is the one-dimensional radiative–convective (RC) model (*86*). These models consider the thermal balance of the atmosphere and have one (vertical) spatial coordinate. The thermal structure of the atmosphere is basically a balance between heating due to the absorption of visible and ultraviolet sunlight by molecules and cooling by emission of radiation from the ground and from gases thermally active in the infrared. Solar heating is determined predominantly by the ozone concentration in the stratosphere and the amount of water vapor in the troposphere. The most important infrared absorbers are H_2O in the troposphere and CO_2 in the stratosphere.

The assumptions of RC models are simple, but their results are reasonably consistent with those of higher dimensionality climate models. In the stratosphere, constituent gases are assumed to be in local radiative equilibrium, i.e., the local radiative heating and cooling rates are assumed equal. In the troposphere, radiative equilibrium would imply a temperature vertical lapse rate (dT/dz) larger than can be supported in stable equilibrium by the air, and convection would develop. Most RC models assume that the tropospheric lapse rate is either fixed ($dT/dz = -6.5$ K/km) or given by the moist adiabatic value (*59, 76, 86*). Typically, the water vapor content of the troposphere is determined by fixing the relative humidity profile.

The RC models calculate the local heating and cooling rates in the atmosphere assuming local thermodynamic equilibrium through a chosen treatment of infrared radiative transport. Some RC models are "broadband," considering only a few spectral regions in the IR, whereas others (e.g., *82, 107*) include more detail and have on the order of 40 IR spectral intervals. Atmospheric transmission functions for the major infrared absorbers (H_2O, CO_2, O_3, CH_4, N_2O) are obtained from band-model fits to the line-by-line tabulated absorption data. Typically, about 20 vertical grid points are used, covering the domain from the surface to the stratopause or above.

The RC models simulate the global mean temperature profile of the atmosphere reasonably well, as shown in Fig. 7 (*82*). They are inexpensive to run and thus can be used to investigate the effect on atmospheric and surface temperatures of various potential perturbations to trace gas concentrations. They can also be coupled to 1-D transport and chemistry models to give a self-consistent model of the chemical and thermal structure of the atmosphere.

B. Model Applications

Atmospheric models are a valuable tool for several types of investigations. First, they provide a diagnostic model for understanding the complex interaction between chemistry and transport for species like ozone. The ozone content of the atmosphere varies markedly with latitude and season. Two-dimensional models especially can be used to simulate these variations, and adjustments in their transport parameters or chemical data base can indicate the sensitivity of the modeled ozone distribution to chemical and transport processes. The stratospheric distributions of other long-lived species (N_2O, H_2, CH_4, CFCs) can also be assessed with the use of 2-D models.

A second use of the models is for direct comparison with the observations of shorter lived trace gases. Where discrepancies occur, they may indicate an inaccuracy or omission in the present understanding of the chemical scheme. The

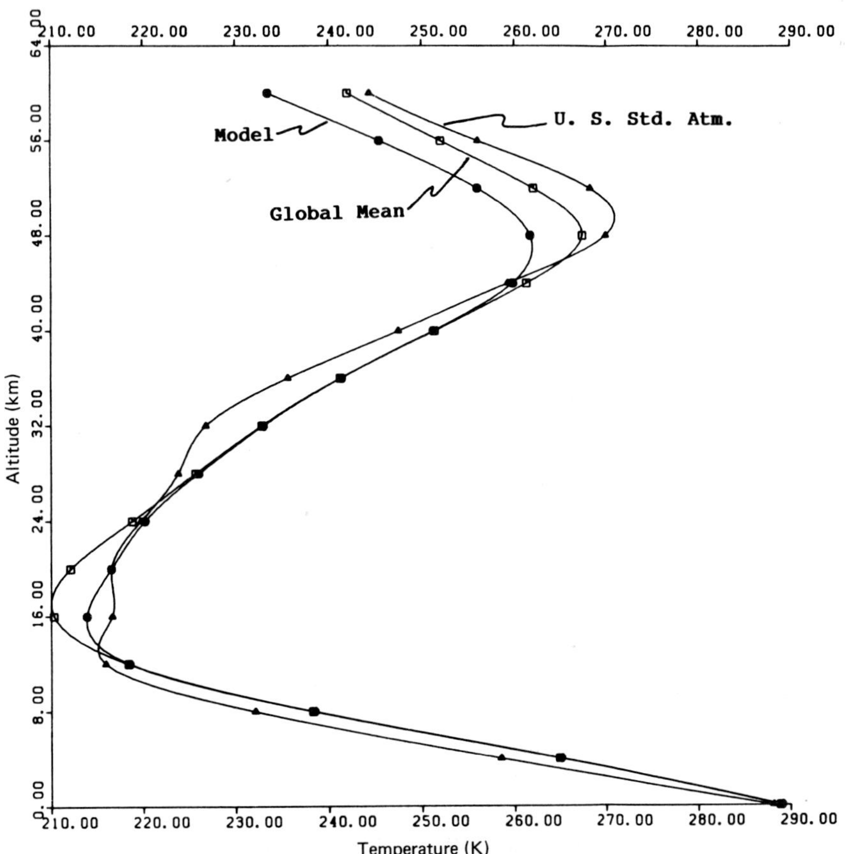

Fig. 7. Vertical atmospheric temperature profiles, degrees Kelvin versus altitude (km). ▲, United States Standard Atmosphere; □, observed global mean temperature; ●, du Pont coupled model calculation.

models can isolate key laboratory or atmospheric measurements that are required to fill gaps in our understanding of stratospheric chemistry and indicate the optimal conditions for detection of trace species.

The most widely known application of models is to investigate the sensitivity of the atmosphere to potential changes in trace gas concentrations that may occur in the future. For example, it is known that the atmospheric CO_2 level is increasing. Models can readily calculate the potential impact, based on current knowledge, of an increased CO_2 concentration on ozone and on the surface temperature. This type of study measuring model sensitivity is important since it will likely be some time before direct measurements (e.g., of surface temperature) can isolate the true atmospheric sensitivity to such changes.

Another potential use of models is to simulate the evolution of the atmosphere in the recent past. Researchers are now accumulating atmospheric measurements with sufficient accuracy that reasonable assumptions can be made regarding the historical temporal evolution of the concentrations of some important source gases (e.g., N_2O, CH_4, CFCs). Using the measured concentration changes over the last 10–20 years, for example, modelers can simulate the expected changes in ozone or surface temperature. Comparison with observations can then be made to estimate the validity of the models and to assess whether such changes continued into the near future could be detected with some level of confidence.

The simple extension of such work to simulate future atmospheric concentrations allows consideration of different scenarios for the effects of man's future activities on the atmosphere. Such assessments should be viewed as a continually changing dynamic process, since both the likely scenarios and model data bases will continually be updated. However, they provide our best tool for synthesizing available knowledge to define the evolution of the atmosphere as it is currently understood. Such efforts at simulating the time dependence of the atmosphere can be placed in perspective by reviewing the adequacy of model representations for the present atmosphere.

Sections III,C–E briefly compare two-dimensional model calculations with available atmospheric measurements. The measurements cited are those critically reviewed in WMO/NASA (*118*) where references and a more complete discussion can be found. Our presentation here is a synopsis of the results given in Miller *et al.* (*68*) and WMO/NASA (*118*).

C. Long-Lived Chemical Species

Chemical species like O_3 and various source gases (N_2O, CFCs, CH_4, and H_2O) have relatively long atmospheric lifetimes. For currently available 1-D or 2-D chemical models, one or more of these species are commonly used to derive or to fit transport parameters. Given that practice, a comparison of calculated concentration with available measurements is primarily a test of the validity of the transport parameterization employed in the model calculation. Since latitudinal and seasonal variations can be significant, the most appropriate comparison for these species can be made with 2-D models. Calculations shown in the figures are from the models of Miller *et al.* (*68*) (du Pont) and Sze *et al.* (*100*) (AER). The chemical reaction rate set used in both models is that recommended in WMO/NASA (*118*).

1. O_3

The most temporally extensive part of the ozone data base is the historical record of columnar ozone from the Dobson station network. The seasonal and latitudinal variability of columnar ozone based on these observations is shown in Fig. 8A. Although data coverage from this network is not truly global because

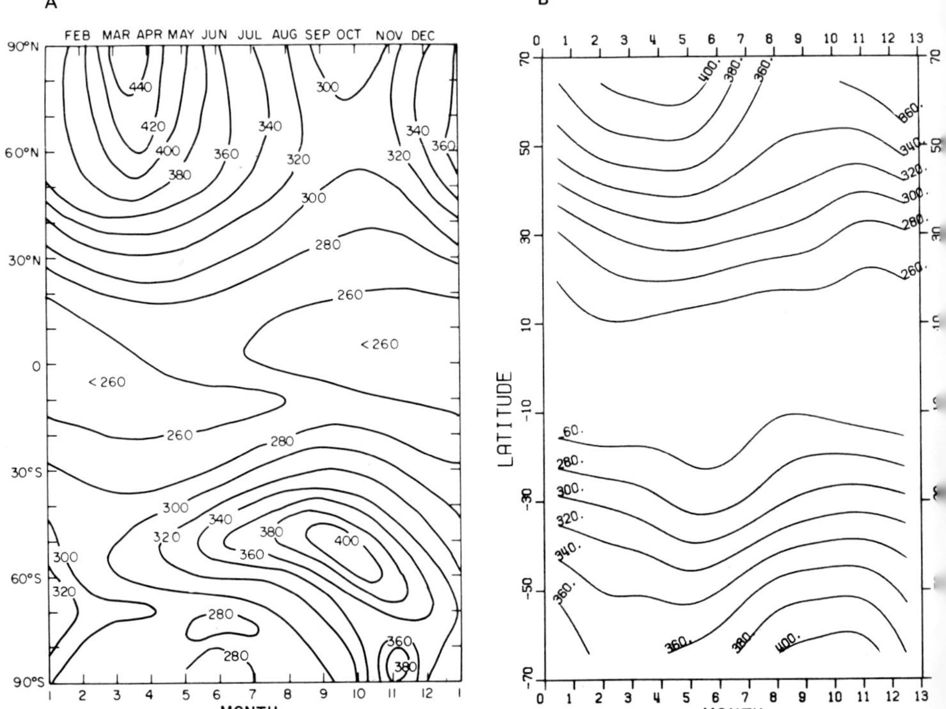

Fig. 8. Total ozone column above the ground (Dobson units) versus latitude and season. (A) Observations, from Dutsch, 1971 (*36*), with permission. (B) du Pont 2-D model, from Hudson *et al.*, 1982 (*51*).

observations have been made predominantly at mid-latitudes of the northern hemisphere, the first 2 years of backscattered ultraviolet (BUV) satellite data analyzed by Hilsenrath *et al.* (*48*) suggest that the principal features of Fig. 8A are realistic. The model calculation in Fig. 8B (*68*) exhibits the degree of agreement with the observed seasonal and latitudinal variation of columnar ozone that is currently achievable.

The vertical concentration profiles of ozone calculated for various latitudes are compared in Fig. 9 with the measurements of Wilcox *et al.* (*113*). The agreement with the observed behavior is generally quite good. Below ~ 15 km, however, the model significantly underestimates the observed ozone abundance at all latitudes. Similar results have been reported in this respect by other 2-D modeling groups (*13, 100, 112*). Inadequate representation of mean transport below ~ 20 km may be responsible. At ~ 30 km, the calculated ozone abundances appear to be slightly higher than the measurements. This feature has also been observed with 1-D models.

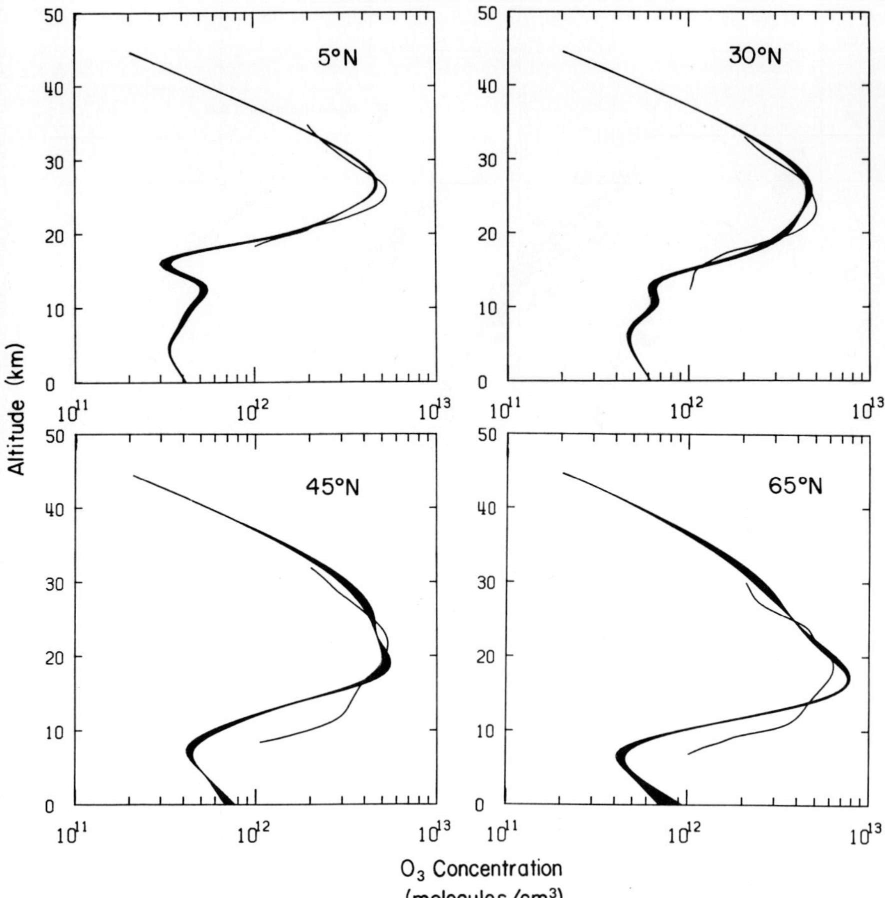

Fig. 9. Springtime vertical ozone profiles at several latitudes. Shaded regions give model calculations; thin curves are observations. (Data from refs. *68* and *113*.)

2. N_2O and CH_4

Figures 10 and 11 show a selection of the available observed data for N_2O and CH_4, respectively, together with the range of model calculations over the period of 1 year in various latitudinal regions. In general, the models give a reasonable simulation of observed N_2O and CH_4.

Both the calculated and the observed profiles decline with altitude more rapidly at high latitudes than in the tropics. This behavior is related to the Hadley circulation, which produces tropical upwelling and corresponding subsidence at high latitudes.

At mid-latitudes, most models overestimate N_2O in the middle stratosphere.

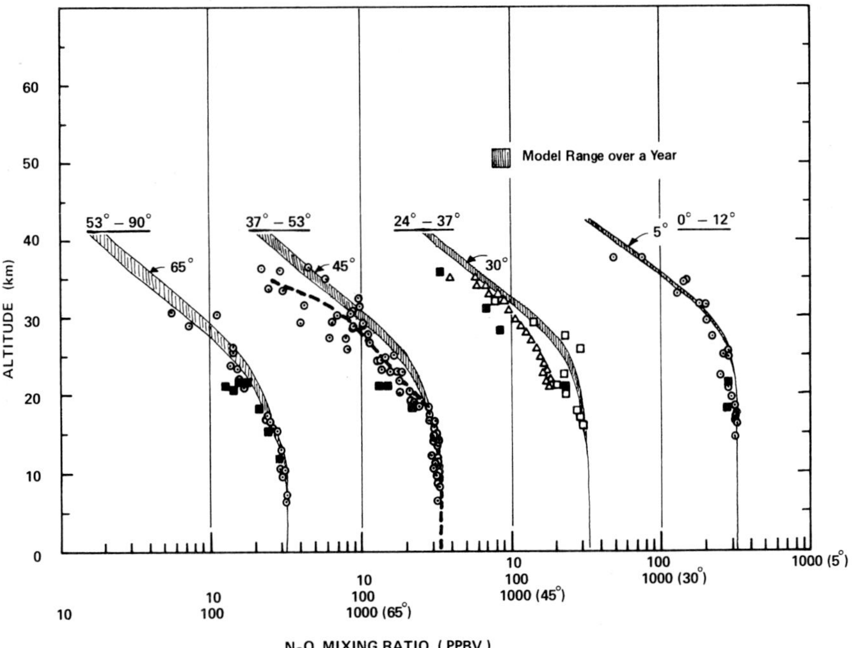

Fig. 10. Observed and calculated vertical mixing ratio profiles for N_2O as a function of altitude. From Miller *et al.* (*68*), with permission.

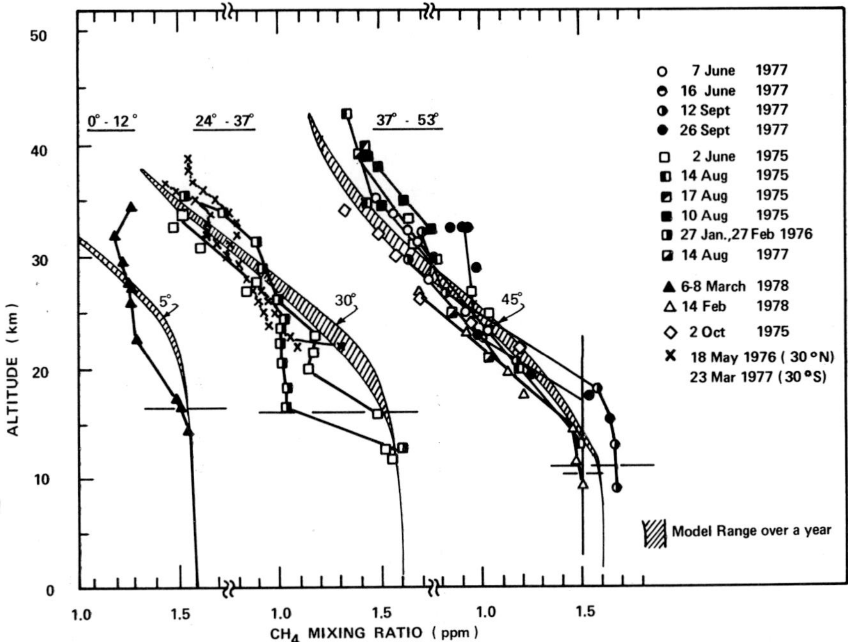

Fig. 11. CH_4 vertical mixing ratio profiles for selected latitude bands. Observations are the points. From Miller *et al.* (*68*), with permission.

For CH_4, the main sink in the atmosphere is reaction with OH radical and thus the calculated profiles for CH_4 depend on the model-calculated OH distribution. Modeling high altitude CH_4 simultaneously with N_2O and CFCs has been a difficult problem, which appears here as an underestimate of the observed CH_4 concentration.

3. *CFCs*

Comparisons between calculated and observed vertical distributions of CFC-11 ($CFCl_3$) and CFC-12 (CF_2Cl_2) in several different latitudinal bands are shown in Figs. 12 and 13, respectively (*68*). Since the concentrations of the CFCs have been increasing with time, the measurements have been scaled to a common date (January 1, 1979) in order to provide a consistent basis for comparison with the model calculations. Both CFC-11 and CFC-12 exhibit latitudinal behavior qualitatively similar to that for N_2O, with the vertical gradient in mixing ratio increasing from the tropics to higher latitudes. For CFC-12, the agreement between the calculated and measured distributions is generally good. For CFC-11, the model overestimates the measured mixing ratios at altitudes above 20 km for all latitudes up to 50°.

A possible cause of the above discrepancies may be an underestimate of the photolysis rates. Both CFC-11 and CFC-12 photolysis depend sensitively on the

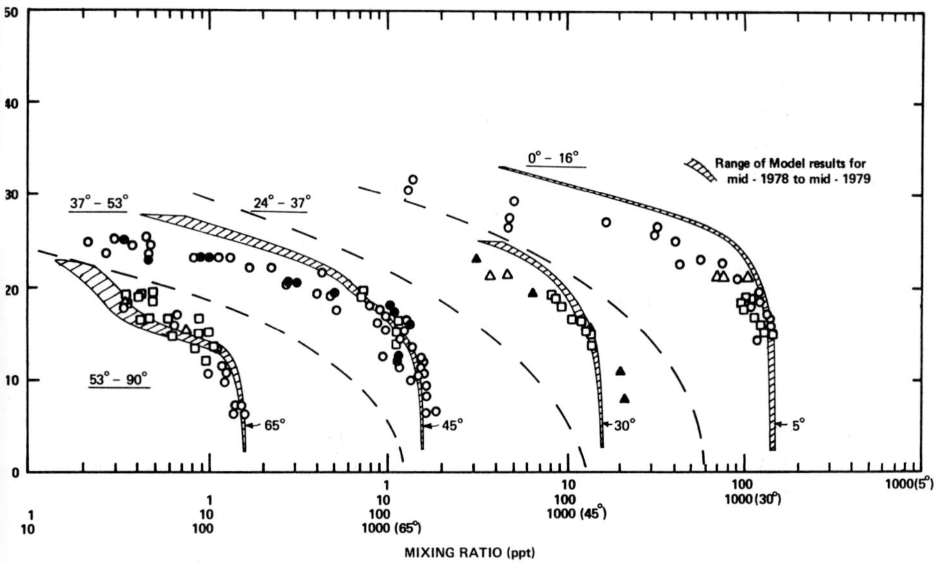

Fig. 12. Observed and calculated vertical mixing ratio profiles for CFC-11 ($CFCl_3$) as a function of latitude. The measurements are scaled to January 1979, based on known release scenarios. From Miller *et al.*, 1981 (*68*), with permission.

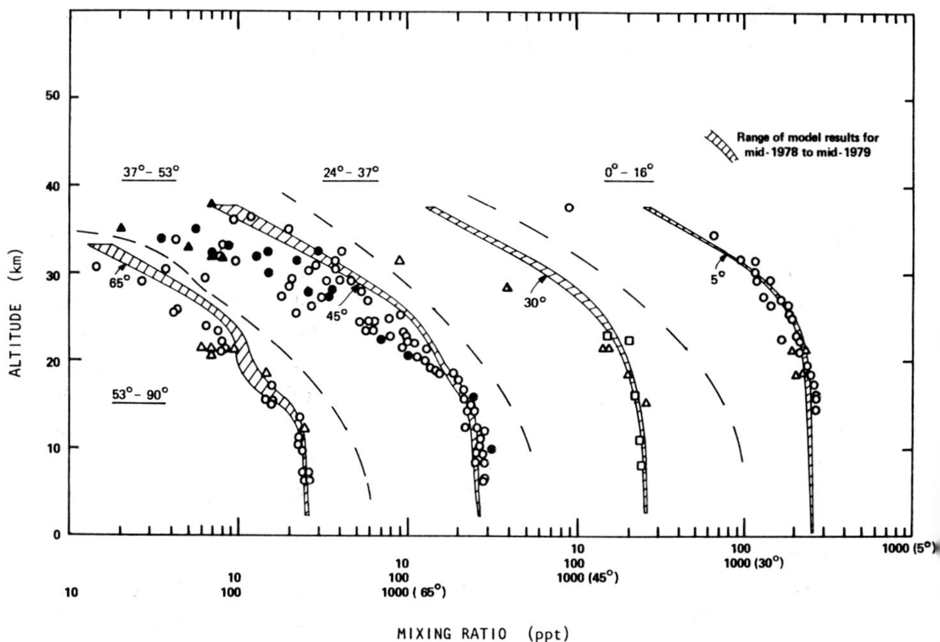

Fig. 13. Observed and calculated vertical mixing ratio profiles for CFC-12 (CF_2Cl_2) as a function of latitude. The measurements are scaled to January 1979. From Miller *et al.*, 1981 (*68*), with permission.

penetration of radiation through the Schumann and Herzberg regions of the molecular oxygen spectrum (around 200 nm). Recent observations (*40*) indicate more penetration of UV in this region than previously calculated. Froidevaux and Yung (*41*) and Brasseur *et al.* (*14*) explored some of the implications for models, and noted significant improvement in the comparison of certain calculated profiles with measured values. With new measurements of these cross sections, it may soon be possible to more successfully model simultaneously the concentration profiles of N_2O, CH_4, and CFCs.

D. Active Radicals and Sink Gases

Once the transport scheme in the model is tested by the comparison of the calculated and observed distributions for the long-lived species, the chemical aspects of the model can be examined in more detail. In the upper stratosphere (above ~ 40 km), chemical processes are more rapid than transport, and a photochemical steady state is generally maintained for the reactive species. Thus, any discrepancy between the observed and calculated abundances of these species would suggest either incompleteness or inaccuracy of the chemical scheme. In the lower stratosphere (below ~ 30 km), however, the physical transport time scales become comparable to those of chemical processes. In this region, a poor comparison between models and measurements may be related to either chemistry or transport difficulties.

Recent model calculations provide general agreement with available measurements for many of the observed species, although some cases indicate inadequacies of our present chemical scheme. The most serious problem, however, is that present comparisons with measurements do little to constrain the models, due to the large variations in observed species concentrations. Continued improvement of measurement precision and a better understanding of natural variations would improve this situation. Another approach receiving increasing attention is the devising of innovative experiments to observe directly groups of chemically coupled species, e.g., through ratio measurements or diurnal variations. Sections III,D,1–6 review the current status of the traditional direct comparisons for the key active radicals (NO, NO_2, OH, HO_2, Cl, ClO) and sink gases (HCl, HNO_3).

1. NO

The vertical profile of NO has been measured at mid-latitudes on numerous occasions. A selection of the measurements is given in Fig. 14 along with

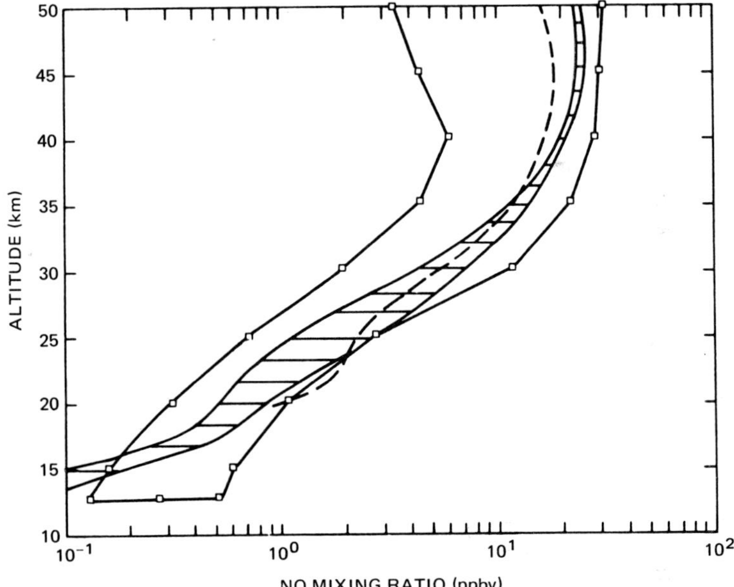

Fig. 14. NO vertical mixing ratio profiles. Observations at mid-latitudes indicated by the band with square points. Striped region gives the seasonal range of daytime average values from du Pont model; dashed curve is from AER model for July noontime. From Hudson *et al.*, 1982 (*51*).

Atmospheric and Environmental Research, Inc. (AER) and du Pont model calculations. The calculations lie toward the higher end of the measurements above 30 km, with a mixing ratio at 50 km of about 20 ppb. This overestimation of NO at high altitudes is also a feature characteristic of 1-D model calculations (*51*). A possible explanation is that the photolysis rate of NO may be underestimated in the model, since in the upper stratosphere and lower mesosphere NO is destroyed by the reaction $NO + h\nu \rightarrow N + O$ followed by $N + NO \rightarrow N_2 + O$.

2. NO_2

Figure 15 presents a comparison of calculated and observed NO_2 mixing ratios at mid-latitudes. The calculated mixing ratios are in good agreement with the measurements from 20 to 40 km.

Seasonal and latitudinal dependence of NO_2 concentration poses a challenging problem for the model simulation. Both Noxon and co-workers (*78*) and Coffey *et al.* (*22*) have measured the NO_2 column abundance as a function of latitude for several seasons. The summer measurements show a monotonic increase with latitude by a factor of about 2 from the equator to 60°N, whereas the winter measurements show a rapid decrease between 45° and 50°N (the so-called Noxon Cliff). The measurements are compared in Fig. 16 with du Pont 2-D model

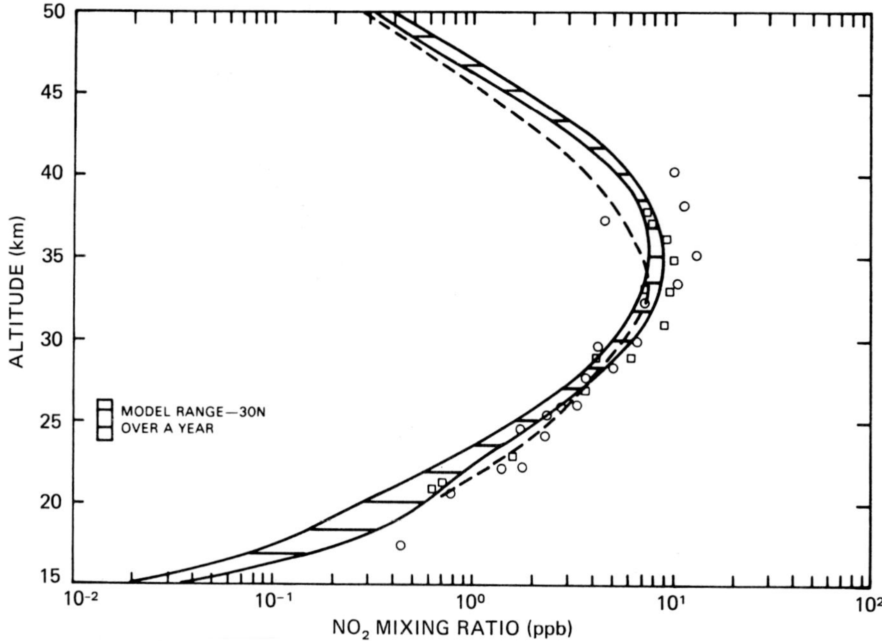

Fig. 15. NO_2 vertical mixing ratio profiles near 30°N. Observations are the points. Model results are as in Fig. 14. From Hudson *et al.*, 1982 (*51*).

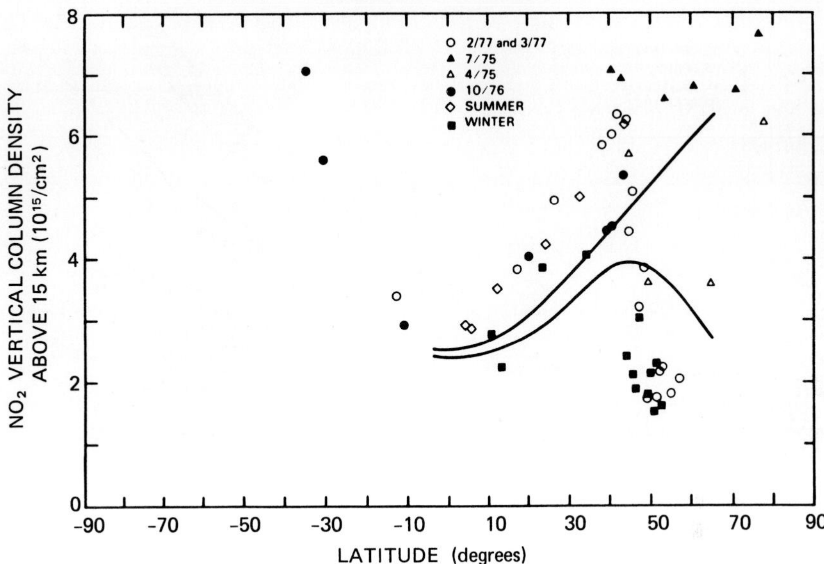

Fig. 16. NO_2 column abundances versus latitude. Data are points. Solid lines are du Pont model calculations for summer (upper) and winter (lower). From Hudson *et al.*, 1982 (*51*).

results. Although the latitudinal resolution in current models is not sufficient to calculate sharp features like the Noxon Cliff, a significant decrease in NO_2 at high latitudes in winter is clearly simulated. Solomon and Garcia (*96*) find a similar feature and ascribe it to the conversion of NO_2 to N_2O_5. A more detailed simulation requires some attention to meteorological factors such as the winter polar vortex.

3. *OH and HO₂*

The measurements for these species to date are scant, and no data are at the present time available below 30 km. Figure 17 illustrates a comparison between the calculated OH profile (daytime average) at 30°N and the available *in situ* measurements (*4*). Two of Anderson's data sets (January 1976, and April 1977) were obtained near sunset, and they have been multiplied by a factor of 2 to allow for diurnal changes from midday. Daytime average column abundances of OH calculated for the region below 55 km are $\sim 6 \times 10^{13}$ molecules/cm^2 at mid-latitudes, with minimal seasonal variations. These results are consistent with those of Burnett and Burnett (*15*), which were obtained using high-resolution solar absorption spectroscopy in the ultraviolet.

The HO_2 mixing ratio between 28 and 36 km at 30°N has been measured on 3 days (*6*). The calculated mixing ratio is somewhat lower than the average of the

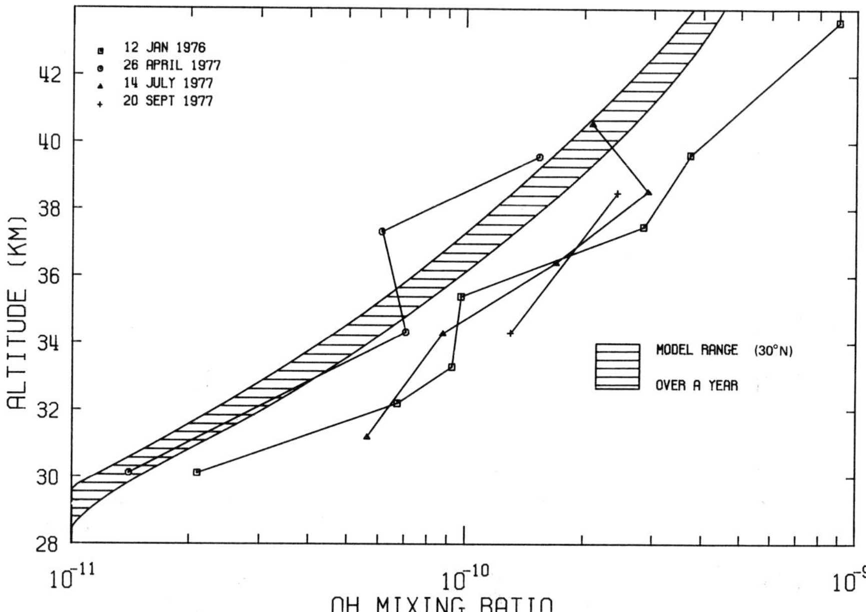

Fig. 17. Vertical profiles of OH mixing ratio near 30°N. Points connected by lines give observations. du Pont model range over the year shown.

measurements, although the observations are too few to permit a meaningful comparison.

4. *Cl and ClO*

Measurements of atomic chlorine in the stratosphere have been quite limited. Considering the uncertainties in the data and the scant observations (between 35 and 40 km only), the agreement with model calculations is thought to be reasonable.

In contrast with Cl, ClO has been measured numerous times by Anderson *et al.* (*5*) using *in situ* resonance fluorescence in balloon launches from Palestine, Texas (32°N). Two of the observations (July 28, 1976 and July 14, 1977) fall clearly outside an envelope that is rather well defined by clustered data with deviations limited to about ± 50% about the observed mean. These two sets of data do not appear representative of the mean distribution of ClO at mid-latitudes; in most comparisons with models, they are not included in the data base defining the mean distribution of ClO. The presence of such high ClO values, even occasionally, remains unexplained.

Recent balloon-borne observations of ClO by Waters *et al.* (*108*), using mm-wave emission techniques, fall into the envelope mentioned above. Further indirect support for this average ClO envelope is provided by the ground-based (mm-

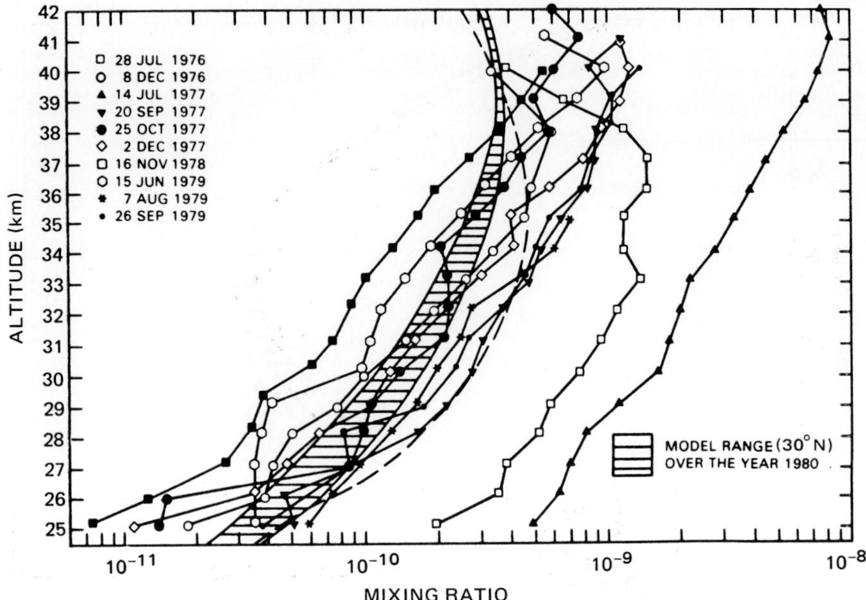

Fig. 18. ClO vertical mixing ratio profiles near 30°N. Connected points are data of Anderson and co-workers (*4, 5*). Banded region gives yearly variation of daytime average from du Pont model. Dashed curve is AER model result for July noontime. From Hudson *et al.*, 1982 (*51*).

wave emission) observations of Parrish *et al.* (*83*). Below 35 km, as shown in Fig. 18, the agreement between the calculated and observed ClO (average) profile is satisfactory. Above 35 km, the calculated values appear to underestimate ClO concentrations, suggesting a current lack of understanding of upper stratospheric chemistry.

5. HNO_3

The observed mid-latitude vertical mixing ratio profile of HNO_3 is shown in Fig. 19, along with recent 2-D model calculations. Above 25 km, the calculations significantly overestimate the measured amount of HNO_3. This overestimate, which is about a factor of 3 at 30–35 km, cannot be easily explained. A part of the discrepancy may be attributable to the model's overestimate of NO in the upper stratosphere, as discussed in Section III,D,1.

The latitudinal variations of HNO_3 obtained by Murcray *et al.* (*70*) and Coffey *et al.* (*22*) are shown in Fig. 20. The rapid variation, with the column density increasing from a minimum of 2×10^{15} molecules/cm^2 at the equator to about 16×10^{15} molecules/cm^2 at 60°N, is well simulated by the models. The dominant contribution to the column density comes from the 15- to 20-km region of the lower stratosphere where transport effects are significant.

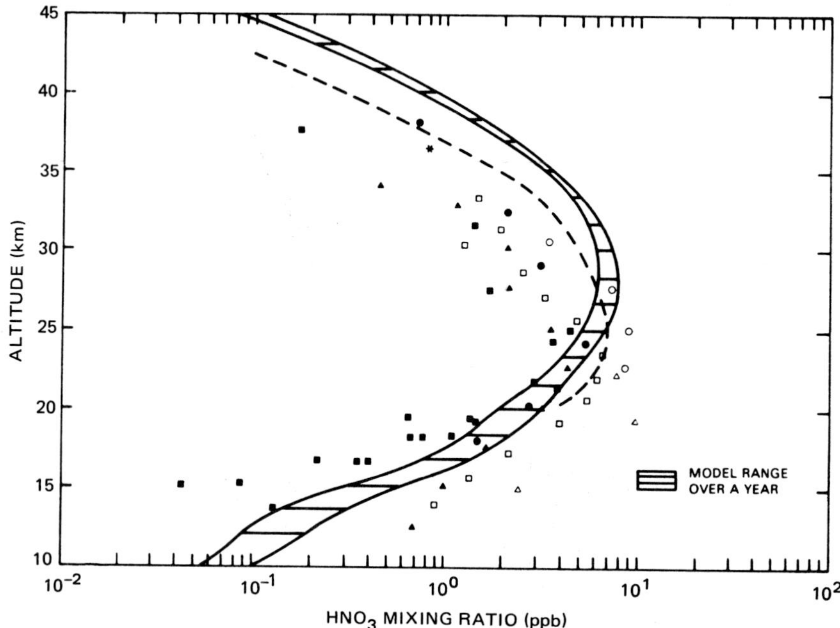

Fig. 19. HNO_3 vertical mixing ratio profiles near 30°N. Observations indicated by points. Striped region gives seasonal range of daytime average values from du Pont model. Dashed curve is from AER model for July, noontime. From Hudson *et al.*, 1982 (*51*).

6. HCl

Figure 21 presents a comparison between the calculated and observed vertical profiles of HCl in the stratosphere for mid-latitudes. Below ~ 25 km, the calculated mixing ratios fall at the high end of the measurements. Above 30 km, the calculated HCl mixing ratio deviates toward the low end of the measurements, similar to the case of the ClO profile in Fig. 18. Some new chemical entities and/or revised sets of kinetic data may be suggested for the Cl_x family, although the existing measurements do not yet provide a sufficiently robust test of the model.

E. Temporary Sink Species

As discussed earlier, HOCl, $ClNO_3$, HO_2NO_2, NO_3, and N_2O_5 are important intermediates in the reaction network (Figs. 1, 2, and 3). They act as temporary sinks, bypassing the major catalytic cycles of O_3 destruction and thus reducing the catalytic efficiency of the active radicals. Observations of these species in the stratosphere would provide further critical tests for the current chemical scheme (*31*). However, they present a particularly difficult analytical problem, because

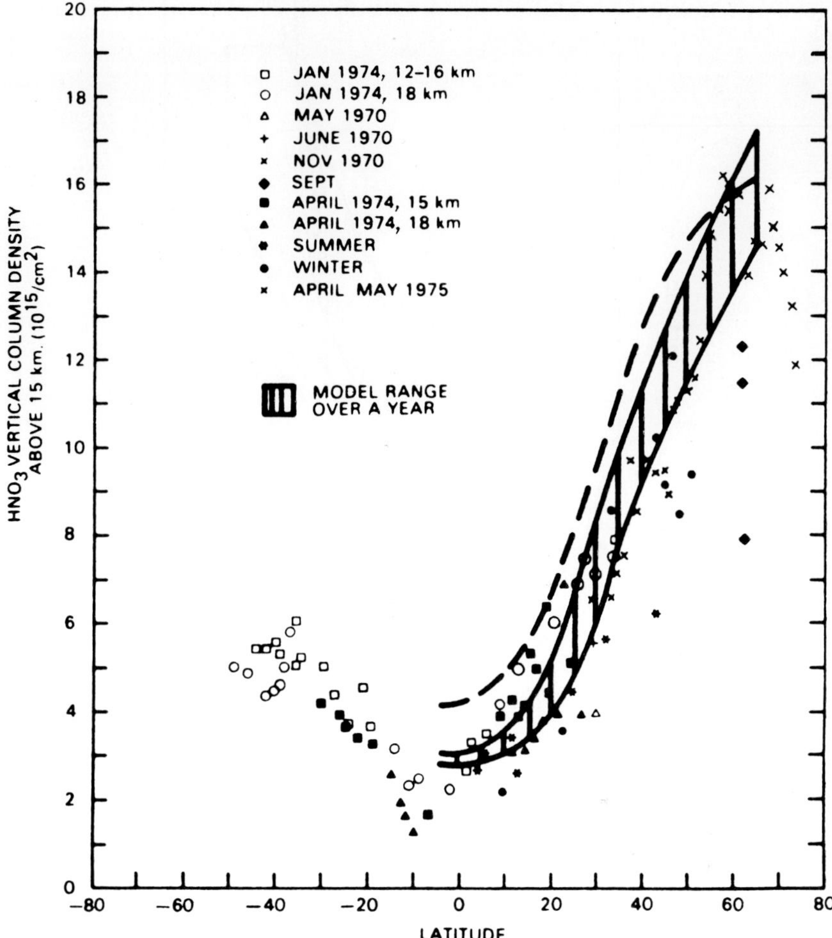

Fig. 20. HNO_3 column abundance versus latitude. Data are points. Vertical banded region gives seasonal variation (upper, winter; lower, summer) from du Pont model. Dashed curve gives AER model results for July. From Hudson *et al.*, 1982 (*51*).

(1) they are large polyatomic molecules that do not possess strong electronic transitions, and (2) their predicted concentrations fall below the detection threshold of long-path IR absorption techniques.

For HOCl and HO_2NO_2, no atmospheric observations are available at the present time. Tentative measurements have been reported for $ClONO_2$ (*71*) and N_2O_5 (*91*). A nighttime profile of NO_3 has been obtained at 43°N latitude from a balloon-borne visible spectrometer (*77*). The modeled vertical profile is consistent with the values observed.

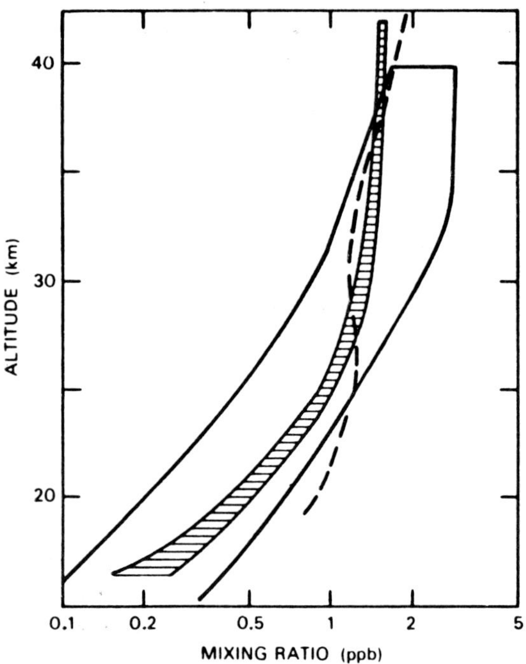

Fig. 21. HCl vertical mixing ratio profiles near 30°N latitude. Outer solid curve is envelope of observations. Model results as in Fig. 19. From Hudson *et al.*, 1982 (*51*).

F. Measurement Programs to Further Test Models

A model can be satisfactorily justified only by surviving restrictive tests using atmospheric measurements. However, as shown in the previous sections, the measured data of the important trace species in the stratosphere are either too sparse in space and time or too variable from one measurement to another to provide significant information. Large deviations among measurements may be due to natural atmospheric variabilities (meteorological origins) and/or errors in measurement techniques. An ambitious program to solve such problems, called the "Balloon Intercomparison Campaign," has just started, sponsored by governmental (NASA; NOAA; NPL, England) and industrial (CMA-FPP) organizations. This international collaboration includes many scientists from Europe, Japan, Canada, and the United States, who measure several important source and sink gases simultaneously with different measurement techniques. The measurements are combined with meteorological monitoring systems. A similar international project, GLOBUS (Global Budget of Stratospheric Trace Constituents), has been planned as part of the Middle Atmospheric Program (MAP). Within a few years, such programs will provide information to permit retrospective evaluation of measurements and more critical tests of the current models.

Another critical test to our model chemistry would be *in situ* simultaneous measurements of chemically related groups of active radicals within the same air mass to see whether the catalytic destruction of O_3 does occur within the stratosphere as the model predicts. A project of this sort headed by J. Anderson at Harvard University is underway, with an initial successful test flight of a balloon-borne system for making repeated measurements during a single flight. This apparatus has the capability to test directly the reproducibility of measurements and the extent of natural variability and cross correlations in the 20- to 40-km altitude region.

Satellite measurements provide data with truly global coverage, a significant advantage over the balloon- or aircraft-borne and ground-based measurement efforts. At the present time, however, only a limited number of abundant trace species, such as O_3, CO_2, H_2O, and NO_2 have been detected. Instruments currently being developed offer potential for several additions to this list.

Measurements of source gases, N_2O, CFCs, CH_4, etc., in the troposphere and accurate knowledge of their origins and their temporal changes are also important. One such measurement program is ALE (Atmospheric Lifetime Experiment), with a global network of six monitoring stations (*84*).

IV. MODEL CALCULATIONS TO CHARACTERIZE POTENTIAL CHANGES

A. Understanding Perturbations

A "perturbation" is a model calculation based on the assumption that one or more minor trace species are changing in concentration, usually due to man's influence on the atmosphere. The implicit assumption in this approach is that there is a well-defined and stable initial or "reference" atmosphere. Ground-level concentrations or source fluxes of long-lived species (e.g., N_2, O_2, H_2O, CO_2, CO, CH_4, CH_3Cl, CH_3CCl_3, CCl_4, N_2O) in the reference atmosphere are fixed to their observed values, and the models are run to a stable solution. For one-dimensional models, the stable solution is a steady state, whereas for two-dimensional models it corresponds to seasonal variations that repeat from one modeled year to the next.

The null hypothesis against which perturbations are traditionally assessed is clearly that the atmospheric composition is a stable state changing only according to natural cyclic variations, e.g., diurnal and seasonal cycles. Against this background, the long-term characteristics of one species are typically changed: the ground level concentration or source flux may be changed (e.g., CO_2, CO, CH_4,

N_2O), an altitude-dependent injection may be specified (SST and conventional aircraft emissions, impulsive cosmic ray events), or a time-dependent ground-level source flux based on the industrial inventory may be specified (CH_3CCl_3, CFCs). When one perturbation is considered in isolation, the other species are treated as in the reference natural atmosphere. This may be accomplished by fixing either the concentrations or the ground-level source fluxes for the other long-lived species. The latter procedure is probably preferable since it allows tropospheric interactions to be modeled.

In a perturbation calculation, the model is typically run either to steady state or for the years of interest (e.g., 1940 to the present to investigate the influence of CFCs). The perturbed atmosphere is compared with the initial reference case, and changes in the chemical and/or thermal structure of the atmosphere are investigated. Ordinarily, the changes of greatest importance, as discussed above, are the impacts of potential man-made perturbations on ozone and on the quantity of ultraviolet light reaching the earth's surface, which is determined by the total vertical column of ozone in the atmosphere. Thus the single quantity of most interest is usually the total ozone column change calculated from a given perturbation scenario.

Perturbation calculations are useful in assessing the current view of the chemical and thermal impacts on the atmosphere of a given potential change in trace gas concentrations. They can be used to gauge the importance of various potential man-made changes, taken individually, on ozone or atmospheric temperatures. The initial view of such calculations (e.g., *73, 74*) was that the different perturbations can adequately be assessed individually. Only recently, however, has it been widely recognized that there are potentially important interactions between the various perturbing species (*75*). Thus, for example, the impact of CFCs on ozone depends sensitively on the atmospheric methane level (*81*). The developing view is that *all* potential perturbations should be considered simultaneously in order to assess the most likely impact on atmospheric ozone (*120*).

It should also be emphasized that changes in chemical reaction rate sets can have a significant impact on perturbation calculations without changing the calculated reference atmosphere by an amount large enough to establish major discrepancies with existing measurements. As discussed in Section III, the stratospheric measurements are currently not sufficiently precise or complete to give detailed tests of model predictions. Perturbation calculations yielding a wide range of predictions can give similar consistency with most stratospheric measurements. The positive detection of several of the temporary reservoir species in the stratosphere (e.g., HO_2NO_2, N_2O_5, $ClONO_2$, HOCl) would help considerably in constraining the range of results for perturbation calculations (*31*).

Given these necessary caveats, in Sections IV,B–E we investigate the individual effects of various potential anthropogenic perturbations. A more comprehensive, simultaneous-perturbations approach is discussed in Section V.

B. Odd Nitrogen Perturbations

One of the first human activities identified as potentially influencing stratospheric ozone was the release of nitrogen oxides from high-flying supersonic transports (SSTs) (*76a*). The suggestion that SSTs might deplete the ozone layer (*53*) led to a concentrated effort to understand stratospheric transport and photochemistry (*21*).

In the natural stratosphere, ozone destruction between the tropopause and the stratopause is dominated by a catalytic cycle involving NO_x, with the rate-limiting step currently thought to be the reaction

$$NO_2 + O \rightarrow NO + O_2 \tag{33}$$

Any source contributing additional NO_2 to the stratosphere enhances the rate of NO_x-catalyzed ozone destruction. The expected levels of NO_x emission from a projected fleet of on the order of 100 SSTs are large enough to influence the stratospheric NO_x budget.

A standard scenario for estimating SST emissions is the release of a steady source, 2×10^8 molecules/cm^2/sec, at an altitude of either 17 or 20 km. Given constant NO_x emissions, a modeled steady state is reached within a few years, because the stratospheric lifetime of NO_x is relatively short. The calculated change in the total ozone column obtained using this scenario has varied considerably as refinements were made in the structure and chemical data base of the models, as illustrated by the Lawrence Livermore National Laboratory (LLNL) model results shown in Table III (*51*).

Note from Table III that the calculated ozone change has been sometimes negative and sometimes positive. The calculations indicating a positive ozone change are those in which lower stratospheric ozone destruction has a strong component due to an HO_x-catalyzed cycle; in these cases, added lower stratospheric NO_x behaves as in the adjacent upper troposphere and influences HO_x radical concentrations through family-linking reactions like

$$HO_2 + NO \rightarrow NO_2 + OH \tag{34}$$

In current models with reduced levels of HO_x, NO_x-catalyzed ozone destruction dominates in the lower stratosphere, and the impact of added NO_x is to significantly decrease the ozone column. The changes in the model thus reflect differences in the chemistry-dependent altitude above which NO_x changes from an ozone producer to an ozone remover.

Conventional (nonsupersonic) aircraft typically fly below the tropopause and inject most of their exhaust NO_x into the upper troposphere. In the troposphere, NO_2 and NO enter into catalytic cycles that can produce ozone through the photooxidation of CO and CH_4 (*24, 38, 60*). The time scale for NO_x in the troposphere is short (a few months), so locally enhanced production of O_3 occurs near the altitude (and latitudes) of greatest aircraft NO_x injection.

TABLE III

Changes in the Lawrence Livermore National Laboratory 1-D Model-Calculated Effect on Total Ozone from NO_x Emissions at 17 and 20 km at a Constant Rate of 2×10^8 Molecules/cm²/sec[a]

Date of evaluation	ΔO_3 (%) for injection at 17 km	20 km
Mid-1974	−4.8	−10.2
Early 1975	−5.3	−11.2
Mid-1975	−4.3	−9.8
Mid- to late 1975	−1.8	−5.2
Late 1975	−1.1	−3.5
Mid-1976	−0.7	−2.9
	−1.2	−4.2
Mid- to late 1976	−0.7	−3.3
Mid-1977	−1.3	−4.8
Mid- to late 1977	+2.0	+0.5
Mid 1978	+3.2	+3.6
Early 1979	+2.6	+2.2
Mid-1979 (Case A)	+2.0	+1.1
Early 1980	+1.7	+0.6
October 1980	−0.3	−4.5
December 1980 (Case B)	−0.7	−5.3
May 1981 (Case D)	−2.2	−7.1

[a] From Hudson *et al.*, 1982 (*51*).

Aircraft emission of NO_x became a significant source of stratospheric NO_x during approximately the mid-1950s and has been increasing since that time. It is estimated that, by 1980, aircraft may have increased the total ozone column amount by about 0.5% (*64, 120*). The projected world aircraft fleet by the year 1990, including a modest SST component (*79*), is projected to increase the ozone column density by approximately 1% (Table 3–8 in ref. *51*). The greatest local ozone change through the year 1990—about 20%—is projected to occur at 8–10 km altitude, as shown in Fig. 22.

The calculated tropospheric ozone increase due to aircraft emissions has not been very sensitive to chemical reaction rate changes. Other uncertainties, including the levels of natural tropospheric NO_x and the inaccuracy inherent in one-dimensional treatments of the troposphere (*60*), should, however, be kept in mind.

The primary source of NO_x in the natural stratosphere is N_2O emission from the surface. The possibility that man might increase tropospheric N_2O emissions

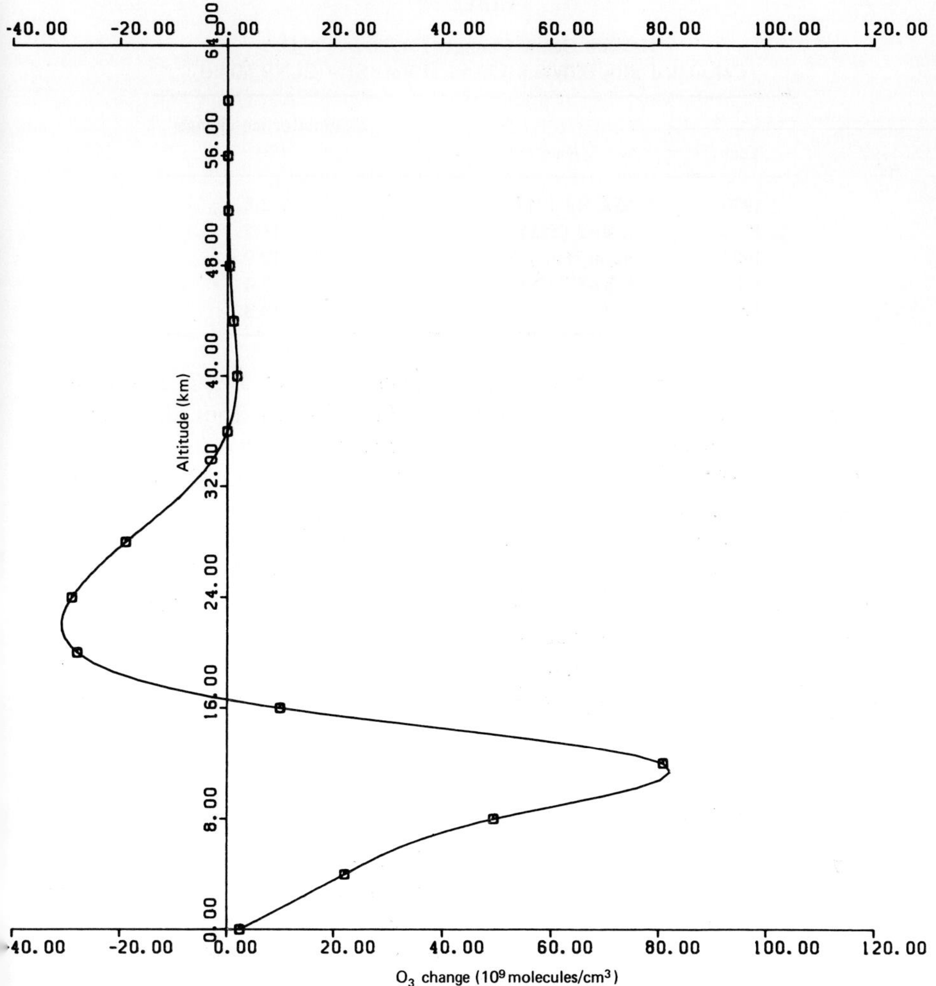

Fig. 22. Vertical ozone concentration profile change due to aircraft emission of NO_x at projected 1990 level. du Pont coupled model.

through the use of "fixed" nitrogen fertilizers (*24*) or internal combustion engines (*110*) has been suggested. Although the source is not yet thoroughly understood, recent detailed measurements indicate that the tropospheric mixing ratio of N_2O is increasing at the rate of 0.2%/year (*109*).

The atmospheric lifetime of N_2O is quite long (~ 100 years), so the effects of an impulsive change in the ground-level N_2O concentration would take several hundred years to develop. Nevertheless, a useful scenario is to consider the response of the atmosphere to a doubling of the ground-level N_2O concentration.

TABLE IV

Column Ozone Changes Due to 2 × N_2O Calculated with Different Chemical Rate Sets (LLNL Model)

Year	Rate set	Column ozone change (%)
1979	NASA RP-1049 (*72*)	−2.8
1981	JPL 81-3 (*55a*)	−11.5
1982	Hudson *et al.*, 1982 (*51*)	−12.9
1982	JPL 82-57 (*55*)	−9.4
1983	Latest	−10.3

By reaction with $O(^1D)$, N_2O gives the primary source of NO_x in a region similar to that of high-altitude SST emissions (near 20 km). Thus, the response of the N_2O increase scenario to changes in the chemical data base is quite similar to that of the SST injection discussed above. Table IV gives LLNL calculations for recent chemical reaction-rate sets.

The vertical profile of the calculated local ozone change due to increased N_2O (+25%) obtained with the du Pont 1-D model is shown in Fig. 23. The model includes thermal feedback effects of the N_2O (see Section IV,E) and JPL 82–57 (*55*) chemistry. The ozone concentration decrease peaks at about 26 km altitude. For 2 × N_2O the calculated column ozone change is −14.2%. The corresponding result from the LLNL 1-D model is −10% (D. J. Wuebbles, private communication, 1983).

C. Halogen Perturbations

Considerable scientific interest has been directed toward assessing the potential impact of man-made halogen compounds, particularly chlorofluorocarbons (CFCs), on stratospheric ozone (*69, 98*). Numerous governmental and international documents discussing the subject have been assembled (*49a, 50, 51, 73–75, 103*).

CFCs are inert, long-lived trace constituents in the atmosphere. No important tropospheric sink for the major fluorocarbons ($CFCl_3$ = CFC-11, CF_2Cl_2 = CFC-12) has been found. Once released from ground level, the CFCs diffuse throughout the troposphere and slowly rise to the stratosphere. Photolysis by UV sunlight occurs in the stratosphere between approximately 20 and 30 km altitude, releasing chlorine atoms.

The estimated atmospheric lifetimes of the CFCs due to stratospheric photolysis are indicated in Table V, along with the observed instantaneous lifetimes obtained with the Atmospheric Lifetime Experiment (*26*). The instantaneous lifetime is the ratio of present total column to present destruction rate, whereas

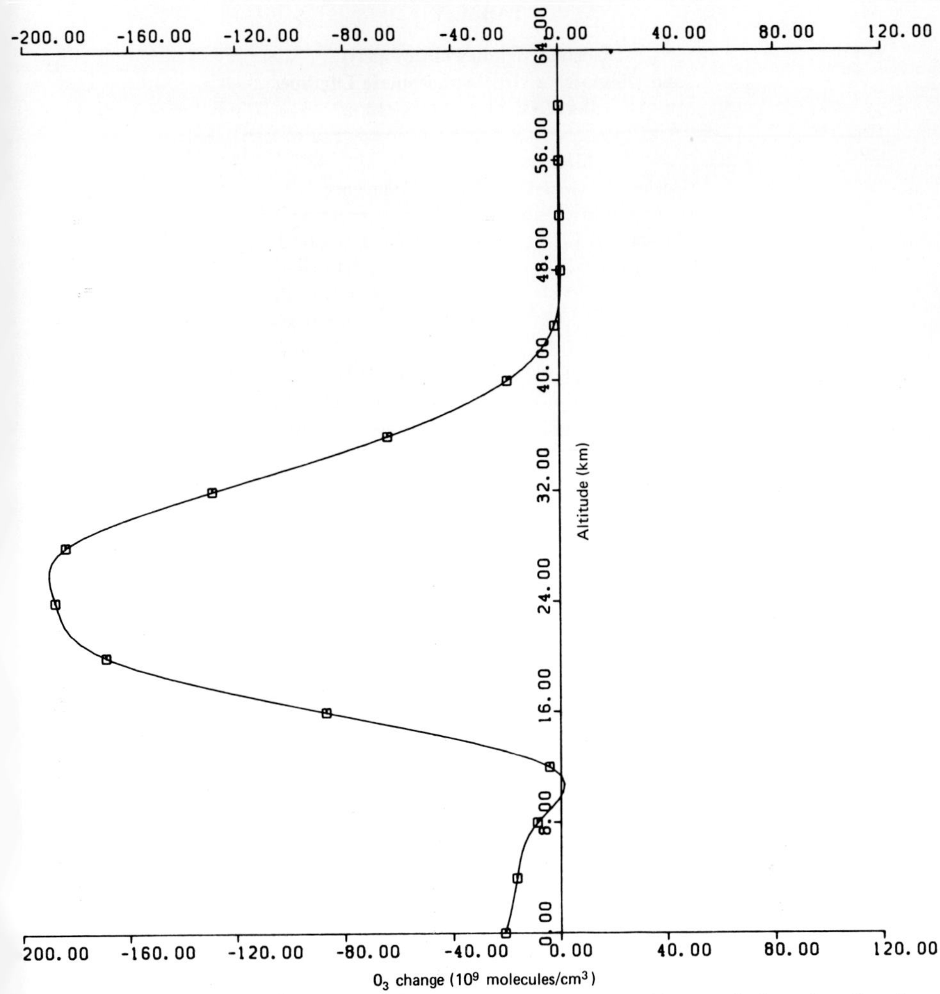

Fig. 23. Vertical ozone concentration profile change due to a 25% increase in the surface flux of N_2O. du Pont coupled model.

the steady-state lifetime is the total column divided by the postulated release rate in the eventual steady state that would be reached in several hundred years. Due to covariances between the species concentrations and photolytic destruction rate constants, both of which are largest near the equator, one-dimensional models may overestimate the CFC lifetime by about 10% (*80*).

Chlorine in the upper stratosphere can destroy ozone through the catalytic cycles discussed in Section II. The contribution of chlorine to the ozone balance in the lower stratosphere has been more controversial. The concentration of ClO,

TABLE V

Calculated[a] (du Pont Models) and Observed[b] (ALE) Atmospheric Lifetimes for CFC-11 and CFC-12

Species	Lifetime (years)	Comments
CFC-11	75	Steady-state, 1-D model
	60	Steady-state, 2-D model
	74	Instantaneous, 2-D model
	78	Instantaneous, observed
CFC-12	140	Steady-state, 1-D model
	120	Steady-state, 2-D model
	140	Instantaneous, 2-D model
	>100	Instantaneous, observed

[a] Owens *et al.* (*80*).
[b] ALE (*26*).

which determines the rate of the catalytic reaction, depends sensitively on the OH radical concentration. No measurements of OH exist in the 20- to 30-km region that is crucial to testing the models. Available ClO measurements extend down to only 25 km. Because of the complex interactions of the various families in this region, chemical rate set revisions have dramatically changed the calculated impact of chlorine on ozone in this region. With NASA RP-1049 (*72*) chemistry, characterized by high OH concentrations, added chlorine was calculated to reduce ozone between 20 and 30 km. With the latest chemical rate recommendations (*55*), a significant ozone increase is indicated in this region. The calculated vertical profile giving the change in local ozone due to continued CFC release is shown in Fig. 24.

Figure 25 indicates in a historic perspective the changing sensitivity of the model-calculated total ozone column in the steady state to a continuous release of CFC-11 and CFC-12 at nominal mid-1970s rates. All these calculations are based on models with a similar chlorine content in the stratosphere (~ 10 ppbv); the changes are due to improvements in model structure and in the chemical reaction rate data base. The calculated steady-state ozone reduction has been larger than 20% (in 1979) and as low as 4% or less (1982). As illustrated in the graph (see also ref. *51*, Table 3-1), different groups obtain similar values for the calculated steady-state ozone depletion. The calculated ozone change due to CFCs alone through the present day is approximately one-fifteenth the steady-state value and is currently estimated by various models to lie in the range 0 to −0.4%.

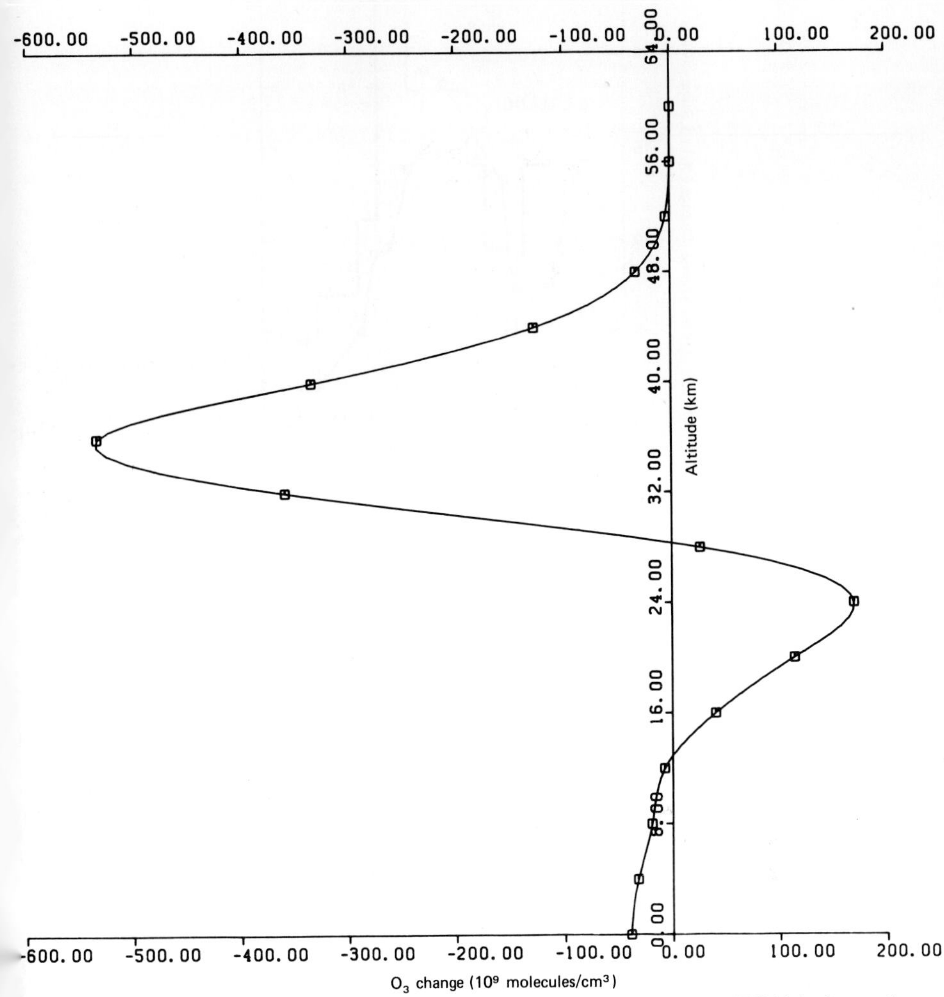

Fig. 24. Vertical ozone concentration profile change due to CFC-11 and CFC-12 in the steady state. du Pont coupled model.

The total ozone above a given area on the earth's surface varies significantly with latitude and season. Two-dimensional models are required to simulate the calculated change in the total ozone column as a function of latitude and season. With recent chemical rate sets, the modeled percentage ozone depletion due to CFCs is only a weak function of latitude and season, as indicated in Fig. 26 (*97*). Earlier calculations, using chemical rate sets giving substantial ozone depletion in the lower stratosphere, showed considerably larger variations (*85*).

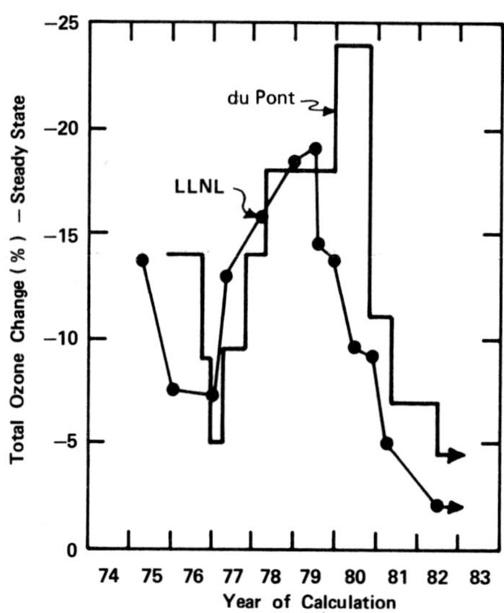

Fig. 25. Calculated ozone change in the steady state due to CFCs from the LLNL and du Pont models versus year during which the calculation was made.

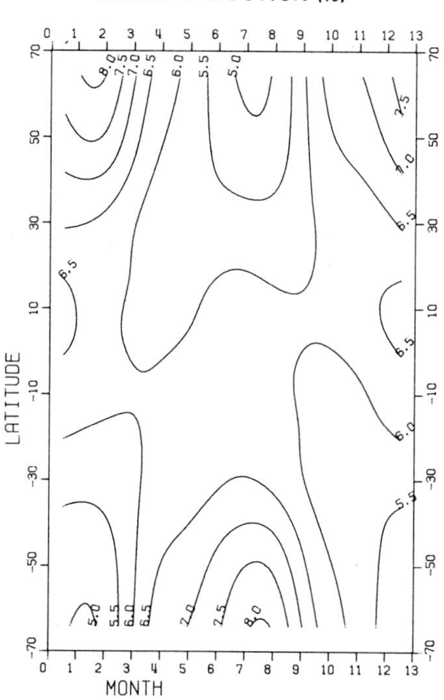

Fig. 26. Latitude–season contour plot of calculated steady-state percentage ozone reduction due to continued release of CFC-11 and CFC-12 at mid-1970s release rates. From Steed *et al.*, 1982 (*97*), with permission.

TABLE VI

Calculated Column Ozone Reduction and Relative Depletion for Various Chlorocarbons[a]

Species	Release rate (10^6 kg/year)	LLNL[b] ΔO_3 (% steady state)	LLNL[b] O_3 depletion potential	du Pont ΔO_3 (% steady state)	du Pont O_3 depletion potential
CFC-11 ($CFCl_3$)	272	1.95	1.00[c]	2.39	1.00[c]
CFC-12 (CF_2Cl_2)	338	2.08	0.86	2.67	0.90
CFC-113 ($C_2F_3Cl_3$)	91	0.52	0.80	0.66	0.82
CFC-114 ($C_2F_4Cl_2$)	18	0.08	0.60	0.10	0.61
CFC-115 (C_2F_5Cl)	4.5	0.01	0.32	0.01	0.35
HCFC-22[d] ($CHClF_2$)	72	0.03	0.05	0.02	0.03
CH_3CCl_3[d]	476	0.51	0.14	0.38	0.09
CCl_4	82	0.65	1.10	0.69	0.96
Total		5.83		6.92	

[a] According to WMO/NASA 1982 (*118*) chemistry.
[b] Wuebbles, 1981 (*119*).
[c] Reference standard by definition.
[d] Contains C—H bond.

Although most attention has been concentrated on CFC-11 and CFC-12, several other chlorocarbons have been considered as possible sources of stratospheric chlorine. Table VI lists the estimated annual releases of several such chlorocarbons (*51*). The calculated steady-state ozone depletion (based on WMO/NASA 1982 chemistry, *51*) and the relative per-pound ozone depletion potentials of these chlorocarbons are also given. About two-thirds of the calculated ozone reduction in steady state is due to CFC-11 and CFC-12, with minor additions due to CCl_4, CFC-113, and CH_3CCl_3. The contributions of CFC-114, CFC-115, and HCFC-22 are negligible at currently estimated release rates.

In terms of its contribution to the calculated ozone depletion to the present day, CCl_4 is relatively more important, accounting for more than either CFC-11 or CFC-12 alone (*29*). This is due to the long release history of CCl_4, which has been produced in large quantities beginning early in the century. For the other species, in contrast, production approached current levels only during the 1970s.

The potential importance of a bromine-catalyzed ozone-destroying process similar to that sketched above for chlorine has been suggested by Wofsy *et al.* (*115*) and Yung *et al.* (*122*). No known mechanism can account for significant bromine sources that might reach the stratosphere, since brominated chemicals tend to be more reactive than the chlorinated analogs. The estimated bromine budget of the troposphere is only about 5% that of chlorine.

D. Other Chemical Perturbations

The possibility that SSTs or the Space Shuttle could influence ozone by directly injecting water vapor into the stratosphere has been suggested (*102*). Water vapor is the dominant source of stratospheric HO_x, through the reaction

$$H_2O + O(^1D) \rightarrow 2OH \tag{35}$$

Although water exists in greater than trace amounts ($\sim$ 1%) in the troposphere, the stratosphere is very dry ($\sim$ 5 ppmv). Although the true mechanism for limited water vapor transport is not fully understood, it appears that the "cold trap" near the tropopause effectively freezes out most H_2O vapor (*36a*). Estimates from CIAP (*21*) suggest that aircraft do not inject enough H_2O into the stratosphere to make a significant impact on ozone.

It has been apparent for some time that atmospheric CO_2 levels have been increasing at a substantial rate, and the consensus is that the increase is due to the burning of fossil fuels (see, e.g., *76, 116*). Although CO_2 does not have a direct chemical effect on ozone, its greenhouse effect (see Section IV,E) can influence ozone indirectly through changes in the thermal structure of the atmosphere.

The effect of a doubled atmospheric CO_2 level on the total ozone column is an increase of about 3%, as illustrated by recent coupled one-dimensional model results shown in Table VII. The ozone increase is caused by the cooling of the stratosphere (see Fig. 28 below), resulting in slower ozone loss since the catalytic destruction cycles slow down with decreasing temperature. The ozone increase due to CO_2 occurs throughout the mid-stratosphere ($\sim$ 25–40 km), as shown in Fig. 27.

An increase in carbon monoxide (CO) may accompany the CO_2 increase. The atmospheric distribution shows much higher CO concentration in the northern hemisphere, suggesting an anthropogenic source, although, due to the shorter

TABLE VII

Calculated Changes Due to Doubled Carbon Dioxide: Coupled 1-D Models

References	Column ozone change (%)	Surface temperature change (K)	Chemistry
Callis *et al.* (*17*)	+3.4	+2.6	JPL 81-3 (*55a*)
Wang *et al.* (*107*)	+2.5	+2.0	WMO/NASA 1982 (*118*)
Wuebbles *et al.* (*120*)	+5.9	+2.0[a]	WMO/NASA 1982 (*118*)
	+5.5	+2.0[a]	JPL 82-57 (*55*)
Owens *et al.* (*82*)	+2.9	+1.7	JPL 82-57 (*55*)

[a] Assumed.

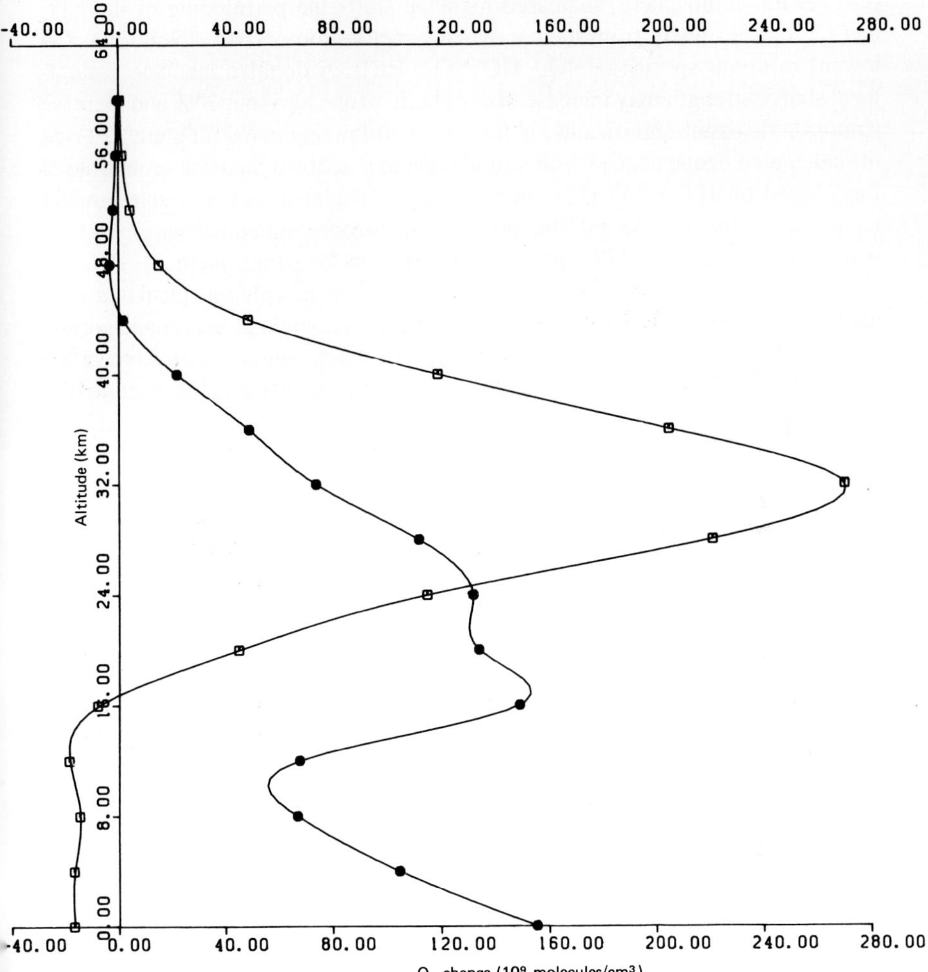

Fig. 27. Vertical ozone concentration profile changes. Curve with open squares (□) due to doubled atmospheric CO_2 level. Curve with filled circles (●) due to doubled CH_4 ground-level source flux. du Pont coupled model.

chemical lifetime of CO, no secular trend has been observed. CO is chemically important, particularly in controlling the concentration of the scavenging OH radical in the troposphere (*60, 99*). Increased CO would be expected to increase O_3 in the troposphere and lower stratosphere (*17*).

Recent measurements (*10, 22a, 39*) indicate that the ground-level concentration of methane has been increasing at the rate of 1.5%/year for at least the last several years. In the troposphere, methane oxidation yields ozone as one product

(*60*). In the stratosphere, increased methane shifts the partitioning of the NO_x and Cl_x species away from the catalytically active forms (NO_2, NO, ClO, Cl) toward reservoir species (HNO_3, HO_2NO_2, HCl). A doubled methane ground-level source strength may increase tropospheric ozone by about 30% and increase stratospheric ozone significantly in the 15- to 30-km region (*81*). Figure 27 gives the calculated ozone change with altitude due to a doubled methane ground-level flux, based on JPL 82-57 (*55*) chemistry and calculated with a coupled model including thermal feedbacks. The modeled total ozone column change is +4.6%, with about two-thirds of the increase occurring in the stratosphere.

Table VIII summarizes the changes in the total ozone column calculated with the present coupled du Pont model for several of the potential man-made perturbations discussed above. (Also given is the calculated column ozone change for the combination of all the perturbations combined, as discussed in Section V.)

E. Greenhouse Effects

As discussed in Section II, the earth's surface temperature is determined by a balance of incoming (visible and ultraviolet) solar radiation and outgoing (infrared) thermal radiation. The average solar flux at the earth's orbit is $F_0 = 1380$ W/cm²/sec. The surface atmospheric temperature T_e is given approximately by

$$(1/4)\,(1 - A)\,F_0 = \tau\sigma T_e^4 \tag{36}$$

where $A = 0.3$ (the average global albedo or reflectance of solar radiation), τ is the effective transmission of the atmosphere in the infrared, and σ is the

TABLE VIII

Calculated Total Ozone Changes for Various Potential Man-Made Perturbations[a]

Perturbation	Column ozone change (%)[b]
Aircraft	+0.5 (+0.4)
1.25 × N_2O	−3.9 (−4.2)
CFCs	−5.7 (−4.7)
2 × CO_2	+2.9 (0.0)
2 × CH_4	+4.3 (+4.3)
All above combined	+1.5

[a] du Pont coupled 1-D model using JPL 82-57 (*55*) chemistry.

[b] Values in parentheses give chemical effect only, assuming fixed temperature.

Stefan–Boltzmann constant. The factor of 1/4 is due to the fraction of daylight hours (1/2) multiplied by the solid–angle effects [(area of circle)/(area of hemisphere) = 1/2]. The left-hand side is the net solar flux, and the right-hand side is the net escaping infrared flux. For the earth's atmosphere, models indicate τ = 0.66, and the surface temperature implied by Eq. (36) is 283 K, close to the observed global mean surface temperature (289 K).

A gas that decreases the infrared transmission (τ) will increase the surface temperature. Equation (36) is only an illustrative approximation, and detailed infrared radiation transport calculations must be made to assess quantitatively the thermal effects of trace gases. But the thermal effects of gases will clearly be most important if they have strong infrared absorption bands in the "window" near 10 μm and thus can reduce the net atmospheric transmissivity. Gases that absorb at 30 μm are radiatively less important, for example, because the earth's surface is already shielded from space by strong water vapor absorption at that wavelength.

Human activities may influence surface temperatures both by changing CO_2 and O_3 and by increasing the concentration of other trace gases that absorb IR strongly in the window region. The largest potential effect is due to the well-established CO_2 increase, which is due primarily to the burning of fossil fuels. The atmospheric CO_2 concentration has been increasing at a rate of 0.3%/year since 1960 (*51, 76*). It has been estimated that a doubling of the atmospheric CO_2 level will occur by the middle of the twenty-first century and that it could cause an average global surface temperature increase of about 3 K.

The direct thermal effect of CO_2 alone, as calculated with one-dimensional radiative/convective models, is a surface temperature increase of about 1.4 K. Various feedback mechanisms act to enhance this value. The increased surface temperature is expected to increase tropospheric water vapor, since 3-D GCM calculations show that the relative humidity tends to stay approximately constant (*64a*). The H_2O feedback raises the surface temperature by an additional 0.5 K. Other feedbacks include the effects of surface temperature on cloud amount and on surface albedo. The latter feedback is expected to be positive, since increased temperatures would melt the more reflective polar ice, reducing the surface albedo. The direct thermal effect on the albedo of land masses and the indirect effects on the hydrological cycle are subjects of considerable current uncertainty and research.

The thermal effects of various trace gases will be discussed using a one-dimensional model that couples a transport and chemistry model to an RC model (*82*). The RC model has its tropospheric temperature lapse rate (dT/dz) fixed at the global mean value, which is similar to the moist adiabatic value, with fixed cloud cover at 6 km altitude. The change in the modeled surface temperature due to a doubled CO_2 concentration is +1.7 K, similar to other RC models with these conditions (*45, 59, 76*). The surface temperature changes to be discussed depend

TABLE IX

Calculated Surface Temperature Changes (K) Due to Steady-State Perturbations from Various RC Models[a]

References	CFCs	$2 \times CH_4$	$2 \times N_2O$	$2 \times CO_2$
Ramanathan (*87*)	+0.85 (0.61)			+2.8[b] (2.0)
Donner and Ramanathan (*33*)		+0.30 (0.30)	+0.33 (0.33)	+2.0[b] (2.0)
Wang *et al.* (*105*)	+0.54 (0.54)	+0.28 (0.28)	+0.44 (0.44)	+2.0 (2.0)
Lacis *et al.* (*58*)	+0.68 (0.47)	+0.26 (0.18)	+0.65 (0.45)	+2.90 (2.0)
Owens *et al.* (*82*)	+0.35 (0.42)	+0.23 (0.28)	+0.29 (0.35)	+1.67 (2.0)

[a] Values in parentheses scaled to 2.0 K for $2 \times CO_2$.
[b] Estimated.

primarily on the infrared absorption properties of the trace gases and are not sensitively dependent on the chemical reaction rate set used. The vertical distribution of the calculated temperature changes, however, depends sensitively on the effect that the associated chemical changes have on ozone.

To a first approximation, the thermal response of the troposphere in an RC model is determined by the infrared flux change at the tropopause. Thus, the *ratio* of a calculated surface temperature change from a given perturbation to that

TABLE X

Surface Temperature Changes for Various Potential Man-Made Perturbations Calculated from the du Pont Coupled 1-D Model

Perturbation	Temperature change (K)[a]	Fraction of total[a]
Aircraft	+0.02 (+0.00)	0.01 (0.00)
$1.25 \times N_2O$	+0.04 (+0.07)	0.01 (0.03)
CFCs	+0.45 (+0.35)	0.16 (0.15)
$2 \times CO_2$	+1.69 (+1.67)	0.60 (0.72)
$2 \times CH_4$	+0.60 (+0.23)	0.23 (0.10)
All above combined	+2.82 (+2.32)	1.00 (1.00)

[a] Values in parentheses represent thermal effects only, with fixed species vertical profiles. For these cases, quoted factors are perturbations in concentrations (not fluxes).

due to doubled CO_2 is less sensitive to the tropospheric feedback assumptions of the model than is the absolute change. The surface temperature changes obtained with several RC models, excluding chemical feedbacks, are given in Table IX for perturbations in the atmospheric concentrations of CFCs, CH_4, N_2O, and CO_2. The CFC concentrations (~4 ppbv total) correspond to the eventual steady state, given present release rates. The numbers in parentheses have been *scaled* to a sensitivity of +2.0 K for doubled CO_2. The direct (scaled) thermal contributions to the greenhouse effect are surface temperature increases of 0.4–0.6 K for

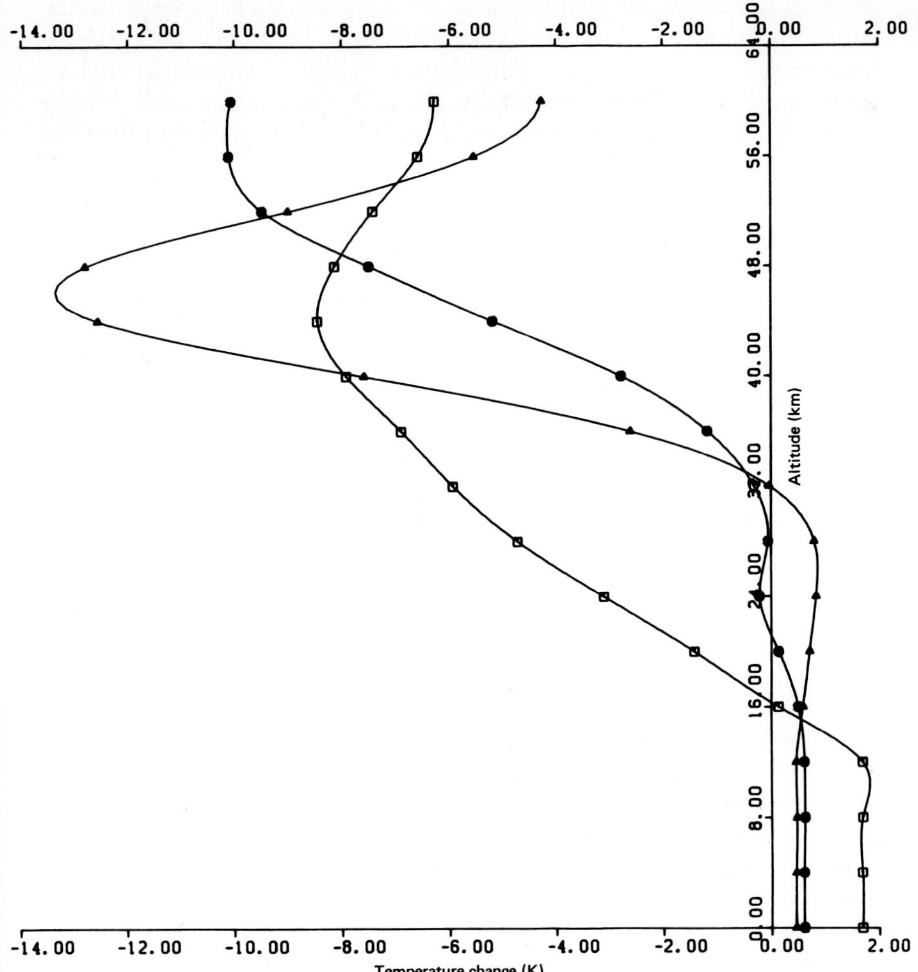

Fig. 28. Vertical temperature change profiles due to various perturbations. □, Doubled CO_2; ●, doubled CH_4 ground-level flux; ▲, CFC release in the steady state. du Pont coupled model.

CFCs in the steady state, 0.2–0.3 K for doubled CH_4, and 0.3–0.4 K for doubled N_2O. It should be noted that the projected time scales for such steady-state changes could be quite different, given the measured present trends and estimated future emissions.

Table X gives the modeled change in the surface temperature due to various individual potential perturbations (along with the total due to all five sources), calculated with du Pont's coupled 1-D model, including chemical feedback. The levels of the perturbations (of ground-level *fluxes*) were chosen to represent the atmospheric content approximately 100 years in the future, if current trends are projected (see ref. *82* for a discussion of the scenarios). As in previous calculations (*76*), the thermal effect of CO_2 accounts for about two-thirds of the calculated surface temperature change. The thermal impacts of aircraft and N_2O are small, and together they give less than a ± 1 K calculated temperature change throughout the atmosphere. Given the differences in the RC model structures and the contribution of chemical feedback, our results for the surface temperature change are similar to those discussed previously for CFCs (*16, 45, 87, 105*) and for N_2O (*33, 45, 105*). Tropospheric and lower-stratospheric ozone increases in our model give an enhanced thermal impact for CH_4.

Figure 28 gives the calculated vertical temperature changes due to each of the three individual perturbations ($2 \times CO_2$, CFCs, and $2 \times CH_4$). The calculated changes due to CO_2 are predominantly due to increased infrared opacity, while the stratospheric temperature changes due to CFCs and CH_4 include the dominant indirect effects of a modified ozone distribution. The combined thermal effect of the five perturbations in Table X is nearly additive. Near the stratopause ($\sim$ 50 km), for all five perturbations combined, a temperature decrease of more than 20 K is calculated.

V. A COMPREHENSIVE APPROACH—MULTIPLE PERTURBATIONS

A. Multiple-Perturbation Scenarios

The extensive study of atmospheric physical and chemical processes generated by concern over individual human activities such as CFC release or SST emissions has not only broadened the list of human activities under consideration as discussed in the preceeding section; the research also has identified important connections among those activities. Calculations summarized by NASA in 1979 (*50*) showed that the combined effect of CFCs and a doubling of N_2O flux was not simply the sum of the individual effects. Chemically, this result could be expected on the basis of the numerous reactions now known to couple the ClO_x, NO_x, and HO_x families in the stratosphere (see Section II). For example, in the case of N_2O and CFCs, the additional NO_2 arising from N_2O decomposition serves to tie up a larger fraction of ClO as chlorine nitrate, reducing the combined impact.

The modeler's goal of predicting the impact of a given perturbation, however, now requires a working knowledge of other emissions that could impact ozone. The calculation of individual perturbations (as discussed in Section IV) can at best indicate how a single emission would alter an otherwise stable atmosphere. This approach implicitly assumes a reference atmosphere in which all emissions other than the one under consideration have reached steady state. The calculated ozone change is then the difference between the modeled ozone concentration in the "ambient" or initial atmosphere (in which all emissions have reached a dynamic equilibrium) and in an atmosphere with the single specified change.

It has become clear that several natural and anthropogenic emissions may be changing in magnitude, and that the reference stable atmosphere and predictions based on it lack necessary realism. Any assessment of a given chemical emission seeks to define how the atmosphere (including that emission) might evolve, and to compare that to the situation without the given emission. The stable reference atmosphere discussed above is inappropriate for comparison if the real atmosphere is actually evolving due to other causes. A new approach is developing wherein the reference atmosphere is redefined to be the atmosphere *expected to evolve based on current estimates of future atmospheric emissions.* A comparison of this "anticipated atmosphere" with the "ambient atmosphere" will define variations in ozone, temperature, etc., that might be expected overall. The contribution of any one factor can then be defined by removing that factor from the "anticipated atmosphere," recalculating, and finally comparing the result with the calculated "anticipated atmosphere."

A series of these more realistic calculations can better define the individual contributions of various emissions, providing the basis for any planning necessary to preserve important atmospheric characteristics like the total ozone concentration. On the other hand, calculations based on the individual perturbations described previously give misleading results.

The results of recent calculations for the combined interaction of different perturbations are summarized in Table XI. In every case, the results are not simply additive. The current best information on any individual effect, therefore, becomes apparent only in the context of a full calculation combining all anticipated man-made changes.

Although the emission resulting from man's future activities cannot be estimated accurately, one can combine the available knowledge to give a scenario for a proposed "anticipated atmosphere." This multiple emissions scenario is predominantly a composite of the individual emission scenarios discussed in Section IV. For chlorofluorocarbons, constant release at current emission rates is assumed by most authors for CFCs-11, -12, -113, -114, -115, and -22. Due to their small production rates, the influence of the fluorocarbons other than

TABLE XI

Pairwise Interactions of Various Perturbations on Total Ozone[a]

Perturbations		Calculated column ozone change (%)			
A	B	(A + B)	(A) + (B)	A only	B only
2 × N_2O	CFCs	−15.3	−20.3	−13.6	−6.7
2 × CH_4	CFCs	−1.6	−3.2	+3.5	−6.7
2 × N_2O	2 × CH_4	−8.1	−10.1	−13.6	+3.5

[a] du Pont 1-D model using WMO/NASA 1982 chemistry.

CFC-11 and CFC-12 on ozone is small (see Table VI). Because of slow rates of transport and photolysis, the atmosphere would begin to approach steady state with respect to constant CFC emissions only after more than 100 years. Chlorocarbons other than CFCs are also commonly included at current production rates, although the atmosphere is much nearer to steady state with respect to these emissions due to their shorter atmospheric lifetimes.

Other changes in atmospheric concentrations of major source gases are expected to occur over different time scales. Nitrous oxide (N_2O) concentrations have increased over the past several years at ~ 0.2% per year, and all the time-dependent calculations discussed below employ that rate of increase. For the reference point 100 years from now, Owens *et al.* (*82*) chose a 25% increase in N_2O flux to the atmosphere.

The common assumption for aircraft emissions employs the projections of Oliver *et al.* (*79*) through the year 1990 with constant emissions at the 1990 level thereafter. Carbon dioxide in recent years has been increasing quite rapidly, at ~ 0.3–0.5% per year. For time-dependent calculations, modelers have recently adopted the fossil fuel projections by Rotty and Marland (*91a*), assuming an airborne fraction of 0.5 (*120*). These lead to a doubling on the 100-year time scale.

An additional perturbant included by many modelers is methane (CH_4). Recent rates of increase have been estimated at 1–2% per year (*39*), doubling with a 50- to 100-year time scale. However, the source(s) of the recently observed changes in the atmospheric methane concentration is not known, and modelers have usually chosen scenarios both with and without future methane changes.

B. Steady-State Multiple Scenario Calculations

As discussed above, the complex interactions between CFCs, N_2O, and CH_4 in the atmosphere yield highly nonlinear combinations of individual perturbations. As illustrated by the modeling results shown in Table XI (*81*), obtained with a chemistry–transport model employing WMO/NASA (*118*) chemistry, the impacts of the separate perturbations on the total column cannot simply be added to estimate the result of a combined perturbation. The incremental impact of CFCs in the steady state, for example, is a total ozone change of −1.7% in the 2 × N_2O + CFC scenario and −5.1% in the 2 × CH_4 + CFC scenario, whereas the CFC scenario alone changes ozone by −6.7%.

The nonlinear interaction between the steady-state CFC and 2 × CO_2 scenarios on the ozone column, calculated with coupled radiative–convective and transport–chemistry models, is shown in Table XII. The "latest" chemistry is that recommended in JPL 82-57 (*55*), with the fast rate of $ClONO_2$ formation and reduced oxygen photolysis cross sections from 200–220 nm. The most recent calculations suggest that doubled carbon dioxide plus CFCs in the steady state leave the total ozone column essentially unchanged.

A recent set of steady-state calculations by Owens *et al.* (*82*) is based on the levels of atmospheric gases that may exist approximately 100 years into the future, if current trends persist. The scenario is that given in Tables VIII and X above: aircraft + 1.25 × N_2O + CFCs + 2 × CH_4. Figure 29 gives the vertical profile of the local ozone concentration change based on this simultaneous perturbation. The figure shows both the absolute concentration change (molecules/cm^3), which indicates the contribution to the total column, and the local percentage change, which shows the magnitude of local effects.

The integrated column ozone change calculated for this scenario is +1.8%. The important aspect of this and similar time-dependent calculations to be discussed below is the near constancy of the total vertical column of ozone. Consequently, the major concern over a change in the ground-level UV-B flux has been reduced.

As Fig. 29 indicates, the relatively small calculated changes in total ozone arise from a cancellation of effects: significant decreases in local ozone above ~ 28 km and comparable increases below that altitude. Thus, in effect, several

TABLE XII

Calculated Total Ozone Change Due to Simultaneous Perturbation: CFCs + 2 × CO_2

	Calculated column ozone change (%)[a]			
	LLNL model		du Pont model	
Perturbation	(1)	(2)	(3)	(4)
CFCs	−5.1	−2.0	−5.7	−4.0
2 × CO_2	+5.9	+4.7	+2.9	+2.6
CFCs + 2 × CO_2	−0.6	+1.8	−3.5	−2.4

[a] (1) Wuebbles *et al.* (*120*), WMO/NASA (*118*) chemistry; (2) D. J. Wuebbles, private communication (1983), latest chemistry; (3) Owens *et al.* (*82*), JPL 82-57 (*55*) chemistry; (4) latest chemistry.

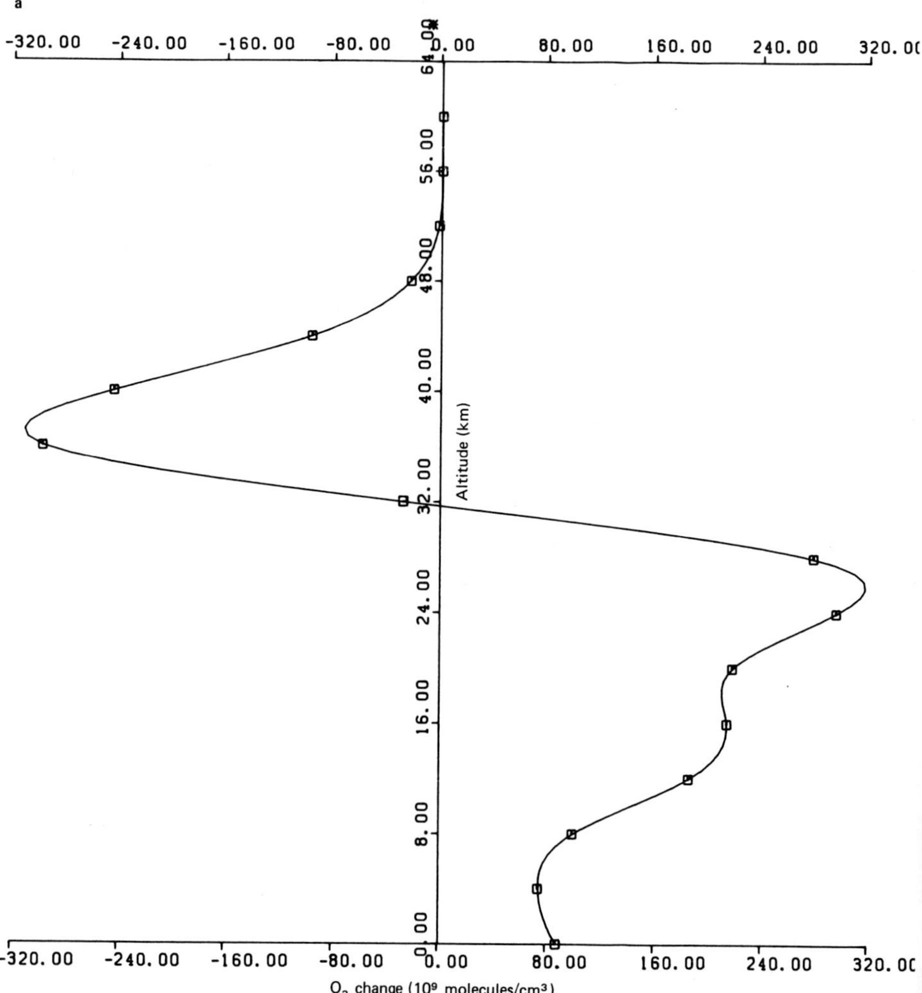

Fig. 29. Vertical ozone profile changes due to the simultaneous steady-state multiple scenario $2 \times CO_2 + 2 \times CH_4$ + aircraft + $1.25 \times N_2O$ + CFCs. (a) Change of local concentration. (b) Local percentage change. du Pont coupled model.

percent of the overhead ozone is calculated to shift from higher to lower altitudes as a result of projected trace gas emissions at ground level. The importance of such a redistribution of ozone requires a rigorous quantitative assessment.

So far as the ultraviolet flux in the troposphere is concerned, the location of ozone has essentially no impact; it depends only on the ozone column density. However, as discussed earlier, the radiative properties of ozone can have consid-

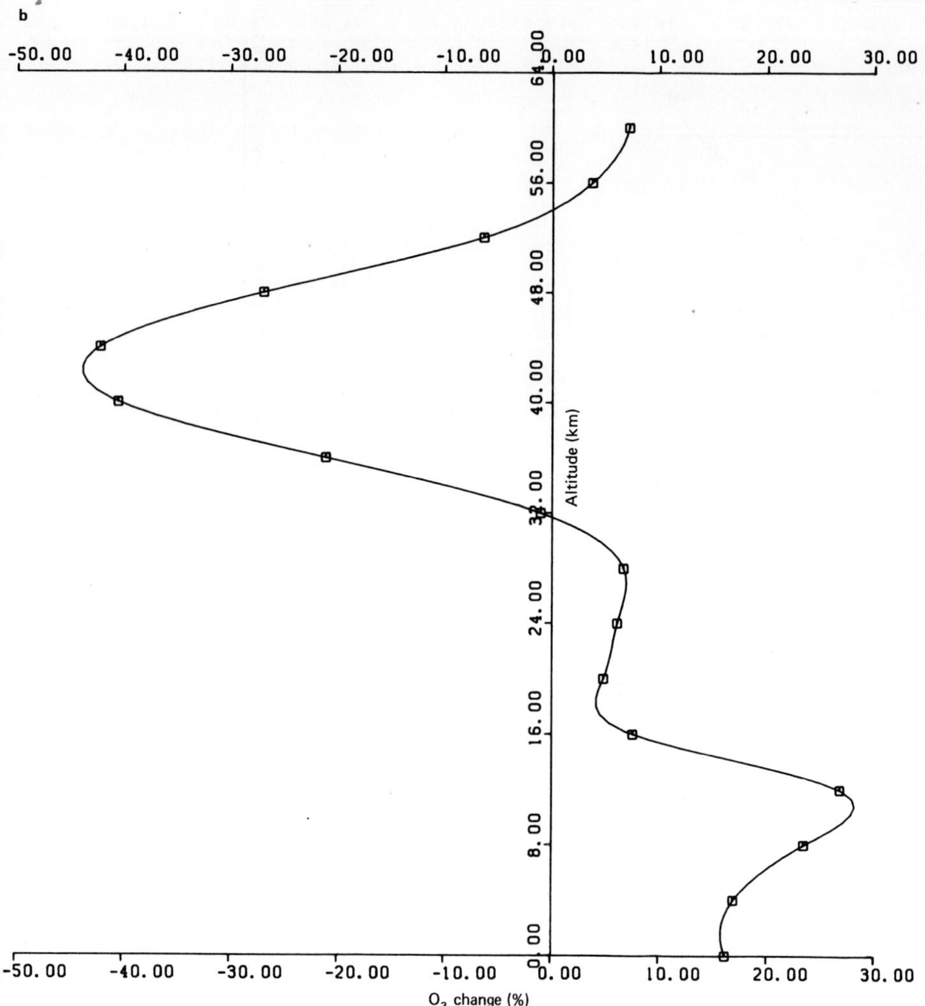

erable influence on temperatures throughout the atmosphere. The temperature inversion (i.e., the increase in temperature with increasing altitude) that defines the stratosphere as distinct from the troposphere below and the mesosphere above results almost entirely from heating by ozone through its strong ultraviolet absorption. Likewise, ozone infrared radiative properties exert a strong influence on atmospheric and surface temperature through the greenhouse effect (see *76, 82, 106*). These temperature changes in turn feed back to the temperature-dependent chemistry in a way that has been included in the state-of-the-art

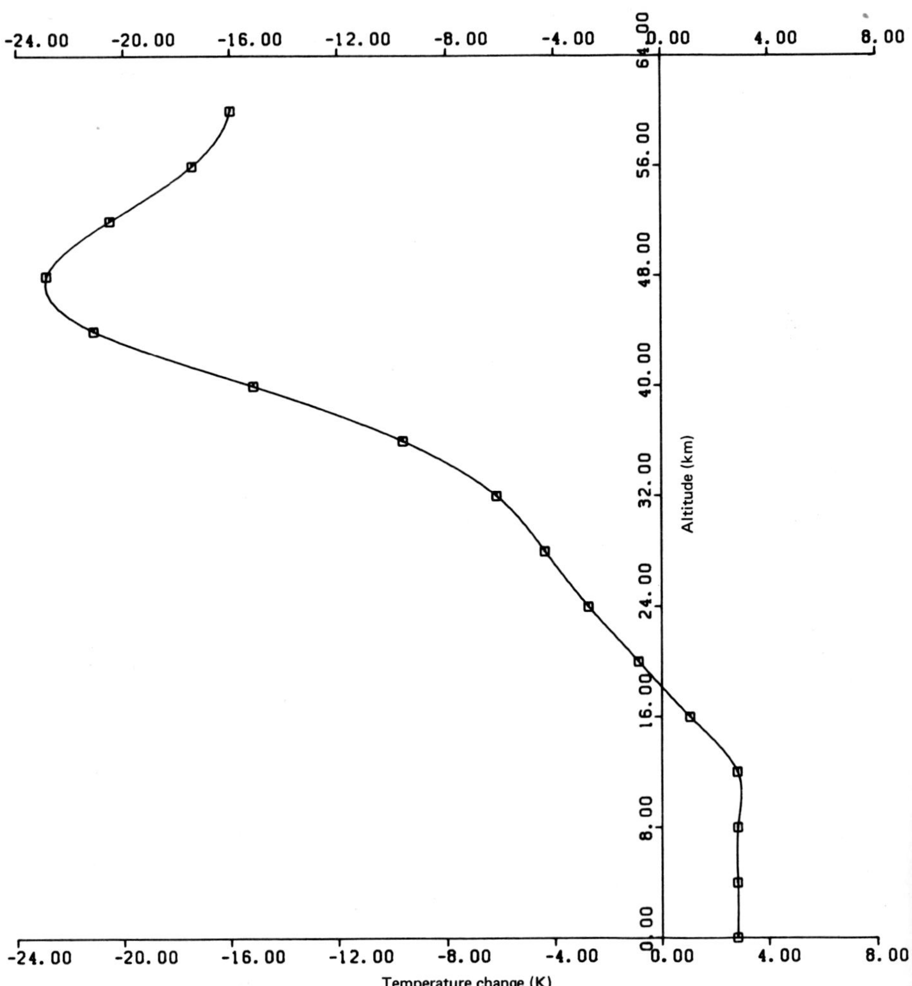

Fig. 30. Vertical temperature change due to the same simultaneous steady-state multiple scenario as in Fig. 29.

atmospheric models. The consequent changes in surface temperature can also be assessed with such models.

The altitude profile of calculated temperature changes from the model of Owens *et al.* (*82*) is given in Fig. 30. The maximum change of about 20 K occurs near the stratopause (~ 50 km). Because temperature gradients are closely related to air movement, the dynamics of the stratosphere might be altered somewhat by such changes. A large enough change could conceivably couple with the troposphere to produce effects on climate at the surface. However, an understanding of atmospheric dynamics adequate to predict these consequences does not currently exist. This area represents one of the major needs for future research. The possibility of changes in dynamics further complicates the issue through its potential feedback on chemistry. Some groups are now attempting to develop models that fully integrate three-dimensional dynamics in a self-consistent way with chemistry and radiation.

Current uncertainties notwithstanding, calculation of an "anticipated atmosphere" permits a more realistic approach to evaluating the effects of an individual perturbation. The results of Owens *et al.* (*82*) in Table XIII, for example, demonstrate the effects of excluding methane changes from the anticipated atmosphere. The calculated difference in total ozone is seen to be 6.0% more ozone when the methane change ($2 \times CH_4$) is included. When methane is considered alone, however, the same increase in the methane source flux leads to a calculated ozone increase of only 4.3%.

Figures 31 and 32 show, respectively, the altitude profiles of ozone change and temperature change for the two different assessments of the impact of methane. Each figure gives (●), the effect of a doubling of the methane concentration alone on the ambient atmosphere, and (□) the difference between $2 \times CH_4$ and $1 \times CH_4$ in the "anticipated atmosphere" (CFCs, $2 \times CO_2$, aircraft, $1.25 \times N_2O$). The difference between the two curves is due to the nonlinear interactions between methane and the ClO_x and NO_x species produced by the CFCs and N_2O.

TABLE XIII

Effect of Excluding a Methane Increase from the "Anticipated Atmosphere"[a]

	Calculated changes from ambient atmosphere	
Scenario	Ozone column (%)	Surface temperature (K)
Aircraft		
+ $1.25 \times N_2O$		
+ CFCs		
+ $2 \times CO_2$	−4.5	+2.25
Above		
+ $2 \times CH_4$	+1.5	+2.82
Difference	+6.0	+0.57
$2 \times CH_4$ alone	+4.3	+0.60

[a] Based on du Pont coupled 1-D model with JPL 82-57 (*55*) chemistry.

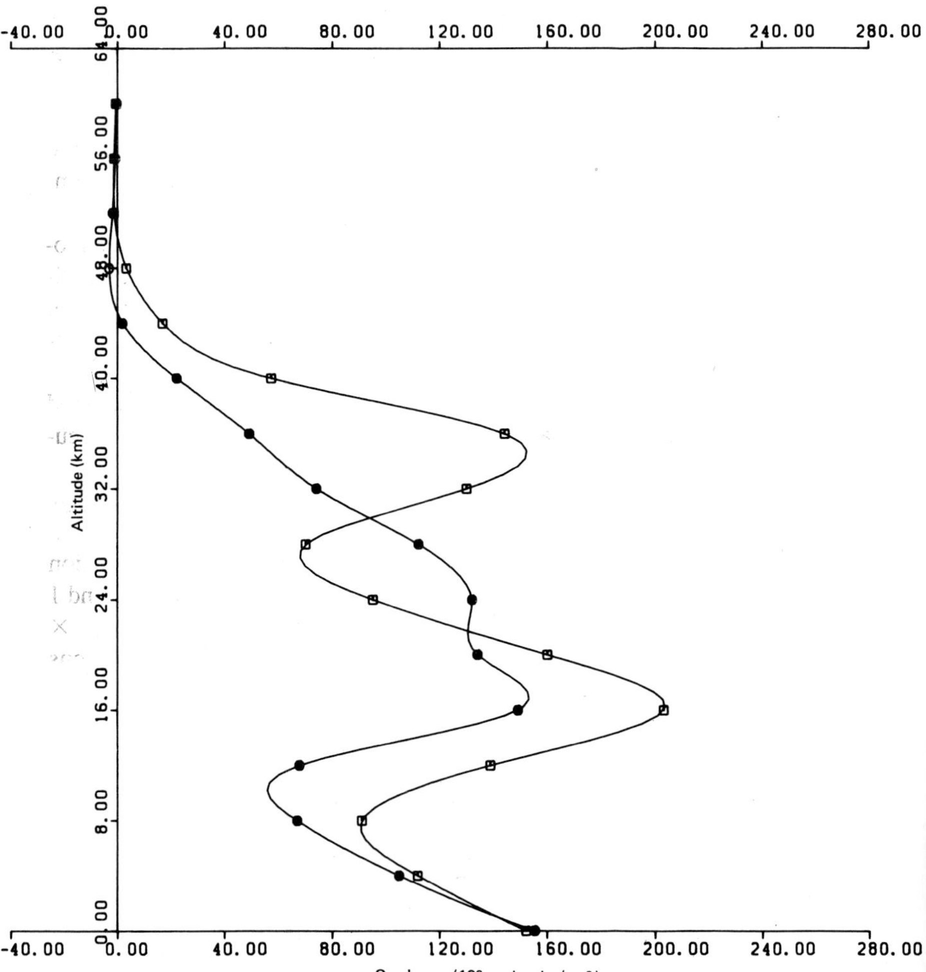

Fig. 31. Comparison of two methods for assessing the effect of a 2 × CH_4 perturbation on the vertical ozone profile. (●) Doubled CH_4 flux versus a stable "initial" atmosphere. (□) Difference between 2 × CH_4 and 1 × CH_4 in the "anticipated atmosphere" with simultaneous CO_2, aircraft, N_2O, and CFC changes. du Pont coupled model.

A similar approach can be used for other perturbations in order to assess, for instance, the significance of an uncertainty in future emissions levels. Such assessments are constrained by the gaps in current knowledge and the high degree of uncertainty involved in many of the model parameters. Nonetheless, this approach provides a framework that is thus more relevant than earlier (single perturbation) approaches.

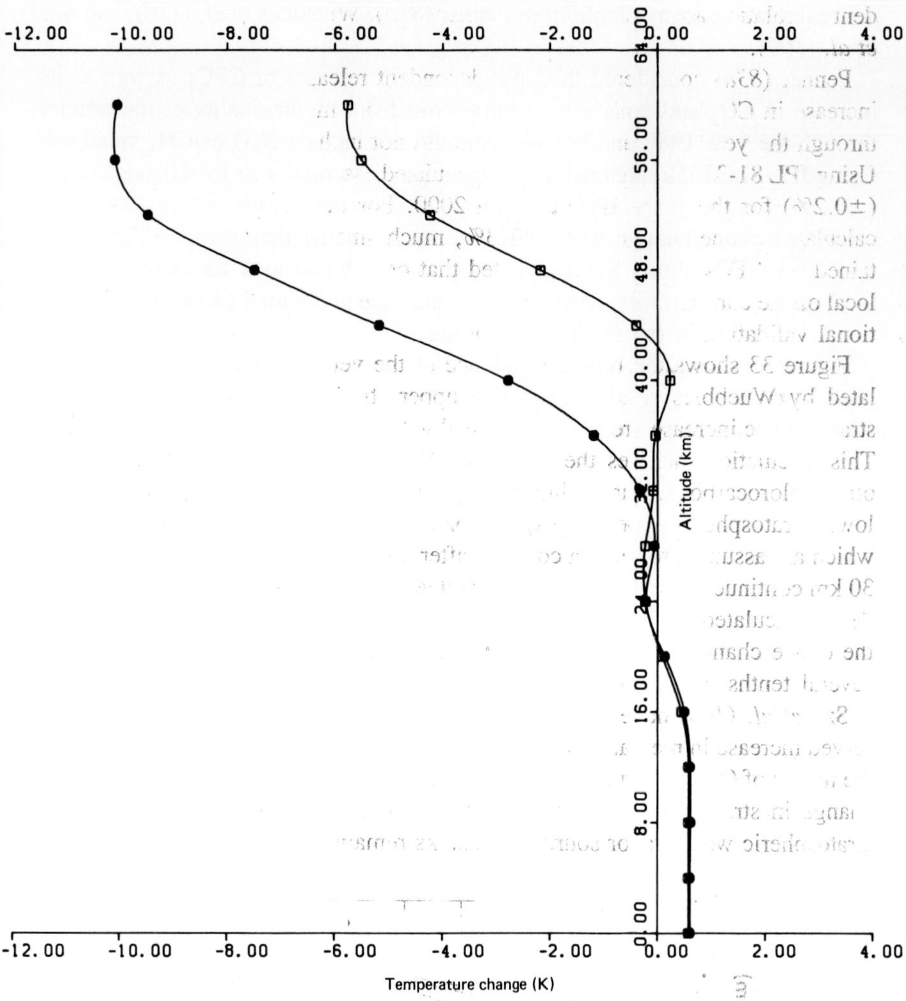

Fig. 32. Comparison of two methods for assessing the effect of a 2 × CH_4 perturbation on the vertical temperature profile. Symbols as in Fig. 31.

C. Time-Dependent Multiple Scenario Calculations

For simplicity, the discussion in Section V,B concentrated on steady-state type calculations for a projected atmosphere with source gas fluxes as they might exist approximately a century in the future. In reality, of course, the time scales for changes in the various emissions can be quite different. Thus, a more complete assessment must examine the possible upward or downward excursions in ozone concentration and temperature during the interim. Examples of such time-dependent calculation are available from Penner (*83a*), Wuebbles *et al.* (*120*), and Sze *et al.* (*100*).

Penner (*83a*) considered the time-dependent releases of CFCs as well as an increase in CO_2 and emissions from aircraft from the beginning of the century through the year 1980 and beyond. She did not include N_2O or CH_4 increases. Using JPL 81-3 (*55a*) chemistry, she calculated essentially no total ozone change ($\pm 0.2\%$) for the years 1960 through 2000. For the decade of the 1970s, the calculated ozone change was -0.13%, much smaller than that (-0.73%) obtained for CFCs alone. She suggested that one should look for changes in the local ozone concentration near 40 km, rather than in the total ozone, for observational validation of the model calculations.

Figure 33 shows the time dependence of the vertical ozone profile as calculated by Wuebbles *et al.* (*120*). The upper stratospheric decrease and lower stratospheric increase are seen to proceed with time at somewhat different rates. This calculation includes the effects of N_2O, CO_2, aircraft NO_x, CFCs, and other chlorocarbons, but excludes CH_4. Thus, the dominant contributor to the lower stratosphere–upper troposphere ozone increase is aircraft NO_x emissions, which are assumed to remain constant after 1990. A small ozone increase below 30 km continues from that time onward due primarily to CFCs. The time dependence calculated by Wuebbles *et al.* for total ozone is given in Fig. 34. Despite the ozone changes as a function of altitude, the total is seen to remain within several tenths of one percent of the level calculated for today's atmosphere.

Sze *et al.* (*100*) have performed a similar calculation including also the observed increase in methane, projected to continue at 1.0% per year. In contrast to the model of Owens *et al.* (*82*), methane decomposition in this model leads to no change in stratospheric water vapor, which is fixed. (A full understanding of stratospheric water vapor sources and sinks remains an elusive goal.) With this

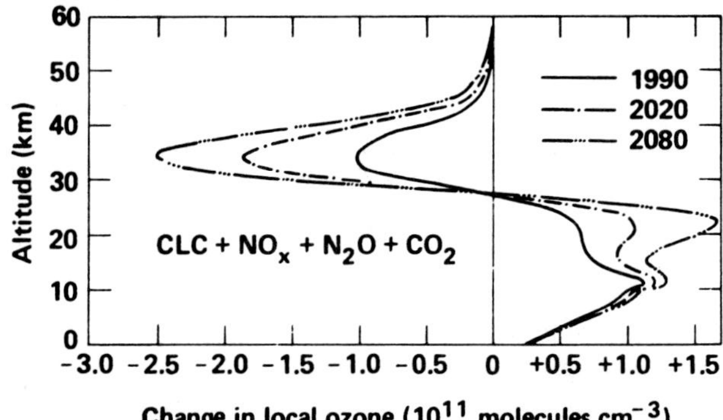

Fig. 33. Calculated absolute change in local ozone at various times relative to 1911 background atmosphere for perturbation containing all anthropogenic sources (except methane). From Wuebbles *et al.*, 1983 (*120*), with permission.

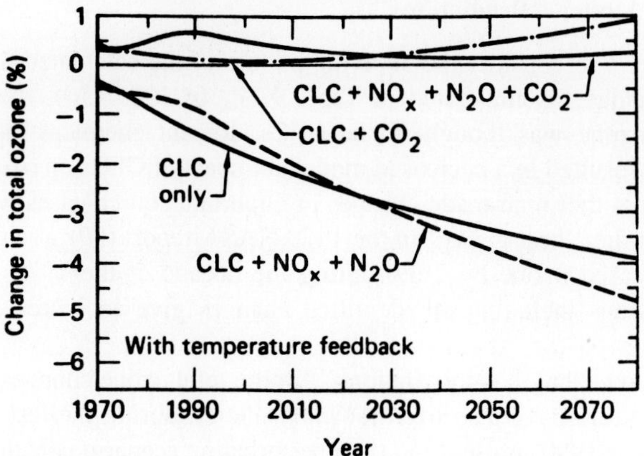

Fig. 34. Calculated changes in total ozone as a function of time relative to 1911 background atmosphere for various combinations of anthropogenic perturbation scenarios. From Wuebbles *et al.*, 1983 (*120*), with permission.

assumption, methane would not contribute to the HO_x budget in the upper stratosphere, and would have a positive effect on ozone concentration in the region. In an application of the "anticipated atmospheric" approach, Sze *et al.* consider both constant CFC emissions indefinitely and an increase of 3% per year from 1980 to 2000, followed by constant emissions thereafter. Results from both calculations, presented in Fig. 35, project increased ozone throughout the time period considered.

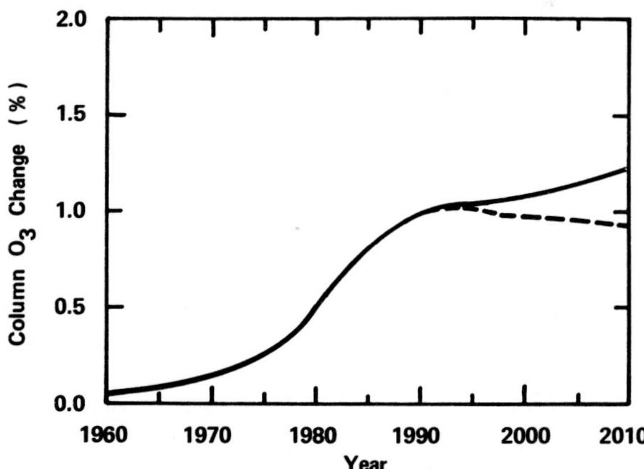

Fig. 35. Calculated percentage change in ozone column compared to the year 1950 as a function of time due to combined effect of CO_2, CH_4, aircraft, N_2O, and CFCs. Solid curve, constant CFC emissions; dashed curve, CFC emissions increase 3% per year from the year 1980 to 2000. From Sze *et al.*, 1983 (*100*), with permission.

D. Limits of Model "Predictions"

It is instructive to consider the evolving picture of the calculated time-dependent ozone change, as illustrated in Table XIV. In 1979 (*50*), the dominant influence on ozone was thought to be CFCs. Recent chemical reaction rate revisions have resulted in a decreased modeled impact of CFCs on ozone, while, simultaneously, other man-made changes in important source gases were identified. Although the "best guess" in the 1979 NASA report (*50*) was that human activities decreased ozone by 2.0% during the decade of the 1970s, the most recent calculations including all identified changes give an increase of about 0.3%.

Figure 36 shows that the "predictions" for the total ozone changes during the next 10 to 30 years have also been radically altered during the last few years. Through the year 1990, present multiple perturbation scenario calculations indicate an ozone *increase* of about 1%, whereas in 1979 the expectation was more than a 4% *decrease*. Recent model calculations suggest that the (presumably) man-made changes to several important stratospheric source gases during the last decade have had little impact on the total amount of ozone above the earth's surface.

The variety of results discussed here gives some idea of the range of increases or decreases in ozone that might arise over the next several decades based on our current understanding, but that range may well be misleading as an indicator of the confidence that can be placed in these results. When a future state of the atmosphere is calculated, researchers must obviously ask the important question: How reliable is the model as a predictor? The reliability depends on the uncertainties involved both in the model structure (physical and chemical schemes) and in the necessary model input information (e.g., scenarios of future changes in source gases). Some uncertainties, like rate constants of chemical reactions

TABLE XIV

Total Ozone Column Change: 1970 to 1980

Vintage of calculation	CFCs only (%)	All man-made effects (%)
1979[a]	−2.0	−2.0 (CFCs)
1981[b]	−0.7	−0.1 (CFCs + CO_2 + aircraft)
1982[c]	−0.5	+0.1 (CFCs + CO_2 + aircraft + N_2O)
1983[d]	−0.3	+0.3 (CFCs + CO_2 + aircraft + N_2O + CH_4)

[a] Reference *50*.
[b] Reference *83a*.
[c] Reference *120*.
[d] References *82* and *100*.

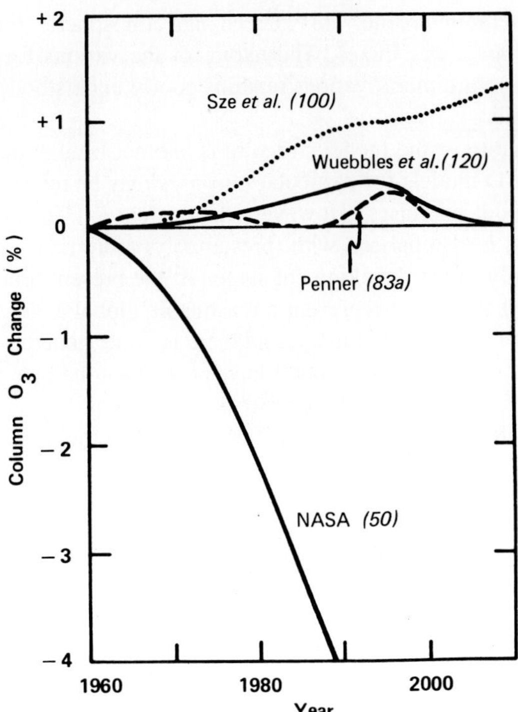

Fig. 36. Summary of calculated ozone column changes due to anthropogenic activities versus time. Scenarios are: NASA, CFC; Penner, CFC + aircraft + CO_2; Wuebbles *et al.*, CFC + aircraft + CO_2 + N_2O; Sze *et al.*, CFC + aircraft + CO_2 + N_2O + CH_4.

employed in the model, can in principle be defined in a quantitative fashion. In practice, other types of uncertainty (e.g., missing chemical mechanisms and future courses of source species emissions) cannot be quantified.

Over the past several years the chemical scheme has undoubtedly been greatly improved, but its current adequacy remains an open question. The available tests of model accuracy have not yet been sufficient to constrain the models to a satisfactory degree. As to the overall model structure, the fact that the current models simulate the observed trace species of the present atmosphere fairly well is encouraging and at least one of the necessary conditions for reliability. However, the lack of a true sufficiency or completeness test, as documented not least by the lack of measurements for the temporary reservoir species, still precludes a claim of model "validity."

Concerning physical processes, the models have become more sophisticated, taking into account the coupling of thermal effects and chemistry. The increase in CO_2 discussed in previous sections is expected to cause modifications to the atmospheric temperature profile. This may change atmospheric transport, as well as tropospheric source gas fluxes. The nature of the various feedbacks due to climate and temperature modifications remains poorly understood, especially for long-term predictions.

The dimensionality of the model structure is another limitation. Some justification of use of 1-D models for particular purposes may be made on the basis of available 2-D model exercises. However, the current 2-D models themselves must be examined in comparison with (presumably) more realistic 3-D models, which are now only in the development stage. At the present time, the state-of-the-art 1-D models appear to represent a reasonable global average of the stratospheric transport and photochemistry, and there is rather general agreement that calculations made with 1- or 2-D models have at least a good qualitative description of atmospheric transport and chemistry.

Large uncertainties in future scenarios for source gas emissions of CH_4, H_2O, N_2O, CO_2, CFCs, etc. continue to exist and greatly limit our ability to accurately forecast future ozone modifications.

It is clear, therefore, that a great deal more basic work is required in all aspects of the subject before detailed quantitative predictions can be made with confidence. This basic research on models and their inputs must in turn be supplemented by atmospheric measurements designed to meaningfully test the model results.

VI. OZONE MEASUREMENTS

The major advances in theoretical understanding of production, transport, and destruction of atmospheric ozone have been paralleled by a continuation of measurement efforts dating back to Dobson's work in the 1930s. The evolution of such research has included the development of new techniques for rocket-, balloon-, and satellite-borne instruments. Since 1976, statisticians have introduced new methods into the analysis of available measurements in an effort to recognize changes that might be attributable to man's activities. That work, in turn, has spurred efforts to better understand the natural variability of ozone and to refine further the accuracy of the measurements.

A. Total Ozone

The principal quantity of concern, historically, has been the integrated total amount of ozone in a vertical column above the earth's surface. It is this total column ozone amount that determines the ultraviolet flux in the UV-B region from 290 to 320 nm. Both ground-based and satellite techniques are used for measuring total ozone.

The longest continuous series of measurements available is that of observations made with the Dobson spectrophotometer, starting as early as 1925 at some stations. Figure 37 depicts currently operating stations that have 15 or more years of data. This network is augmented by use of the M-83 filter photometer, the locations of which comprise most of the "other" sites marked in Fig. 37. In a search for long-term trends or variations, the geographical representativeness of such systems must be assessed. The concentration of stations in the Northern Hemisphere mid-latitudes and the absence of measurements over ocean areas leaves open the possibility of systematic ozone variations (e.g., standing waves), which might be lost or misconstrued in any effort to average results from individual stations to construct a global climatology of ozone.

The obvious, if difficult, solution to global coverage is the creation of a continuous system of well-intercalibrated satellite measurements. The principal satellite systems providing observations of total ozone data thus far are the

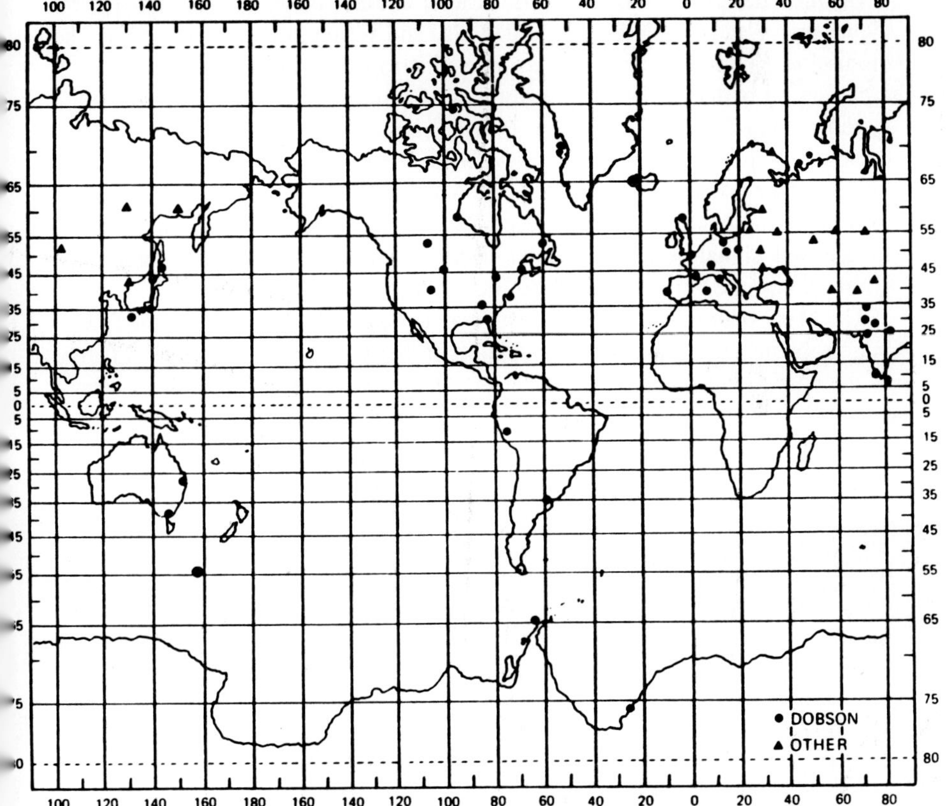

Fig. 37. Global network of ground-based total ozone observing stations with 15 or more years of observations. From Hudson *et al.*, 1982 (*51*).

Backscattered Ultraviolet (BUV, flown on Nimbus 4, and measuring from 1970–1977), the Solar Backscattered Ultraviolet (SBUV, Nimbus 7, 1978–present), and the Total Ozone Mapping Spectrometer (TOMS, Nimbus 7, 1978–present). Others include the Infrared Interferometer Spectrometer System (IRIS), the High Resolution Infrared Radiometer System (HIRS), and the Multichannel Filter Radiometer (MFR). From within this group of instruments there exist measurements throughout the period from 1969 to present; however, the individual instruments often provide no more than 1–2 years of data. Moreover, the extensive coverage of the Nimbus 4 BUV instrument suffers from concerns over instrumental drift and uneven global coverage. The inability to recalibrate a deployed satellite system necessitates careful attention to sufficient overlap of successive instruments and an accurate method for ground truth verification, neither of which has been completely accomplished to date. Thus, analysis of these data for trends suffers some potentially serious drawbacks.

Detailed descriptions of the various total column ozone measurement techniques are provided in NASA RP-1049 (*72*) and WMO Report No. 9 (*117*). The discussion is updated with more recent results from Hudson *et al.* (*51*), including a discussion of the types and magnitudes of errors associated with each of the systems. Major concerns with the Dobson measurements include possible systematic bias associated with absorption coefficient standards, anomalous absorption by other species (notably SO_2, for which concentration may be increasing over many stations), corrections necessary for stratospheric aerosol contributions to detected signal, effects of solar cycle variations on the measurement algorithm, and the contribution of tropospheric ozone.

The M-83 photometer is sensitive to standardization problems with the filters, bias problems due to the broad bandwidth of the filters, and the empirical nature of the ozone determination from the raw data. Furthermore, there have been a number of improvements in design and calibration procedures that have improved data quality, but at the expense of continuity necessary for elucidation of long-term trends.

Of the satellite methods, the BUV currently provides the longest data base. Those data, however, suffer from a known instrument degradation as evidenced by a trend in the average difference between BUV and Dobson ozone observations during the lifetime of the satellite (*51*). During the later years of the satellite's operation, a weak power source also limited the amount of data collected, resulting in less than complete global coverage, unfortunately negating one of the key advantages of a satellite measuring system. These problems appear to have been resolved with the incorporation of several improvements in the SBUV system on Nimbus 7. Provided that sufficient attention is paid to continuity and ground truth, the SBUV data are expected to be the beginning of a long-term global record that will be sufficient to indicate accurately both natural and unusual trends and variations in total ozone.

The lack of global coverage in ground-based observation systems and the limited length of satellite measurement records combine to make difficult the estimation of actual long-term variations in ozone. As discussed in the introduction, shorter term (e.g., seasonal, annual, daily) variations are reasonably well understood, but the complications of proper global averaging and real or apparent (measurement-associated) random fluctuations contribute in significant ways to the uncertainty in any derived trend.

The earliest form of analysis consisted of graphically presenting the data with associated error bars and assessing the trend by linear regression or piecewise linear least-squares fitting techniques [e.g., Komhyr *et al.* (*57*); Johnston *et al.* (*54*); Angell and Korshover (*1–3*); London and Oltmans (*61*); and Birrer (*9*)]. A typical example of such analysis is given in Fig. 38, in which the world average demonstrates the conclusion of Johnston *et al.* (*54*) that total ozone decreased during the early 1960s and increased throughout the remainder of the decade. Angell and Korshover (*3*), who compiled the data in Fig. 38, have concluded that there is no evidence of a decrease in the 1970s that could be attributed to man-made causes. That search for a change, however, is hampered by the difficulty of compiling a representative global average. Angell and Korshover compiled the world average by using a 1, 3, 4, 3, 1 weighting of north polar, north mid-latitude, tropical, south mid-latitude, and south polar data in an attempt to avoid giving too much weight to the heavy concentration of measuring stations in the northern mid-latitudes. This consideration, however, leads to a broadening of the uncertainty limits over those obtained from a simple average of available data.

The other complication in a search for long-term trends is identification of any natural cyclic variations with a long period. For example, the world data in Fig. 38 show maxima in 1958, 1970, and 1979, all of which were years of sunspot maxima. The lack of adequate measurements of solar ultraviolet flux variations over a sunspot cycle has inhibited theoretical verification of such a change, as has the relative shortness of the historical record relative to the 11-year solar cycle.

Shorter term variations in ozone are well known, including daily fluctuations, the annual cycle, and the quasi-biennial oscillation. Recent analyses have suppressed such variations by considering a moving 10-year average or decadal trend (*51*). In such an analysis, it is important to consider not only the trend for a given 10-year period, but also to note its variation. Thus a -1% change during 1969–1978, although indicative of a possible decline in ozone, was reduced to -0.3% for the decade 1970–1979, indicating perhaps a longer term cyclic variation rather than a consistent decline attributable to man's activities (*51*).

A more general approach recently adopted by statisticians involves fitting the observed ozone values to a computer model calculated curve of the historical ozone trend. The statistical fit includes terms for cyclic variations. In addition, before being fit, the raw data are typically deseasonalized by one of several

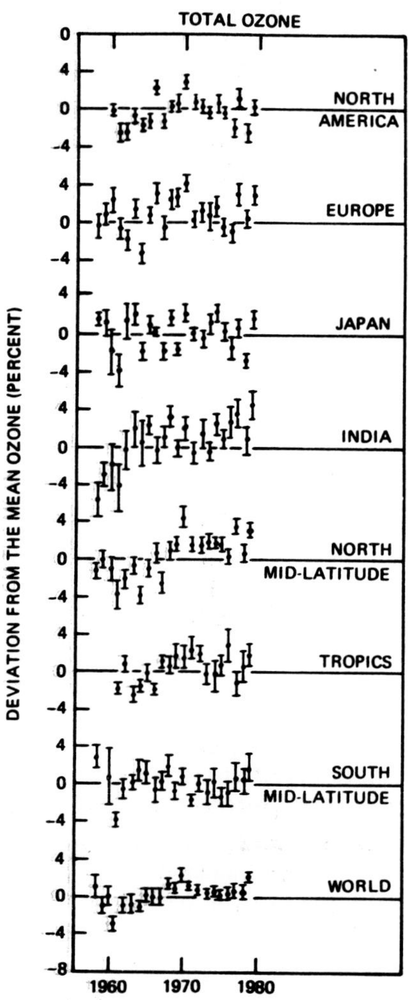

Fig. 38. Variations in the annual total ozone amount for different continents and the world. From WMO, 1981 (*117*).

methods, e.g., by subtracting the average value for, say, all Februarys from the value for February in each year, and so on. The data are typically analyzed station by station to produce a statistical model of the time series of ozone measurements for each station. The model consists of a long-term ozone average, a seasonal component, perhaps other cylical components, and a deterministic historical trend term. The regression coefficient for the trend term then gives the magnitude of the ozone variation that is consistent with the shape of the deterministic trend.

Because most anthropogenic changes are postulated to have reached sufficient magnitude to generate observable changes in ozone since about 1970, the deterministic shape tested by statisticians is commonly taken to be a constant total O_3 amount prior to 1970 with a linear decrease in subsequent years. Using this form, Reinsel *et al.* (*89*) have examined 36 Dobson station records from 1958 through 1979 and concluded that the change since 1970 is a statistically nonsignificant increase of $+0.79 \pm 1.30\%$. St. John *et al.* (*93*) use a smooth curve from model calculations by Miller *et al.* (*67*) rather than the piecewise linear form and consider only the 14 stations with nearly complete records. Their result is a change of $+1.2 \pm 1.2\%$.

Bloomfield *et al.* (*11*) perform a similar analysis using Fourier-transformed time series of measurement data. They argue that this frequency domain analysis can provide a better understanding of the various contributions to trend uncertainty. Their determined trend, based on 36 stations, is similar: $+0.44 \pm 1.04\%$. In general, all of these results rely on statistical weights to determine the appropriate average trend value for the stations considered. In that sense, the observed trend is strictly that for the observed locations only, and not necessarily a global trend. Reinsel *et al.* (*90*) have investigated the global representativeness of their results by examining the Nimbus 4 satellite data over the period April 1970 through April 1977. They found that the mean and standard deviation of trend estimates for the entire data set (gridded into 468 blocks over the globe) were in close agreement with those from the subset of the data taken in the 36 blocks that include the Dobson stations used in their earlier analysis. Thus, within an uncertainty of $\pm 0.6\%$ and for the time period considered, the 36 Dobson locations are thought to be representative of the global trend. A. Miller *et al.* (*65*) have found less consistency with ground based data using different regions and time periods. One may infer that a conclusive determination of global trend with small uncertainty limits would be made easier with a long-term satellite data set.

The consistent theme from analyses of total ozone measurements is that no decrease has taken place during the decade beginning in 1970, and possibly a small increase has occurred. This "no change" result is, of course, not inconsistent with the multiple perturbation model calculations discussed in Section V. As this is at best a null verification of a null hypothesis, however, the agreement should not be given much significance. Currently, both modelers and statisticians are seeking examples of calculated changes that can in principle be tested by measurements. As part of that effort, both groups have looked toward the behavior of ozone in specific regions of the atmosphere.

B. Vertical Ozone Concentration Profile

The dominant physical and chemical processes vary significantly with altitude in the stratosphere. Consequently, the calculated impact on ozone of various perturbing substances also varies, sometimes leading to increases at some altitudes with partially offsetting decreases at others. This is especially true of multiple-perturbation calculations, in which the net effect on total ozone is calculated to be minimal, despite rather significant changes in particular altitude regions. A related point is that the theoretical local percentage change may be quite large in regions containing little of the total ozone, improving the opportunity for detecting such a change if it occurs.

Changes near the lower extent of the ozone layer, such as those calculated for methane and aircraft NO_x emissions, can be quite large. However, this region is also strongly affected by weather patterns, leading to rather large short-term fluctuations and compounding the problem of analysis for a trend. At the upper part of the ozone layer, between 30 and 50 km altitude, the CFC perturbation is calculated to have its largest impact. Models also show contributions from methane and N_2O to ozone changes in this region. The remoteness from tropospheric circulation disturbances and the general dominance of chemistry over transport make this an ideal region to search for a partial test of the photochemical theory with ozone measurements.

The primary source of historical measurements of the vertical concentration profile for ozone involves the application of the Umkehr inversion technique to measurements made with the Dobson spectrophotometer. Because these measurements are time consuming—they require intermittent zenith sky observations while the sun is between the horizon and 30° elevation—and because reliable measurements can be taken only on clear days, the data set is considerably smaller than that for total ozone. The clear sky requirement also introduces a fair weather bias that may introduce a systematic error in trend detection. Most other potential errors in Dobson total ozone data are shared by Umkehr measurements, along with uncertainties in the inversion algorithm, unique calibration sensitivities, and susceptibility to optical effects of tropospheric and stratospheric aerosols (*51*). The technique provides its most accurate data near the ozone maximum, and random measurement errors tend to be largest in the lower and higher "Umkehr layers" (near the bottom and top of the stratosphere).

Profile measurements up to about 35 km have been taken continuously for the last 10 to 15 years at 11 stations using ozone sonde balloons. These are very unevenly distributed globally, and hence provide a rather inadequate data base for trend analysis, but are useful for cross comparison with Umkehr and satellite data.

Most of the satellite total ozone instruments also contain instruments for measuring the vertical ozone profile. In some cases, the same instrument is used for both. Both BUV and SBUV techniques provide profile data and, as for total ozone, provide the most extensive measurement set to date. Infrared techniques include the Limb Radiance Inversion Radiometer (LRIR, on Nimbus 6, June 1975–January 1976) and the Limb Infrared Monitor of the Stratosphere (LIMS,

Nimbus 7, October 1978–May 1979). The Stratospheric Aerosol and Gas Experiment [SAGE, Stratospheric Aerosol Measurement (SAM-2), February 1979–present] provides two solar occultation profiles per day, as will the future SAGE II. Other planned projects include SBUV-2, to be flown on Tiros-N, and new instruments on the future Upper Atmosphere Research Satellite (UARS).

Just as with total ozone, however, the analysis of these satellite data for small trends requires careful attention to a number of criteria (*51*): (1) greatest interval of temporal coverage, (2) similarity of observational technique, (3) similarity of inversion algorithms, (4) continuing assessment/instrument performance from ground truth measurements, (5) observational continuity, (6) independence from recalibration in orbit by indirect methods, and (7) availability of the data sets. Only the BUV/SBUV combination provides sufficient duration and similarity of technique to merit serious consideration in this regard. In addition to the uncertainties and probable instrument degradation discussed above, the assumption of constant solar spectral irradiance through the 250- to 340-nm wavelength region is critical for the BUV instrument in that any variation could cause a wavelength-dependent change in instrument response and, hence, an altitude-dependent trend in derived ozone profiles. In fact, this consideration may be related to the observed drift noted earlier. The SBUV instrument has been designed to overcome some of these difficulties. The diffuser plate is better protected against degradation, and the instrument operation includes a mode for directly measuring the solar spectral irradiance between 160 and 400 nm.

Analyses of ozone profile data similar to those for total ozone have been performed by many of the same groups that have studied total ozone. Angell and Korshover (see *51*) concluded that there is no evidence of a trend between 32 and 46 km (Umkehr layers 7, 8, 9) during the years 1970 to 1979, using data from most of the operating Umkehr stations (20). Statistical analyses of data from 12 stations covering the years 1956–1978 show changes in levels 7, 8, and 9 of $-0.1 \pm 1.8\%$ (G. Reinsel, private communication, 1983) and $+8.3 \pm 5.0\%$ (*12*). D. S. St. John (private communication, 1983) analyzed 1956–1979 data for 11 stations and found a change of $+1.1 \pm 2.5\%$. The Reinsel and St. John analyses differ from the Bloomfield *et al.* analysis (*12*) primarily by including in the regression a term allowing for step-function changes in procedure at four of the stations. Without that term, results from all three groups are similar.

Results obtained to date provide encouragement for continuing to refine these analyses as a detection scheme for any changes that may be taking place. More recent work centers around the use of time series of auxiliary variables in the regression. Such factors as temperature, solar cycle variations, meteorological parameters, and atmospheric transmittance have been included in searches for correlations with apparently random fluctuations in ozone concentrations. The atmospheric transmittance data are especially interesting in view of the expected impact of atmospheric aerosols on measured ozone concentrations. Time-dependent variations in the stratospheric aerosol content in particular may introduce biases that should be removed before trends can adequately be determined. Unfortunately, few aerosol data exist for most Umkehr sites, and aerosol content varies considerably over the globe.

Analyses of satellite data are also in progress, though no long-term studies are yet published. As with total ozone, the global coverage of the satellites promises to provide key information on the average variations in ozone as the data set becomes long enough to begin extracting trends.

Because for all the data the confidence limits remain broad, it is difficult to compare results with model calculations. However, virtually all model calculations indicate decreases in ozone at ~ 40 km over the past decade which have not so far appeared in the trend analyses of profile data.

VII. PROSPECTS FOF FUTURE RESEARCH

Progress in understanding stratospheric ozone has been exceptionally rapid during the past decade. As is often the case, however, much of the newly derived knowledge has opened up further areas of research that have yet to be fully addressed.

The chemistry of the stratosphere is without a doubt much better understood as a result of recent research in kinetics and photochemistry. Nevertheless, the model-calculated ozone concentration is highly sensitive to the details of the chemical reaction rate data, including temperature and pressure dependencies. Due primarily to experimental difficulties in their measurement, the reaction product distributions have not been determined for many reactions. This necessitates making assumptions about likely products—some of which have already been shown to be incorrect. Equally important is the use of modern direct experimental techniques to remeasure reaction rates which have been assumed to be fairly well determined by earlier, more indirect methods. Dramatic surprises such as those observed in the last few years (e.g., $OH + HNO_3$) often make significant differences in the calculation of past and future ozone changes due to human influences.

The physics of atmospheric motions, particularly the development of a more acceptable a priori method for describing transport in one- and two-dimensional models, remains an important area of research. Lacking such a justification, transport parameters are "derived" from fits with measurements of long-lived tracers. This procedure introduces measurement- and chemistry-based uncertainties into the model characterization and may not adequately describe transport over the shorter time scales of importance to the reactive free radicals.

The importance of calculating, rather than assuming and fixing, the temperature distribution in the atmosphere has become evident with the realization that temperature changes may be coupled with changes in ozone and other trace gases. The importance of these temperature effects lies both in the feedback with local chemistry and in the impact of these changes on the earth's surface temperature. Much work remains in suitably simulating the radiative processes in the atmosphere and their interplay with the global tropospheric and stratospheric winds.

During the next few years, the key to a better overall understanding of the atmosphere will probably lie in the area of field measurements. It is clear that currently available measurements do not sufficiently constrain the models to provide confidence in their assessments of historical or future changes in ozone. During the past few years, many changes in the chemical reaction data base used in the models have strongly influenced calculations of ozone perturbations without severely affecting the degree of agreement between models and measurements. Often, chemical data base changes improve the model predictions for some species at the expense of others.

Much of the future effort in field measurements can be expected to lie in developing new strategies of planned experiments. For example, current intercomparison studies of several techniques used to measure the same chemical species should begin to clarify the extent to which differences in reported measurements are due to instrumentation or to natural variability. Another effort will be to make truly informative combinations of measurements of photochemically related species. Using models for guidance, it should be possible to identify relationships (e.g., species concentration ratios) that can be tested directly and are not subject to large uncertainties in unmeasured parameters. In this way, perhaps, the model's data base can be broken down into testable segments. Success of the detailed tests will yield increasing confidence as the systematic verification of the models continues. Alternatively, discrepancies will lead to more directly identified areas of model weakness.

It is therefore likely that model development will be even more strongly coupled to measurement programs. An area of intensified research will undoubtedly be the development of a fully coupled three-dimensional general circulation model with a complete treatment of atmospheric chemistry. Undoubtedly, this is a goal in order to provide much of the information on chemical/dynamic couplings that may be important in the atmosphere. It is also important to understand not only average changes that might be anticipated but also how they would be distributed over the globe. Finally, only a general circulation model can address the question of whether the existing lower dimensionality models have inherent limits or biases associated with their method of "averaging down" from three to fewer dimensions.

Model calculations and advances in one- and two-dimensional model structure will, of course, continue in parallel with these developments. The recent attention to multiple perturbations has led to the conclusion that only a simultaneous treatment of all suspected past or future changes of the major trace gases can hope to give an accurate picture of atmospheric ozone or temperature changes. This is especially important in evaluating proposed alterations of human activities. The impact of any single activity that influences stratospheric trace gas concentrations can most accurately be evaluated against a background in which all other suspected perturbations continue. For example, assessing the impact of CFCs alone on an otherwise idealized stable atmosphere can be quite misleading. This realization places an important new demand on research efforts: all likely changes in man-made emissions that could impact atmospheric ozone and temperature must be identified early and projected into the future as accurately as possible. Clearly, the importance of accurate and diverse tropospheric monitoring programs for all important atmospheric trace gases cannot be overemphasized.

Ozone itself remains the final check of stratospheric research. Monitoring of total ozone, the ozone profile, and the temperature profile will be an integral part of ongoing research efforts. Analysis of these measurements can place strong limits on the magnitude of changes predicted by model calculations. At the same time, such analyses provide an independent determination of whether ozone is in fact changing, thus providing both a safeguard during a period of continued research and an impetus to further stimulate that research.

The question of potential ozone modification as a result of human activities—whether through fossil fuel combustion, agricultural practices, aircraft emissions, or chemical manufacturing—remains an open one. All of these activities, and perhaps several of which we remain unaware, can potentially modify stratospheric ozone and temperature. It is not yet known whether such source gas changes have resulted in small total ozone or surface temperature perturbations. The raising of this question, however, has stimulated remarkable developments in this still rapidly expanding scientific field. Mankind has been made aware of possible inadvertent modification of the ozone layer. However, measurements to date have established no long-term trends in ozone, and most recent assessments indicate that a decrease in the total amount of stratospheric ozone has not occurred and is not projected to occur if current trends in human activities are continued into the future.

References

1. Angell, J. K., and Korshover, J. (1973). Quasi-biennial and long-term fluctuations in total ozone. *Mon. Weather Rev.* **101**, 426.
2. Angell, J. K., and Korshover, J. (1976). Global analysis of recent total ozone fluctuations. *Mon. Weather Rev.* **104**, 63.

3. Angell, J. L., and Korshover, J. (1978). Global ozone variations: An update into 1976. *Mon. Weather Rev.* **106,** 725.
4. Anderson, J. G., Margitan, J. J., and Stedman, D. H. (1977). Atomic chlorine and the chlorine monoxide radical in the stratosphere: Three *in situ* observations. *Science* **198,** 501–503.
5. Anderson, J. G., Grassl, H. J., Shetter, R. E., and Margitan, J. J. (1980). Stratospheric free chlorine measured by balloon-borne *in situ* resonance fluorescence. *J. Geophys. Res.* **85,** 2869–2887.
6. Anderson, J. G., Grassl, H. J., Shetter, R. E., and Margitan, J. J. (1981). HO_2 in the stratosphere: Three *in situ* observations. *Geophys. Res. Lett.* **8,** 289–292.
7. Andrews, D. G., and McIntyre, M. E. (1978). An exact theory of nonlinear waves in a Lagrangian-mean flow. *J. Fluid Mech.* **89,** 609.
8. Arnold, F., Viggiano, A. A., and Schlager, H. (1982). Implications for trace gases and aerosols of large negative ion clusters in the stratosphere. *Nature (London)* **297,** 371.
9. Birrer, W. H. (1974). Some critical remarks on trend analysis of total ozone data. *Pure Appl. Geophys.* **112,** 523.
10. Blake, D. R., Mayer, E. W., Tyler, S. C., Mikade, Y., Montague, D. C., and Rowland, F. S. (1982). Global increase in atmospheric methane concentrations between 1978 and 1980. *Geophys. Res. Lett.* **9,** 477.
11. Bloomfield, P., Thompson, M. L., Watson, G. S., and Zeger, S. (1981). "Frequency Domain Estimation of Trends in Stratospheric Ozone," Tech. Rep. 182. Dept. of Statistics, Princeton Univ., Princeton.
12. Bloomfield, P., Thompson, M. L., and Zeger, S. (1982). A statistical analysis of Umkehr measurements of 32–46 km ozone. *J. Appl. Meteorol.* **21,** 1828.
13. Borucki, W. J., and Eberstein, I. J. (1980). Comparison of the Nimbus-4 BUV Ozone Data with the Ames Two-Dimensional Model," NASA Tech. Memo. 81207. NASA, Ames, California.
14. Brasseur, G., De Rudder, A., and Simon, P. C. (1983). Implication for stratospheric composition of a reduced absorption cross section in the Herzberg continuum of molecular oxygen. *Geophys. Res. Lett.* **10,** 20.
15. Burnett, C. R., and Burnett, E. B. (1979). Spectroscopic measurements of the vertical abundance of hydroxyl (OH) in the earth's atmosphere (abstr.). *EOS, Trans. Am. Geophys. Union* **60,** 336.
16. Callis, L. B., and Natarajan, M. (1981). Atmospheric carbon dioxide and chlorofluoromethanes: Combined effects on stratospheric ozone, temperature, and surface temperature. *Geophys. Res. Lett.* **8,** 587.
17. Callis, L. B., Natarajan, M., and Boughner, R. E. (1983). On the relationship between the greenhouse effect, atmospheric photochemistry, and species distribution. *J. Geophys. Res.* **88,** 1401.
19. Chang, J. S., Hindmarsh, A. C., and Madsen, N. K. (1974). Simulation of chemical kinetics transport in the stratosphere. *In* "Stiff Differential Systems" (R. A. Willoughby, ed.), p. 51. Plenum, New York.
20. Chapman, S. (1930). On ozone and atomic oxygen in the upper atmosphere. *Phil. Mag., Ser. 7* **10,** 369.
20a. Chemical Manufacturers Association (CMA) (1982). "World Production and Release of Chlorofluorocarbons 11 and 12 through 1981." CMA, Washington, D.C.
21. CIAP (1975). "The Effects of Stratospheric Pollution by Aircraft" (A. J. Grobecker, S. C. Coroniti, and R. H. Cannon, Jr., eds.), DOT–TST–75–50. Dept. of Transport., Washington, D.C.
22. Coffey, M. T., Mankin, G., and Goldman, A. (1981). Simultaneous spectroscopic determina-

tion of the latitudinal, seasonal, and diurnal variation of stratospheric N_2O, NO, NO_2, and HNO_3. *J. Geophys. Res.* **86,** 7331–7341.

22a. Craig, H., and Chou, C. C. (1982). Methane: The record in polar ice cores. *J. Geophys. Res.* **9,** 1221–1224.

23. Crutzen, P. J. (1969). Determination of parameters appearing in the "dry" and "wet" photochemical theories for ozone in the stratosphere. *Tellus* **21,** 368.

24. Crutzen, P. J. (1974). Photochemical reactions initiated by and influencing ozone in unpolluted tropospheric air. *Tellus* **26,** 47–57.

25. Cunnold, D., Alyea, F. N., Phillip, N., and Prinn, R. G. (1975). A three-dimension dynamical-chemical model of atmospheric ozone. *J. Atmos. Sci.* **32,** 170.

26. Cunnold, D., Prinn, R., Rasmussen, R. A., Simmonds, P. G., Alyea, F. N., Cordelino, C. A., Crawford, A. J., Fraser, P., and Rosen, R. (1983). The Atmospheric Lifetime Experiment 3, Lifetime methodology and application to three years of $CFCl_3$ data, and 4, Results for CF_2Cl_2 based on three years of data. *J. Geophys. Res.* **88,** 8379–8414.

27. DeMore, W. B., and Yung, Y. L. (1982). Catalytic processes in the atmospheres of Earth and Venus. *Science* **217,** 1209.

28. DeMore, W. B., Watson, R. T., Golden, D. M., Hampson, R. F., Kurylo, M., Howard, C. J., Molina, M. J., and Ravishankara, A. R., eds. (1982). "Chemical Kinetics and Photochemical Data for Use in Stratospheric Modeling," Eval. No. 5, Jet Propulsion Lab. Publication 82–57. Jet Propulsion Lab., Pasadena, California.

29. Derwent, R. G., and Curtis, A. R., (1981). "Stratospheric Ozone Depletion Estimates for Global Halocarbon Usage Estimated by the Linear Superposition of Concentrations from Individual Halocarbons," AERE Rep. R10168. HM Stationery Office, London.

30. Derwent, R. G., and Eggleton, E. J. (1981). Two-dimensional model studies of methyl chloroform in the troposphere. *Q. J. R. Meterol. Soc.* **107,** 231.

31. Derwent, R. G., and Eggleton, E. J. (1981). On the validation of one-dimensional CFC-ozone depletion models. *Nature (London)* **293,** 387.

32. Dobson, G. M. B. (1931). A photoelectric spectrophotometer for measuring the amount of atmospheric ozone. *Proc. Phys. Soc.* **43,** 324.

33. Donner, L., and Ramanathan, V. (1980). Methane and nitrous oxide: Their effects on the terrestrial climate. *J. Atmos. Sci.* **37,** 119.

34. Dunkerton, T. (1978). On the mean meridional mass motions of the stratosphere and mesosphere. *J. Atmos. Sci.* **35,** 2325.

35. Dutsch, H. U. (1968). The photochemistry of stratospheric ozone. *Q. J. R. Meterol. Soc.* **94,** 483.

36. Dutsch, H. (1971). Photochemistry of atmospheric ozone. *Adv. Geophys.* **15,** 219–322.

36a. Ellsaesser, H. W., Harries, J. E., Kley, D., Penndorf, R. (1980). Stratospheric H_2O. *Planet. Space Sci.* **28,** 827–835.

37. Filkin, D. L., and Schaffers, W. J. (1972). "Numerical Solution of Simultaneous Differential Equations by the Method of Lines." SIAM-SIGNUM, Austin, Texas.

38. Fishman, J., Ramanathan, V., Crutzen, P. J., and Liu, S. C. (1979). Tropospheric ozone and climate. *Nature (London)* **282,** 818.

39. Fraser, P. J., Khalil, M. A. K., Rasmussen, R. A., and Crawford, A. J. (1981). Trends in atmospheric methane in the southern hemisphere. *Geophys. Res. Lett.* **8,** 1063.

40. Frederick, J. E., and Mentall, J. E. (1982). Solar irradiance in the stratosphere: Implications for the Herzberg continuum absorption of O_2. *Geophys. Res. Lett.* **9,** 461.

41. Froidevaux, L., and Yung, Y. L. (1982). Radiation and chemistry in the stratosphere: Sensitivity to O_2 absorption cross sections in the Herzberg continuum. *Geophys. Res. Lett.* **9,** 854.

42. Garcia, R. R., and Solomon, S. (1983). A numerical model of the zonally averaged dynamical and chemical structure of the middle atmosphere. *J. Geophys. Res.* **88,** 1379.

43. Gear, C. W. (1971). "Numerical Initial Value Problems in Ordinary Differential Equations." Prentice-Hall, New York.

43a. Goody, R. M. (1964). "Atmospheric Radiation, Vol. 1, Theoretical Basis." Oxford Univ. Press (Clarendon), London and New York.

44. Gotz, F. W. P., and Dobson, G. M. B. (1928). Observations of the height of the ozone in the upper atmosphere. *Proc. R. Soc. London. Ser. A* **120,** 251.

45. Hansen, J., Johnson, D., Lacis, A., Lebedeff, S., Lee, P., Rind, D., and Russell, G. (1981). Climatic impact of increasing atmospheric carbon dioxide. *Science* **213,** 957.

46. Hartley, W. N. (1881). On the absorption of solar rays by atmospheric ozone. *J. Chem. Soc.* **39,** 111.

47. Hessvedt, E. (1969). A photochemical model for the ozone layer. *Ann. Geophys.* **25,** 99.

48. Hilsenrath, E., and Schlesinger, B. M. (1979). "The Seasonal and Interannual Variability of Total Ozone as Revealed by the BUV Nimbus-4 Experiment," NASA HC-A17/MF-A01, 4th NASA Weather and Climate Program Sci. Rev., 277 (N79–2063311–417). NASA, Washington, D.C.

49. Holton, J. R. (1981). An advective model for two-dimensional transport of stratospheric trace species. *J. Geophys. Res.* **86,** 11989.

49a. Hudson, R. D., ed. (1977). "Chlorofluoromethanes and the Stratosphere." NASA Ref. Publ. 1010. NASA, Washington, D.C.

50. Hudson, R. D., and Reed, E. I., eds. (1979). "The Stratosphere: Present and Future, " NASA Ref. Publ. 1049. NASA, Washington, D.C.

51. Hudson, R., Alpert, J., Bojkov, R., Chang, J., DeMore, W., Ehhalt, D., Frederick, J., Geller, M., Holton, J., London, J., McGee, T., Reed, E., Stolarski, R., Sundararaman, N., and Watson, R., eds. (1982). "The Stratosphere 1981: Theory and Measurements, " World Meteorol. Org. Global Ozone Research and Monitoring Proj. Rep. No. 11. World Meteorol. Org., Geneva.

52. Hunt, B. G., (1966). Photochemistry of ozone in a moist atmosphere. *J. Geophys. Res.* **71,** 1385.

53. Johnston, H. S. (1971). Reduction of stratospheric ozone by nitrogen oxide catalysts from supersonic transport exhaust. *Science* **173,** 517.

54. Johnston, H., Whitten, G., and Birks, J. (1973). Effects of nuclear explosions on stratospheric nitric oxide and ozone. *J. Geophys. Res.* **78,** 6107.

55. JPL 82–57: see ref. *28*.

55a. JPL 81–3 (1981). "Chemical Kinetics and Photochemical Data for Use in Stratospheric Modelling, " Eval. No. 4: NASA Panel for Data Evaluation, JPL Publ. 81–3. Jet Propulsion Laboratory, Pasadena, California.

56. Keeling, C. D., Bacastow, R. B., Bainbridge, A. E., Ekdahl, E. A., Jr., Guenther, P. R., Waterman, L. S., and Chin, J. F. S. (1976). Atmospheric carbon dioxide variations at Mauna Loa Observatory, Hawaii. *Tellus* **28,** 538.

57. Komhyr, W. D., Barratt, E. W., Slocum, G., and Weickmann, H. K. (1971). Atmospheric total ozone increase during the 1960s. *Nature (London)* **232,** 390.

58. Lacis, A. A., Hansen, J., Lee, P., Mitchell, T., and Lebedeff, S. (1981). Greenhouse effects of trace gases, 1970–1980. *Geophys. Res. Lett.* **8,** 1035.

59. Lindzen, R. S., Hou, A. Y., and Farrell, B. F. (1982). The role of convective model choice in calculating the climate impact of doubling CO_2. *J. Atmos. Sci.* **39,** 1189.

60. Logan, J. A., Prather, M. J., Wofsy, S. C., and McElroy, M. B. (1981). Tropospheric chemistry: A global perspective. *J. Geophys. Res.* **86,** 7210.

61. London, J., and Oltmans, S. J. (1978/1979). The global distribution of long-term total ozone variations during the period 1957–1975. *Pure Appl. Geophys.* **117,** 345.

62. Louis, J. F. (1974). "A Two-Dimensional Transport Model of the Atmosphere, " Ph.D. Dissertation, Univ. of Colorado, Boulder.
63. Luther, F. M., and Gelinas, R. J. (1976). Effect of molecular multiple scattering and surface albedo on atmospheric photodissociation rates. *J. Geophys. Res.* **81,** 1125.
64. Luther, F. M., Chang, J. S., Duewer, W. H., Penner, J. E., Tarp, R. L., and Wuebbles, D. J. (1979). "Potential Environmental Effects of Aircraft Emissions, " Lawrence Livermore Laboratory, UCRL-52861. Lawrence Livermore Laboratory, Livermore, California.
64a. Manabe, S., and Wetherald, R. T. (1976). Thermal equilibrium of the atmosphere with a given distribution of relative humidity." *J. Atmos. Sci.* **24,** 241–259.
65. Miller, A. J., Nagatani, R. M., Rogers, T. E., Fleig, A. J., Heath, D. F. and Kaveeshwar, V. G. (1982). Total ozone variations 1970–1974 using Backscattered Ultraviolet (BUV) and ground-based observations. *J. Appl. Meteorol.* **21,** 621–630.
66. Miller, C., Meakin, P., Franks, R. G. E., and Jesson, J. P. (1978). The fluorocarbon-ozone theory. V. One-dimensional modeling of the atmosphere. The base case. *Atmos. Environ.* **12,** 2481.
67. Miller, C., Steed, J. M., Filkin, D. L., and Jesson, J. P. (1981). The fluorocarbon ozone theory. 7. One-dimensional modeling—An assessment of anthropogenic perturbations. *Atmos. Environ.* **15,** 729.
68. Miller, C., Filkin, D. L., Owens, A. J., Steed, J. M., and Jesson, J. P. (1981). A two-dimensional model of stratospheric chemistry and transport. *J. Geophys. Res.* **86,** 12039.
69. Molina, M. J., and Rowland, F. S. (1974). Stratospheric sink for chlorofluoromethanes: Chlorine atom catalyzed destruction of ozone. *Nature (London)* **249,** 810.
70. Murcray, D. G., Barker, D. B., Brooks, J. N., Goldman, A., and Williams, W. J. (1975). Seasonal and latitudinal variation of the stratospheric concentrations of HNO_3. *Geophys. Res. Lett.* **2,** 234–224.
71. Murcray, D. G., Goldman, A., Murcray, F. H., Murcray, F. T., and Williams, W. J. (1979). Stratospheric distribution of $ClONO_2$. *Geophys. Res. Lett.* **6,** 857–859.
72. NASA RP–1049, 1979: see ref. *50.*
72a. NASA RP–1010 (1977). See ref. *49a.*
73. National Academy of Sciences (1976). "Halocarbons: Effects on Stratospheric Ozone." National Research Council, Washington, D.C.
74. National Academy of Sciences (1979). "Stratospheric Ozone Depletion by Halocarbons: Chemistry and Transport." National Research Council, Washington, D.C.
75. National Academy of Sciences (1982). "Causes and Effects of Stratospheric Ozone Reduction: An Update." Environ. Studies Board, National Research Council, Washington, D.C.
76. National Academy of Sciences (1982). "Carbon Dioxide and Climate: A Second Assessment." CO_2/Climate Review Panel, National Research Council, Washington, D.C.
76a. National Academy of Sciences (1975). "Environmental Impact of Stratospheric Flight." Natl. Acad. Sci., Washington, D.C.
77. Naudet, J. P., Huguenin, D., Rigamel, P., and Cariolle, D. (1981). Stratospheric observations of NO_3 and its experimental and theoretical distribution between 20 and 40 km. *Planet. Space Sci.* **29,** 707.
78. Noxon, J. F. (1979). Stratospheric NO_2. 2. Global behavior. *J. Geophys. Res.* **84,** 5067.
79. Oliver, R. C., Bauer, E., Hidalgo, H., Gardner, K. A., and Wasylkiwsky, W. (1977). "Aircraft Emissions: Potential Effects on Ozone and Climate," Dept. of Transport. Rep. FAA–EQ–77–3. U.S. Dept. Transport., Washington, D.C.
80. Owens, A. J., Steed, J. M., Miller, C., Filkin, D. L., and Jesson, J. P. (1982). The atmospheric lifetimes of CFC-11 and CFC-12. *Geophys. Res. Lett.* **9,** 700.
81. Owens, A. J., Steed, J. M., Filkin, D. L., Miller, C., and Jesson, J. P. (1982). The potential effects of increased methane on atmospheric ozone. *Geophys. Res. Lett.* **9,** 1105.
82. Owens, A. J., Hales, C. H., Filkin, D. L., Miller, C., Yokozeki, A., and Jesson, J. P. (1983).

A coupled one-dimensional radiative-convective, transport-chemistry model of the atmosphere. I. Model structure and steady-state perturbation calculations (submitted).

83. Parrish, A., De Zafra, R. L., Solomon, P. M., Barrett, J. W., and Carson, E. R. (1981). Chlorine oxide in the stratospheric ozone layer: Ground-based detection and measurement. *Science* **211,** 1158–1161.

83a. Penner, J. E. (1982). Trend prediction for O_3: An analysis of model uncertainty with comparison to detection thresholds. *Atmos. Environ.* **16,** 1109–1115.

84. Prinn, R. G., Rasmussen, R. A., Simmonds, P. G., Alyea, F. N., Cunnold, D. M., Lane, B. C., Cardelino, C. A., and Crawford, A. J. (1983). The Atmospheric Lifetime Experiment. 5, Results for CH_3CCl_3 based on three years of data. *J. Geophys. Res.* **88,** 8415–8426.

85. Pyle, J. A. (1980). A calculation of the possible depletion of ozone by chlorofluorocarbons using a two-dimensional model. *Pure Appl. Geophys.* **118,** 355.

86. Ramanathan, V., and Coakley, J. A., Jr. (1978). Climate modeling through radiative-convective models. *Rev. Geophys. Space Phys.* **16,** 465.

87. Ramanathan, V. (1975). Greenhouse effect due to chlorofluorocarbons: Climatic implications. *Science* **190,** 50.

88. Reed, R. J., and German, K. E. (1965). A contribution to the problem of stratospheric diffusion by large-scale mixing. *Mon. Weather Rev.* **93,** 313.

89. Reinsel, G., Tiao, G. C., Wang, M. N., Lewis, R., and Nychka, D. (1981). Statistical analysis of stratospheric ozone data for detection of trends. *Atmos. Environ.* **15,** 1569.

90. Reinsel, G., Tiao, G. C., and Lewis, R. (1981). A statistical analysis of total ozone data from the BUV satellite experiment. *J. Atmos. Sci.* **39,** 418.

91. Roscoe, H. K. (1982). Tentative observation of stratospheric N_2O_5. *Geophys. Res. Lett.* **9,** 901–902.

91a. Rotty, R. M., and Marland, G. (1980). "Constraints on Carbon Dioxide Production from Fossil Fuel Use," Institute for Energy Analysis, Rep. No. ORAU/IEA–80–9. Inst. for Energy Analysis, Oak Ridge, Tennessee.

92. Rowland, F. S., Rogers, P. J., and Montague, D. C. (1983). Ozone depletion by the ClO_x chain: Potential importance of ultraviolet photolysis of NaCl and the stratospheric cycles of gaseous sodium. *J. Geophys. Res.* (in press).

93. St. John, D. S., Bailey, S. P., Fellner, W. H., Minor, J. M., and Snee, R. D. (1981). Time series search for trend in total ozone measurements. *J. Geophys. Res.* **86,** 7299.

94. Schonbein, C. F. (1840). Recherches sur la nature de l'odeur qui se manifeste dans certaines actions chimiques. *C. R. Hebd. Seances Acad. Sci.* **10,** 706.

95. Simmonds, P. G., Alyea, F. N., Cardelino, C. A., Crawford, A. J., Cunnold, D. M., Lane, B. C., Lovelock, J. E., Prinn, R. G., and Rasmussen, R. A. (1983). The Atmospheric Lifetime Experiment. 6. *J. Geophys. Res.* **88,** 8427–8441.

96. Solomon, S., and Garcia, R. R. (1983). On the distribution of nitrogen dioxide in the high latitude stratosphere. *J. Geophys. Res.* **88,** 5229–5239.

97. Steed, J. M., Owens, A. J., Miller, C., Filkin, D. L., and Jesson, J. P. (1982). Two-dimensional modelling of potential ozone perturbation by chlorofluorocarbons. *Nature (London)* **295,** 308.

98. Stolarski, R. S., and Cicerone, R. J. (1974). Stratospheric chlorine: A possible sink for ozone. *Can. J. Chem.* **52,** 1610.

99. Sze, N. D. (1977). Anthropogenic CO emissions: Implications for the atmospheric CO-OH-CH_4 cycle. *Science* **195,** 673.

100. Sze, N. D., Ko, M. K. W., Livshits, M., Wang, W. C., Ryan, P. B., Specht, R. E., McElroy, M. B., and Wofsy, S. C. (1983). "Annual Report on the Atomospheric Chemistry, Radiation and Dynamics Program." Atmos. and Environ. Res., Inc., Cambridge, Massachusetts.

101. Turco, R. P., and Whitten, R. C. (1978). A note on the diurnal averaging of aeronomical models. *J. Atmos. Terr. Phys.* **40,** 12.
102. Turco, R. P., Toon, O. B., Pollack, J. B., Whitten, R. C., Poppoff, I. G., and Hamill, P. (1980). Stratospheric aerosol modification by supersonic transport and space shuttle operations—climate implications. *J. Appl. Meteorol.* **19,** 78.
103. U. K. Dept. of Environ. (1979). "Chlorofluorocarbons and Their Effect on Stratospheric Ozone, " Pollution Paper No. 15. HM Stationery Office, London.
104. Varga, R. S. (1962). "Matrix Iterative Analysis." Prentice-Hall, New York.
105. Wang, W. C., Lacis, A. A., Mo, T., and Hansen, J. E. (1976). Greenhouse effects due to man-made perturbations of trace gases. *Science* **194,** 685.
106. Wang, W. C., Ko, M., and Sze, N. D. (1980). "Atmospheric CFCs, N_2O, CH_4, and O_3: Coupled Effects on the Earth's Climate." Quadrennial Int. Ozone Symp., IAMAP, Boulder, Colorado, August.
107. Wang, W. C., Ko, M. K. W., Sze, N. D., and Luther, F. M., (1981). "Influence of infrared band absorption treatments on the calculated O_3 increase due to CO_2." Amer. Geophys. Union Meeting, San Francisco, December.
108. Waters, J. W., Hardy, J. C., Jarnot, R. F., and Pickett, H. M., (1981). Chlorine monoxide radical, ozone and hydrogen peroxide: Stratospheric measurements by microwave limb sounding. *Science* **214,** 61–64.
109. Weiss, R. F. (1981). The temporal and spatial distribution of tropospheric nitrous oxide. *J. Geophys. Res.* **86,** 7185.
110. Weiss, R. F., and Craig, H. (1976). Production of atmospheric nitrous oxide by combustion. *Geophys. Res. Lett.* **3,** 751.
111. Weiss, R. F., Keeling, C. D., and Craig, H. (1981). The determination of tropospheric nitrous oxide. *J. Geophys. Res.* **86,** 7197.
112. Widhopf, G. F., and Glatt, L. (1979). "Two-Dimensional Description of the Natural Atmosphere Including Active Water Vapor Modeling and Potential Perturbations Due to NO_x and HO_x Aircraft Emissions, " FAA–EE–79–07. Federal Aviation Admin., Washington, D.C.
113. Wilcox, R. W., Nastrom, G. D., and Belmont, A. D. (1977). Periodic variations of total ozone and its vertical distribution. *J. Appl. Meteorol.* **16,** 290–298.
114. Wofsy, S. C., and McElroy, M. B. (1974). HO_x, NO_x, and ClO_x: Their role in atmospheric photochemistry. *Can. J. Chem.* **52,** 1582.
115. Wofsy, S. C., McElroy, M. B., and Yung, Y. L. (1975). The chemistry of atmospheric bromine. *Geophys. Res. Lett.* **2,** 215.
116. World Meteorol. Org. (WMO) (1979). "Report of the First Session of the CAS Working Group on Atmospheric Carbon Dioxide, " WMO Project on Res. and Monitoring of Atmos. CO_2, Rep. No. 2. WMO, Geneva, Switzerland.
117. World Meteorol. Org. (WMO) (1981). "Assessment of Performance Characteristics of Various Ozone Observing Systems, " Rep. of meeting of experts, Boulder, Colorado, July 1980, WMO Global Ozone Recording and Monitoring Proj. Rep. No. 9. WMO, Geneva, Switzerland.
118. WMO/NASA 1982: see ref. *51.*
119. Wuebbles, D. J. (1981). "The Relative Efficiency of a Number of Halocarbons for Destroying Stratospheric Ozone, " Lawrence Livermore National Laboratory, UCJD 18924. Lawrence Livermore Natl. Lab., Livermore, California.
120. Wuebbles, D. J., Luther, F. M., and Penner, J. E. (1983). Effects of coupled anthropogenic perturbations on stratospheric ozone. *J. Geophys. Res.* **88,** 1444.
121. Yung, Y. L. (1976). A numerical method for calculating the mean intensity in an inhomogeneous Rayleigh scattering atmosphere. *J. Quant. Spectrosc. Radiat. Trans.* **16,** 775.
122. Yung, Y. L., Pinot, P. J., Watson, R. T., and Sander, S. P. (1979). Atmospheric bromine and ozone perturbations in the lower stratosphere. *J. Atmos. Sci.* **37,** 379.

Overview of Health Effects of Formaldehyde[1]

Andrew G. Ulsamer

United States Consumer Product Safety Commission
Bethesda, Maryland

James R. Beall

United States Department of Energy
Germantown, Maryland

Han K. Kang

United States Veterans Administration
Washington, D.C.

James A. Frazier

National Research Council
Washington, D.C.

I. Introduction 338
II. Regulatory Activities 340
III. Chemical Properties 341
  A. Chemical Forms 341
  B. Chemical Reactivity 342
IV. Sources and Exposure 343
V. Metabolism 348
VI. General Toxicology 353
VII. Hypersensitization 359
VIII. Teratogenic and Reproductive Effects 361
  A. Inhalation Studies 361
  B. Dermal Studies 364
  C. Ingestion Studies 364

[1]The opinions expressed in this article are those of the authors and do not necessarily reflect official positions or policies of the Consumer Product Safety Commission, Department of Energy, Veterans Administration, or the National Research Council.

HAZARD ASSESSMENT OF CHEMICALS:
Current Developments, Vol. 3

ISBN 0-12-312403-4

D. Injection Studies 366
E. *In Vitro* Studies 367
IX. Genetic Effects 367
X. Carcinogenicity 369
XI. Epidemiology 375
A. Controlled Human Exposure Studies 375
B. Case Reports 376
C. Cross-Sectional Studies 378
D. Cohort and Case-Control Studies 383
XII. Summary and Conclusions 389
References 392

I. INTRODUCTION

Formaldehyde (H_2CO or HCHO),[2] first prepared by Butlerov in 1859, is today one of the most widely used chemicals in commercial production (*170*). In 1981, about 5.9 billion pounds of a 37% aqueous solution of formaldehyde (formalin) was produced in the United States. Approximately one-half of the annual production is used in the manufacture of urea–formaldehyde (UF) and phenol–formaldehyde (PF) resins for bonding of pressed wood products, especially plywood and particle board. Lesser amounts of UF resins are used in permanent press fabrics and, until recently, for UF foam insulation (UFFI). For a more detailed discussion of the chemistry, uses, and history of development of UF resins, see Meyer (*160*).

Formaldehyde is released by some wood and other products containing UF and related resins (*170*). Such formaldehyde initially may be present in small amounts in the manufactured product or may subsequently result from the hydrolysis of the UF resin bonding or coating the wood or other products. Wood

[2]Abbreviations: BCME, bis(chloromethyl) ether; CNS, central nervous system; $CV_{\%}$, closing volume expressed as percentage of the vital capacity; DEN, diethylnitrosamine; DNA, deoxyribonucleic acid; $FEF_{25-75\%}$, same as MMEF; $FEV_{1.0}$, forced expired volume in 1 sec; $FEV_{\%}$, ($FEV_{1.0}$ $\div$ FVC) $\times$ 100; FH_4, tetrahydrofolic acid; FVC, forced vital capacity; GSH, glutathione; HCHO, formaldehyde; HMPA, hexamethyl phosphoramide; HMT, hexamethylenetetramine; HR, hexamethylenetetramine–resorcinol; LC_{50}, median lethal concentration; LD_{50}, median lethal dose; $MEF_{50\%}$, maximum expiratory flow rate at 50% FVC; MMEF, maximum mid-expiratory flow calculated as the mean forced expiratory flow during the middle half of FVC; MMF, same as MMEF; NAD, nicotinamide adenine dinucleotide; PF, phenol–formaldehyde; PMR, proportionate mortality ratio; RE, reticuloendothelial; RNA, ribonucleic acid; SMR, standardized mortality ratio; TBA, tumor-bearing animals; TPA, tetradecanoylphorbol acetate; TLV, threshold limit value; TWA, time-weighted average; UF, urea–formaldehyde; UFFI, urea–formaldehyde foam insulation.

products containing UF resins emit more formaldehyde than wood products containing PF resins. Formaldehyde may be released also from UFFI, a product used for retrofitting insulation primarily in older structures (*45*). UFFI was made on site by mixing the liquid reactants and spraying them into small holes in the walls. Success with UFFI varied due to a number of factors that could not be predictably controlled. For example, mixing conditions of the liquid reactants and the temperature both were critical factors affecting the quality and at times resulted in poor formation of foam and release of unreacted formaldehyde. Even when the mixing conditions were ideal, some formaldehyde was released. The formaldehyde thus released permeated the walls into the living spaces wherein the occupants were exposed to concentrations that varied widely. Increasing numbers of consumers of formaldehyde-releasing products complained to both state and federal agencies about formaldehyde causing irritation of the eyes, nose, throat, and skin, and about persistent cough, dizziness, nausea, and headaches.

Many of the data on the noncarcinogenic effects of formaldehyde have been reviewed in the National Research Council's two reports (*169, 170*): one prepared by the Committee on Toxicology (1980) for the Consumer Product Safety Commission (CPSC) assessing the adverse health effects of formaldehyde (*169*), and the other by the Committee on Formaldehyde and Other Aldehydes in 1981 assessing the health and certain environmental effects for the Environmental Protection Agency (EPA) (*170*). Both these studies extensively reviewed the published literature on formaldehyde and discussed the irritation, the sensitization, and the other data on adverse health effects, but found a lack of adequate data to assess the risks of carcinogenicity. The reports pointed out that a substantial portion of the people exposed might react adversely even to low concentrations, and recommended that exposure be kept at the lowest practical concentration in indoor residential air.

In October 1979, concerns about the adverse health effects from exposure to formaldehyde grew considerably when the Formaldehyde Institute announced preliminary findings from the Chemical Industry Institute of Toxicology (CIIT) study of carcinogenicity in rats and mice experimentally exposed to formaldehyde gas. These findings, which are reviewed in more detail in Section X, showed that formaldehyde produced nasal squamous cell carcinomas in animals exposed to 14.3 ppm.[3]

Data on the mutagenicity, teratogenicity, carcinogenicity, and other chronic effects of formaldehyde were reviewed by the Federal Panel on Formaldehyde (a panel of 16 senior scientists established by CPSC and other agencies with the cooperation of the National Toxicology Program) (*70*). The panel concluded in

[3]Conversion factor: 1 ppm = 0.82 mg/m^3.

its report that formaldehyde has been demonstrated to be mutagenic and carcinogenic under laboratory conditions and should be presumed to pose a cancer risk to humans, although data were not available for direct assessment in exposed humans. Similar conclusions were reached by the National Institute for Occupational Safety and Health (NIOSH) (*235*) and the International Agency for Research on Cancer (*117*).

This article presents a review of the major findings of the available published research. The published literature on formaldehyde is too voluminous to be covered comprehensively in this overview. It is our intent therefore to characterize to the extent possible the adverse health effects of and the hazards posed by exposure to formaldehyde. To accomplish this objective, we shall discuss some of the relevant chemical reactions, sources of human exposure, its metabolism and metabolic fate, its irritation and sensitization properties, the carcinogenic, mutagenic, teratogenic, and reproductive effects, and what is presently known about epidemiologic studies, some of which are in progress.

II. REGULATORY ACTIVITIES

The state of California has banned the sale of urea–formaldehyde foam insulation (UFFI) unless the free formaldehyde content of the foam is less than 0.01% by weight. The city of Cincinnati, Ohio also has banned the installation of UFFI as have the states of Connecticut and Massachusetts. Installation of the product is banned for schools, nurseries, and certain other institutions in the state of Colorado. The states of New Hampshire and New York require that installers of UFFI warn potential buyers of the potential health effects of formaldehyde. The states of Minnesota and Wisconsin have established standards of 0.5 and 0.4 ppm, respectively, for formaldehyde levels in mobile homes. The state of Texas requires that retailers or manufacturers of mobile homes warn consumers of the potential health effects of formaldehyde. On the national level, installation of UFFI was banned in residences and schools by the CPSC on August 9, 1982. The ban was overturned by the Fifth Circuit Court of Appeals in April 1983, and the Commission's request for reconsideration was denied in June 1983. The Commission's request to the Justice Department to appeal the case to the Supreme Court was turned down in September 1983 despite numerous scientific errors contained in the Fifth Circuit decision.

With regard to other residential standards and guidelines, the American Society of Heating, Refrigeration, and Air Conditioning Engineers (ASHRAE) has recommended that formaldehyde levels not exceed 0.1 ppm. Denmark, The Netherlands, and West Germany have residential standards for formaldehyde of 0.12, 0.10, and 0.10 ppm, respectively.

III. CHEMICAL PROPERTIES

A. Chemical Forms

1. Monomeric Formaldehyde

Formaldehyde is a colorless gas that is usually manufactured by reacting methanol vapor with air in the presence of a catalyst. It is designated by its molecular formula HCHO or the structural formula

$$\begin{matrix} H \\ H \end{matrix} \!\!>\! C = O$$

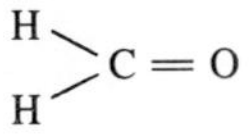

Commercially, it is not available in the monomeric form but is commonly sold as an aqueous solution of from 30 to 56% formaldehyde by weight with from 0.5 to 15% methanol added to prevent polymerization.

It has a characteristic pungent, suffocating odor, and it is highly irritating to exposed membranes of the eyes, nose, and respiratory tract. Some of its physical properties are density, 1.067 (air = 1.000); vapor pressure, 400 mm Hg at 33°C; flash point, 430°C; boiling point, −19°C; and melting point, −118°C. In addition, formaldehyde polymerizes slowly at temperatures below 80–100°C.

2. Trioxane

Formaldehyde is available commercially also as the cyclic trimer trioxane (trioxymethylene), designated by the molecular formula $C_3H_5O_3$ or the structural formula

$$\begin{matrix} & H_2 & \\ & C & \\ O & & O \\ | & & | \\ H_2C & & CH_2 \\ & O & \end{matrix}$$

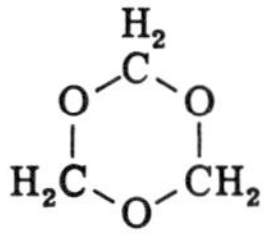

In pure form, trioxane is a colorless crystalline solid that has a nonirritating chloroformlike odor. It boils at 115°C and melts at 61–62°C.

3. Paraformaldehyde

This commercial form of formaldehyde is a colorless solid prepared by condensation of methylene glycol (methanediol) and is designated by the formula $HO—(HCHO)_8—H$. It has the same characteristic odor as monomeric formaldehyde and melts over a wide temperature range (120–170°C). Thus, heating paraformaldehyde on a hot plate releases formaldehyde. This is a procedure commonly used for disinfecting large areas.

B. Chemical Reactivity

Polymerization of double-bonded methylene compounds and of simple methyl derivatives is the principle mechanism by which formaldehyde reacts with other chemicals to form some of the resinous products mentioned elsewhere in this article.

It is beyond the scope of this article to discuss in depth all of the chemical reactions involving formaldehyde. The reactions given below were selected to show some of the more important ways in which formaldehyde reacts with other chemicals.

1. Industrial Preparation

Although cheaper reagents may be used by manufacturers, the general method for preparation of formaldehyde from alcohols (methanol) is oxidation with air or dehydrogenation over hot copper or silver catalyst. For example:

$$CH_3OH \xrightarrow[\Delta]{Cu} HCHO + H_2$$

2. Reaction with Urea

Formaldehyde's high degree of chemical reactivity has been attributed to the direct attachment of the carbonyl carbon to two hydrogen atoms. It is because of this strongly reducing reactivity that it has found widespread commercial use in reactions with urea and resinous substances to produce a wide variety of products. This abbreviated illustration shows the basic reaction used in forming the urea–formaldehyde polymer.

$$HCHO + H_2N{-}\underset{\underset{O}{\|}}{C}{-}NH_2 \longrightarrow HOCH_2{-}\overset{H}{\overset{|}{N}}{-}\underset{\underset{O}{\|}}{C}{-}NH_2 \xrightarrow{HCHO}$$

$$HOCH_2{-}\overset{H}{\overset{|}{N}}{-}\underset{\underset{O}{\|}}{C}{-}\overset{H}{\overset{|}{N}}{-}CH_2OH \xrightarrow{HCHO + \text{urea}}$$

```
            —CH₂—N—
                 |
                 CH₂
                 |
   —N— CH₂— N—C—N—
    |           ||
  O=C           O
    |
   —N— CH₂— N—C—N—
             |  ||
                O
```

3. Reaction with Hydrochloric Acid

This reaction is important owing to the possible formation of the known carcinogen bis(chloromethyl) ether where human exposure to both chemicals may occur.

$$2HCHO + 2HCl \rightarrow ClH_2C{-}O{-}CH_2Cl + H_2O$$

4. Reactions of Biological Interest

The following reactions of formaldehyde are important since they can involve nucleic acids, proteins, and other amine-containing molecules within the cell.

Hydration:

$$HCHO + H_2O \rightarrow CH_2(OH)_2$$

Amines:

$$HCHO + R_2NH \rightarrow R{-}\underset{\displaystyle R}{\underset{|}{N}}{-}CH_2OH$$

$$HCHO + RNH_2 \rightarrow R{-}NH{-}CH_2OH$$

$$HCHO + R{-}\underset{\displaystyle O}{\underset{\|}{C}}{-}NH_2 \rightarrow R{-}\underset{\displaystyle O}{\underset{\|}{C}}{-}NH{-}CH_2OH$$

$$R{-}NH{-}CH_2OH + RNH_2 \rightarrow R{-}NH{-}CH_2{-}NH{-}R + H_2O$$

Acetal:

$$HCHO + ROH \rightarrow R{-}O{-}CH_2OH$$

IV. SOURCES AND EXPOSURE

The sources of formaldehyde to which humans may be exposed have been reviewed extensively (*44, 45, 160, 169, 170, 229, 230*). These sources can be divided into two basic classes: (1) commercial manufacturing processes and products, and (2) natural processes. For 1978, it was estimated that approximately 68% of the total formaldehyde production of $1,580,000 \times 10^3$ kg was due to commercial processes; natural processes accounted for the remainder. All of the formaldehyde produced by natural processes was released into the atmosphere.

The primary sources of formaldehyde released by natural processes are the incomplete combustion of fossil fuels and refuse (65%) and the photochemical oxidation of hydrocarbons released by automotive exhaust. Sources of formaldehyde from combustion processes outdoors include incinerators, refineries, power plants, houses, and businesses, as well as automobile, bus, truck, and jet exhausts. The amount of formaldehyde released in automobile exhaust has been decreasing steadily with the increasing use of catalytic converters. Gas stoves, ovens, and unvented heaters are major indoor combustion processes that are sources of formaldehyde.

Almost all (94%) commercially produced formaldehyde goes into the manufacturing of various resins and plastics. The remaining 6% is used for embalming fluid and tissue fixation, and as a preservative in various products. Over half (55%) of all commercially produced formaldehyde is used to produce urea–formaldehyde and phenol–formaldehyde resins.

Most of the urea–formaldehyde (UF) resins produced (28% of commercial formaldehyde production) are used as bonding agents in the manufacture of products such as interior grades of plywood, particle board, and fiber board, and for paper and textile treating and coating resins, protective coatings, and laminates. Additionally, until banned by the CPSC in August 1982, a small amount of UF resin production was used to make urea–formaldehyde foam insulation for homes, schools, and commercial buildings (*70*). Press time, temperature, and moisture content influence the release of formaldehyde from wood products as do ambient humidity and temperature loading and background levels of formaldehyde. Release of formaldehyde from urea–formaldehyde foam insulation is affected by temperature and humidity, age of chemicals, mixing of components, and other factors.

Phenol–formaldehyde (PF) resins are produced in quantities approximately equal to urea–formaldehyde resins. They are used as adhesives for exterior grades of plywood and particle board, and as friction material, foundry and shell moldings insulation, molding compounds, protective coatings, and laminates. The release of formaldehyde from PF products using PF resins is much less than that from those containing UF resins. Meyer (*160*) has estimated that PF resins are 1000 times more stable than UF resins.

Formaldehyde is used also to produce melamine and acetal resins (11% of commercial formaldehyde production). These resins are used predominantly in plastics and molding compounds; release of formaldehyde from products using these resins is considered negligible.

Other uses of formaldehyde include production of the chemical intermediates pentaerythritol, 1,4-butanediol, trimethylpropane, and hexamethylenetetramine (HMT). Production of these chemicals consumes approximately 20% of commercial formaldehyde. It is uncertain how much formaldehyde is released by products synthesized from these chemicals.

Formaldehyde (or a derivative) is used at low concentration as a disinfectant and preservative in a variety of cosmetics, including shampoos, makeup, eyeshadow, and bubblebath. Formaldehyde is used also for the preservation and hardening of biological specimens and as a topical fungicide. A partial listing of products containing formaldehyde is shown in Table I.

TABLE I

Product Uses of Formaldehyde

Adhesives	Insulation and fiberglass
Concrete	Intermediate chemicals
Cosmetics	Laminates
Deodorants	Leathers
Detergents	Lubricants
Dry cleaning solutions	Mothballs
Dyes	Paints
Embalming fluids	Paper
Explosives	Polishes
Fertilizers	Photographic developing solutions
Fiberboard	Particleboard
Food	Pharmaceuticals
Food packaging materials	Plastics
Friction materials	Plywood
Fuels	Rubber
Fungicides	Textiles
Furniture	Water softening chemicals

Many formaldehyde-containing products have the ability to release formaldehyde during and after manufacture, thereby exposing workers and consumers to the chemical. These data, although not extensive for many exposure settings, indicate that exposure is widespread and can be significant for certain populations. Table II identifies the sources of human exposure to formaldehyde and provides an estimation of the size of each subpopulation, the mean exposure level and standard deviation to which each is exposed, and the duration of exposure in hours per week.

Although data are limited, workers in many different occupations can be exposed to formaldehyde. Formaldehyde production workers, resin production workers, veneer panel production workers, textile workers, embalmers, and pathologists appear to be exposed to higher concentrations of formaldehyde than other types of workers.

The most extensive data bases exist for two types of consumer exposures. These are houses that have been insulated with UFFI and mobile homes.

Houses with UFFI have been found to have significantly higher formaldehyde concentrations than non-UFFI homes (*70*). On the basis of data collected on UFFI installed in panels under near ideal conditions, it has been estimated that UFFI at 25°C can contribute 0.05–0.4 ppm to the formaldehyde burden of the home. When similar panels were tested at 23, 33, and 40°C (temperatures that may be encountered in wall cavities), the average concentration of formaldehyde emitted from the panels increased 6-fold at 33°C and 13-fold at 40°C, compared to that emitted at 23°C. Emission may continue over a period of several years.

TABLE II

Human Exposure to Formaldehyde[a]

Exposure source	Estimated number exposed	Number of observations	Mean exposure level (ppm) ± standard deviation	Estimated duration (h/week)
I. Occupational				
A. Direct production of formaldehyde	420	3	1.34 ± 0.27	40
B. Commercial use of formaldehyde and formaldehyde products				
1. Urea–formaldehyde foam installers	2,000–15,000	42	0.25 ± 0.09	40
2. Manufacturers				
a. Resin producers	2,000–45,000	165	0.31 ± 0.04	40
Urea–formaldehyde foam producers	30–80	28	0.68 ± 0.41	40
b. Molded products producers	Unknown	88	0.23 ± 0.04	40
c. Furniture, production of veneered panels	Unknown	41	0.92 ± 0.36	40
d. Textile producers	360–6,000	84	0.54 ± 0.09	40
e. Textile storage	Unknown	22	0.28 ± 0.08	40
f. Fertilizer producers	500–900	11	0.82 ± 0.35	30
3. Others				
a. Embalmers	70,000	12	1.04 ± 0.36	20
b. Pathologists	12,000	11	2.79 ± 1.45	30

c. Biology instructors	35,000	21	0.32 ± 0.17	20
d. Students				
College/university	1,200,000	21	0.32 ± 0.17	1
Medical school	60,000	7	1.56 ± 0.74	15
High school	Unknown	21	0.32 ± 0.17	1
II. Residential and commercial building levels, use of				
A. Plywood/particle board				
1. Conventional homes				
a. Denmark	Unknown	25	0.53 ± 0.15	100–150
b. United States	Unknown	12	0.28 ± 0.16	100–150
2. Mobile homes	2,200,000	836	0.37 ± 0.02	100–150
B. Urea–formaldehyde foam insulation				
1. Homes	1,750,000	751	0.12 ± 0.02	100–150
2. Shopping center, offices, and stores	Unknown	66	1.17 ± 0.41	40
III. Ambient levels				
A. Air	220,000,000	41	0.02 ± 0.018	168
B. Non-UFFI homes	Unknown	51	0.03 ± 0.004	100–150

[a] From Ulsamer *et al.* (*230*).

Mobile homes, which use particle board and plywood extensively in their construction, are well insulated (but not with UF foam insulation) and are relatively air tight. Mobile homes were found to contain an average of 0.37 ppm of formaldehyde (*230*). Levels in new homes were higher than those in older homes. Emission may continue for a period of years.

V. METABOLISM

Formaldehyde is an important intermediate in the biosynthesis of amino acids, lipids, and nucleotides. It also alkylates nucleic acids and proteins and is converted to formate and CO_2 (Fig. 1). Following acute inhalation exposure, formaldehyde gas is absorbed primarily via the upper respiratory tract in dogs (*63*) and in rats (*223*). Formaldehyde can penetrate into the lower respiratory tract when adsorbed to particulates (*5, 6*).

When rats inhaled [^{14}C]formaldehyde, nasal tissues were found to contain the highest concentration (10- to 100-fold greater than other tissues) of radiolabel; most of the isotope remaining in the body was distributed throughout body tissues (*102, 103*). The chemical form of the radiolabel was not defined but, based on what is known about formaldehyde's chemical reactivity and metabolism, it is unlikely that any of the radiolabel remained as formaldehyde.

Dermal absorption of [^{14}C]formaldehyde has been demonstrated in several species of laboratory animals, including rats and monkeys (*118*), guinea pigs (*236*), and rabbits (*195*). The chemical form of the radiolabel has yet to be determined but preliminary data from *in vitro* diffusion studies using rabbit skin (*100*) indicate that formaldehyde per se cannot be detected enzymatically.

Formaldehyde that enters the body is rapidly metabolized to formate. Intravenous (iv) infusion of formaldehyde into dogs demonstrated that formate levels in blood rapidly increased whereas formaldehyde could be detected only during infusion (*148*). The rapidity of this conversion was demonstrated by the finding that the peak in blood formate concentration occurred within the same time frame, and was of the same magnitude, regardless of whether formaldehyde or sodium formate was infused into dogs. The plasma half-life of formate was also the same following injection of either chemical (between 80 and 90 min). Following infusion of formaldehyde into cynomolgus monkeys, the half-life of the formaldehyde in the blood was estimated to be 1.5 min (*157*). Similar estimates of half-life have been made for cats, guinea pigs, rabbits, and rats (*194*). More recently, Heck (*102*) has shown that [^{14}C]formate distributes similarly to [^{14}C]formaldehyde in rat blood cells and plasma following iv injection, and follows the same decay curve.

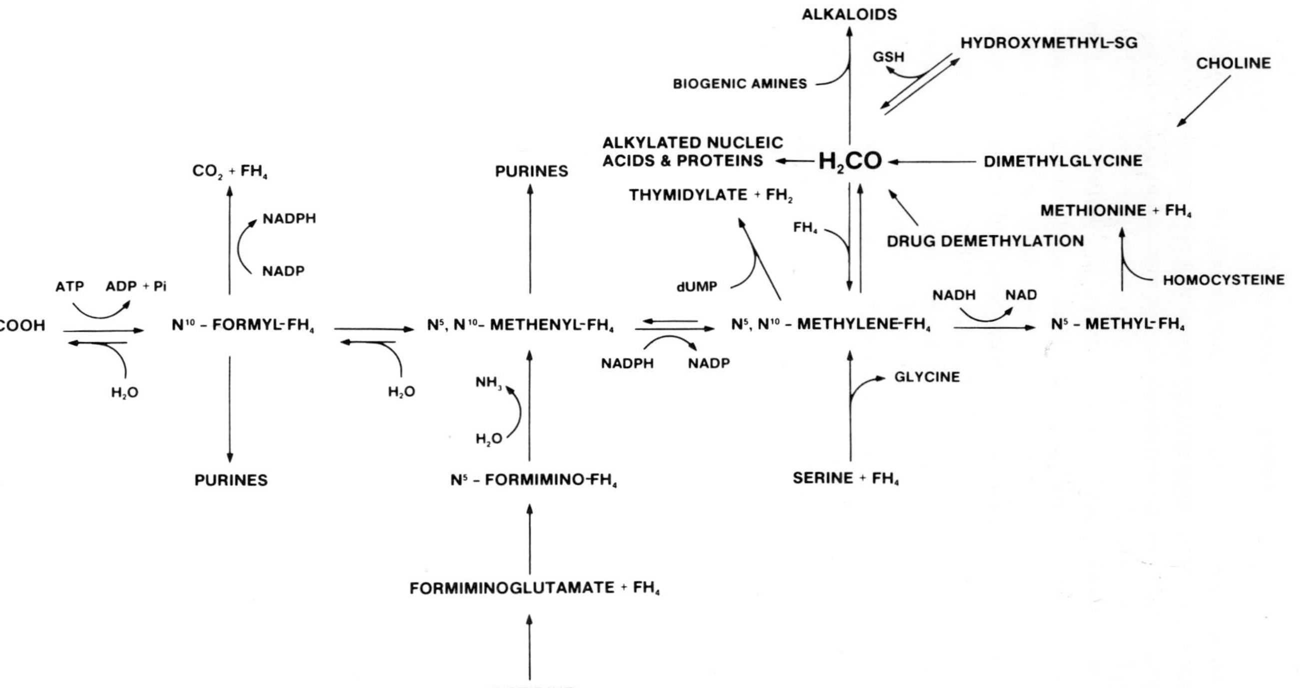

Fig. 1. Formaldehyde interconversions. FH_4, Tetrahydrofolic acid; GSH, glutathione; FH_2, dihydrofolic acid; dUMP, deoxyuridine monophosphate.

Several studies have described the metabolic fate of formaldehyde once it enters the body. DuVigneaud *et al.* (*60*) found that approximately 80% of subcutaneously (sc) administered formaldehyde is converted to CO_2, while a small amount remained in body tissues incorporated into choline. Neely (*172*) administered [^{14}C]formaldehyde intraperitoneally (ip) to rats at doses of 7 and approximately 70 mg/kg. At the higher dose, 82% of the radiolabel was expired as CO_2 after 24–48 h and 14% was recovered in the urine. Most of the radiolabel in the urine was found to be incorporated into methionine and a cysteine adduct with lesser amounts as serine and "formaldehyde." "Formaldehyde" was determined by a chemical method of analysis. Since no radiolabeled peak corresponding to formaldehyde could be detected when a sample of urine was chromatographed, "formaldehyde" may be an artifact resulting from cleavage of a conjugate during chemical analysis. Neither the cysteine adduct nor serine could be detected following administration of the 7 mg/kg dose. The nature of the cysteine adduct was not defined, but Neely found that a chromatographically identical product could be formed by adding formaldehyde to cysteine or to urine. Edwards *et al.* (*61*) identified the cysteine adduct as *N*-formylcysteine in rats and mice following ip administration of [^{14}C]formaldehyde. Methionine and serine were also found in the urine of both species. The reaction of formaldehyde with cysteine occurs nonenzymatically and in preference to that with glutathione (GSH) (*97*). More recently, Mashford and Jones (*155*) demonstrated that in rats administered 4 mg/kg of formaldehyde ip, 82% of the dose was exhaled within 48 h as CO_2, while 5.5% was excreted in the urine. At 40 mg/kg, 78% of the dose was exhaled as CO_2 after 48 h and 11% was excreted in urine. The metabolites in urine were the same at either dose: *N*-(hydroxymethyl)urea, *N,N'*-bis(hydroxymethyl)urea, and formate. The authors postulated that the urea conjugates are formed in the urine by chemical reaction with free formaldehyde, and that 3–5% of the higher dose may therefore have been excreted in the urine as free formaldehyde; no formaldehyde was found in expired air. It is uncertain whether the formaldehyde in the urine is free or exists in the form of a labile conjugate. The excretion of formate contrasts with results obtained by other investigators and may be related to the strain of rat used. When formaldehyde is inhaled by F344 rats (*102*), rather than injected, 40% of the radiolabel is retained in the animal, while 40% is exhaled and 20% appears in the urine. In the nasal mucosa of these rats, RNA contained the greatest amount of radiolabel with a lesser amount in protein and a small amount in DNA (*103*). In WI-38 human fibroblasts exposed to [^{14}C]formaldehyde, most of the radiolabel is incorporated into RNA with lesser amounts in DNA and protein (*190*). The purine bases of the nucleic acids were labeled most heavily. Formate appeared rapidly in the blood and urine of humans exposed to formaldehyde gas (*64*). Einbrodt also found a small amount of formaldehyde in the urine by chemical analysis but this may have resulted from the breakdown of a labile conjugate. Eels *et al.* (*62*) noted a rise in formate in the blood following ingestion of formalin by a 41-year-old woman. In addition to being converted rapidly to formate and CO_2 as well as being incorporated into various body chemicals formaldehyde also can alkylate amino acids, such as cysteine (*107*) and lysine (*228*), proteins (*72, 215*), nucleotides (*107*), and DNA (*36, 72*). The reaction of formaldehyde with DNA to form stable linkages is enhanced by the presence of amino acids (especially lysine) and histones (*215*). Linkages between protein and DNA have been reported in formaldehyde-exposed rats (*101*) and in mouse leukemia L1210 cells (*196*).

The conversion of formaldehyde to formate is catalyzed by formaldehyde dehydrogenase (*231*). This enzyme catalyzes an easily reversible reaction between GSH, formaldehyde, and NAD to yield *S*-formylglutathione. The actual substrate for formaldehyde dehydrogenase is probably the hemimercaptal of formaldehyde and GSH which forms nonenzymatically (*238*). Hydrolysis of *S*-formylglutathione to formic acid and GSH is catalyzed by *S*-formylglutathione hydrolase (*232*). This reaction is described as being very fast, highly specific, and apparently irreversible; it is not inhibited by formate. In human liver, hydrolase activity is present in great excess over formaldehyde dehydrogenase activity. Both enzymes are of cytoplasmic origin in liver (*204, 231, 232*). Human erythrocytes (*148*) and brain, sheep liver, rat brain, kidney, and muscle, rabbit brain, and bovine brain and adrenal (*232*) also can rapidly convert formaldehyde to formate. In human liver, *S*-formylglutathione can be hydrolyzed also by glyoxalase II to formic acid and GSH (*232*). The authors noted that the activity of this enzyme is equivalent to that of formaldehyde dehydrogenase in liver. Formaldehyde is also oxidized to formic acid by a nonspecific aldehyde dehydrogenase and by the tetrahydrofolic acid (FH_4) pathway (*115*). The aldehyde dehydrogenase that oxidizes formaldehyde to formate is found primarily in mitochondria of liver cells; the microsomal aldehyde dehydrogenase has not been found to be reactive with formaldehyde (*133*).

Formaldehyde, as well as formate, is converted to CO_2. This can occur via the FH_4 pathway (Fig. 1) or via deamination of serine (formed from formaldehyde) to form pyruvate, which is then oxidized to CO_2 by the mitochondria. Finally, formate can be converted to CO_2 via catalase, but this pathway is apparently of much less importance than the FH_4 pathway (*157, 181, 244*). Den Engelse *et al.* (*57*) have shown that lung is less efficient than liver in converting formate to CO_2.

Formaldehyde that enters the FH_4 pathway does so by nonenzymatic reaction with FH_4 to form an N^5-carbinolamine (*126*). The carbinolamine proceeds rapidly through an imine to N^5,N^{10}-methylene-FH_4 in a reaction with an equilibrium constant of approximately 10^4 in favor of the formation of the end product (*126, 177*). The equilibrium constant for the dissociation of N^5,N^{10}-methylene-FH_4 has been reported to be 3×10^{-5} (*21*). The reaction is inhibited by thiols, such as 2-mercaptoethanol, which react preferentially with formaldehyde (*126*).

The efficiency of these processes in metabolizing formaldehyde was demonstrated by the infusion studies discussed above (*148, 157, 194*). A more recent study by Heck *et al.* (*104*) has actually quantitated labile formaldehyde in various tissues of F344 rats before and after inhalation of either formaldehyde or chloromethane. The method used by Heck measures both free and bound formaldehyde (the formaldehyde derivatives of both glutathione and FH_4 react) without distinguishing between the two forms. "Labile formaldehyde" levels range from 0.42 μmol/g for nasal mucosa to 0.097 μmol/g for brain; liver contains 0.20 μmol/g. Inhalation of 6 ppm of formaldehyde for 6 h/day for 10 days did not significantly alter the concentration of labile formaldehyde in the nose (the only tissue measured postexposure). A similar finding was made following inhalation of 15 ppm of formaldehyde (*103*). Glutathione levels were likewise unchanged following inhalation of 15 ppm of formaldehyde in this study. Previous work from the same laboratory demonstrated that CO_2 production from inhaled formaldehyde is directly proportional to dose at 0.5 and 13.1 ppm in rats (*86*). When chloromethane was inhaled by rats, the formaldehyde concentrations in liver and testes approximately doubled while increasing sevenfold in brain; no data were given for the nose (*194*). Whether this increase represents formation of a conjugate not seen when formaldehyde itself is inhaled or whether it is related to depletion of GSH as postulated by the author is unclear.

Formaldehyde in tissue can result from a number of sources. The primary source of endogenous formaldehyde is the degradation of serine (*15*) with some contribution from the degradation of other amino acids (see Fig. 1). Oxidative demethylation of *N,N*-dimethylglycine (from choline degradation) also contributes significantly to endogenous formaldehyde. Cytochrome *P*-450-dependent N-demethylation of drugs can contribute additional formaldehyde (*1, 244*). Using aminopyrine as the substrate for the demethylation reaction, Waydhas *et al.* (*244*) found that the rate of formaldehyde oxidation to formate exceeded the rate of formaldehyde production in perfused rat liver by a factor of 12. Other xenobiotics including dihalomethanes (*2*), methanol (*157*), dimethylnitrosamine (*119*), hexamethylphosphoramide (HMPA) (*50*), bis(chloromethyl) ether (BCME) (*213*), dibromoethane (*110*), and dimethylsulfoxide (*131*) lead to the production of formaldehyde. Formaldehyde is also formed *in vitro* in the presence of an amine acceptor, apparently by nonenzymatic breakdown of N^5,N^{10}-methylene-FH_4 (*137, 140, 226*). This reaction produces alkaloids from biogenic amines or drugs *in vitro* and probably *in vivo* (*145*). Formaldehyde resulting from the metabolism of HMPA, and dibromoethane (*50*), dimethylnitrosamine (*119*), and bis(chloromethyl) ether (*213*) may be the active species for these carcinogens. Using rat liver microsomes, formaldehyde production and accumulation could be demonstrated from HMPA (*50*) and dimethylnitrosamine (*119*). The accumulation of formaldehyde in microsomal preparations is not unexpected since microsomes have no detectable aldehyde dehydrogenase of the type capable of oxidizing formaldehyde (*133*), whereas formaldehyde dehydrogenase is primarily cytoplasmic in origin (*231*). Dodd *et al.* (*58*) have shown that labile formaldehyde also accumulated *in vivo* in tissues following inhalation of chloromethane. In contrast, when monkeys were administered methanol (which also metabolized to formaldehyde) by a nasogastric tube, no increase in formaldehyde concentrations could be detected in body tissues (*157*). Alcohol dehydrogenase, which converts methanol to formaldehyde, is cytoplasmic in origin (*185*) as is formaldehyde dehydrogenase. These findings raise the possibility that formaldehyde derived from xenobiotics metabolized by the microsomes may lead to accumulation of formaldehyde conjugates in tissue and pose a greater risk of carcinogenicity than formaldehyde derived from xenobiotics metabolized in the cytoplasm.

Exposure to formaldehyde may also cause additive toxicity to that caused by chemicals that require glutathione for detoxification (*136*). Chemicals with the potential for additive toxicity by this mechanism are numerous and include acetaminophen and corticosteroids (*52, 53, 128, 147*). Exposure to these chemicals may increase during periods when toxic exposure to formaldehyde occurs since they could be used to treat acute symptoms, such as headaches and skin rashes.

VI. GENERAL TOXICOLOGY

The acute toxicity of formaldehyde has been studied in several animal species by different routes of administration. The reported LD_{50} and LC_{50} values are summarized in Table III. These studies indicate that when formaldehyde is

TABLE III

Acute Toxicity

Species/ strains	Medium tested	Route	Observation period	Measure	Reference
Rats	2% solution	po	14 days	LD_{50}, 800 mg/kg	*217*
Guinea pigs	2% solution	po	14 days	LD_{50}, 260 mg/kg	*217*
Albino mice	Vaporized aqueous solution	Respiratory	45 days	LT_{50}, 100 min, (320 ppm)	*19*
Rabbit	Solution	Dermal	—	LD_{50}, 270 mg/kg	*144*
Rat	Solution	iv	—	LD_{50}, 87 mg/kg	*141*
Rat	35.5% solution	sc	—	LD_{50}, 420 mg/kg	*216*
Mice	35.5% solution	sc	—	LD_{50}, 300 mg/kg	*216*
Rat	35.5% solution	Respiratory	—	LC_{50}, 1.0 mg/liter (830 ppm), 30 min	*216*

TABLE IV

Lowest Effective Concentration of Formaldehyde: Human and Animal Controlled Studies[a]

Concentration (ppm)	Length of exposure	Species	Effect	Reference
0.01	5 min	Human	Eye irritation	*206*
0.05–0.06	Minutes	Human	Odor threshold	*71, 159, 237*
0.07	Minutes	Human	Optical chronaxy threshold	*159*
0.08	1.5 months	Rabbit	Changes in evoked potential of optic nerve	*23*
0.08	Minutes	Human	Threshold to affect the functional state of cerebral cortex	*159*
0.2	1 h	Human	Eye, nose, and throat irritation	*192*
0.25	5 h	Human	Dryness of nose and throat, decrease in mucous flow rate	*9*
0.31	1 h	Guinea pig	Increased airway flow resistance, decreased compliance	*5*
0.55	10 min	Rat	Reduction in respiratory rate	*127*
0.83	3 months	Rat	Histologic and histochemical changes in cerebral amygdaloid complex	*24*
0.83	1 min	Human	Altered functional state of cerebral cortex	*71*
0.83	90 days	Rat	Peribronchial and perivascular hyperemia, lymphohistiocytic proliferation in lung, focal hyperplasia and RE system activation in liver and changes in cerebral cortex	*71*
0.83	10 min	Human	Irritation of upper tract and eyes, accelerated breathing, EEG changes such as alpha rhythm enhancement, changes in automatic nervous system	*209*

0.83	10 months over two generations	Rat	Morphological changes in upper respiratory tract, decreased liver weight	*161*
0.83	Continuous, beginning 10–15 days before mating	Rat	Increase in size and number of extramedulary hematopoietic centers, increased epithelial proliferation of common bile duct, increased abnormalities of renal epithelium	*88*
1.4	1 min	Human	Eye sensitivity to light lowered in unacclimated group	*159*
1.67	Continuous or intermittent	Guinea pig, rat	Sensitization (inhalation), leukocytosis, and change in blood cholinesterase	*178*
2	6 h/day, 5 day/week for 18 months	Rat	Epithelial hyperplasia, squamous cell metaplasia of nasal turbinates, rhinitis	*223*
3.8	90 days continuous	Rat, dog, rabbit, monkey, guinea pig	Death in 1/15 rats, some inflammation of lungs in all species	*45a*
4.1	1 h on days 1–19 of gestation	Rat	Increase in threshold of neuromuscular excitability, peripheral white blood cells, decreased hemoglobin and rectal temperature in pregnant animals	*211*
4.2	1 min	Human	Unbearable without respiratory protection	*246*
15.5	10 h	Mouse, rabbit, guinea pig	5/7 mice, 3/5 rabbits, 8/20 guinea pigs dead; closed eyes, slow deep respiration, convulsions	*201*
41.5	1 h/day, 3 day/week for 35 weeks	Mouse	Upper respiratory tract inflammation, basal cell hyperplasia, epithelial stratification, bronchopneumonia	*113*
482	4 h	Rat	LC_{50} (approximate)	*168*

[a] From Gupta *et al.* (*94a*), with permission.

administered by subcutaneous or intravenous injection to rats it is more lethal than when administered orally. This may be because formaldehyde reacts with chyme, thus decreasing the amount available for absorption. The effects observed following exposure of experimental animals and humans to formaldehyde by inhalation include tissue irritation, sensitization, and CNS effects (*169, 170, 233, 234*). These effects, summarized in Tables IV and V [from Gupta *et al.* (*94a*)], include eye, nose, throat, and pulmonary irritation and hyperemia, skin

TABLE V

Human Occupational and Residential Studies: Ranges of Formaldehyde Concentrations Giving Adverse Effects[a]

Concentration (ppm)	Effect	Type of exposure	Reference
0.01–10	Nausea; eye, nose, and throat irritation; headache; vomiting; stomach cramps	Residential	*44*
0.02–4.15	Diarrhea, eye and upper respiratory tract irritation, headache, nausea, vomiting	Residential	*29, 85, 203, 249*
0.09–5.6	Burning of eye and nose; sneezing, coughing, and headaches; 3 out of 7 suffered from asthma or sinus problems	Occupational	*129*
0.3–2.7; mean, 0.68; median, 0.4	Annoying odor, constant prickling of mucous membranes, disturbed sleep, thirst, heavy tearing	Occupational	*212*
0.13–0.45	Burning and stinging of eyes, nose, and throat; headaches	Occupational	*27*
0.83	Loss of olfactory sense, increased upper respiratory disease, subatrophic and hypertrophic alterations in nose and throat, ciliostasis of nasal mucosa, increased absorptive function of nasal mucosa	Occupational (greater than 5 years to less than 10 years)	*250*
0.9–1.6	Itching eyes, dry and sore throats, disturbed sleep, unusual thirst upon awakening in morning	Occupational	*164*
0.9–2.7	Tearing of eyes, irritation of nose and throat	Occupational, 1966	*22*
Unknown	Chronic airway obstruction, respiratory tract and eye irritation, small decrease in pulmonary function during workday and workweek	Occupational	*205*
1.3–3.8	Menstrual disorders, pregnancy complications, low birth weight of offspring	Occupational	*214*
4 or less	Inflammation, reactions of upper respiratory tract, chronic bronchitis, conjunctivitis, and skin changes	Occupational (7-year mean exposure)	*134*

[a] From Gupta *et al.* (*94a*), with permission.

rashes, changes in cerebral cortex, development of headaches, and many other effects.

In many instances, the changes that occur in animals that are exposed to formaldehyde are similar to those that occur in humans who are likewise exposed. This may be illustrated by the effects on the airways. Amdur (*5, 6*) exposed guinea pigs to formaldehyde and found, even at 0.07 ppm in the presence of NaCl particles, significantly increased airway resistance and decreased lung compliance. Murphy *et al.* (*167*) noted similar changes after exposure to higher concentrations of formaldehyde in rats; they also found signs of eye and nasal irritation, dyspnea, and an increase in liver alkaline phosphatase. When exposed to low concentrations of formaldehyde, humans often experience nose and throat irritation (*8, 169, 170*). In humans, pulmonary irritation may be characterized by cough, a feeling of tightness in the chest, and wheezing (signs of bronchial constriction) (*84, 169, 170, 189*). A protective mechanism against the respiratory effects of inhaled formaldehyde appears to exist in mice, which are able to decrease their respiration rate by up to 50% when exposed to formaldehyde (*127*). Barrow (*12*) also reported similar results in mice and to a lesser extent in rats. Formaldehyde is also a severe eye irritant in rabbits (*35*) and in humans (*8, 192*). Significantly, humans experience conjunctival irritation when exposed to as little as 0.20 ppm of formaldehyde alone (*192*) or 0.01 ppm of formaldehyde in artificial smog (*206*). Formaldehyde causes skin irritation in guinea pigs (*39*) and in humans (*87, 186*).

The NAS report "Formaldehyde and Other Aldehydes" (*170*) mentions various effects of formaldehyde on the central nervous system in humans. For example, CNS effects such as thirst, dizziness and apathy, and inability to concentrate have been reported in workers using formaldehyde resin (*169*). Electroencephalographic (EEG) changes have been reported in human subjects exposed to 0.044 ppm of formaldehyde (*71*).

Formaldehyde causes hyperemia or inflammation in liver and kidney in rats (*71, 73, 88, 216*). Microscopically, formaldehyde also causes cloudy swelling, cytoplasmic vacuolization, and necrosis in the liver, and hyperemia, edema, and necrosis in the kidney. Macroscopic changes in the liver have also been produced by formaldehyde. When exposure is repeated over a period of weeks, changes include a mottled appearance and a decrease in liver weight (*13, 77*). Following a single high exposure, liver size may increase (*201*). Similarly, Murphy *et al.* (*167*) found that liver weight (absolute and relative to body weight) increased in rats following a single inhalation exposure to 35 ppm of formaldehyde for 18 h.

The toxic effects on liver that occur in response to high levels of exposure are usually more pronounced and occur more frequently than those caused by lower levels (*13, 73, 77*). Consequently, a general dose–response relationship may exist for organ toxicity caused by formaldehyde. For example, in the study conducted for CIIT, inhalation of formaldehyde caused changes in liver weight as well as microscopic changes in the livers of mice at 14.3 ppm, but caused only significant decreases in relative liver weights at 6 ppm in mice (*13*). Similarly, in the Formaldehyde Institute study, inhalation of 3 ppm of formaldehyde caused a decrease only in liver weight (*77*). At 6 and 12 months, in the CIIT study, hepatic centrilobular vacuolization and necrosis occurred in mice receiving 14.3 ppm but not in the control groups ($p < 0.001$) (*13*). Similar changes occurred in rats that inhaled 0.8 ppm of formaldehyde in a reproductive study (*88, 89*). An abstract of one study reported that dermally applied formaldehyde caused liver changes (*208*).

Transient effects on the hematopoietic system occurred in rats and mice after 6 months of exposure to formaldehyde by inhalation (*13*). These effects were reflected by statistically significant decreases in (1) reticulocytes in female mice exposed to 2.1, 5.6, or 14.3 ppm; (2) mean corpuscular hemoglobin in male and female rats exposed to 14.3 ppm of formaldehyde; and (3) mean corpuscular hemoglobin concentration in male rats exposed to 2.1, 5.6, or 14.3 ppm of HCHO (*13*). Male and female rats had significant ($p < 0.05$) increases in mean corpuscular hemoglobin, mean corpuscular hemoglobin concentration, and myeloid to erythroid ratios after 13 weeks of exposure by inhalation to 12.7 ppm of formaldehyde (*163*). This could indicate myeloid hyperplasia or erythroid hypoplasia.

Microscopic examination of the tissues in this study (which involved three exposure concentrations: 4.0, 12.7, and 38.6 ppm) revealed several lesions that resulted from exposure to 38.6 ppm of formaldehyde (*163*). In rats, the lesions included ulceration and necrosis of the nasal turbinates and trachea, congestion and hemorrhage in the lungs, congestion of hepatic sinusoids, and cytoplasmic vacuolation and congestion of the adrenal cortex. Some of these microscopic changes could have resulted from secondary effects of formaldehyde. Formaldehyde-related changes in mice included necrosis of the nasal turbinates and trachea, and pulmonary congestion and hemorrhage.

The effects of subchronic exposure to formaldehyde have also been examined in a study in which mice were exposed to formaldehyde by inhalation (*139*). Five groups of 10 male and 10 female mice each were exposed for 6 h/day, 5 days/week for 13 weeks to concentrations of either 2, 4, 10, 20, or 40 ppm of formaldehyde. Although the study was designed only to help establish exposure levels for a subsequent chronic toxicity study, which was never done, it produced some interesting findings. At 4 ppm and above, a dose-related increase in squamous metaplasia of the nasal cavity developed. At 10 ppm and above, epithelial hyperplasia, squamous metaplasia, and inflammation of the trachea also developed. In addition to these changes, at 40 ppm bronchial inflammation, epithelial hyperplasia, metaplasia, and granulation were observed. There was also an 80%

mortality rate at 40 ppm. The study revealed a significant depression of weight in both sexes at 20 and 40 ppm.

Interestingly, a few systemic effects at 40 ppm were sex related. These included ovarian involution and endometrial atrophy and a decrease in the liver to body weight ratio in female mice. These findings are particularly interesting because Shumilina (*214*) reported that several menstrual and reproductive alterations occurred in women who were exposed to formaldehyde during their work. (Details of Shumilina's findings are presented in Section XI.) Both sexes showed atrophy and necrosis of the thymus.

This study shows that 65 days of exposure to formaldehyde with weekly recovery periods of 48 h will produce significant changes in the upper respiratory tract at 4.0 ppm and serious systemic effects and death at 40 ppm (*13*).

In a study conducted for Biodynamics, the subchronic toxicity of formaldehyde was studied under contract by the Synthetic Organic Chemical Manufacturers' Association (*77*). They exposed rats, hamsters, and monkeys to concentrations of 0.2, 1.0, or 3.0 ppm for 22 h a day, 7 days a week, for 26 weeks. The summary table of microscopic findings in rats exposed to 0.2 or 1.0 ppm of formaldehyde revealed that albuminous degeneration of hepatocytes, hyperplasia of the bile duct, and focal hemorrhage developed in the livers of several of them. These changes did not occur in the concurrent control group. Four of the five rats exposed to 1.0 ppm and whose organs were subjected to microscopic examination also had hepatic necrosis. The necrosis did not occur in the other groups. At 3.0 ppm, the liver weight and the liver to body weight ratio were significantly decreased in rats, but necrosis of the liver was not seen.

VII. HYPERSENSITIZATION

Formaldehyde solution and probably gas, as well as polymers containing formaldehyde, have induced and elicited hypersensitivity reactions in humans. Dermal reactions may follow dermal or inhalation exposure and may be immediate or delayed in nature. Immediate reactions are characterized by urticaria, while the more common delayed reactions lead to erythema, edema, and vesiculation. Respiratory reactions result from exposure to airborne formaldehyde and are characterized by rhinitis or asthma. Asthmatic responses may be immediate or late, with present data indicating that at least late reactions to formaldehyde are of immunologic origin.

Early work by Horsfall (*112*) demonstrated that formaldehyde produced delayed contact dermatitis in a sensitive patient when tested by immersion of the hand in formaldehyde solution as low as 0.2 ppm. When the patient inhaled formaldehyde through a mouthpiece, a delayed dermal reaction was also produced. Rostenberg *et al.* (*197*) confirmed Horsfall's finding that formaldehyde could cause a delayed contact dermatitis. These investigators studied nurses who developed dermatitis from repeatedly handling thermometers sterilized in 10% formaldehyde. Positive reactions were elicited by patch tests with 0.5% formaldehyde. More recent studies involving dermatologic patients from many countries, including the United States, have shown that 1–7% of these patients were sensitive to 2% formaldehyde by patch testing (*34, 47, 81, 105, 198, 199, 222*). Approximately 5–8% of subjects without dermatologic complaints (selected from the San Francisco area) became sensitized following dermal exposure to formaldehyde in concentrations of 0.37–3.7% and challenge concentrations of 0.3% (*154*). Later work by Jordan (*125*) showed that positive delayed reactions were found to occur in dermatologic patients by patch testing with 30 ppm of formaldehyde (four of nine patients). However, spraying 28 ppm of formaldehyde in water on exposed skin did not produce a positive reaction.

Positive dermal sensitization reactions to formaldehyde have been caused by many products including textiles (*17, 47, 114*), paper (*20, 75*), cleaning agents (*80, 132*), coolants (*10, 96*), nail hardeners (*47, 162, 183*), photographic chemicals (*47, 75*), and embalming fluid (*47*). Delayed contact dermatitis reactions have also been produced by resins containing formaldehyde. These include melamine–formaldehyde (*79, 146b, 146c*), urea–formaldehyde (*111*), and phenol–formaldehyde resins (*47, 49, 78, 149, 150*). Immediate dermal reactions to formaldehyde or products containing formaldehyde have also been reported (*74, 106, 125, 156*).

Exposure to formaldehyde vapor has produced rhinitis in exposed individuals (*205, 240*), as well as bronchial asthma (*3, 14, 108, 174, 188*). The development of bronchial asthma following exposure to formaldehyde vapor is perhaps best demonstrated by the study of Hendrick and Lane (*108*). They included five staff members of a hemodialysis unit, two of whom developed wheezing, chest tightness, and cough after several months of repeated exposure to formaldehyde. Symptoms were delayed and worsened at night. Various hematological changes including eosinophilia were also noted. Provocative inhalation tests, in which either 25 or 10% formalin was painted on a board in a chamber, produced similar late asthmatic symptoms in these individuals. Later work (*109*) showed that formaldehyde concentrations during these exposures approximated 5 and 3 ppm, respectively. The several-month exposure period required for development of symptoms, the delayed onset and recurrent nocturnal pattern of the asthma, and the development of eosinophilia are all consistent with an immunological reaction as opposed to an irritation reaction. A late asthmatic reaction has also been reported (*3*) to occur in a painter who was exposed to 2 ppm of formaldehyde in a provocative inhalation test. This individual initially experienced rhinitis and then asthma while spraying paint later found to contain formaldehyde.

Inhaled formaldehyde vapor can also produce an immediate reaction in exposed individuals (*14*). In another case, Frigas *et al.* (*82*) reported an immediate response following provocative inhalation tests with pulverized urea–formaldehyde foam insulation: the response did not occur with formaldehyde gas delivered through a face mask. Aluminum oxide dust did not produce a reaction in this patient who had developed asthma following insulation of her house with urea–formaldehyde foam insulation. The immunological nature of the immediate reactions is more open to question since IgE antibodies have not been isolated from exposed humans as they have in some cases for other chemical allergens such as isocyanates (*16*) and trimellitic anhydride (*184*). Demonstration of asthmatic responses at low formaldehyde exposures may depend upon the presence of particulates. Respiratory symptoms have also developed in workers exposed to hexamethylenetetramine–resorcinol (HR) resin (*84*). More recently, work by Frigas *et al.* (*82*) has shown that 37 other individuals with respiratory symptoms, following formaldehyde exposure at home or on the job, did not develop asthma following inhalation of formaldehyde gas in compressed breathing air. Since previous investigators (*3, 108, 188*) used room air rather than compressed breathing air, the presence of naturally occurring particulates in the room air used by these previous investigators may have allowed formaldehyde to penetrate into the lungs. It is known that particulates aid formaldehyde in reaching the lower respiratory tract (*5, 6, 139*).

VIII. TERATOGENIC AND REPRODUCTIVE EFFECTS

The potential of formaldehyde to interfere with embryonic and fetal development has been reviewed previously (*70*). Since then, studies on the toxicity of formaldehyde have been completed that provide additional information about the reproductive and teratogenic effects of this chemical. This section discusses the main findings of the earlier studies and relates them to the results of more recent research.

A. Inhalation Studies

In a series of four publications beginning in 1968, Gofmekler and colleagues reported on the toxic and teratogenic effects of formaldehyde (*88–90, 191*). All of these publications appear to be based on an experiment in which 36 female rats (12 per group) were exposed to 0, 0.01, or 0.83 ppm of formaldehyde from 10 to 14 days before impregnation through gestation. Three male rats per dose level were also exposed for 6 to 10 days before mating.

In 1968, Gofmekler (*88*) reported the effects of formaldehyde on fertility, fetal weights, and organ weights. At 0.01 and 0.83 ppm, formaldehyde increased the duration of gestation by 14–15% as well as the average body weight of offspring and their heart, adrenal, and kidney weights. In contrast, the liver and lungs from pups in the treated groups weighed less than those from the control pups. This finding likely represents direct or indirect effects of formaldehyde on developing fetal lung and liver. Although Gofmekler reported a decrease in the litter size, data in the article show that exposed groups had average litter sizes of 19.6 and 17.3 pups, as compared to the control value of 11.2, which is nearer a ''normal'' size. For these calculations, Gofmekler apparently assumed that all females in each group became pregnant.

Gofmekler *et al.* (*89*) published additional results related to the effects of HCHO (0.01 and 0.83 ppm) on the developing embryo [identical data were reported by Pushkina *et al.* (*191*)]. Significant decreases in ascorbic acid concentrations occurred in the whole embryo and in the maternal liver at both dose levels. A significant increase in the ascorbic acid concentration in liver occurred in offspring from dams exposed to 0.01 ppm only. The lack of a similar change in the group exposed to 0.83 ppm, however, raises questions about the significance of this finding. RNA concentrations in maternal livers were greater at both dose levels than they were in controls. RNA concentrations of fetal brain were similar in control and treated groups. DNA content was significantly lower in maternal and fetal liver in both treated groups and control animals. The authors concluded that formaldehyde ''significantly inhibited the synthesis of nucleic acids.'' However, because the RNA concentration in the liver increased as the DNA concentration decreased, this conclusion is not completely supported by the data in the article.

The above publication (*89*) also describes microscopic changes in the liver, kidneys, and other organs of fetuses from dams exposed to 0.01 or 0.83 ppm of formaldehyde. Changes in the liver included an increased proliferation of epithelial cells in the bile duct and segmented forms in the hepatic sinusoids. Changes in the kidney included renal epithelial cells with polymorphic nuclei, casts in the lumina of some tubules, and functional alterations in the renal tubule apparatus. Also, exposure to 0.83 ppm decreased myocardial glycogen, involuted thymic lymphoid tissue, and disintegrated lymphocytes. Histologically, the testes of adult males exposed to formaldehyde were similar to those of the controls. In contrast, formaldehyde inhalation by pregnant dams did not cause macroscopically discernible changes in embryonic or fetal development; there were no terata.

Sheveleva (*211*) studied the teratogenic potential of formaldehyde in pregnant albino rats. The dams were exposed by inhalation to 0.004, 0.0004, or 0.0 ppm of formaldehyde for 4 h each day on days 1 through 19 of gestation. Fifteen females per group were killed on day 20, while six were kept to obtain progeny. Exposure to 0.004 ppm of formaldehyde decreased neuromuscular excitability, spontaneous mobility, rectal temperature, and hemoglobin concentration in the dams.

On day 20, the number of preimplantation deaths was higher in both groups exposed to formaldehyde than it was in the controls. The number of live fetuses was approximately the same in all groups. If added together, these data indicate that the number of zygotes was *greater* in dams that were exposed to formaldehyde. If that is correct, the data provide indirect support for the findings of Gofmekler (*88*) showing an increase in litter size. Until more research is done, the potential effect of HCHO on litter size will remain uncertain. In Sheveleva's study, no external malformations were observed in the offspring that were removed by hysterotomy. This finding is similar to that of Gofmekler and his colleagues (*88, 89*) as well as other researchers (*152, 179, 180*).

On day 22, six dams from each group delivered offspring; all progeny appeared to be normal at birth (*211*). At 1 month postnatal, the female offspring from control dams were larger than the female offspring from treated dams. For male progeny, the opposite was true. At 1 month, the spontaneous mobility of progeny from treated dams was less than that of control progeny. By 2 months, the hemoglobin and leukocyte concentrations were decreased in progeny of dams exposed to formaldehyde, but not in a dose-related manner (*211*).

Guseva (*95*) measured the nucleic acid content in the testes of rats exposed to formaldehyde. During a 6-month exposure period, three groups of male rats received formaldehyde orally and by inhalation as follows: Group 1, 0.1 mg/liter in drinking water and 0.4 ppm by inhalation; Group 2, 0.01 mg/liter in drinking water and 0.2 ppm by inhalation; Group 3, 0.005 mg/liter in drinking water and 0.1 ppm by inhalation. Group 4 served as untreated controls. Exposure in the drinking water was continuous. Simultaneous exposure to formaldehyde by both routes occurred five times per week for 4 h each time. Reproductive function was evaluated by pairing each treated male with two virgin untreated females and evaluating the resulting pregnancies. On day 20, an unspecified number of pregnant females were killed and their offspring were removed and examined. The remaining dams were allowed to produce offspring. Guseva did not report the results of the examination of the fetuses. The number and weight of newborn rats were recorded. Observations of their subsequent development extended over 1 month. The time of eye opening and other developmental indices were recorded for the offspring of males in Groups 1 and 3 only.

There was no effect of formaldehyde on the weight of the fetuses or the size of the litters. The offspring were morphologically normal at birth and developed normally thereafter. Gonadotropin levels were not significantly different between males in the control group and those in the treatment groups. However, the amount of nucleic acid in the testes of males exposed to 0.4 and 0.2 ppm of formaldehyde was significantly less than the amount in the testes of the controls (*95*).

Sanotskii *et al.* (*202*) studied the effects of formaldehyde on reproduction in an unspecified strain and number of albino rats. They exposed groups of pregnant and nonpregnant rats to 0, 0.4, or 0.5 ppm of formaldehyde for 4 h per day for 20 days. Nonpregnant rats responded more to the effects of formaldehyde than did pregnant ones. In nonpregnant rats, exposure to formaldehyde at 5.0 ppm altered renal function by decreasing daily diuresis and urinary chlorides, and increasing urinary protein concentrations. The increase in concentration of protein in urine may have simply reflected decreased urinary output. Altered hepatic function was manifested by a decrease in urinary excretion of hippuric acid. At 5 ppm only, blood hemoglobin decreased in the pregnant rats. This finding supports the findings by CIIT that exposure to 2 ppm of formaldehyde and above in rats decreases mean corpuscular volume and mean corpuscular hemoglobin concentration after 6 months of exposure (*13*). Exposure to 0.4 ppm did not affect the parameters that were estimated for either pregnant or nonpregnant rats.

B. Dermal Studies

We found one teratology study of formaldehyde in which exposure was by dermal application. In a pilot study, Overman (*179*) applied formalin to the denuded back of pregnant hamsters for 2 h per day on days 7–11 of gestation. This treatment resulted in a potentially meaningful increase in resorptions and birth defects. To determine if the changes were significant, he repeated the study using larger numbers of animals. In this latter study, exposure to HCHO did not affect the survival or development of the offspring of hamsters (*180*).

C. Ingestion Studies

Although human exposure to formaldehyde occurs most commonly by the respiratory and dermal routes, it may occur by the ingestion of formaldehyde-based preservatives. One teratogenic study of HCHO following oral ingestion has been done. Marks *et al.* (*152*) intubated pregnant albino mice on days 6–15 of gestation with 0, 74, 148, or 185 mg/kg/day. On day 19, the mice were killed and the offspring were examined. At 185 mg/kg, HCHO was toxic to 22 of 34 pregnant mice. At 74 mg/kg, there was a significant decrease in average weight gain during pregnancy. Treatment with HCHO did not result in malformed offspring. Because formaldehyde reacts with or binds to chyme and intestinal contents, as well as to tissue, the amount of HCHO that test animals are exposed to following ingestion is unknown.

Hexamethylenetetramine (HMT), an antimicrobial food additive as well as a medication used to treat chronic bladder infections, degrades to formaldehyde and ammonia in an acid medium or in the presence of protein (*93*). Reproductive studies using orally administered HMT have produced information that helps us to understand the potential effects of formaldehyde.

In 1970, Della Porta *et al.* (*55*) reported on the effects of orally administered HMT in rats. Females and males were given 1% HMT in the drinking water, beginning when the rats were 8 weeks old. Two weeks later, the animals were mated, and treatment of the females was continued during pregnancy and nursing. A group of 24 male and 24 female progeny was randomly selected for continued exposure to HCHO until they were 20 weeks old. Groups of 12 untreated dams and 48 pups were used as controls.

Treated females and control females produced 124 and 118 offspring, respectively. The progeny of the treated dams were not malformed although mean body weights of the treated males and females were significantly less than those of controls. The weights remained depressed for up to 9 weeks for males and up to 13 weeks for females before becoming comparable to those of the controls. When the offspring were autopsied at 22 weeks, no macroscopic or microscopic lesions were seen. Body weights and organ weights (liver, kidneys, spleen, thymus, pituitary, adrenals, and testes) of offspring were similar for treated and control groups.

A second experiment reported in this article involved exposure of rats to HMT in the drinking water over three generations. F_1 and F_2 animals were given HMT until week 40 postnatal; F_3 animals were given HMT until week 20. F_1, F_2, and F_3 animals were observed for 130 weeks; survivors were sacrificed at 3 years of age. The survival rates of all the generations of offspring were not affected by HCHO. Mean body weights obtained during the experiments showed no significant differences between control and treated groups.

One year later, Natvig *et al.* (*171*) reported findings similar to those of Della Porta *et al.* (*55*) when they gave HMT in the feed to Wistar rats. Male and female rats were fed a diet containing either 0.0 or 0.16% HMT starting at 2 months of age and continuing for 3 months, then mated with group mates. Their offspring were fed the same diet. The offspring were weighed at 7 and 15 weeks, measured for voluntary muscle activity at 6 weeks, and killed when they were 123 days old. There were no detectable differences between rats in the test and control groups. The fertility of the treated animals was similar to that of control animals. The offspring from both groups had similar muscular activity, body weights, general health, and organ weights.

Formaldehyde administered in the diet had no effect on reproduction in beagles (*116*). From 4 days after mating to day 56 of pregnancy, pregnant bitches were fed concentrations of 600 or 1250 ppm HMT, or of 125 or 375 ppm formaldehyde in the diet. Control dogs ate unadulterated chow. Neither formaldehyde nor HMT affected the pregnancy rate. Maternal body weights increased normally during pregnancy in all groups. The duration of gestation was not affected by formaldehyde or HMT. Mean litter sizes were within the normal range for all groups. The group that received 1250 ppm of HMT had a greater percentage of stillborn pups than any other group; this was due mainly to one litter in which seven of nine pups were dead. The stillborn pups were not malformed. At 1250 ppm of HMT, there were some signs of neonatal toxicity.

During the first month after parturition, the pups from bitches given 1250 ppm HMT grew less than normal. The retarded growth coincided with increased neonatal mortality. Consequently, the percentage of pups that survived to weaning was lower than it was in the other groups. Nevertheless, pups that survived for up to 9 months exhibited normal behavior, appearance, mobility, and muscular coordination.

D. Injection Studies

Palkovits and Mitro (*182*) studied formalin-induced stress in neonatal rats. They injected one group of newborn Wistar rats with 0.02 ml of 2% formaldehyde ip once on the day of birth. A second group was injected daily for the first 4 days after birth. Control animals were untreated. All neonates were decapitated 24 h after the last injections.

In the neonates injected for 4 days, degenerative cellular atrophy occurred in the ventromedial arcuate of the hypothalamus. Single injections of formaldehyde did not cause degenerative changes but did cause decreased cellular activity in the medial field of the ventromedial nucleus and in the arcuate nucleus and an accumulation of granules in the neuronal cytoplasm. In both groups, formaldehyde injections increased nuclear volume in the adrenals. These changes indicate that the hypothalamus of the neonatal rat is sensitive to corticoid feedback induced by formaldehyde administration.

Cohen (*38*) studied the response to formaldehyde injection of fetal rats by measuring ascorbic acid levels in the adrenals. The first fetus from each litter served as the control. During a hysterotomy, approximately 6 μl/g body weight of 2% formaldehyde was injected sc into one or more litter mates. Fetuses were injected with formaldehyde at either 18.5, 19.5, 20.5, or 21.5 days of gestation. Injections of formaldehyde at 20.5 days of gestation resulted in decreased ascorbic acid levels in the adrenals. Injections on other days of gestation did not cause this response in fetuses.

Conner *et al.* (*42*) studied the contragestational properties of formaldehyde. On day 3 or 7 after mating, 0.05 ml of 40, 20, 10, 7, 3.5, 2.0, 0.5, 0.05, or 0.0005% formaldehyde was instilled into one uterine horn and 0.09% saline into the other horn (control) of pregnant Sprague–Dawley rats. All solutions of formaldehyde also contained 12–15% methanol. On day 15, dams were sacrificed.

Injections of 40 and 20% formaldehyde produced maternal toxicity and death. Injections of 7.0 through 0.5% on day 3 terminated most pregnancies. When these concentrations were injected on day 7, most pregnancies continued. The authors concluded that the contragestational properties of formaldehyde were similar to those of other protein denaturing agents, including ethanol, methanol, and silver nitrate. Because methanol solutions alone were not tested, contragestational effects of methanol could not be clearly distinguished from those of formaldehyde.

E. *In Vitro* Studies

Johnson (*122*) used an *in vitro* assay involving hydra to evaluate the teratogenic potential of formaldehyde. The minimal concentration of formaldehyde that was toxic to the adult hydra was also teratogenic. More importantly, the maximal concentration of formaldehyde that was not toxic to the adult hydra was also not teratogenic. This *in vitro* assay system with hydra has been used to evaluate numerous chemicals (*123*). It accurately predicts the ratio of teratogenic doses to maternally toxic doses for several mammalian species *in vivo*. From the results of this assay, a chemical would probably not be expected to cause terata at an exposure that was not also toxic to the adult. Such a prediction is consistent with results of *in vivo* assays (*152, 180*).

IX. GENETIC EFFECTS

Formaldehyde has been found to be mutagenic to viruses, *Escherichia coli, Pseudomonas fluorescens, Salmonella typhimurium,* and to strains of yeast, fungi, *Drosophila,* grasshopper, and mammalian cells (*11, 70, 117, 225*). It produces gene mutations and chromosomal aberrations, including deficiencies, duplications, inversions, and translocations. In most experiments, although the results were positive, dose–response relationships were difficult to demonstrate (*70*). In the presence of other mutagens, such as X rays, ultraviolet radiation, and hydrogen peroxide, formaldehyde increases the frequencies of observed mutants. In *E. coli* and *Saccharomyces cerevisiae,* the lethal and mutagenic effects of formaldehyde are greater in the test systems using excision repair-deficient strains than in those with normal repair mechanisms (*69*). The mutations or DNA damage caused by formaldehyde may be related to its ability to cause crosslinks in nucleic acids (*36*).

The recent work of Temcharoen and Thilly (*225*) showed a positive relationship between the time and concentration of formaldehyde exposure and the mutagenic and toxic effects observed in *S. typhimurium* (strain TM 677) *in vitro,* both with and without rat liver microsomes. Connor *et al.* (*43*) recently found positive results using formalin in the Ames assay, thus confirming the mutagenic effects of formaldehyde. They also found the mutagenicity expressed over a narrow range of exposure concentrations.

In vivo assays using mammalian cells showed that formaldehyde induces sister-chromatid exchange in hamster ovary cells and in human lymphocytes (*175*). Formaldehyde also causes cell transformation in the mouse BALB/c 3T3 cells (*117*). Brusick (*32*) used the BALB/c 3T3 cell to demonstrate that formaldehyde acts as both an initiator and a promoter of cell transformation. Other data indicate that formaldehyde can initiate C3H/10T1/2 cell transformation with tetradecanoylphorbol acetate (TPA) as a promoter and that formaldehyde can act as a promoter in C3H/10T1/2 cells initiated with *N*-methyl-*N'*-nitro-*N*-nitrosoguanidine (*26*).

Whole animal systems have been used to evaluate the mutagenic potential of formaldehyde. It caused chromosome breaks in the spermatocytes of the grasshopper (*151*) and mutations in early larval spermatocytes of *Drosophila melanogaster* (*11*). Formaldehyde did not induce dominant lethal mutations in mice (*66*). More recently, Fontignie-Houbrechts (*76*) reported increased dominant lethality during weeks 1 and 3 after treatments of male mice with 50 mg/kg (ip) of formaldehyde; the effect was marginal, however, and may not have been treatment related. At all matings in this study, treated and control dams averaged less than 1.0 resorption each. To reveal statistical differences between control and treatment groups, preimplantation losses were added to postimplantation losses. The accurate determination of preimplantation losses, however, depends upon an accurate counting of corpora lutea, which is difficult and subject to a 10–15% error.

Data are becoming available that demonstrate that formaldehyde induces mutagenic changes in human cells *in vitro* and possibly in humans themselves. Goldmacher and Thilly (*91*) grew lymphoblast TK6 cells from a human donor *in vitro* and exposed them to formaldehyde for 2 h. At 4.6 ppm (150 μM), formaldehyde induced a significant number of mutations in the cells. The minimal concentration of formaldehyde that induced a detectable number of mutations was 4.0 ppm. Between 4.0 and 4.6 ppm, there seemed to be a simple linear dose–response relationship between the concentration of formaldehyde used and the number of mutations induced. Data on other chemicals that have been tested using the same system indicate that a simple linear dose–response relation should exist for formaldehyde at even lower concentrations. Thus, according to the authors, 10 exposures at 0.2 ppm would probably cause a mutagenic response of similar magnitude to that caused by 1 exposure at 2.0 ppm.

Evidence is beginning to appear showing that chromosomal effects observed *in vitro* following formaldehyde exposure could also occur in human leukocytes *in vivo*. Preliminary data were recently obtained on eight medical students who were exposed intermittently to about 1.0 ppm of formaldehyde during a 10-week anatomy course (*221*). As compared with control students, the students who were exposed to formaldehyde had an increase in sister-chromatid exchange rates in the chromosomes. If additional research validates these preliminary data, the findings could provide important insight into the potential chromosomal effects of formaldehyde on humans.

X. CARCINOGENICITY

Over the past 30 years, numerous animal studies have been reported in the literature concerning the carcinogenicity of formaldehyde. These include studies conducted with rats (*4, 13, 56, 242, 243*), mice (*13, 56, 113*), hamsters (*51*), and rabbits (*166*) by various exposure routes. With the exception of a few recent studies (*4, 13*), the interpretation of earlier studies is complicated by a variety of limitations relating to the extent of histopathology, dose, duration, number of animals tested and survived, lack of controls, route of administration, and chemical form tested.

By far the two most important carcinogenesis studies conducted to date with formaldehyde are the recently completed Chemical Industry Institute of Toxicology study (*13*) and the New York University (NYU) study reported by Albert *et al.* (*4*).

In the recent CIIT study (*13*), F344 rats and B6C3F_1 mice were exposed by inhalation to 0, 2, 5.6, or 14.3 ppm of formaldehyde for 6 h/day, 5 days/week for up to 2 years. Initially, 240 animals (120 males and 120 females) were exposed at each level. Randomly selected animals were sacrificed at 6, 12, 18, and 24 months. Additional rats from each group were also sacrificed at 27 or 30 months of the experiment. Over 50% of the rats exposed to 14.3 ppm of formaldehyde experienced an early unscheduled death, many due to squamous cell carcinoma of the nasal cavity, whereas in other exposure groups proportions of unscheduled early deaths ranged from 13 to 22% of the total number of rats exposed. Approximately 40 tissues were evaluated histopathologically from all animals except the 2- and 5.6-ppm animals sacrificed at 6 and 12 months. All gross lesions were histopathologically examined, and, for the nasal cavity examinations, multiple sections were evaluated.

At about month 12 of the study, the first nasal cancer was noted in a rat exposed to 14.3 ppm formaldehyde. By month 18, it was reported that 37 rats exposed to 14.3 ppm of formaldehyde had nasal cancer: 28 squamous cell carcinomas and 1 spindle cell sarcoma among 44 rats dead or moribund, and 8 squamous cell carcinomas among 40 rats sacrificed at 18 months (*224*). Other dose-related changes observed in rats at this time were squamous metaplasia and dysplasia of the nasal mucosa. After 24 months of exposure, the number of rats with nasal cancers at 14.3 ppm increased to 108: 103 squamous cell carcinoma, 2 nasal carcinoma, 2 undifferentiated carcinoma or sarcoma, 1 carcinosarcoma. Of the rats exposed to 5.6 ppm, two had squamous cell carcinoma of the nasal cavity. Tracheal metaplasia and bone marrow hyperplasia were also observed in rats exposed to 14.3 ppm. In addition, in all three formaldehyde-exposed groups, polypoid adenomas in the nasal cavity were observed. The incidence of the adenomas among animals that survived until 24 months appeared to increase in a dose-related manner (*13*).

In mice, squamous cell carcinoma occurred in the nasal cavities of two males at 14.3 ppm formaldehyde exposure in the CIIT study. In addition, a significant increase in the incidences of some of the nonneoplastic lesions (epithelial dysplasia, squamous cell metaplasia) was observed in mice exposed to 5.6 or 14.3 ppm formaldehyde. Mortality was substantial in all groups of male mice, but this was primarily attributed to fighting for dominance among the group-caged male mice. Wounds inflicted by cage mates resulted in infection and subsequent death.

In light of the extremely low spontaneous incidence rate of this type of cancer (*235*), and the high incidence of the same type of cancer in rats exposed to formaldehyde, the nasal cancers in mice are believed to be related to formaldehyde exposure. It should be noted that mice were able to decrease their rate and volume of respiration such that the dose of formaldehyde received by mice at 15 ppm was approximately equivalent to that received by rats at 5.6 ppm (*12*). Thus the carcinogenic response of these two species may be very similar.

The CITT study protocol and sections of nasal cavities and other tissues from exposed and control rats sacrificed at 6, 12, and 18 months of the study were reviewed by a panel of pathologists formed by the Interagency Regulatory Liaison Group (*25*). All tissues of exposed (14.3 ppm) and control mice sacrificed at 6 and 12 months also were examined by the panel. The members of the panel generally concurred with the observations, diagnoses, and interpretations of the CIIT pathologists (*25*). It was found, after histopathological analysis of the nasal cavities of formaldehyde-exposed animals, that no ulceration occurred in the nasal cavities at 6 or more months.

The methodology for formaldehyde generation and measurement was also reviewed by a panel of experts, which agreed that "the Battelle approach to formaldehyde vapor generation was a suitable adoption of accepted methods and principles and, therefore, it was sound and based upon the best available technology. The same type of assessment applied to the chamber air monitoring system, which also combined two well-established procedures" (*94*).

Two experiments conducted at NYU produced findings similar to those reported by CIIT (*4*). In the first experiment, 99 male Sprague–Dawley rats were exposed to a mixture of formaldehyde and hydrogen chloride (HCl) at concentrations of 14.7 and 10.6 ppm, respectively, for 6 h/day, 5 days/week for life. Groups of 50 air sham-exposed and 50 untreated rats were used as controls. Histologic sections were taken from the nasal cavity, larynx, trachea, pulmonary lobes, liver, bladder, kidney, spleen, and other organs with gross pathologic alterations. Bis(chloromethyl) ether (BCME) levels in the exposure chamber were too low to measure but were estimated to average about 1.0 ppb. Of the 99 rats exposed to the gaseous formaldehyde and hydrogen chloride, 28 developed nasal tumors: 25 squamous cell carcinomas and 3 papillomas. The first carcinoma was seen at 223 days. No tumors were observed in the respiratory tract of the controls.

Formaldehyde and HCl can combine to form BCME. BCME can cause lung and nasal cancer in rats upon inhalation. The most common type of nasal cancer induced by BCME in rats was esthesioneuroepithelioma (a tumor of the nerve tissue) and not squamous cell carcinoma as actually observed (*138, 142*). It was unlikely that BCME was involved in the development of the nasal cancer observed in this experiment because (1) it normally induces esthesioneuroepithelioma and (2) its concentration in the exposure chamber (1.0 ppb) was estimated to be far below that which previously produced carcinoma in rats. This experiment appears to support the findings of the CIIT study but was complicated by the presence of minute amounts of BCME and by the unknown effects of HCl alone or in combination with appropriate control groups.

In the second experiment, 100 male Sprague–Dawley rats were used in each of the following exposure groups: (1) gaseous mixture group as in Experiment I; (2) combined exposure to HCHO and HCl, in which the two gases were not premixed at high concentrations but fed separately into the inlet air supply of the exposure chamber; (3) formaldehyde alone; (4) hydrogen chloride alone; and (5) air sham-exposed controls. Formaldehyde concentrations ranged from 14.1 to 14.3 ppm, while HCl concentrations in groups ranged from 9.5 to 10.2 ppm. No BCME measurements had been made at the time of the reporting. The experiment had been in progress for 588 days at the time of the report. Therefore, the tumor data reported include only nasal lesions in rats that produced grossly evident nasal swelling. The number of nasal cancers in each group is as follows: Group 1, 12; Group 2, 6; Group 3, 10; Group 4, 0; Group 5, 0. Final results are yet to be reported. A significantly greater degree of irritation was observed in rats exposed to formaldehyde plus HCl as opposed to rats exposed to formaldehyde alone.

The results of the CIIT and the NYU studies provide adequate evidence that formaldehyde gas is carcinogenic in two strains of rats. In addition, formaldehyde appears to have induced nasal cancer in $B6C3F_1$ mice.

The carcinogenic potential of formaldehyde was tested also in hamsters and mice in combination with the known chemical carcinogens diethylnitrosamine (*51*) and coal tar (*113*). Dalbey (*51*) studied the potential carcinogenicity of formaldehyde and the possible tumor-promoting activity of formaldehyde in hamsters. In the first experiment, male Syrian golden hamsters were exposed to

10 ppm formaldehyde for 5 h/day, 5 days/week for life. Histopathologic examinations were made on two sections of the nasal turbinates, larynx, trachea, and lung; nasal sections were not consistently cut. No tumors were observed in histologic sections of respiratory tract tissues in either control or treated animals. Both hyperplastic and metaplastic areas were observed in the nasal epithelium of 5% of the hamsters exposed to formaldehyde, whereas none was observed in control animals. Survival in both treated and control groups was very poor: over 40% of the animals died within 80 weeks of the study, and over 80% of the animals died within 100 weeks.

The second experiment involved groups of male Syrian golden hamsters. The first group was exposed to 20 ppm formaldehyde for 5 h/day, 1 day/week for life. The second group received injections of 0.5 mg of diethylnitrosamine (DEN) once weekly for 10 weeks. The third group was exposed to 30 ppm formaldehyde 48 h prior to each of 10 weekly DEN injections, followed by weekly HCHO exposures for lifetime, but beginning 2 weeks after the last DEN injection. Survival of hamsters in all groups was again poor: over 40% of the animals died within 60 weeks of the study. No tumors were observed in untreated animals or in those receiving only formaldehyde. Tumors in one or more sites in the respiratory tract were observed in 77% of DEN-treated controls. Lifetime exposure to formaldehyde either prior to or after DEN injection did not significantly increase the number of tumor-bearing animals (TBA) above those DEN-only controls. However, the ratio of the number of tracheal tumors/TBA was almost doubled in the group given formaldehyde prior to each DEN injection over DEN-only controls. The author suggests that under these experimental conditions formaldehyde may enhance the carcinogenicity of DEN in the respiratory tract.

In the study reported by Horton *et al.* (*113*), groups of 42–60 C3H mice were exposed to coal tar aerosol and/or formaldehyde at concentrations of 40, 80, and 160 ppm for three 1-hour periods/week for 35 weeks. The 160-ppm group was exposed for only 4 weeks because of toxicity. Mice that survived 35 weeks at 40 ppm were subsequently exposed to 122 ppm of formaldehyde for another 35 weeks. Survival after 1 year was poor in all groups. There is no mention of histopathological evaluation of nasal tissues, so presumably no grossly visible tumors were observed. Coal tar-exposed mice developed lung cancer. However, in formaldehyde-exposed mice, no tumors were reported in lungs and trachea. The major limitations of this study for assessing the carcinogenic potential of formaldehyde are that too few animals survived beyond 1 year, exposures were too short, and histopathology was not adequately reported.

The carcinogenic potential of formaldehyde has also been tested by a variety of other routes of exposure, including oral (*56*), subcutaneous injection (*242, 243*), application to the buccal mucosa (*166*), and skin painting (*135, 219*).

Because of the limitations in the study design and lack of detailed description of study protocols, these studies could not provide firm evidence regarding formaldehyde carcinogenicity in animals. Notwithstanding the limitations, some of the studies suggest that formaldehyde may be carcinogenic in tissues other than nasal epithelium and in other species. These studies are discussed below.

Hexamethylenetetramine (HMT) is a urinary tract antiseptic that owes its activity to formaldehyde. HMT decomposes *in vivo* to generate formaldehyde and ammonia. The following two studies utilized HMT and are relevant for the evaluation of the carcinogenic potential of formaldehyde.

Della Porta *et al.* (*56*) administered HMT to the drinking water of CTM, SWR, or C3Hf mice at 1.25–12.5 g/kg body weight/day for up to 60 weeks while Wistar rats received HMT at 1.5–2.5 g/kg body weight/day for 104 weeks. No treatment-related tumors were observed either in mice or in rats.

In the second experiment, Watanabe and Sugimoto (*243*) injected rats subcutaneously with 1–2 ml of a 9–40% solution of HMT once a week until tumors developed. Of the 20 treated rats, 8 developed tumors: 7 sarcomas at the site of injection and 1 adenoma. Subcutaneous injection of formic acid, a metabolite of formaldehyde, did not induce tumors. Watanabe *et al.* (*242*) also injected rats subcutaneously with 0.4% formalin (1 ml/week for 15 months). Of the 10 rats treated, 4 developed sarcomas: 2 in the skin of the injection site, 1 in the liver, and 1 in the peritoneal cavity. The studies indicate that subcutaneous injection of formalin and HMT induced tumors. However, it is not certain what role the repeated injury to the subcutaneous tissue may have played in the induction of the sarcomas, even though results of formic acid injection were negative.

The other study suggesting formaldehyde-induced tumors at the site of application was reported by Mueller *et al.* (*166*). Rabbits were fitted with oral cavity tanks designed to continuously expose the palate to 3% formalin with minimal mechanical irritation. Six rabbits were exposed to formalin, 4 rabbits were fitted with oral tanks that did not contain formalin, and 10 rabbits served as controls. Each exposure lasted to 90 min and was repeated five times per week for a period of 10 months. Animals were sacrificed at 1 month after the last exposure. Of six rabbits treated with formalin, two developed grossly visible leukoplakias that, according to the authors, showed histological features of carcinoma *in situ*. In animals that were fitted with tanks without formalin, no lesions were apparent.

Two skin initation/promotion studies in mice were reported recently. Krivanek *et al.* (*135*) tested formaldehyde for its ability to irritate and/or promote skin tumorigenesis in CD-1 female mice. No treatment-related nodules were observed. A similar study by Spangler and Ward (*219*) also produced negative results. Preliminary data from the first 48 weeks of an ongoing 78-week study with female Sencar mice show no treatment-related tumors. It should be noted that in both studies it is uncertain as to how much of the highly volatile formaldehyde applied to mouse skin is actually being absorbed, and how much could penetrate the stratum corneum to reach a target.

Since formaldehyde causes cancer in experimental animals, the question of mechanism becomes important. Formaldehyde most likely acts through a genotoxic mechanism, although its ability to act as a promoter may play a role in the expression of its carcinogenicity. The genotoxic properties of formaldehyde are clearly indicated by its mutagenicity in viruses, bacteria, insects, and cultured mammalian cells (including human cells); it can also initiate mammalian cell transformation. Swenberg *et al.* (*223*) reported that exposure to 6 or 15 ppm of formaldehyde (but not to 2 ppm) for 6 h/day for 3 days increased cell turnover in the nasal cavity of rats. The increased cell turnover observed may in turn increase the likelihood of DNA damage becoming fixed, thus leading to tumor development. To what extent, however, increased cell turnover occurs beyond this very brief period is unknown. It can be surmised that squamous cells are more resistant to the toxic effects of formaldehyde and thus would tend to be less likely to undergo increased turnover.

There is also evidence that formaldehyde can act as a promoting agent. As Dalbey (*51*) reported, exposure of hamsters to 30 ppm of formaldehyde 2 days prior to each of 10 weekly injections of a known carcinogen, DEN, resulted in an increased number of tracheal adenomas per tumor-bearing animal as compared to hamsters given DEN only, whereas no tumors were found in the group treated with only formaldehyde. Additionally, it has been shown that in cell transformation assays formaldehyde can promote cells initiated by *N*-methyl-*N'*-nitro-*N*-nitrosoguanidine, and formaldehyde-initiated cells can be promoted by tetradecanoylphorbol acetate. Thus, formaldehyde may exert its carcinogenic effect by one mechanism or a combination of mechanisms.

Some suggest that the cellular response to formaldehyde irritation is the direct and necessary precursor to development of cancer of the nasal cavity. Studies, however, have shown that not all chemicals capable of inducing epithelial hyperplasia, a specific irritant effect, cause cancer. For example, in the NYU study reported by Albert *et al.* (*4*), formaldehyde with or without irritant hydrochloric acid vapor produced similar numbers of nasal cancers, although the degree of irritation with hydrochloric acid was greater than with formaldehyde alone. Furthermore, nasal cancer has not been observed in rats exposed to hydrochloric acid alone. In Dalbey's study of hamsters, hyperplastic and metaplastic areas were observed in the nasal epithelium of 5% of animals exposed to formaldehyde and none in controls, but no tumors were observed in those hamsters receiving formaldehyde. The Federal panel also considered the same issue and concluded that there was no evidence that "irritation," or induction of epithelial hyperplasia, is sufficient to account for the formaldehyde carcinogenicity (*70*). However, the panel did recognize that the induction of epithelial hyperplasia may contribute, to some extent, to cancer activity by enhancing stages of carcinogenesis such as tumor promotion or tumor growth.

XI. EPIDEMIOLOGY

Information on the acute and chronic health effects of formaldehyde in humans comes largely from (1) controlled human exposure studies, (2) case reports of individuals who were exposed to formaldehyde, and (3) cross-sectional studies in which measurement of exposure (formaldehyde) and effects (prevalence of symptoms, signs, and disease) were made at the same time among individuals repeatedly exposed to formaldehyde in residential or occupational settings. In addition, limited information on the carcinogenicity of formaldehyde is available from cohort and case-control studies in which the ascertainment of exposure and effects relate to two different points in time.

A. Controlled Human Exposure Studies

Several controlled experiments with increasing concentrations of airborne formaldehyde have been conducted on healthy volunteers to investigate its acute effects.

Andersen (*8*) studied the effects of formaldehyde on airway function, comfort perception, and learning capacity in 16 healthy young men exposed for 5 h to either 0.25, 0.42, 0.83, or 1.6 ppm formaldehyde. The exposures occurred in a climate chamber at 23°C and 50% relative air humidity. No significant changes in pulmonary functions (vital capacity $FEV_{1.0}$, $FEF_{25-75\%}$) or in performance on mathematical tests were reported between the control period and the period of exposure to formaldehyde. However, there was a significant reduction in mucociliary function at all concentrations except for 0.83 ppm. Subjective perception of discomfort, namely eye irritation and dryness in the nose and throat, was reported even at the lowest exposure, and the number of complaints increased with increasing formaldehyde levels. After exposure to 0.25, 0.42, 0.83, and 1.6 ppm formaldehyde, 3, 5, 15, and 15, respectively, of the 16 volunteers had complained of eye irritation and dryness of the nose and throat. The author concluded that the formaldehyde concentrations should be lower than 0.25 ppm in order for a 5-h exposure not to cause mucous membrane irritation and reduction in the natural clearance mechanisms of the mucous membranes.

In a second study, two separate experiments were conducted using a 30 m^3 climatic chamber (*245*). In the first experiment, 33 healthy students (24 men and 9 women) were exposed for 37 min to formaldehyde concentrations continuously rising to a maximum of 3.2 ppm. Every 5 min the subjects filled out questionnaires and their eye blinking rates were measured. In the second experiment, 48 healthy students (35 men and 13 women) were divided into four groups and subjected to five exposures of 1.5 min each and varying formaldehyde concentrations (0, 1, 2, 3, and 4 ppm). Between two exposures, the subject could recover for 8 min in the well-ventilated room. In both experiments, eye, nose, and throat irritation increased as a function of increasing formaldehyde concentration. The authors concluded that, on the average, significant changes in physiological parameters and comfort perception occurred at the following concentrations of formaldehyde: eye irritation, 1.2 ppm; nose irritation, 1.2 ppm; throat irritation, 2.1 ppm; annoyance (desire to leave the room), 1.2 ppm; eye blinking rate, 1.7 ppm. They also reported that at 2.1 ppm of formaldehyde exposure, 10% of the subjects experienced moderate eye irritation, 7% of the subjects had strong or very strong eye irritation, 33% of the subjects exhibited a doubling of eye blinking rate, and 20% expressed a desire to leave the room. In light of these observations, the authors suggested that a TLV of 2 ppm might be too high.

In a third study (*192*), a number of volunteers (5 or 10 at each exposure level) were exposed to 0, 0.1, 0.2, 0.5, 1.0, 3.0, or 5.0 ppm of formaldehyde for 1 h. Predominant complaints were eye, nose, and throat irritation, tear flow, nasal secretion, and awareness of objectionable odor. The total sum of complaints (frequency times intensity) was clearly dose dependent and at 0.2 ppm and above the responses differed significantly from control values.

These experimental studies demonstrate that upper respiratory tract irritation and eye irritation occur at formaldehyde concentrations of 0.2 ppm and above. The subjects of the experiments were healthy young adults, who may be less susceptible to the irritant effects of formaldehyde than people with allergy, children, and the elderly who are already suffering from respiratory tract illness. Also, the duration of exposure was short compared to consumer and industrial exposure.

On the basis of the studies by Andersen (*8*) and Weber-Tschopp *et al.* (*245*) and other available animal and human studies, the National Academy of Sciences concluded that there is no population threshold for the irritant effects of formaldehyde (*169*).

B. Case Reports

Numerous case reports of ill health associated with formaldehyde are available. Contact urticaria was described in a 28-year-old woman who worked as a carver and model setter in a factory in which leather dresses were manufactured (*106*). The leather contained small amounts of formaldehyde. She had urticaria almost daily (severely on the hands, with occasional edema of the lips) during the work week. During weekends and vacations, when she did not come into contact with leather, there was no evidence of urticaria.

A second case of contact urticaria was reported by Lindskov (*146*). A 26-year-old female who worked in a pathology laboratory for 8 years suffered daily outbreaks of urticaria of the face, neck, forearms, and dorsa of the hands. This occurred whenever she fixed tissue specimens. The rashes developed 15 min after she started working at the fume cupboard and disappeared a few hours after work was finished. The tissues were fixed in 10 and 20% solutions of buffered formaldehyde. The woman was transferred to other work in the laboratory and the urticaria disappeared. Similarly, Harris (*99*) described four persons who developed acute papulovesicular eczema following contact with urea–formaldehyde resins. The condition persisted until the workers were reassigned to areas without formaldehyde.

Sakula (*200*) reported a case of acute respiratory distress in a hospital laboratory technician following exposure to formalin. Severe bronchial asthma followed the slightest inhalation of formalin vapor, but the worker was free from attacks on weekends and holidays. A case of pneumonitis following heavy exposure to formaldehyde by inhalation has been reported (*189*). The case involved a 27-year-old neurology resident who spent 15 h exposed to high concentrations of formaldehyde vapor during preparation of brain specimens for student demonstrations. The following week, after only 2 h spent at the same activity, he developed acute respiratory distress including progressive dyspnea and chest tightness over a period of 15 h. Chest X rays showed increased interstitial markings with early edema. Decreased pulmonary function as measured by FVC, $FEV_{1.0}$, and MMEF was also noted on day 2 and day 24 after the onset of symptoms. This is said to be the first report describing a clinical picture of acute pneumonitis in man following formalin inhalation.

A number of employees of a dress shop reported burning, stinging eyes, nose and throat irritation, and headaches (*25*). Formaldehyde was found to off-gas from wrinkle-proof apparel and its concentrations in the shop ranged from 0.13 to 0.45 ppm. Similar symptoms were reported among workers involved in a paper conditioning process. The workers processed wood pulp paper which was previously treated with urea–formaldehyde or melamine–formaldehyde resin for shrinkage control. Air samples collected in the breathing zone of the workers revealed formaldehyde concentrations ranging from 0.9 to 1.6 ppm (*164*).

The United States Consumer Product Safety Commission has received over 3000 complaints involving formaldehyde vapor released from building materials: about 2000 involve urea–formaldehyde foam insulation and the remainder involve plywood, particle board, paneling, and other wood products. Predominant symptoms reported in the complaints were nausea, eye, nose, and throat irritation, headache, vomiting, and stomach cramps. Exposure data compiled by the CPSC on average formaldehyde levels in homes are as follows: homes without UFFI, 0.03 ppm; homes with UFFI, 0.12 ppm; mobile homes, 0.38 ppm; ambient air, 0.01 ppm.

C. Cross-Sectional Studies

Numerous cross-sectional studies of workers, volunteers, or residents exposed to formaldehyde reported health problems. Adverse health effects associated with formaldehyde exposure include eye, nose, and throat irritation, sneezing, shortness of breath, sleeplessness, tight chest, nausea, and excess phlegm (*77, 85, 129, 192, 203, 205*).

A recent study of Wisconsin mobile home residents reported that some of the above symptoms were significantly associated with the level of formaldehyde in the homes (*7*). Residents of 137 randomly selected mobile homes were enrolled for a 6-month prospective double-blind observation. Each month the residents filled out health questionnaires and after about 6 months the residents filled out a more detailed comprehensive health questionnaire and were asked to attend one of three clinics for a physical examination. Spirometry and single-breath diffusing capacity tests were also performed on all the participating residents over the age of 12. Formaldehyde levels in the homes were measured every month. Prevalence of burning eyes increased significantly with increasing mean formaldehyde levels in the homes. However, the presence of cough was not associated with the level of formaldehyde in the homes. The prevalence of clinical signs of irritation correlated with increasing mean formaldehyde levels in the homes: <0.4 ppm, 10%; >0.4 ppm, 24%; >0.8 ppm, 56%. The mean formaldehyde level in the homes of those who had clinical signs of irritation was 0.7 ± .3 ppm, while, for those without signs, the mean level was 0.4 ± 0.3 ppm. This difference was statistically significant. Results of spirometry were not associated with formaldehyde levels.

Similar effects were reported at even lower concentrations among German school children (*33*). The children were exposed to formaldehyde from urea–formaldehyde resin used for panels, acoustic ceilings, and school furniture. The mean formaldehyde concentrations ranged from 0.13 to 0.57 ppm. The study group (n = 1594) had a significant increase in upper respiratory irritation, eye irritation, and functional disturbances (headache, lack of concentration, dizziness, nausea) compared to a control group (n = 497). A substantial reduction of symptoms (71%) was reported 8 months after removal of the formaldehyde emission sources.

The irritant effect of formaldehyde on the upper respiratory tract and eyes was also reported in a recent study of New Zealand workers (*18*). The exposure group consisted of 110 workers employed in particle board manufacturing plants (n = 30), furniture manufacturing (n = 18), pathology laboratories (n = 28), chemical manufacture (n = 13), fiberglass products (n = 14), and other industry (n = 7). Formaldehyde levels, when measured in a few workplaces, ranged from 0.1 to 2.4 ppm; but total aldehyde levels for most workplaces were generally less than 1.0 ppm. A control group consisted of 56 government employees who were free of known formaldehyde exposure. Significant differences between the two groups in the prevalence of eye, nose, and throat irritation were reported. However, prevalence of lower respiratory tract symptoms with the exception of breathlessness, was not significantly different between the two groups.

An outbreak of hemolytic anemia among patients on hemodialysis was described in a recent report (*176*). The outbreak occurred shortly after a new system using filters impregnated with formaldehyde resins was installed. When the filters were removed, hematocrit values returned to previous levels. The severity and incidence of some responses were related to the concentration of exposure.

Other health effects attributed to formaldehyde from cross-sectional studies include respiratory problems, dermatitis, neurologic difficulties, and menstrual and reproductive disorders. Schoenberg and Mitchell (*205*) studied five groups of employees from a filter manufacturing plant to determine adverse effects of exposure to phenolic (phenol–formaldehyde) resin fumes. Groups of workers currently exposed to phenolic resins showed an excess of chronic cough and/or phlegm when compared to previously exposed workers or ''never-on-line'' workers who had never been production-line workers or supervisors. In addition, after adjustments were made for differences in total cigarette consumption, workers on the present production line for more than 5 years had a significantly lower $FEV_{1.0}/FVC$ and $MEF_{50\%}/FVC$ ratio ($p < 0.05$) than the never-on-line group. These results suggest that long-term exposure to phenol–formaldehyde resin fumes may lead to chronic airway obstruction. No systematic measurement of formaldehyde concentration was made during this study but, based on measurements by others, levels of formaldehyde were estimated to be in the range of 0.4 to 0.8 ppm. Exceptionally high levels (8.8–13.5 ppm) could occasionally occur when cross-current fans were turned off. In this plant, even never-on-line workers were occasionally exposed to resin fumes, which may explain the high prevalence of acute symptoms such as eye irritation (80%), nose irritation (53%), and lower respiratory tract symptoms (47%) among these workers. As noted by the authors, the limitations of this study include small numbers of exposed subjects, probable formaldehyde exposure among the never-on-line workers, and the potential for selective bias commonly associated with cross-sectional studies. In addition, the possible role of the parent resins, phenol, and other exposure from the industrial process prevent a clear determination that long-term exposure to formaldehyde may lead to chronic airway obstruction.

In a study of rubber workers exposed to hexamethylenetetramine–resorcinol (HR) resins, Gamble *et al.* (*84*) found more self-reported symptoms (itch, rash, cough, chest tightness, burning eyes, running nose, and persistent cough and phlegm) among HR-exposed workers than among nonexposed workers. Contrary to the previous findings of Schoenberg and Mitchell (*205*), there were no differences in lung function between HR-exposed workers and nonexposed workers. There were, however, significant differences in lung function measurements before and after the regular work shift for HR-exposed workers, but not among the nonexposed workers. The resin investigated in this study was composed of resorcinol as the phenol donor and hexamethylenetetramine (HMT) as both a formaldehyde donor and a catalyst. There was no association between decreases in lung function and ambient levels of resorcinol, formaldehyde, hydrogen cyanide, or ammonia. Mean concentrations of formaldehyde were 0.05, 0.01, and 0.04 ppm for HR-exposed, non-HR-exposed, and control groups, respectively. The decrease in pulmonary function was related to the quantity of respirable particulates obtained from personal samples. Chemical analysis of particulates, however, was not performed.

In a study of 73 workers exposed to phenolic (phenol–formaldehyde) resin dust in the textile industry, Sparks and Peters (*219a*) reported a statistically significant acute drop in $FEV_{1.0}$ and FVC over the shift in garment-line workers exposed to the phenolic resin dust. Workers exposed only to processed cotton dust did not show a significant drop in $FEV_{1.0}$ and FVC over the work shift. The limitations of this study include no measurement of formaldehyde levels, small number of study subjects, a high rate of absence and refusal, mixed-dust exposure, and use of respirators. In spite of these limitations, the study suggests the possible role of formaldehyde in inducing chronic obstructive effects on lung function.

A significant reduction in $FEV_{1.0}$ and other pulmonary functions after a day of work was reported among 47 workers exposed to formaldehyde (*4a*). Exposure to formaldehyde (mean, 0.36 ppm; range, 0.04–1.25 ppm) occurred in the area where sawdust and wood chips were cemented together under high pressure in the process of manufacturing chipboard. Another 20 workers from the same plant (at the carpentry works), but not exposed to formaldehyde or other agents known to irritate the lung, were also examined. In addition to the usual symptoms such as irritation of eyes, nose, and throat, workers exposed to formaldehyde displayed a significant reduction in lung function irrespective of their smoking status: $FEV_{1.0}$ decreased by an average of 0.17 liter, $FEV_{\%}$ decreased by 2.4%, MMF decreased by 0.39 liter/sec, and $CV_{\%}$ increased by 3.4%. These findings are consistent with signs of airway obstruction, which apparently subsides over the weekend of nonexposure, as evidenced by the normal lung function on Monday morning before work.

A study of 199 workers involved in the manufacture and processing of formaldehyde did not show any significant difference in lung function, nor did the workers demonstrate abnormal lung X-ray or blood biochemical parameters as compared to a control group consisting of 91 steel construction workers (*92*). The majority of formaldehyde workers in this study was exposed to less than 0.2 ppm formaldehyde. Under this exposure condition, the study could not demonstrate that the workers suffered from chronic impairment of health.

Eight cases of occupational asthma (three smokers and five nonsmokers) were reported among 28 members of the nursing staff of a hemodialysis unit where formalin was used to sterilize the artificial kidney machine (*108*). Recurrent episodes of productive cough accompanied by wheeze were a prominent feature and, for five persons, attacks had extended over the previous 3 years. Inhalation provocation tests were performed on the five subjects with histories of recurrent attacks of wheezing. In two of these subjects, the test resulted in asthmatic attacks like those experienced at work. Peak expiratory flow rates fell approximately 50% and wheezing began 2 and 3 h after exposure to formalin and lasted for 10 h to 10 days. Three of the five subjects had no respiratory reaction to inhalation of formaldehyde similar to that experienced in the dialysis unit. Two of the five had no symptoms, and one developed conjunctivitis. This latter patient developed redness, weeping, and sensations of grittiness of the eyes when heavily exposed to formalin. In the absence of symptoms and exposure, there was no apparent reduction in lung function. The authors suggested that although formalin may not have been the etiologic agent in all cases, it may have increased the susceptibility to other agents, which could, perhaps, explain the high incidence of bronchitislike symptoms. In the absence of a comparison group, an alternative explanation of the asthma being attributable to chance alone still exists. However, the explanation is unlikely because of the high proportion of the staff that developed the symptoms and because of the positive responses observed after the inhalation provocation test.

Formaldehyde-related asthma and dermatitis were also reported by Kerfoot and Mooney (*129*). A survey of six Detroit area funeral homes conducted by the authors showed that embalmers were generally exposed to formaldehyde at mean levels ranging from 0.25 to 1.39 ppm, with a total range of 0.09–5.26 ppm. They experienced acute toxic effects including eye and nose burning, sneezing, coughing, and headaches. Asthma or sinus problems were reported by three out of seven morticians. In addition, two workers experienced dermatitis, with one case being so severe that the worker discontinued working for a period of time until he recovered. Embalming agents contain formaldehyde as well as a variety of other chemicals such as tissue moisturizers, smooth muscle relaxants, bleaches, an auxiliary antiseptic agent (phenol), dyes, buffers, wetting agents, water conditioners and/or anticoagulants, perfumes and odor suppressors, and vehicles (methanol, ethanol, and glycerin). In light of the possible mixed exposure to a variety of the above chemicals during embalming, and the lack of an appropriate control group in the study, the relationship of formaldehyde exposure among embalmers to development of asthma and dermatitis remains inconclusive.

In a mail survey of 20 funeral homes in Los Angeles, 57 of 80 embalmers responded (*187*). Nine (16%) reported symptoms compatible with acute bronchitis and 17 (30%) were considered to have chronic bronchitis. The 31 asymptomatics, however, had worked longer than the bronchitics (18 vs 11 years). In the absence of a control group and in light of the possible mixed exposure to other chemicals, these findings are, at best, only suggestive of the role of formaldehyde in development of bronchitis.

Engle and Calnan (*65*) reported an outbreak of dermatitis in a car factory. A total of 50 cases of dermatitis was observed in 3 years (1962–1965) among 150 employees who handled rubber weather strips coated with phenol–formaldehyde resins. The workers who developed dermatitis had been exposed to the adhesives containing phenol–formaldehyde resins for from 1 day to 2 years before the onset of the eruption, with an average period of contact of 17 weeks. The average duration of the eruption was 12 weeks; however, in three cases it persisted for up to 2.5 years. The eruption was generally an erythematous vesicular rash of the fingers and hands. Three materials were handled by these employees: (1) the rubber weather strips, used for some years, (2) adhesives A and B introduced 4 years before in 1962 and supplied by the same manufacturers, and (3) toluene, which was the solvent used to activate the adhesive. The rubber weather strips alone were ruled out as a cause because they came from various suppliers and had not changed in composition for a long time: toluene would not be expected to cause sensitization. Among the 29 patch-tested dermatitis patients, 4 (14%) gave a weak reaction to phenol alone, while 65% had a positive reaction to the adhesive resins. It is, therefore, probable that formaldehyde in the resins was a causal agent.

Outbreaks of dermatitis in several industries using formaldehyde resins were reported by Schwartz *et al.* (*207*). In a factory in which plywood was laminated, 600 cases of dermatitis were reported among about 800 workers during the first 6 months of operation. In a second reported outbreak, over 40 workers out of a total of 100 developed dermatitis in a factory in which tool handles were made from laminated glass fabric and phenol–formaldehyde resins. Although no unexposed group was available for comparison, the high proportion of formaldehyde-exposed workers developing dermatitis is quite impressive.

In a hemodialysis unit in which formalin was used as a sterilant, 6 of 13 staff members developed dermatitis within 3 weeks (*218*). Four of the six gave positive patch tests to 3% formalin. It was not clear why only the hemodialysis unit was affected since other units also used formalin. The author speculated that it might be due to the use of a detergent that lowered the resistance of the skin to formaldehyde vapors and to the high temperature and concentration of formalin in the preparation room.

Shumilina (*214*) reported a high incidence of menstrual and reproductive function disorders among 446 women workers (130 finishers and 316 inspectors) exposed to urea–formaldehyde resins. Formaldehyde concentrations of 1.2–3.6 ppm were often found in the finishers' work area of the fabric trim shop, while levels from 0.04 to 0.06 ppm occurred in the inspectors' work area. A group of 200 saleswomen not exposed to formaldehyde was used for comparison. The reproductive disorders reported to be more common among those exposed, primarily in finishers, included menstrual disorders, increased complications during pregnancy, and a higher percentage of neonates with low birth weights. The role of formaldehyde in the development of these disorders is uncertain, however, because of the lack of information on the work environment and the socioeconomic status of the study and control groups. In addition, many of these disorders are known to be associated with physical and mental stress, personal habits (alcohol, cigarette, and caffeine consumption). nutritional status, and other factors related to the socioeconomic status of women. More recently, Olsen and Døssing have reported that female workers in mobile home day care centers experienced significant increases in irritation of mucous membranes, tiredness, headaches, and menstrual irregularities when compared to control workers (*175a*). These workers were exposed to a mean formaldehyde concentration of 0.43 ppm whereas control workers were exposed to 0.06 ppm. The findings reported by Shumilina and by Olsen and Døssing are intriguing and indicate a need for further studies in this area.

Neshkov and Nosko (*173*) reported a high incidence of sexual dysfunction among male workers employed in a plant producing glass fiber-reinforced plastic. These workers were exposed to vapors of phenol, formaldehyde, aniline, epichlorohydrin, styrene, and a combination of glass fiber and glass-reinforced plastic dust. The levels of each of these chemicals were 1.5 times the respective maximum permissible level (0.4 ppm) in 63% of the samples. Among the 143 workers examined, 58 (40.5%) had psychoneurologic and sexual complaints. Sexual complaints included a diminution of libido, premature ejaculation, a weakening erection, and decrease in the satisfaction derived from an orgasm. Analysis of the sexual complaints revealed a direct relationship to the duration of employment at the plant. Testicular dysfunction among the workers was also reported, including decreased volume and increased viscosity of ejaculate and decreased number of spermatozoids. The authors concluded that these sexual dysfunctions were the results of complex toxic effects of chemicals on the cells of the cortex and those of the subcortical brain stem structures, which participate in the regulation of sexual function. It is impossible to determine how much of the sexual dysfunction might be attributable to formaldehyde since appropriate control groups were not included and these workers were exposed to a variety of toxic chemicals in addition to formaldehyde. Of particular interest are reports of epichlorohydrin-induced sterility in animals (*46, 124*).

D. Cohort and Case-Control Studies

Epidemiologic studies available for review in these categories involved two classes of workers: (1) those from occupations or industries in which formaldehyde exposure may have occurred, and (2) those from occupations or industries in which formaldehyde exposure has definitely occurred.

Studies involving the first class of workers were not designed to evaluate the role of formaldehyde in the development of disease or cancer. Although the studies did not specifically specify exposure to formaldehyde, they were included because of the likelihood of their exposure to formaldehyde.

Moss and Lee (*165*) have reported elevated risks of oral and pharyngeal cancers among male textile workers. They reported 77% excess deaths due to these cancers compared with the general male population of Wales and England. These textile workers may have been exposed to formaldehyde, since formaldehyde has been widely used in the textile industry for producing creaseproof, crushproof, and shrinkproof fabrics. No environmental measurements of formaldehyde, however, were made in this study. A recent NIOSH survey showed concentrations of formaldehyde in textile plants in the United States ranging from 0.1 to 1.4 ppm. Bross *et al.* (*30*) reported a significantly increased risk of nasal cancer among textile workers and shoemakers. Textile workers also had a significantly elevated risk of cancer of the pancreas and stomach. No information on formaldehyde concentrations was reported.

Decoufle (*54*) has reported significantly increased risks of cancer of the buccal cavity, pharynx, and larynx among male leather workers. Bladder cancer and malignant lymphomas were also associated with increased risks among male and female employees in the leather industry. The cancer risk for these employees was calculated relative to the risk for those who were in clerical positions. A variety of chemicals including formaldehyde, azo dyes, chromium compounds, and tanning extracts has been used in the production of leather goods. Some of these chemicals are carcinogenic in animals. Industrial hygiene surveys of a calfskin tannery in the United States showed concentrations of formaldehyde in the finishing department ranging from 1 to 8.6 ppm.

In contrast to the above studies of industrial workers, many studies of medical personnel, i.e., physicians, pathologists, and laboratory technicians, did not indicate cancer hazards. Medical personnel, particularly pathologists and certain laboratory technicians, are very likely to be exposed to formaldehyde. Doll and Peto (*59*) studied mortality rates of a total of 20,540 British male doctors by their specialties. The follow-up was from 1952–1971, and the number of observed deaths was compared to the expected deaths calculated from the general male population. No excess deaths from cancers of either lung, mouth, or esophagus, or from other cancers were reported among the 853 physicians classified as laboratory scientists (pathologists and biochemists).

Harrington and Shannon (*98*) studied the mortality pattern of pathologists and medical laboratory technicians in the United Kingdom. Mortality patterns of a total of 2079 pathologists alive and active at some time between 1955 and 1973, and a total of 12,944 medical technicians who registered between 1963 and

1973, were followed up to the end of 1973 and compared with those of the population of England and Wales or Scotland. During the study period, 156 pathologists and 154 medical technicians died. No excess deaths from cancer of the lung, bronchus, trachea, digesive tract and peritoneum, or bladder were reported in either group. However, in male pathologists in England and Wales, a statistically significant increase in lymphatic and hematopoietic neoplasms was observed (8 observed vs 3.3 expected, $p < 0.01$). The mortality data were not analyzed for nasal or pharyngeal cancers.

Jensen and Andersen (*121*) reported a case-control study of lung cancer risk among Danish physicians. Information on the medical specialty of 84 physicians who died from lung cancer was compared with the information for 252 controls. The controls were also physicians and were matched with the cases by age, sex, and survival, at least until the time of lung cancer development. Physicians who specialized in pathology, including forensic medicine and anatomy, did not show an unusual risk of lung cancer death. In fact, none of the 84 lung cancer victims were pathologists. Furthermore, previous employment at some time during their professional careers in pathology, forensic medicine, or anatomy was not associated with the increased lung cancer risk. Jensen (*120*) reported earlier that during the period 1943–1976 three cases of cancer of the nasal cavities, sinuses, and nasopharynx were observed in Danish doctors. However, none of these three physicians had ever worked in a pathology department or as an anatomist.

In all these studies of industrial workers and medical personnel, actual exposure to formaldehyde is not known. The lack of information concerning the number of persons actually exposed to formaldehyde and the level of their exposure, in addition to confounding exposures to other chemicals, makes interpretation of these reports regarding formaldehyde difficult.

The second category of studies includes investigations designed specifically to evaluate health hazards of formaldehyde among workers who were employed in areas where formaldehyde was produced or used. Since 1980, several mortality studies of workers who had been potentially exposed to formaldehyde have been reported. Wong (*248*) studied a cohort of 2026 white males who were ever employed through 1977 at a chemical plant in which formaldehyde and other chemicals were produced. The plant was built in the early 1940s and, in addition to formaldehyde, handled a variety of substances, such as oxygenated hydrocarbons, benzene, asbestos, and inorganic and organic pigments. The cohort was followed through December 31, 1977. A total of 146 deaths was observed. Vital status was not ascertained for 51 workers. Death certificates were obtained for all but 10 of the individuals known to have died. The number of expected deaths was calculated based on mortality rates for white males in the United States. Mortality from cancer at all sites for the entire cohort was not different from that expected (37 vs 36.5). Excess SMR values were observed for cancer of the prostate (4 vs 1.31) and brain (3 vs 1.61), and for Hodgkins disease (2 vs 0.83).

But none of the increased SMR values was statistically significant. Limiting the analysis to deaths that occurred more than 30 years after first employment in the plant, the author found a statistically significant excess of deaths from prostatic cancer (4 vs 0.93). There were no deaths observed from cancers of the nose or nasal sinuses. Analysis of data by the proportionate mortality ratio (PMR) revealed significantly elevated PMR values for all cancer (PMR = 148), cancer of the bone (PMR = 625), and cancer of the prostate (PMR = 367). The role of formaldehyde in the development of cancer is difficult to determine from this study for a number of reasons. First, the study includes all white males who were ever employed at the plant between the early 1940s and December 1977. But no information is given as to what proportion of these workers was actually exposed to formaldehyde and for how long. Second, only a small number of individuals had achieved 20+ years of latency, limiting the chance of identifying a carcinogenic risk from exposure to formaldehyde. Third, the size of the cohort is too small to detect any risk of nasal cancer even if excess nasal cancer was actually induced by formaldehyde: there was only an 8% probability of detecting a threefold increase in risk of death from nasal cancer.

Marsh (*153*) conducted a proportional mortality analysis on 136 deaths occurring between 1950 and 1976 among male workers who had worked for at least 1 month in areas of the plant where exposure to formaldehyde as well as to other chemicals would have occurred. Only 36 decedents had spent the largest portion of their employment in plant areas where significant formaldehyde exposure could have occurred routinely. The number of observed deaths was compared to expected numbers calculated by applying the cause-specific proportional mortality of United States white and nonwhite males to the total number of white and nonwhite deaths in the study group after adjusting for age and time period. In general, the PMRs for all cancer and specific cancer sites were not significantly different than expected, with the exception of cancers of the digestive organs and peritoneum. When analyzed by age at death, the youngest of the three age groups (45, 45–64, 65+) of white male formaldehyde-exposed workers indicated excess mortality from cancer of the digestive tract (PMR = 413, 2 observed deaths). Among those white males exposed for a total of less than 5 years and with more than 20 years from onset of exposure, the PMR for cancer of the digestive tract was 320 (5 observed deaths). No sinonasal cancer deaths were observed in the study. The mortality data were not analyzed separately for pharyngeal cancer. This study also is inadequate to assess the carcinogenicity of formaldehyde in humans because (1) very limited information on formaldehyde exposure is available, (2) a very small number of decedents ($n = 36$) could have had actual exposure to formaldehyde for a significant portion of their employment, and (3) the small number of deaths gives the study limited power to detect increases in mortality from nasal cancer: there was only a 7% probability of detecting a threefold increase in risk of death from nasal cancer.

A study of the same plant population as above, however, followed further through 1980 by another group of investigators showed different results. In a recent meeting, Liebling *et al.* (*145a*) reported a PMR analysis of those who died between January 1, 1976 and December 31, 1980. Among 211 decedents identified during this 5-year period, 24 workers appeared on seniority lists from work areas with known formaldehyde exposure. An analysis based on the 24 deaths indicated statistically elevated PMRs for buccal and pharyngeal cancer (2 observed, PMR = 870). The worker who died of pharyngeal cancer had a squamous cell cancer, which is the same histologic type as developed in experimental animals exposed to formaldehyde.

A team of NIOSH investigators has reported a case of squamous cell carcinoma of the nasal cavity in a 57-year-old worker who had 25 years of occupational exposure to low concentrations of formaldehyde in the textile finishing industry (*95a*). A recent survey at a fabric finishing plant in the United States indicated that formaldehyde concentrations in air ranged from 0.2 to 1.2 ppm; exposures may have been higher in the past when more highly concentrated solutions were used (*64a*). Workers involved in a textile finishing operation also are likely to be exposed to sodium hypochlorite, as well as to dyes, dye carriers, and antifoaming agents. Cancers of the nasal cavity and sinuses are very rare tumors. The annual incidence among United States males is approximately 8 per million for a lifetime risk among United States males of about 6 per 10,000—about the same as the lifetime risk for breast cancer in males. As much as the reports of nasal and pharyngeal cancer cases in workers exposed to formaldehyde may help to generate an etiologic hypothesis, they would hardly be expected to provide evidence of causation or lack of causation.

Walrath and Fraumeni (*241*) studied the proportion of cancer deaths in a group of 1106 deceased embalmers who had been licensed to practice embalming in New York State between 1902 and 1979. Formaldehyde has been the main preservative in commercial embalming fluids and its use in embalming fluids is required by various state laws in the United States. The number of observed deaths due to specific causes was compared to the expected numbers calculated by the same method described in the Marsh study. The PMR value for all cancers was slightly elevated. However, the study shows an unusually high PMR for death from skin cancer among white male embalmers. The excess was greater among those licensed for more than 35 years than among those licensed for less than 35 years (PMRs of 196 vs 354), among those licensed after age 30 than those licensed before age 30 (PMRs of 151 vs 424), and among those licensed for embalming only than those licensed for both embalming and funeral directing (PMRs of 337 vs 178). The proportionate mortality for kidney and brain cancers also was elevated among white males licensed only for embalming. No excess mortality from cancers of the respiratory tract was reported. Despite the limitations of the PMR study, a high magnitude of excess PMRs as well as an upward trend by years since first licensed and the type of license suggest a role of formaldehyde in the development of skin cancer among embalmers.

A case-control study of du Pont chemical plant employees who had died from cancer was recently reported (*69*). A total of 481 cancer deaths from 1957 to 1979 among male employees at eight formaldehyde-manufacturing and -using plants was compared to an equal number of du Pont employees who were known to have not died from cancer (controls). Each cancer case was matched to a control for age (± 3 years), plant location, sex, pay class (wage or salary), and adjusted service date (± 3 years). Information on work histories for both cases and controls was obtained from personnel records, medical records, or co-worker interviews. The potential for exposure to formaldehyde for each job was estimated on the basis of several factors, including job descriptions, air-monitoring data, and statements by employees concerning odor or sensory irritation. The data were analyzed by tumor site, latent period, duration of exposure, exposure level and frequency, cumulative exposure index, age and year of death, and age and year of first exposure. Analyses of lung cancer deaths were adjusted for subjects' cigarette smoking habits. The study did not show a significantly elevated relative risk of cancer in any of the above analyses. No nasal cancer deaths were observed. The authors concluded that cancer mortality rates for formaldehyde-exposed du Pont workers were no higher than the rates among nonexposed co-workers. They further state that airborne formaldehyde levels of 1 ppm TWA and a 2 ppm ceiling provide adequate worker protection. This study should be considered preliminary, however, until certain limitations are accounted for. First, a major limitation of the study is the possibility of observation bias: all controls were identified from lists of active employees, whereas all cases were dead when identified. Second, the cases and controls may have been overmatched on factors related to exposure of interest, limiting the study's ability to detect differences in the degree of exposure between cases and controls. Third, for most of the site-specific cancers studied, the number of workers who died from these cancers was too small to reliably detect moderately elevated relative risks of cancer resulting from formaldehyde exposure.

In a recent meeting, Levine (*143*) reported a mortality study of 1477 male undertakers who were first licensed in Ontario, Canada between 1928 and 1957. The cohort was followed through December 31, 1977. A total of 337 undertakers was known to have died. Vital status of 209 undertakers was not ascertained. All undertakers were potentially subjected to formaldehyde exposure during embalming. The number of expected deaths was calculated based on mortality rates for United States white males, because death rates for Canadian males were not yet available in suitable computer format. Deaths from all causes (337 observed vs 370.1 expected) and all cancers (60 observed vs 67.7 expected) were not significantly higher than expected. No deaths due to nasal cancer or skin cancer were observed. Mortality from cancer of the prostate (2 observed vs 3.1 expected), kidney (1 observed vs 1.8 expected), and lung (19 observed and 21.3 expected) was less than expected. Cirrhosis of the liver was the only cause of death reported to be significantly in excess (18 observed vs 10.5 expected). The author speculated that the excess for cirrhosis of the liver might be attributable to increased alcoholism among the undertakers. However, other deaths related to alcoholism, such as deaths from suicide or accident, were less than expected. Whether the cirrhosis may have been due to formaldehyde or its metabolite in the liver cannot be determined at this time. No analyses of the mortality data by magnitude of formaldehyde exposure were presented. It should be noted that the undertakers also included funeral directors who had no or minimum contact with bodies or the chemicals used to treat them. The embalmers' exposure to formaldehyde was reported to be intermittent (a few hours/week) and low (less than 1 ppm) when it occurred. The embalmers in small towns may experience no exposure at all during some years. Until the mortality data are fully analyzed and published, the findings should be considered preliminary.

XII. SUMMARY AND CONCLUSIONS

As this and other reviews indicate, formaldehyde is a ubiquitous chemical. Exposure can occur in homes, workplaces, and the general environment. Small amounts of formaldehyde can cause adverse health effects and, since many of those exposed to it may not perceive or recognize its odor, the sense of smell cannot be relied upon to give warning against its possible adverse health effects.

At present, the ambient air concentrations of formaldehyde appear to be best characterized in mobile homes, in homes containing urea–formaldehyde foam insulation, and in some workplaces. Data on other sources of exposure are more limited. Formaldehyde that is present in homes probably poses the greatest potential health hazard, since people of all ages and states of health spend most of each day in their residences. Concentrations of formaldehyde in indoor air almost always exceed those in outdoor air. Products most likely to emit formaldehyde into the indoor air appear to be urea–formaldehyde resin-bonded wood products, such as particle board and plywood, and, until recently, urea–formaldehyde foam insulation. The relative contributions of the various unvented combustion sources in the home require additional characterization. More data are needed to define all of the factors involved in formaldehyde release from resins.

Formaldehyde is a very reactive chemical and readily combines with DNA, RNA, protein, amino acids, and a variety of other organic chemicals in the body. Conjugation of formaldehyde with low-molecular-weight biochemicals can facilitate intracellular reactions with DNA. Some animal carcinogens may act in a similar manner following metabolism to formaldehyde. Because endogenously formed formaldehyde is apparently metabolized very rapidly, it would not be expected to react with cellular macromolecules in the same way or to the same degree as exogenous formaldehyde. Nor would it be expected to pose the same health hazards.

The metabolic reactions of formaldehyde in humans and animals are similar. It has been found that body tissues can rapidly metabolize exogenous formaldehyde even at high concentrations, although some of this formaldehyde can still react with tissues and cellular macromolecules before being metabolized.

Formaldehyde is an irritant via all routes of exposure and can cause local damage to the eyes, respiratory tract, and skin when inhaled. The effects of formaldehyde may not be limited only to these tissues, since it forms conjugates with biological chemicals that may affect tissues that are remote from the respiratory tract. Thus, in some studies, there is evidence showing that formaldehyde causes adverse effects in tissues and organ systems that are not directly exposed, including the central nervous and hematopoietic systems, as well as the kidneys, adrenals, and liver. Effects in liver appear to be generally dose related, although, at present, the mechanisms by which these effects occur are unknown.

In the case of liver exposure, formaldehyde also may enhance the toxicity of chemicals and drugs that require glutathione for hepatic detoxification. Drugs requiring glutathione include aspirin, acetaminophen, and corticosteroids, all of which could be used to treat symptomatic effects resulting from exposure to formaldehyde, such as nausea and headaches and dermal irritation. Additional work is needed in this area.

Some individuals have become sensitized to formaldehyde following dermal contact and, perhaps, following inhalation exposure. Dermal contact with formaldehyde elicits both immediate and delayed reactions that are immunological in nature. Repeated exposure of susceptible individuals to formaldehyde via inhalation causes asthmatic reactions, which may be either immediate or delayed. The delayed reactions appear to be immunological in nature. Since asthmatic reactions involve the lower respiratory tract, elicitation of these reactions may depend upon the presence of airborne particulates to carry formaldehyde into the low respiratory tract.

The currently available data do not show that the embryo is unusually sensitive to formaldehyde nor is there any information to show that formaldehyde is teratogenic in rodents when administered orally or applied dermally in nontoxic amounts to the dams. Also, the *in vitro* data do not provide any evidence to support the conclusion that formaldehyde causes terata at exposure concentrations that are not toxic to the adult.

Inhalation of formaldehyde has caused fetotoxic effects but not teratogenic effects. Further studies of formaldehyde exposure by inhalation are needed to elucidate the meaning of these changes. Limited evidence suggests that formaldehyde may affect the menstrual cycle and perhaps reproduction in women repeatedly exposed. Additional work is needed to validate these findings.

Formaldehyde causes mutation and chromosomal aberrations in a wide variety of bacteria, yeasts, fungi, and insects in several test systems, as well as in some human cell test systems with and without metabolic activation. These effects are enhanced in excision repair-deficient test systems. In some testing systems, formaldehyde interacts with other mutagens resulting in a greater effect than either would cause alone. Formaldehyde has been shown to cause cell transformation in BALB/c 3T3 cells and to act as both an initiator and a promoter in C3H/10T1/2 cells. Chromosomal breaks and mutations have not been found in whole animal studies. A likely explanation for this latter finding may be that an insufficient dose reaches the target tissues. At present, there are very limited data indicating possible adverse effects on chromosomes in humans from formaldehyde exposure. Additional studies are needed to characterize whether there are any such effects directly attributable to formaldehyde.

Most of the carcinogenic studies on formaldehyde in animals done in the past have had certain limitations that made unclear the meaningful interpretation of the data. However, the results of recent long-term experiments in rats and mice conducted by Battelle for CIIT and studies by NYU both provide acceptable evidence that exposure to formaldehyde by inhalation causes squamous cell carcinomas in the nasal tissues of rats and mice. In addition, dose-related metaplastic and dysplastic changes were noted in the experimental animals, as well as adenomas of the nose at exposures as low as 2.0 ppm (the lowest dose tested). Formaldehyde may also cause cancer at other sites of exposure.

Several studies are now under way that may help to define the body defense mechanisms against the effects of formaldehyde, such as decreases in rate of respiration, tidal volume, or nasal mucous flow. The current data are not sufficient for a clear understanding of how effective such mechanisms actually are.

Under controlled exposure conditions, formaldehyde irritates the eyes, nose, and throat in healthy humans at concentrations as low as 0.2 ppm. The proportion of exposed individuals who experience any irritation and the degree of the irritation experienced increase as the concentration and duration of exposure increase. Effects would be expected to be more severe and to occur at lower concentrations in subpopulations that include the elderly or the infirm under conditions of chronic exposure.

Exposure to formaldehyde in homes and in workplaces may result in signs and symptoms attributed to the exposure, such as headache, dermatitis, chronic airway obstruction, and menstrual, reproductive, and sexual dysfunction, in addition to irritation of the eyes, skin, and mucous membranes of the nose and throat. The results of investigations that show these health effects are usually those given in case reports or cross-sectional studies. Many of these studies lacked appropriate controls and/or environmental exposure measurements, making it difficult to clearly establish formaldehyde as the causal agent. However, there has been some validation of the many independent reports of respiratory system disorders and of dermatitis in persons who are exposed to formaldehyde in a variety of environmental situations. Certainly not all of the reports have been or can be validated; however, the findings from many of them strongly indicate that formaldehyde is the major contributor to the observed adverse health effects. The reports of reproductive disorders among women who are exposed to formaldehyde are provocative. Although two independent epidemiological studies have shown that formaldehyde affects human menstrual cycles, there is still a question of uncertainty about these data that precludes making a firm conclusion. However, the implications of these studies are serious and far-reaching, owing to the ubiquitousness of the potential for exposure. The need for further research on the reproductive effects of formaldehyde in humans is both urgent and crucial.

Several epidemiological studies have been designed to evaluate the relationship between formaldehyde exposure and cancer in humans. These studies revealed significant increases in cancer of the prostate, digestive system, and upper respiratory tract associated with exposure to formaldehyde. All of these studies have limitations in design, methodology, sample size, information on exposure, patient follow-up, selection bias, or overmatching and exposure–effect associations. They do not provide definitive evidence upon which to evaluate the carcinogenicity of formaldehyde in humans. These limitations should be corrected in the design of future epidemiological studies of formaldehyde. In the interim, formaldehyde should be regarded as posing a carcinogenic risk to humans.

Concern over the known and potential health effects of formaldehyde has led to certain regulatory actions against products containing formaldehyde (UFFI and mobile homes) by various states, a city, and a federal agency (the Consumer Product Safety Commission). Standards to limit residential exposure to formaldehyde have been issued by various European countries. More recently, ASHRAE has recommended that formaldehyde levels in buildings not exceed 0.01 ppm.

References

1. Aebi, H., and Roggen, G. (1959). *Pharm. Acta Helv.* **33,** 413–428.
2. Ahmed, A. E., and Anders, M. W. (1978). *Biochem. Pharmacol.* **27,** 2021–2025.
3. Alanko, K., Keskinen, H., and Saarinen, L. (1977). *Duodecim* **93,** 306–318.
4. Albert, R. E., Sellakumer, A. R., Laskin, S., Kuschner, M., and Nelson, N. (1982). *J. Natl. Cancer Inst.* **68,** 597–603.

4a. Alexandersson, R., Hedenstierna, G., and Kalmodin-Hedman, B. (1982). *Arch. Environ. Health.* **37,** 279–284.

5. Amdur, M. O. (1959). *Int. J. Air Pollut.* **1,** 170–183.
6. Amdur, M. O. (1960). *Int. J. Air Pollut.* **3,** 201–220.
7. Anderson, H. A., Hanrahan, L. P., Dally, K. A., and Rankin, J. (1981). "Irritant Symptomology, Clinical Observations and Formaldehyde Exposure Among Wisconsin Mobile Home

Residents.'' Paper presented at the International Symposium on Indoor Air Pollution, Health and Energy Conservation, Amherst, Massachusetts, Oct. 13–16, 1981.

8. Andersen, I. (1979). *In* ''Indoor Climate: Effects on Human Comfort, Performance and Health in Residential, Commercial, and Light-Industry Buildings. Proc. of the First International Indoor Climate Symposium, Copenhagen, Aug. 30–Sept. 1, 1978'' (P. O. Fanger and O. Valbjorn, eds.), pp. 65–87. Danish Building Res. Inst., Copenhagen.
9. Anderson, I., Lundquist, G. R., and Molhave, L. (1979). *Ugeskr. Laeg.* **141,** 966–971.
10. Angelini, G., and Meneghini, C. L. (1977). *Contact Dermatitis* **3,** 219–220.
11. Auerbach, C., Moutschen-Dahmen, M., and Moutschen, J. (1977). *Mutat. Res.* **39,** 317–362.
12. Barrow, C. S., Steinhagen, W. H., and Chang, J. C. F. (1983). *In* ''Formaldehyde Toxicity'' (J. E. Gibson, ed.), pp. 16–24. Hemisphere Publ., New York.
13. Battelle Columbus Laboratory (1981). ''A Chronic Inhalation Toxicology Study in Rats and Mice Exposed to Formaldehyde: Final Report Prepared for the Chemical Industry Institute of Toxicology,'' Vols. 1–4, CIIT Docket No. 10922. Chem. Ind. Instit. Toxicol., Raleigh, North Carolina.

13a. Battelle Northwest Laboratory (1981). ''Subchronic Study Report on Formaldehyde,'' pp. 1–41. Tracor Jitco, Inc., Rockville, Maryland.

14. Baur, X., and Fruhmann, G. (1979). *Prax. Klin. Pneumol.* **33** (Suppl. 1), 317–322.
15. Benkovic, S. A. (1980). *Annu. Rev. Biochem.* **49,** 227–251.
16. Bernstein, I. L. (1982). *J. Allergy Clin. Immunol.* **70,** 24–31.
17. Berrens, L., Young, E., and Jansen, L. (1964). *Br. J. Dermatol.* **76,** 110–115.
18. Bierre, T. H., Tunnicliff, M. J., and Darby, F. W. (1981). ''A Review of Occupational Health Problems Associated with Formaldehyde Exposure: Report of the Department of Health.'' New Zealand Department of Health, Southern Regional Occupational Health Unit, Wellington.
19. Bitron, M., and Aharon, E. (1978). *Am. Ind. Hyg. Assoc. J.* **39,** 129–138.
20. Black, H. (1971). *Contact Dermatitis Newsletter* **10,** 162, 242.
21. Blakley, R. L. (1960). *Biochem. J.* **74,** 71–82.
22. Blejer, H. P., and Miller, B. H. (1966). ''Occupational Health Report of Formaldehyde Concentrations and Effects on Workers at the Bayly Manufacturing Company, Visalia, California,'' Study Rep. No. S–1806. State of California Health and Welfare Agency, Dept. of Public Health, Bureau of Occupational Health, Los Angeles.
23. Bokina, A. I., Eksler, N. D., Semenenko, A. D., and Merkur'yeva, R. Y. (1976). *Environ. Health Perspect.* **13,** 37–42.
24. Bonashevskaya, T. I. (1973). *Arkh, Anat. Gistol. Embriol.* **65,** 56–59.
25. Boorman, G., Gupta, K. C., Hahn, F., Kleinerman, J., Reznik, G., and Stinson, S. (1980). *In* ''Report of the Pathology Group Review of the CIIT Studies on Formaldehyde Exposures in Rodents.'' Prepared for the U.S. Interagency Regulatory Liaison Group, Task Group on Formaldehyde, Washington, D.C.
26. Boreiko, C. (1982). *CIIT Activities* **2,** 3–5.
27. Bourne, H. G., Jr., and Seferian, S. (1959). *Ind. Med. Surg.* **28,** 232–233.
28. Breysse, P. A. (1961). *Occupational Health Newsletter,* Univ. of Washington, March issue.
29. Breysse, P. A. (1977). *Environ. Health Safety News* **26,** 1–6.
30. Bross, I. D. J., Viadama, E., and Houton, L. (1978). *Arch. Environ. Health* **33,** 300–307.
31. Brusick, D. J. (1982). ''*In Vivo* Mutagenicity Studies on Formaldehyde Vapors: Final Report Submitted to Formaldehyde Institute.'' Litton Bionetics, Kensington, Maryland.
32. Brusick, D. J., Jagannath, D., Myhr, A., and Stetka, D. (1980). *Environ. Mutagen.* **2,** 253.
33. Burdach, S., and Wechselberg, K. (1980). *Fortschr. Med.* **98,** 379–384.
34. Burry, J. N., Kirk, J., Reid, J. G., and Turner, T. (1980). *Med. J. Aust.* **1,** 183–184.
35. Carpenter, C. P., and Smyth, H. F., Jr. (1946). *Am. J. Ophthalmol.* **29,** 1363–1372.
36. Chaw, Y. M., Crane, E., Lange, P., and Shapiro, R. (1980). *Biochemistry* **19,** 5525–5531.

37. Clive, D., and Spector, J. F. S. (1975). *Mutat. Res.* **31,** 17–29.
38. Cohen, A. (1972). *C. R. Hebd. Seances Acad. Sci.* **275,** 921–924.
39. Colburn, C. W. (1970). E. I. du Pont de Nemours and Co., Rep. No. 1013. Haskell Laboratory for Toxicology and Industrial Medicine, Wilmington, Delaware.
40. Commonwealth of Massachusetts, Department of Public Health. (1979). "Regulations Concerning Hazardous Substances from Statement of Governor Edward J. King at News Conference Concerning Urea–Formaldehyde Foam, November 1, 1979." Department of Public Health, Boston.
41. Connecticut Department of Consumer Protection (1979). "Report on Urea–Formaldehyde Foam Insulation, " April 1979. Department of Consumer Protection, Hartford.
42. Conner, E. A., Blake, D. A., Parmley, T. H., Burnett, L. S., and King, T. M. (1976). *Contraception* **13,** 571–582.
43. Connor, T. H., Barrie, M. D., Theiss, J. C., Matney, T. S., and Ward, J. B. (1984). *Mutat. Res. Lett.* (in press).
44. Consumer Product Safety Commission (1980). "Urea–Formaldehyde Foam Insulation: Regulation Options Briefing Paper." CPSC, Washington, D.C.
45. Consumer Product Safety Commission (1982). "Ban of Urea–Formaldehyde Foam Insulation." *Fed. Regist.* **47,** 11366–14421.
45a. Coon, R. A., Jones, R. A., Jenkin, C. R., Jr., and Siegel, J. (1970). *Toxicol. Appl. Pharmacol.* **16,** 646–655.
46. Cooper, E. R. A., Jones, A. R., and Jackson, H. (1974). *J. Reprod. Fertil.* **38,** 379.
47. Cronin, E. (1980). "Contact Dermatitis." Churchill, London.
48. Cronin, E. (1963). *Br. J. Dermatol.* **75,** 267–273.
49. Cronin, E. (1982). *Semin. Dermatol.* **1,** 33–42.
50. Dahl, A. R., Hadley, W. M., Hahn, F. F., Benson, J. M., and McClellan, R. O. (1982). *Science* **216,** 57–59.
51. Dalbey, W. (1982). *Toxicology* **24,** 9–14.
52. Davis, D., Potter, W., Jollow, D., and Mitchell, J. (1974). *Life Sci.* **14,** 2099–2109.
53. Davis, M., Ideo, G., Harrison, N. G., and Williams, R. (1975). *Chin. Soc. Mol. Med.* **49,** 495–502.
54. Decoufle, P. (1979). *Arch. Environ. Health* **34,** 33–37.
55. Della Porta, G., Cabral, J. R., and Parmiani, G. (1970). *Tumori* **56,** 325–334.
56. Della Porta, G., Colnaghi, M. I., and Parmiani, G. (1968). *Food Cosmet. Toxicol.* **6,** 707–715.
57. Den Engelse, L., Gebbink, L. M., and Emmelot, P. (1975). *Chem.-Biol. Interact.* **11,** 535–544.
58. Dodd, D. E., Bus, J. S., and Barrow, C. S. (1982). *Toxicol. Appl. Pharmacol.* **62,** 228–236.
59. Doll, R., and Peto, R. (1977). *Br. Med. J.* **1,** 1433–1436.
60. DuVigneaud, V., Vergy, W. G., and Wilson, J. E. (1950). *J. Am. Chem. Soc.* **72,** 2819–2820.
61. Edwards, K., Jackson, H., and Jones, A. R. (1970). *Biochem. Pharmacol.* **19,** 1791–1795.
62. Eels, J. T., McMartin, K. E., Black, K., Virayotha, V., Tisabell, R. H., and Tephly, T. R. (1981). *J. Am. Med. Assoc.* **246,** 1237–1238.
63. Egle, J. L. (1972). *Arch. Environ. Health* **25,** 119–124.
64. Einbrodt, H. J., Prajsnar, D., and Erpenbeck, J. (1976). *Zentralbl. Arbeitsmed. Arbeitsschutz Prophyl.* **26,** 154–158.
64a. Elliott, L., Stayner, L., and Blade, L. (1982). "The NIOSH Industrywide Study of Formaldehyde." Presented at the American Industrial Hygiene Conference, Cincinnati, Ohio, June 6–11, 1982.
65. Engel, H. O., and Calnan, C. D. (1966). *Br. J. Ind. Med.* **23,** 62–66.

66. Epstein, S. S., Arnold, E., Andrea, J., Bass, W., and Bishop, Y. (1972). *Toxicol. Appl. Pharmacol.* **23,** 288–325.
67. Fajen, J. M. (1981). Unpublished report. Prepared for Monograph No. 25. Int. Agency Res. Cancer.
68. Fassbinder, W., Frei, U., and Koch, K. M. (1979). *Klin. Wochenschr.* **57,** 673–679.
69. Fayerweather, W. E., Pell, S., and Bender, J. R. (1982). Case control study of cancer deaths in du Pont workers with potential exposure to Formaldehyde. *In* ''Formaldehyde: Toxicology—Epidemiology—Mechanisms'' (J. J. Clary, J. E. Gibson, and R. S. Waritz, eds.). Dekker, New York.
70. Federal Panel. (1982). Report of the Federal Panel on Formaldehyde. *Environ. Health Perspec.* **43,** 139–168.
71. Fel'dman, Y. G., and Bonashevskaya, T. I. (1971). *Hyg. Sanit.* **36,** 174–180.
72. Feldman, M. Y. (1973). *Prog. Nucleic Acid Res. Molec. Biol.* **13,** 1–49.
73. Fischer, M. H. (1905). *J. Exper. Med.* **6,** 487–518.
74. Fisher, A. A. (1973). ''Contact Dermatitis,'' 2nd Ed., pp. 273–277. Lea & Febiger, Philadelphia.
75. Fisher, A. A. (1976). *Cutis* **17,** 665–667.
76. Fontignie-Houbrechts, N. (1981). *Mutat. Res.* **88,** 109–114.
77. Formaldehyde Institute (1982). ''A 26-Week Inhalation Toxicity Study of Formaldehyde in Monkeys, Rats, and Hamsters,'' Proj. No. 79–7259. Formaldehyde Instit., Scarsborough, New York.
78. Foussereau, J., Cavalier, C., and Selig, D. (1976). *Contact Dermatitis* **2,** 254–258.
79. Fregert, S. (1981). *Contact Dermatitis* **7,** 56.
80. Fregert, S., and Gruvberger, B. (1980). *Contact Dermatitis* **6,** 366.
81. Fregert, S., Hjorth, N., Magnusson, B., Bandmann, H. J., Calnan, C. D., Cronin, E., Malten, K., Meneghini, C. L., Pirila, V., and Wilkinson, D. S. (1969). *Trans. St. John's Hosp. Dermatol. Soc.* **55,** 17.
82. Frigas, E., Filley, W. V., and Reed, C. E. (1981). *Chest* **79,** 706–707.
83. Frisch, K. S., and Saunders, J. H. (1973). ''Plastic Foams,'' Vol. 2, pp. 675–700. Dekker, New York.
84. Gamble, J. F., McMichael, A. J., and Battigelli., M. (1976). *Am. Ind. Hyg. Assoc. J.* **37,** 499–513.
85. Garry, V. F., Oatman, L., Pleus, R., and Gray, D. (1980). *Minn. Med.* **63,** 107–111.
86. Gibson, J. E. (1981). Presentation to Consumer Products Safety Commission, Washington, D.C., Dec. 15, 1981.
87. Glass, W. I. (1961). *N. Z. Med. J.* **60,** 423–427.
88. Gofmekler, V. A. (1968). *Hyg. Sanit.* **33,** 327–332.
89. Gofmekler, V. A., and Bonashevskaya, T. I. (1969). *Hyg. Sanit.* **34,** 266–268.
90. Gofmekler, V. A., Pushkina, N. N., and Klevtsova, G. N. (1968). *Hyg. Sanit.* **33,** 112–116.
91. Goldmacher, U. S., and Thilly, W. G. (1983). *Mutat. Res.* **116,** 417–422.
92. Goldmann, P., Hey, W., Strassburger, U., Fleig, I., and Theiss, A. M. (1982). Formaldehyde morbidity study. *Medichem. Newsletter,* The Hague, Jan.–Feb., 1982.
93. Goodman, L. S., and Gilman, A., eds. (1975). ''The Pharmacological Basis of Therapeutics,'' 5th Ed. Macmillan, New York.
94. Gralla, E. J., Heck, H. d'A., Hrubesh, L. W., and Meadows, G. W. (1980). ''A Report of the Review of Formaldehyde Exposure Made by the CIIT *ad hoc* Analytical Chemistry Team Held at the Battelle Memorial Laboratories,'' CIIT Docket No. 62620. CIIT, Raleigh, North Carolina.

94a. Gupta, K. C., Ulsamer, A. G., and Pruss, P. W. (1982). *Environ. Int.* **8,** 349–358.

95. Guseva, V. A. (1972). *Hyg. Sanit.* **10,** 102–103.
95a. Halperin, W. E., Goodman, M., Stayner, L., Elliott, L. J., Keenlyside, R. A., and Landrigan, P. J. (1983). *J. Am. Med. Assoc.* **249,** 510–512.
96. Hamann, K. (1980). *Contact Dermatitis* **6,** 446.
97. Hemminki, K. (1981). *Toxicol. Lett.* **9,** 161–164.
98. Harrington, J. M., and Shannon, H. S., (1975). *Br. Med. J.* **II,** 329–332.
99. Harris, D. K. (1953). *Br. J. Ind. Med.* **10,** 255–268.
100. Harris, R. (1982). Conversion of formaldehyde to formate by rabbit skin *in vitro.* Personal communication.
101. Heck, H. d'A, Chin, T. Y., and Schmitz, M. C. (1983). *In* "Formaldehyde Toxicity" (J. E. Gibson, ed.), pp. 26–36. Hemisphere Publ., New York.
102. Heck, H. d'A. (1982). *CIIT Activities* **2,** 3–7.
103. Heck, H. d'A. (1982). Reaction of formaldehyde in rat nasal mucosa. *In* "Formaldehyde: Toxicology—Epidemiology—Mechanisms" (J. J. Clary, J. E. Gibson, and R. S. Waritz, eds.). Dekker, New York.
104. Heck, H. d'A., White, E. L., and Casanova-Schmitym, M. (1982). *Biomed. Mass Spectrom.* **9,** 347–353.
105. Hegyi, E. (1979). *Allerg. Immunol.* **25,** 104–115.
106. Helander, L. (1977). *Arch. Dermatol.* **113,** 1433.
107. Hemminki, K. (1982). *Toxicol. Lett.* **11,** 1–6.
108. Hendrick, D. J., and Lane, D. J. (1977). *Br. J. Ind. Med.* **34,** 11–18.
109. Hendrick, D. J., Rando, R. J., Lane, D. J., and Morris, M. J. (1982). *J. Occup. Med.* **24,** 893–897.
110. Hill, O. L., Shih, T., Johnston, T. P., and Struck, R. F. (1978). *Cancer Res.* **38,** 2438–2442.
111. Hjorth, N., and Fregert, S. (1967). *Contact Dermatitis Newsletter* **2,** 18.
112. Horsfall, F. L. (1934). *J. Immunol.* **27,** 569–581.
113. Horton, A. W., Tye, R., and Stemmer, K. L. (1963). *J. Natl. Cancer Inst.* **30,** 31–43.
114. Hovding, G. (1961). *Acta Derm.-Vernereol.* **41,** 194–200.
115. Huennekens, F. M., and Osborn, M. J. (1959). *Adv. Enzymol.* **2,** 369–446.
116. Hurni, H., and Ohder, H. (1973). *Food Cosmet. Toxicol.* **11,** 459–462.
117. IARC (1982). *IARC Monogr. Eval. Carcinog. Risk Chem. Humans* **29,** 346–389.
118. Jeffcoat, A. R., Chasalow, F., Feldman, D. B., and Marr, H. (1983). *In* "Formaldehyde Toxicity" (J. E. Gibson, ed.), pp. 38–49. Hemisphere Publ., New York.
119. Jensen, D. E., Lotlikar, P. D., and Magee, P. N. (1981). *Carcinogenesis* **2,** 349–354.
120. Jensen, O. M. (1980). *Lancet* II(8192), 480–481.
121. Jensen, O. M., and Andersen, S. K. (1982). *Lancet* II(8277), 913.
122. Johnson, M. (1982). Personal communication.
123. Johnson, E. M., and Gabel, B. E. G. (1982). *J. Am. Coll. Toxicol.* **1,** 57–71.
124. Jones, A. R., Davis, P., Edwards, K., and Jackson, H. (1969). *Nature (London)* **244,** 83.
125. Jordan, W. R., Jr., Sherman, W. T., and King, S. E. (1979). *J. Am. Acad. Dermatol.* **1,** 44–48.
126. Kallen, R. G., and Jencks, W. P. (1966). *J. Biol. Chem.* **241,** 5851–5863.
127. Kane, L. E., and Alarie, Y. (1977). *Am. Ind. Hyg. Assoc. J.* **38,** 509–522.
128. Kaplowitz, N., Kuhlenkamp, J., Goldstein, L., and Reeve, J. (1980). *J. Pharmacol. Exp. Ther.* **212,** 240–245.
129. Kerfoot, E. J., and Mooney, T. F. (1975). *Am. Ind. Hyg. Assoc. J.* **36,** 533–537.
130. Kitchens, J. F., Casner, R. E., Edwards, G. S., Harvard, W. E., III, and Macri, B. J. (1976). "Investigation of Selected Potential Environmental Contaminants. Formaldehyde," EPA Rep. 560/2–76/009. U.S. Environ. Prot. Agency, Washington, D.C. NTIS PB–256 839.
131. Klein, S. M., Cohen, G., and Cederbaum, A. I. (1981). *Biochemistry* **20,** 6006–6012.

132. Kleinhans, D., and Dayss, U. (1980). *Dermatosen Beruf Umwelt* **28,** 101–103.
133. Koivulo, T., and Koivusala, M. (1975). *Biochim. Biophys. Acta* **397,** 9–23.
134. Kratochvil, I. (1971). *Prac Lek.* **23,** 374–375.
135. Krivanek, N. D., McAlack, J. W., and Chromey, N. C. (1982). Skin initiation/promotion study with formaldehyde in CD-1 Mice. *In* ''Formaldehyde: Toxicology—Epidemiology—Mechanisms'' (J. J. Clary, J. E. Gibson, and R. S. Waritz, eds.). Dekker, New York.
136. Ku, R. H., and Billings, R. (1982). *Pharmacologist* **24,** 97.
137. Kucharczyk, N., Yang, J., and Sofia, R. D. (1979). *Biochem. Pharmacol.* **28,** 2219–2222.
138. Kuschner, M., Laskin, S., Drew, R. T., Capiello, V., and Nelson, N. (1975). *Arch. Environ. Health* **30,** 73–77.
139. LaBelle, C. W., Long, J. E., and Christofano, E. E. (1955). *AMA Arch. Ind. Health* **11,** 297.
140. Laduron, P., and Leyson, J. E. (1975). *Biochem. Pharmacol.* **24,** 929–933.
141. Langecker, H. (1954). *Naunyn-Schmiedebergs Arch. Exp. Pathol. Pharmakol.* **22,** 166–170.
142. Laskin, S., Kuschner, M., Drew, R. P., Capiello, V. P., and Nelson, N. (1971). *Arch. Environ. Health* **23,** 135–136.
143. Levine, R. J. (1982). Mortality of Ontario undertakers: A first report. *In* ''Formaldehyde: Toxicology—Epidemiology—Mechanisms'' (J. J. Clary, J. E. Gibson, and R. S. Waritz eds.). Dekker, New York.
144. Lewis, R. J., Sr., and Tatken, R. L. (1980). ''Registry of Toxic Effects of Chemical Substances,'' 1979 Ed., p. 695. Natl. Inst. Occup. Safety and Health, Washington, D.C.
145. Leysen, J., and Laduron, P. (1974). *FEBS Lett.* **47,** 299–303.
145a. Liebling, T., Rosenman, K., Pastides, H., and Lemeshows, S. (1982). *Am. J. Epidemiol.* **116,** 570.
146. Lindskov, R. (1982). *Contact Dermatitis* **8,** 333–334.
146a. Logan, W. S., and Perry, A. O. (1972). *Arch. Dermatol.* **106,** 717–721.
146b. Logan, W. S., and Perry, A. O. (1973). *Clin. Orthop.* **90,** 150–152.
147. Malone, A., Louis, A., Benedetti, M., Schneider, M., Smith, R. L., Kreher, L., and Lam, R. (1975). *Biochim. Soc. Trans.* **3,** 730–732.
148. Malorny, G., Rietbroch, N., and Schneider, M. (1965). *Naunyn-Schmiedebergs Arch. Exp. Pathol. Pharmakol.* **250,** 419–436.
149. Malten, K. E. (1958). *Dermatologica* **117,** 103–109.
150. Malten, K. E. (1975). *Contact Dermatitis* **1,** 181–182.
151. Manna, G. K., and Parida, B. B. (1967). *J. Cytol. Genet.* **1,** 88–91.
152. Marks, T. A., Worthy, W. C., and Staples, R. E. (1980). *Teratology* **22,** 51–58.
153. Marsh, G. M. (1983). *In* ''Formaldehyde Toxicity'' (J. E. Gibson, ed.), pp. 237–254. Hemisphere Publ., New York.
154. Marzulli, F. N., and Maibach, H. I. (1974). *Food Cosmet. Toxicol.* **12,** 219–227.
155. Mashford, P. M., and Jones, A. R. (1982). *Xenobiotics* **12,** 119–124.
156. McDaniel, W., and Marks, J. (1979). *Arch. Dermatol.* **115,** 628.
157. McMartin, K. E., Martin-Amat, G., Noken, P. R., and Tephly, T. R. (1979). *Biochem. Pharmacol.* **28,** 645–649.
158. McMartin, K. E., Martin-Amat, G., Makar, A. B., and Tephly, T. R. (1977). *J. Pharmacol. Exp. Ther.* **20,** 564–572.
159. Melekhina, V. P. (1960). *USSR Literature on Air Pollution and Related Occupational Diseases* **3,** 135.
160. Meyer, B. (1979). ''Urea–Formaldehyde Resins.'' Addison-Wesley, Reading, Massachusetts.
161. Misiakiewicz, Z., Szulinska, G., Chyba, A., and Czyz, E. (1977). *Rocz. Panstw. Zakl. Hig.* **28,** 99–106.
162. Mitchell, J. C. (1981). *Contact Dermatitis* **7,** 173.
163. Mitchell, R. I., Pavkov, K. L., Persing, R. L., and Holzworth, D. A. (1979). ''Final Report

on a 90-day Inhalation Toxicology Study in Rats and Mice Exposed to Formaldehyde.'' Battelle, Columbus, Ohio.
164. Morrill, E. E., Jr. (1961). *Air Cond. Heat. Vent.* **58,** 94–95.
165. Moss, E., and Lee, W. R. (1974). *Br. J. Ind. Med.* **31,** 224–232.
166. Mueller, R., Raabe, G., and Schumann, D. (1978). *Exp. Pathol.* **16,** 36–42.
167. Murphy, S. D., Davis, H. V., and Zaratzian, V. I. (1964). *Toxicol. Appl. Pharmacol.* **6,** 520–528.
168. Nagornyi, P. A., Sudakova, Z. A., and Shchablenko, S. M. (1979). *Gig. Tr. Prof. Zabol.* **1,** 27–30.
169. National Research Council (1980). ''Formaldehyde: An Assessment of Its Health Effects,'' Report of the Committee on Toxicology. Nat. Acad. Sci., Washington, D.C.
170. National Research Council (1981). ''Formaldehyde and Other Aldehydes,'' Report of the Committee on Aldehydes. Nat. Acad. Sci., Washington, D.C.
171. Natvig, H., Andersen, J., and Rasmussen, E. W. (1971). *Food Cosmet. Toxicol.* **9,** 491–500.
172. Neely, W. B. (1964). *Biochem. Pharmacol.* **13,** 1137–1142.
173. Neshkov, N. S., and Nosko, A. M. (1976). *Gig. Tr. Prof. Zabol.* **12,** 92–94.
174. Noceto, J. B., and Laffont, H. (1962). *Arch. Mal. Prof. Med. Trav. Secur. Soc.* **23,** 314–316.
175. Obe, G., and Beek, B. (1979). *Drug Alcohol Depend.* **4,** 91–94.
175a. Olsen, J. H., and Døssing, M. (1982). *Am. Ind. Hyg. Assoc. J.* **43,** 366–370.
176. Orringer, E. P., and Mattern, W. D. (1976). *N. Engl. J. Med.* **294,** 1416–1420.
177. Osborn, M. J., Tabbert, P. T., and Huennekens, F. M. (1960). *J. Am. Chem. Soc.* **82,** 4921–4927.
178. Ostapovich, I. K. (1975). *Gig. Sanit.* **1,** 9–13.
179. Overman, D. (1981). ''Testing for Percutaneous Embryotoxicity of Laboratory Reagents in the Hamster.'' Poster session presentation at the Teratology Society 21st Annual Meeting, Stanford University, Stanford, California, June 21–25, 1981.
180. Overman, D. (1982). Personal communication.
181. Palese, M., and Tephly, T. R. (1975). *J. Toxicol. Environ. Health* **1,** 13–24.
182. Palkovits, M., and Mitro, A. (1968). *Gen. Comp. Endocrinol.* **10,** 253–262.
183. Paltzik, R. L., and Enscoe, I. (1980). *Cutis* **25,** 647–648.
184. Patterson, R., Zeiss, C. R., and Pruzansky, J. J. (1982). *J. Allergy Clin. Immunol.* **70,** 19–23.
185. Pietruszko, R. (1980). *Isozymes: Curr. Top. Biol. Med. Res.* **4,** 107–130.
186. Pirila, V., and Kilpio, O. (1949). *Ann. Med. Intern. Fenn.* **38,** 38–51.
187. Plunkett, E. R., and Barbela, T. (1977). *Am. Ind. Hyg. Assoc. J.* **38,** 61–62.
188. Popa, V., Delescu, D., Stanescu, D., and Gavrileseu, N. (1969). *Dis. Chest.* **56,** 395–404.
189. Porter, J. A. H. (1975). *Lancet* **7935,** 603–604.
190. Pruett, J. J., Scheuenstuhl, H., Michaeli, D., and Nevo, Z. (1980). *Arch. Environ. Health* **35,** 15–20.
191. Pushkina, N. N., Gofmekler, A., and Klevtsova, G. N. (1968). *Bull. Exp. Biol. Med.* **66,** 868–870.
192. Rader, J. (1974). ''Irritative Effects of Formaldehyde in Laboratory Halls.'' Ph.D. Dissertation. Institute for Pharmacology and Toxicology, University of Würzburg, Würzburg, Federal Republic of Germany.
193. Ragan, D. L., and Boreiko, C. J. (1981). *Cancer Lett.* **13,** 325–331.
194. Rietbrock, N. (1969). *Naunyn-Schmiedebergs Arch. Exp. Pathol. Pharmakol.* **263,** 88–105.
195. Robbins, J. D., and Norred, W. P. (1980). ''Bioavailability in Rabbits of Formaldehyde from Durable Press Textiles,'' Report to CPSC, Contract No. CPSC–IAG–80–1397. U.S. Dept. Agriculture, Athens, Georgia.
196. Ross, W. E., and Shipley, N. (1980). *Mutat. Res.* **79,** 277–283.

197. Rostenberg, A., Bairstow, B., and Luter, T. (1952). *J. Invest. Dermatol.* **19,** 459–462.
198. Rudner, E. J. (1975). *Contact Dermatitis* **1,** 277–280.
199. Rudner, E. J., Clendenning, W. E., Esptein, E., Fisher, A., *et al.* (1973). *Arch. Dermatol.* **108,** 537–540.
200. Sakula, A. (1975). *Lancet* **7939,** 816.
201. Salem, H., and Cullumbine, H. (1960). *Toxicol. Appl. Pharmacol.* **2,** 183–187.
202. Sanotskii, I. V., Fomenko, V. N., Sheveleva, G. A., Sal'nikova, L. S., Nakoryakova, M. V., and Pavlova, T. Y. (1957). *Gig. Tr. Prof. Zabol.* **1,** 25–28.
203. Sardinas, A. V., Most, R. S., Guilietti, M. A., and Honchar, P. (1979). *J. Environ. Health* **41,** 270–272.
204. Savenije-Chapel, E. M., and Noordhock, J. (1980). *Biochem. Pharmacol.* **29,** 2023–2029.
205. Schoenberg, J. B., and Mitchell, C. A. (1975). *Arch. Environ. Health* **3,** 574–577.
206. Schuck, E. A., Stephens, E. R., and Middleton, J. T. (1966). *Arch. Environ. Health* **13,** 570–575.
207. Schwartz, L., Peck, S. M., and Dunn, J. E. (1943). *Public Health Rep.* **58,** 899–904.
208. Searle, C. E. (1968). *Annu. Rep.—Cancer Res. Campaign* **46,** 246–247.
209. Sgibnev, A. K. (1968). *Gig. Tr.Prof. Zabol.* **12,** 20.
210. Shelabrick, J. E., and Steadman, T. R. (1979). ''Product/Industry Profile and Related Analysis on Formaldehyde and Formaldehyde-Containing Consumer Products.'' Battelle, Columbus, Ohio.
211. Sheveleva, G. A. (1971). *Toksikol. Nov. Prom. Khim. Veshchestv* **12,** 78–86.
212. Shipkovitz, H. D. (1968). ''Formaldehyde Vapor Emissions in the Permanent-Press Fabrics Industry,'' Rep. No. TR-52. U.S. DHEW, PHS, Consumer Protection and Environmental Health Service, Environmental Control Administration, Cincinnati, Ohio.
213. Shooter, K. V. (1975). *Chem.-Biol. Interact.* **11,** 575–588.
214. Shumilina, A. V. (1975). *Gig. Tr. Prof. Zabol.* **12,** 181–221.
215. Siomin, Y. A., Simonov, V. V., and Povrenny, A. M. (1973). *Biochim. Biophys. Acta* **331,** 27–32.
216. Skog, E. (1950). *Acta Pharmacol. Toxicol.* **6,** 299–318.
217. Smyth, H. F., Seaton, J., and Fischer, L. (1941). *J. Ind. Hyg. Toxicol.* **3,** 259–268.
218. Sneddon, I. B. (1968). *Br. Med. J.* **1,** 183–184.
219. Spangler, E. F., and Ward, J. M. (1982). Skin initiation/promotion study with formaldehyde in Sencar mice. *In* ''Formaldehyde: Toxicology—Epidemiology—Mechanisms'' (J. J. Clary, J. E. Gibson, and R. S. Waritz, eds.). Dekker, New York.
219a. Sparks, P. J., and Peter, J. M. (1980). *Int. Arch. Occup. Environ. Health* **45,** 221–229.
220. Spassowski, M. (1965). *Bull. Hyg.* **40,** 807–808.
221. Spear, R. (1982). *New Physician* **6,** 17.
222. Sugai, T., and Yamamoto, S. (1980). *Contact Dermatitis* **6,** 154.
223. Swenberg, J. A., Gross, E. A., Martin, J., and Popp, J. A. (1983). *In* ''Formaldehyde Toxicity'' (J. E. Gibson, ed.), pp. 132–146. Hemisphere Publ., New York.
223a. Swenberg, J. A., Gross, E. A., Randall, H. W., and Barrow, C. S. (1982). The effect of formaldehyde exposure on cytotoxicity and cell proliferation. *In* ''Formaldehyde: Toxicology—Epidemiology—Mechanisms'' (J. J. Clary, J. E. Gibson, and R. S. Waritz, eds.). Dekker, New York.
224. Swenberg, J. A., Kerns, W. D., Mitchell, R. J., Gralla, E. J., and K. L. Pavkov, (1980). *Cancer Res.* **40,** 3398–3402.
225. Temcharoen, P., and Thilly, W. G. (1983). *Mutat. Res.* **119,** 89–93.
226. Thorndike, J., and Beck, W. S. (1977). *Cancer Res.* **37,** 1125–1139.
227. Tou, J. C., Kallos, G. J. (1976). *Anal. Chem.* **48,** 958–963.

228. Tyihak, E., Trezl, L., and Rusznak, I. (1980). *Pharmazie* **35,** 18–20.
229. Ulsamer, A. G., Dailey, R. L., Beall, J. R., Kang, H. K., Curry, M., and May, J. (1981). ''Integrated Risk Assessment: Formaldehyde.'' Interagency Regulatory Liason Task Group on Formaldehyde, Washington, D.C.
230. Ulsamer, A. G., Gupta, K. C., Cohn, M. S., and Preuss, P. W. (1982). ''Formaldehyde in Indoor Air: Toxicity and Risk.'' Paper presented at the 75th annual meeting of the Air Pollution Control Association, New Orleans, June 20–25, 1982.
231. Uotila, L., and Koivusalo, M. (1974). *J. Biol. Chem.* **249,** 7653–7663.
232. Uotila, L., and Koivusalo, M. (1974). *J. Biol. Chem.* **249,** 7664–7672.
233. U.S. Department of Health, Education, and Welfare (1976). ''Criteria for a Recommended Standard . . . Occupational Exposure to Formaldehyde.'' U.S. DHEW (NIOSH) Publ. No. 77–126. U.S. Govt. Printing Office, Washington, D.C.
234. U.S. Department of Health, Education, and Welfare. (1976). ''Irritant Effects of Industrial Chemicals: Formaldehyde.'' U.S. DHEW (NIOSH) Publ. No. 77–117. U.S. Govt. Printing Office, Washington, D.C.
235. U.S. Department of Health, Education, and Welfare. (1981). ''Formaldehyde—Evidence of Carcinogenicity.'' NIOSH Curr. Intell. Bull. No. 34. U.S. Natl. Inst. Occupational Safety and Health, Washington, D.C.
236. Usdin, V. R., and Arnold, G. B. (1979). ''Transfer of Formaldehyde to Guinea Pig Skin.'' Prepared for Cotton, Inc., Contract No. 78–391. Gillette Res. Inst., Rockville, Maryland.
237. Wahren, J. (1980). ''Swedish Indoor Air Standards.'' Presentation at the CPSC Technical Workshop on Formaldehyde, Gaithersburg, Maryland, April 9, 1980.
238. Wald, G., Greenblatt, C., and Brown, P. K. (1953). *Fed. Proc.* **12,** 285–186.
239. Walker, J. F. (1964). Formaldehyde, 3rd ed. *ACS Monogr.*
240. Wallenstein, G., and Rebohle, E. (1976). *Allerg. Immunol.* **22,** 287–295.
241. Walrath, J., and Fraumeni, J. F., Jr. (1983). *In* ''Formaldehyde Toxicity'' (J. E. Gibson, ed.), pp. 227–235. Hemisphere Publ., New York.
242. Watanabe, F., Matsunaga, T., Soejima, T., and Iwata, Y. (1954). *Gann* **45,** 451–452.
243. Watanabe, F., and Sugimoto, S. (1955). *Gann* **46,** 365–367.
244. Waydhas, C., Weigl, K., and Sies, H. (1978). *Eur. J. Biochem.* **9,** 143–150.
245. Weber-Tschopp, A., Fischer, T., and Granjean, E. (1977). *Int. Arch. Occup. Environ. Health* **39,** 207–218.
246. Wiley, H. W. (1908). ''General Results of the Investigations Showing the Effect of Formaldehyde upon Digestion and Health,'' Circ. 42. U.S. Dept. of Agriculture, Bur. of Chemistry, Washington, D.C.
247. Wisconsin Department of Industry, Labor, and Human Relations (1980). ''Report and Standard: Proposed Final Indoor Ambient Air Quality Standard for Mobile Homes as Amended Through the Public Hearing Process.'' Dept. of Industry, Labor, and Human Relations, Madison.
248. Wong, O. (1983). *In* ''Formaldehyde Toxicity'' (J. E. Gibson, ed.), pp. 256–271. Hemisphere Publ., New York.
249. Woodbury, M. A., and Zenz, C. (1980). ''Formaldehyde Vapor Problems in Homes from Chipboard and Foam Insulation.'' Paper presented to the Wisconsin Department of Health, Education, and Welfare, Committee to Coordinate Toxicology and Related Programs.
250. Yefremov, G. G. (1970). *Zh. Ushn, Nos. Gorl. Bolezn.* **30,** 11–15.

Chlorinated Ethanes: Sources, Distribution, Environmental Impact, and Health Effects

Hans Konietzko

Institute of Occupational and Social Medicine
University of Mainz
Mainz, Federal Republic of Germany

I. General Remarks 402
A. Technical and Economic Data 402
B. Environmental and Biological Data 403
II. Monochloroethane (CH_3CH_2Cl) 408
A. Technical and Economic Data 408
B. Environmental and Biological Data 409
III. 1,1-Dichloroethane (CH_3CHCl_2) 411
A. Technical and Economic Data 411
B. Environmental and Biological Data 412
IV. 1,2-Dichloroethane (CH_2ClCH_2Cl) 414
A. Technical and Economic Data 414
B. Environmental and Biological Data 415
V. 1,1,1-Trichloroethane (CH_3CCl_3) 420
A. Technical and Economic Data 420
B. Environmental and Biological Data 422
VI. 1,1,2-Trichloroethane ($CHCl_2CH_2Cl$) 426
A. Technical and Economic Data 426
B. Environmental and Biological Data 428
VII. 1,1,1,2-Tetrachloroethane (CH_2ClCCl_3) 429
A. Technical and Economic Data 429
B. Environmental and Biological Data 430
VIII. 1,1,2,2-Tetrachloroethane ($CHCl_2CHCl_2$) 431
A. Technical and Economic Data 431
B. Environmental and Biological Data 433
IX. Pentachloroethane ($CHCl_2CCl_3$) 435
A. Technical and Economic Data 435
B. Environmental and Biological Data 436

HAZARD ASSESSMENT OF CHEMICALS:
Current Developments, Vol. 3

ISBN 0-12-312403-4

X. Hexachloroethane (CCl_3CCl_3) 438
A. Technical and Economic Data.... 438
B. Environmental and Biological Data 439
XI. General Conclusions 441
References 442

I. GENERAL REMARKS

The chlorinated ethanes include monochloroethane, 1,1-dichloroethane, 1,2-dichloroethane, 1,1,1-trichloroethane, 1,1,2-trichloroethane, 1,1,1,2-tetrachloroethane, 1,1,2,2-tetrachloroethane, pentachloroethane, and hexachloroethane. All chlorinated ethanes are technological products and have not been detected as natural products.

A. Technical and Economic Data

1. History

Most chlorinated ethanes were first synthesized during the nineteenth century; large-scale production, however, began around the middle of the twentieth century. Since exact information is important for the assessment of the environmental hazard, the available data have been included in each section.

2. Properties

Although all the chlorinated ethanes have a similar chemical structure, their physical and chemical properties differ slightly, depending on the number and position of the chlorine atoms. Increasing numbers of chlorine atoms raise the boiling point and heat of vaporization per mole, increase density, viscosity, and surface tension, lower the heat of combustion, and decrease solubility and inflammability. The boiling points and densities of asymmetric compounds are lower than those of their symmetric isomers. All chlorinated ethanes are readily soluble in organic solvents and only slightly miscible in water; they form azeotropes with both (*190*).

3. Production

Chloroethanes were produced from acetylene until it became too expensive. For the last 20 years, production has been ethylene based. An excess of hydrogen chloride is produced in these new technologies. The problem of environmental load can be satisfactorily solved from a technical point of view by appropriate production processes (oxychlorination and hydrochlorination) and by residue utilization and incineration plants (*190, 206*).

4. *Impurities*

A few chlorinated ethanes are sensitive to physical and chemical influences, i.e., they are oxidized by light and oxygen and form hydrogen chloride, phosgene, carbon oxides, and acetyl chloride. Chlorinated acetic acids formed in a wet environment are highly corrosive. Chloroethanes, therefore, must be stabilized prior to technical application. The most important stabilizing agents are amines, phenols, alcohols, and terpene. Special stabilization properties are required for each field of application. Since most producers have developed special stabilizing systems for their products, it is impossible to provide any general data on the type and amount of stabilizing agent. The most common stabilizing agents have been included in each section. The degree of purity of chlorinated ethanes varies considerably, depending on the compound, producer, and technical requirements. As a rule, the higher the degree of purity, the greater the stability of the chlorinated ethane (*190, 206*).

5. *Use*

All chlorinated ethanes are, or were at one time, important to various degrees for technology. Chlorinated ethanes have been produced in immense quantities. The estimated total United States production for 1976 was almost 9 billion pounds (*19*). At present, they are used almost exclusively as initial and intermediate products for chemical synthesis processes and as solvents for cleaning and extraction procedures. Other fields of application are of no real importance; they, however, have been included in each section.

B. Environmental and Biological Data

1. *Environmental Load*

Chlorinated ethanes are released into the environment during production, storage, transport, consumption, and disposal. Most escape into the air in the form of vapor, the rest into wastewater. Human exposure occurs directly through air and water and indirectly through animals and plants. Measurements of atmospheric concentrations are available for only a few chlorinated ethanes; these measurements have been cited in the relevant sections. The tropospheric half-life of chlorinated ethanes, estimated on the basis of their reactions with hydroxyl radicals, ranges from 18 to 380 days, i.e., considerably shorter than the half-life of carbon tetrachloride or Freons (8×10^4 days, *43*). Through photooxidation of chlorinated ethanes, phosgene and hydrogen chloride as well as monochloroacetyl chloride, dichloroacetyl chloride, and trichloroacetyl chloride can be detected under laboratory conditions. It is conceivable that these compounds are formed in the higher layers of the stratosphere and affect the ozone layers there (*174*).

Since chlorinated ethanes are heavier than water and poorly absorbed by soil particles, they quickly penetrate the thick soil layer and accumulate in the groundwater layer. Parts of them also may evaporate from the water surface. Under experimental conditions at the 1-ppm level in agitated water, the evaporation half-life is about 30 min for all chlorinated ethanes. On the whole, the bioconcentration of chlorinated ethanes is low, but it increases as the chlorine content increases. Neely *et al.* (*122a*) have shown that bioaccumulation is related to the octanol/water partition coefficient of the compound. Log(octanol/water partition coefficient) and other measurement data on the bioconcentration and biomagnification in environmental systems, when available, were included in the relevant sections. All reported information tends to indicate that, in contrast to other halogenated hydrocarbons, such as pesticides and Freons, the chlorinated ethanes do not constitute a global threat to the environment. Since their impact is limited to the immediate environment, it is estimable and controllable. Hazards for man have been demonstrated at production and processing sites, particularly the workplace; these hazards are discussed in detail in each section.

At the workplace, chlorinated ethanes are inhaled primarily as vapor; percutaneous absorption is less important. Ingestion can occur either accidentally or intentionally. The chlorinated ethanes themselves, however, are not the only hazard; their decomposition products also are dangerous. In the presence of hot metal and open flames, e.g., welding equipment and lighted cigarettes, every chlorinated ethane is capable of forming phosgene. High concentrations of phosgene, a dangerous irritant gas, can induce pulmonary edema with fatal outcome.

Increased misuse of chlorinated ethanes by solvent sniffers has been reported. The number of undetected cases is estimated to be high, but exact statistics are not available (*8*).

2. *Pharmacokinetics*

All chlorinated ethanes are lipophilic. The usual mode of intake is by inhalation of vapor. Absorption through the skin or ingestion through drinking water or food is less important. The inhaled dose depends on the chlorinated ethane concentration in the respiratory air and the length of exposure. Additional factors such as the respiratory minute volume can modify this process. Given constant exposure, vapor uptake continues until the concentration in external air, respiratory air, and body organs has reached a steady state. The distribution rate in the body depends on organ perfusion and solubility. In regard to the perfusion rate, the following three distribution compartments have been described: (1) internal organs (good perfusion), (2) muscle tissue (moderate perfusion), and (3) fatty and connective tissue (poor perfusion). Concentrations within these compartments are primarily governed by the partition coefficients between the individual organs. The blood/air partition coefficients increase in direct relation to the length of the carbon chain: they are markedly higher for isomers with a high boiling point (1,2-dichloroethane, 1,1,2-trichloroethane, 1,1,2,2-tetrachloroethane). The oil/blood partition coefficients react similarly (*5, 155*). Chlorinated ethanes with high partition coefficients are therefore distributed in the body faster and exhaled slower than chlorinated ethanes with lower partition coefficients. The partition coefficients have been included in each section.

Exhalation, however, is not the only elimination route. Biochemical mechanisms of elimination are also operative in parallel to these predominantly physical processes, in which inhaled chlorinated ethanes are released unaltered. Through these biochemical mechanisms, the structures of lipophilic chlorinated ethanes are altered so that these ethanes are capable of coupling with endogenous substances from the intermediary metabolism and thereby rendered water soluble and excretable. Metabolism occurs primarily in the liver. It does not lead to detoxification in the case of every compound; relatively harmless initial substances can also be activated to toxic metabolites. Compared to some of the aromatic chlorinated hydrocarbons used worldwide as pesticides, the biological half-life of chlorinated ethanes is short. The accumulation of higher concentrations in the human body or in experimental animals has not been detected.

3. Toxic Effects

The toxic effects of chlorinated ethanes can be derived from their pharmacokinetics. In man as well as in experimental animals, a distinction is made between nonspecific and specific effects; both are essentially dose dependent.

a. Nonspecific Effects. Nonspecific means that the organism reacts uniformly and homogeneously, independent of the chemical structure of the chlorinated ethane but dependent on the dose. These reactions include irritation of skin and mucosa, CNS[1] irritation or paralysis, and sensitization of the myocardial conduction system.

The skin, as a rule, is only reddened. Since chronic skin contact with liquid chlorinated ethanes is poorly tolerated, serious skin damage is usually avoided. Independent of this direct irritation, long-term exposure to vapor drys out the skin, making it susceptible to various allergenic substances in the occupational and private sphere. Direct allergization by chlorinated ethanes is rare.

The effect on the central nervous system can best be understood as a prolonged

[1]Abbreviations: CNS, central nervous system; DNA, deoxyribonucleic acid; ECD, electron capture detector; FID, flame ionization detector; GC, gas chromatography; GC/MS, gas chromatography/mass spectrometry; LC, lethal concentration; LD, lethal dose; MAK, maximale Arbeitsplatzkonzentration = maximum workplace concentration; MCC, maximum ceiling concentration; RNA, ribonucleic acid; TLV, threshold limit value; TLV–TWA, threshold limit value, time weighted average; TLV–STEL, threshold limit value, short-term exposure limit.

preanesthetic state that seldom progresses to an anesthetic state. The symptomatology is ambiguous, uncharacteristic, and often misleading. Depending on the concentration and the length of exposure, affected individuals will complain of mild discomfort, loss of appetite, pain in the stomach, dizziness, fatigue, drunkeness, nausea, and vomiting. Disorders of balance and motor ataxias resembling alcohol intoxication develop in the advanced stage. Continued intake of chlorinated ethanes results in anesthesia and eventually fatal respiratory paralysis. Chlorinated ethanes must be fat soluble to affect the central nervous system. The exact mechanism of action, however, is unknown. Analogous studies using chemically inert anesthetics tend to indicate that the effect is produced primarily by physical changes at the synapses. Chemical reactions apparently do not occur. This means that the nonspecific effects on the central nervous system decrease in direct relation to the elimination; the effects subside completely within a few hours after exposure has been discontinued.

Chlorinated ethanes, like all halogenated hydrocarbons, can sensitize the myocardial conduction system to sympathetic impulses and sympathomimetics, the consequence being cardiac dysrhythmias. Depending on the dose, single or multiple extrasystoles have been observed; rare doses of fatal ventricular fibrillation have even been reported. The pathogenesis of the sensitization process is unknown.

b. Specific Effects. Specific means that chlorinated ethanes preferentially damage certain organs in dependency on their chemical structure. Analogous to the effect of other chlorinated hydrocarbons, the specific organ effect is largely a consequence of metabolism. The type and extent of the metabolic transformation or the metabolite determine whether organ damage will occur and which organs will be affected. This explains, for example, the varying degrees of hepatotoxicity.

c. Carcinogenicity. Estimates of the cancer risk can be inferred from epidemiological studies and case reports or from animal experiments. These studies are, however, elaborate and expensive, and therefore only few such studies have been carried out with chlorinated ethanes. For example, a carcinogenicity test may take as long as 3 years and cost as much as $200,000 (*39, 101*). In view of this, short-term tests have been frequently used as screening procedures. These are based on the fact that most organic carcinogens initially affect DNA, i.e., the genetic material. Such substances, therefore, can be both carcinogenic and mutagenic. Short-term tests attempt to establish DNA damage at the molecular, cellular, and multicellular level. At present, the most important tests are the *Salmonella typhimurium* plate-incorporation assay, X-linked recessive lethal test in *Drosophila melanogaster,* unscheduled DNA synthesis, and *in vitro* transformation test. If the carcinogenic metabolites are also to be included in the study, these tests must be combined with metabolic activation systems, usually rat liver microsomes (*4, 131, 137, 154, 195*). The correspondence of comparative studies on the carcinogenicity and mutagenicity of known carcinogens ranges from 88 to 94% (*137*). A dose–response relationship cannot be inferred from animal tests nor from *in vitro* short-term tests. It therefore is impossible to cite reliable maximum permissible carcinogenic concentrations for man.

No correlation exists between carcinogenicity and teratotoxicity or embryotoxicity (*205*). For practical reasons, however, teratogenic findings, insofar as they are known, have been included in the respective ''Carcinogenicity'' sections.

4. *Analytic Methods*

Gas chromatography (GC) and gas chromatography/mass spectrometry (GS/MS) are the most reliable methods for detecting chlorinated ethanes in air, water, and biological material. Because of its low sensitivity, infrared spectroscopy is suitable only for higher concentrations. The rapid development of analytic techniques makes it impossible to recommend any particular methods for general use. Additional information on analytic techniques has been included in some sections.

5. *Standards of Permissible Exposure*

Standards for occupational exposure are threshold limit values (TLV). These values consider only the concentration in the air and the intake through the lungs and skin, i.e., normal conditions at the workplace. Because of individual differences in sensitivity, they are not to be understood as sharp lines of demarcation between harmless and dangerous concentrations. The following values are important for chlorinated ethanes: threshold limit value, time weighted average (TLV–TWA, mean concentration for an 8-h day in a 40-h week), threshold limit value, short-term exposure limit (TLV–STEL, maximum concentration for up to 15 min exposure) (*185*). The German MAK value corresponds approximately to the TLV–TWA (*102*).

The United States Environmental Protection Agency (EPA) has published standards for the protection of both aquatic and human life for many contaminants (*170*). These standards are chosen to protect both the health of the aquatic organisms and their suitability for human consumption. Because of the differing toxicity in both media, separate criteria are given for freshwater and for saltwater. The 24 h average concentration and the maximum ceiling concentration (MCC), which can not be exceeded, are cited for both criteria.

The limit values for the protection of human health are calculated from the possible contaminant intake through drinking water and eating fish.

II. MONOCHLOROETHANE (CH_3CH_2Cl)[2]

A. Technical and Economic Data

1. History

Monochloroethane, the oldest known chloroethane, was first synthesized in the fifteenth century. Large-scale production began around 1930 when the production of tetraethyllead increased (*190, 206*).

2. Physical Properties

Monochloroethane is a colorless, thin liquid with a burning taste and pleasant ethereal odor that forms a gas at room temperature. It is readily soluble in many organic solvents and is highly flammable (*190, 206*). Its physical properties are summarized in Table I.

3. Chemical Properties

Monochloroethane hydrolyzes or oxidizes gradually at room temperature. It dehydrochlorinates to ethylene at temperatures above 400°C. Phosgene and hydrogen chloride are formed as combustion products. Monochloroethane is the most reactive substance among the chloroethanes and therefore is the starting material for many syntheses (*190, 206*).

4. Production and Transport

Monochloroethane is produced almost exclusively by free radical chlorination of ethane or hydrochlorination of ethylenes. The other production processes are of no practical importance.

The technical degree of purity determined with gas chromatography is 99.9%.

Monochloroethane is stored and transported in compression-proof steel tanks or canisters with a reflective coating to protect it against sunlight. Stabilizing agents are unnecessary (*190*).

5. Use

The annual United States production in 1976 was 670 million pounds. Approximately 85% was used to produce tetraethyllead for antiknock compounds. It is reasonable to assume that the production of monochloroethane will decline in the next few years due to regulations limiting the lead content in gasoline or prohibiting the sale of leaded gasoline. In the United States, alkyllead consumption dropped from 660 to 548 million pounds between 1973 and 1975.

In 1973, 13 million pounds was used to produce ethyl cellulose and ethylhydroxyethyl cellulose. An additional 10 million pounds was used for the

[2]Synonyms: chloroethane, ethyl chloride, hydrochloric ether, chloroethyl, muriatic ether.

TABLE I

Physical Properties of Monochloroethane

Property	Value
Molecular weight	64.52
Density	924 kg/m^3 (0°C)
Vapor density	2.76 kg/m^3 (20°C)
Vapor pressure	403 mbar (−10°C)
	623 mbar (0°C)
	929 mbar (10°C)
	1342 mbar (20°C)
Melting point	−138°C
Boiling point	12°C
Spontaneous ignition temperature	517°C
Explosion limit with air	3.6–14.8 vol%
Solubility in water	0.455 mass% (0°C)
Solubility of water in monochloroethane	0.07 mass% (0°C)

synthesis of dyes and chemicals, refrigerants, extracting agents in temperature-sensitive odoriferous substances, solvents in the plastics industry, adhesive agents, and inks, as well as refrigeration anesthetics (*37, 190*).

B. Environmental and Biological Data

1. Environmental Load

Monochloroethane does not occur as a natural product. Production and utilization can result in contamination of the environment. However, information concerning its levels in air, water, soil, and food is not available. The half-life of monochloroethane in the troposphere is estimated to be 18 days (*43*).

2. Workplace Hazard

The production, storage, transport, and usage of monochloroethane can be hazardous. The United States National Institute for Occupational Safety and Health (NIOSH) estimates that approximately 113,000 workers in the United States are exposed to monochloroethane (*121*). The risk in the production, storage, and transport, however, is slight. The free use of high doses (i.e., refrigeration anesthesia in medical interventions) is more dangerous. Recent studies indicate an increase in the use of monochloroethane as drugs (*65*).

3. Pharmacokinetics

Monochloroethane is rapidly absorbed through the lungs. Approximately 75% is bound to cell constituents in the blood and 25% to plasma (*80*). The Ostwald solubility coefficient (blood/air) is 2.5; the oil/blood partition coefficient is 960 (*15*). The highest concentration is found in perirenal fatty tissue and the lowest in cerebrospinal fluid. The concentration in the brain is twice that of the blood. Unaltered monochloroethane is exhaled rapidly and completely. Slight metabolism to ethanol after dechlorination could be detected only after high anesthetic doses (*31, 204*).

4. *Toxic Effects*

a. Animal Tests. No reactions were observed in guinea pigs after 13 h inhalation of 10,000 ppm; however, abnormal behavior, histologic alterations in the liver and kidneys, and, in the presence of an open flame, severe pulmonary irritation were observed at concentrations above 20,000 ppm (*125, 157, 158*). The narcotic concentration was approximately 18,000 ppm for cold-blooded animals and began at 36,000 ppm for rodents (*41, 86*). At concentrations above 100,000 ppm, all animals were anesthetized (*31, 157, 158*). Conclusive animal tests on chronic toxicity are not available.

b. Effects in Man. The acute irritant effect on skin and mucosa is slight; the narcotic effect, however, is pronounced. Mild subjective complaints are to be expected after 20 min exposure to concentrations above 13,000 ppm (*24*). Signs of inebriation and impaired coordination can be observed at concentrations above 20,000 and above 25,000 ppm, respectively. The anesthesia stage is noticeable at 36,000 ppm (*24, 158*). Respiratory arrest has been observed at 60,000 ppm (*2*). Isolated deaths have been reported after short anesthesia. Since monochloroethane is rapidly absorbed and eliminated, it was once used as a short anesthetic. It has not been used as a general anesthetic, however, for more than 50 years because of its poor controllability, its low narcotic range, and the danger of overdosage. The observations presented here were published 30 to 40 years ago; current information is not available.

Since monochloroethane is not metabolized, specific organ effects are highly unlikely. One case of reversible cerebellar dysfunction in a 28-year-old woman who used monochloroethane as a narcotic for several months was reported (*65*). Occupational intoxication following chronic inhalation has not been described and is highly unlikely, since the fire hazard and low explosion limits eliminate the possibility of higher concentrations at the workplace. Epidemiologic studies on larger samples are not available.

Permissible exposure standards are presented in Table II.

c. Carcinogenicity. Studies on the carcinogenicity, mutagenicity, and teratogenicity of monochloroethane are not available.

5. *Analysis*

Special methods for analyzing the monochloroethane concentration in air, water, or biological material have not been reported.

TABLE II

Standards of Permissible Exposure for Monochloroethane

Index	Permissible exposure [ppm (mg/m³)]	Reference
TLV–TWA	1000 (2600)	*185*
TLV–STEL	1250 (3250)	*185*
MAK	1000 (2600)	*102*
EPA criteria	Not established	*170*

III. 1,1-DICHLOROETHANE (CH_3CHCl_2)[3]

A. Technical and Economic Data

1. Physical Properties

1,1-Dichloroethane is a colorless liquid with a sweet chloroformlike smell. It is miscible with all liquid chlorinated hydrocarbons and with most organic solvents (*55a, 190, 206*). Its physical properties are summarized in Table III.

2. Chemical Properties

1,1-Dichloroethane is extremely stable under dry conditions and at room temperature. Hydrogen chloride splits off when it is warmed. In the presence of catalytic amounts of chlorine and iron and at temperatures above 150°C, it dehydrochlorinates to vinyl chloride (*190*).

3. Production

The most important industrial production process is the bonding of hydrogen chloride to vinyl chloride in the liquid or gas phase (*190*).

4. Use

Information on the annual United States production is not available. Large amounts of 1,1-dichloroethane are apparently neither produced nor imported. Estimates of European and Japanese production are also not available. Most of

[3]Synonyms: α-dichloroethane, ethylidene chloride, ethylene dichloride, chlorinated hydrochloric ether.

TABLE III

Physical Properties of 1,1-Dichloroethane

Property	Value
Molecular weight	98.97
Density	1176 kg/m³
Melting point	−97°C
Boiling point	57°C
Vapor pressure	93 mbar (0°C)
	153 mbar (10°C)
	242 mbar (20°C)
	369 mbar (30°C)
Explosion limit in air	5.9–15.9 vol%
Solubility in water	0.5 mass%
Solubility of water in 1,1-dichloroethane	0.009 mass% (20°C)
Log(octanol/water partition coefficient)	1.79

the 1,1-dichloroethane produced is processed to 1,1,1-trichloroethane. Some is used as solvent for raw rubber or silicone grease and as extracting agent for temperature-sensitive food (*19, 190, 206*).

B. Environmental and Biological Data

1. Environmental Load

1,1-Dichloroethane does not occur as a natural product. Environmental hazards can arise in the production, further utilization, and disposal. However, measurements of its levels in air, water, soil, and food are not available. The half-life in the troposphere is estimated to be 31 days (*43*). The low production excludes the possibility of a major environmental hazard.

2. Workplace Hazard

Production, storage, transport, and further utilization can result in a health hazard. NIOSH estimated that approximately 4600 workers in the United States are exposed to 1,1-dichloroethane (*121*).

3. Pharmacokinetics

Nothing is known about the intake, distribution, metabolism, and elimination of 1,1-dichloroethane and its possible metabolites. The formation of acetic acid has been speculated (*204*). The blood/air and oil/blood partition coefficients are 4.7 and 40, respectively (*155*).

4. *Toxic Effects*

a. Animal Tests. For mice, the acute narcotic concentration is 8000–10,000 ppm; the lethal concentration is 17,300 ppm (*93, 114, 171*). The acute oral LD_{50} is 14.1 g/kg body weight (*173*).

Specific organ effects were studied in two animal experiments. Whereas rats, guinea pigs, rabbits, and cats showed no clinical signs of damage after 3 months inhalation of 500 ppm, kidney damage was demonstrable in cats after a subsequent increase to 1000 ppm. No pathologic alterations were demonstrable in rats, guinea pigs, rabbits, and dogs exposed to 500 and 1000 ppm for more than 6 months. Kidney and liver damage, however, were observed after a considerable increase in the dose (*67*).

b. Effects in Man. 1,1-Dichloroethane was used as an anesthetic at the beginning of the twentieth century. Information on concentration and complications during anesthesia is not available. 1,1-Dichloroethane-induced occupational diseases are unknown.

Standards of permissible exposure are summarized in Table IV.

c. Carcinogenicity. Animal experiments on carcinogenicity have yielded negative findings. However, due to the high death rate of the experimental animals, a definitive evaluation will not be possible until more studies are available (*120*).

No embryotoxic or fetotoxic effects were observed after exposure of pregnant Sprague–Dawley rats to subnarcotic concentrations for 7 h a day during days 6–15 of gestation (*162*).

5. *Analysis*

Special methods for analyzing 1,1-dichloroethane concentration in air, water, or biological material have not been reported.

TABLE IV

Standards of Permissible Exposure for 1,1-Dichloroethane

Index	Permissible exposure [ppm (mg/m^3)]	Reference
TLV–TWA	200 (810)	*185*
TLV–STEL	250 (1010)	*185*
MAK	100 (400)	*102*
EPA criteria	Not established	—

IV. 1,2-DICHLOROETHANE (CH_2ClCH_2Cl)[4]

A. Technical and Economic Data

1. History

1,2-Dichloroethane was first synthesized in 1795 (*28*); large-scale production began in 1960.

2. Physical Properties

1,2-Dichloroethane is a water-white, colorless, slightly flammable liquid which smells like chloroform and is readily soluble in most organic solvents (*56, 137a, 190*). Its physical properties are summarized in Table V.

3. Chemical Properties

In the dry state, pure 1,2-dichloroethane is stable at higher temperatures. At temperatures above 600°C, it breaks down to vinyl chloride, hydrogen chloride, and acetylene. It is decomposed by light, air, and moisture; hydrogen chloride is given off. Photooxidation can result in the formation of phosgene, carbon monoxide, and hydrogen chloride (*56, 190*).

4. Production and Transport

1,2-Dichloroethane is produced by adding chlorine to ethylene via direct chlorination using ferric chloride and other metallic chlorides as catalysts or via oxychlorination, in which chlorine is formed by oxidizing hydrogen chloride in the presence of cuprous catalysts. As a result of the increased importance of 1,2-dichloroethane, several modifications have been developed that can be applied in the gas or the liquid phase. The basic requirement for optimal yield with few by-products is that the raw materials, particularly ethylene, be free of impurities. Waste gases (e.g., nitrogen dioxide, carbon monoxide, carbon dioxide, small amounts of ethylene, 1,1-dichloroethane) are formed only during oxychlorination with air. The environmental load of 1,2-dichloroethane with the oxychlorination procedure is twice as high as that with direct chlorination (*49, 56, 190*).

1,2-Dichloroethane usually has the least impurities among the chloroethanes, and contains only traces of 1,1,2-trichloroethane. At a water concentration of 50 ppm, 1,2-dichloroethane is so stable that it can be transported without stabilizing agents. As a rule, however, commercial 1,2-dichloroethane is stabilized with 0.1% alkylamines. It can be stored and transported in iron and V2A steel tanks, but not in aluminum drums (*56, 190*).

[4]Synonyms: ethylene chloride, ethylene dichloride, dichloroethylene, *sym*-dichloroethane, α,β-dichloroethane, glycol dichloride.

TABLE V

Physical Properties of 1,2-Dichloroethane

Property	Value
Molecular weight	98.97
Density	1281 kg/m^3
Vapor pressure	33 mbar (0°C)
	85 mbar (20°C)
	320 mbar (50°C)
	930 mbar (80°C)
Melting point	−35°C
Boiling point	83°C (1 bar)
Spontaneous ignition temperature	440°C
Explosion limit in air	6.2–16.0 vol% (20°C, 1.013 bar)
Solubility in water	0.873 mass% (0°C)
Solubility of water in 1,2-dichloroethane	0.16 mass% (20°C)
Log(octanol/water partition coefficient)	1.48

5. *Use*

The annual worldwide production of 1,2-dichloroethane is estimated to be 51 billion pounds. In the United States alone, production increased from 510 million pounds to 8 billion pounds (i.e., 16-fold), between 1955 and 1972; the production in 1979 was estimated to be 15 billion pounds. 1,2-Dichloroethane heads the list of organic chlorine derivatives produced in the United States and is the fifteenth most produced chemical. Approximately 83% is immediately processed by the producer. At least 90% is used for the synthesis of vinyl chloride; the rest is used for the production of 1,1,1-trichloroethane, ethyleneamine, tetrachloroethylene, trichloroethylene, and dichloroethylene. 1,2-Dichloroethane is used as a lead scavenger agent in gasoline and, occasionally, as a fungicide (*36, 49, 56, 190*). Until a few years ago, antirheumatic ointments containing as much as 80% 1,2-dichloroethane were available on the market (*150, 202*).

B. Environmental and Biological Data

1. *Environmental Load*

1,2-Dichloroethane does not occur as a natural product. Production, storage, disposal, and further processing can result in an environmental hazard.

With a total production of 11 billion pounds, the environmental load for 1,2-dichloroethane in 1977 was estimated to be 378 million pounds. As a result of vinyl chloride production alone, 100 million pounds was released into the air, 11 million pounds leaked into wastewater, and 60 million pounds was disposed of as industrial waste. Accurate information on the use of the other 207 million pounds (e.g., as gasoline additive or pesticide) is not available (*36, 49*). 1,2-Dichloroethane cannot be detected in automobile exhaust (*57*).

The half-life of 1,2-dichloroethane in the atmosphere is estimated to be several weeks or months (*129*). Attempts to detect 1,2-dichloroethane in the atmosphere have been unsuccessful. Concentrations in the vicinity of production plants, however, were between 0.1 and 186 ppb. The farther away the measuring points were from the production plant, the lower were the values, i.e., within the parts per trillion (ppt) range (*89, 130*). On the basis of the measurement results and the population density in the vicinity of production plants, it has been estimated that approximately 12.5 million people are exposed to annual mean 1,2-dichloroethane concentrations of 0.01–10 ppb; only 300,000, however, are exposed to concentrations exceeding 1 ppb (*79, 179*). Concentrations of 0.01–9.4 ppb were obtained in large Japanese cities (*127*).

Wastewater and waste are contaminated by 1,2-dichloroethane-bearing tar. These complex tar compounds are composed predominantly of aliphatic halogenated hydrocarbons, which can include 33% 1,2-dichloroethane or 1,1,2-trichloroethane, and 0.06% vinyl chloride (*74*). Because of its high vapor pressure, 1,2-dichloroethane, insofar as it is not firmly bound in tars, volatilizes. The remainder is highly stable and slightly soluble in water; 1 part 1,2-dichloroethane dissolves in approximately 120 parts water. Degradation is probably gradual. Exact data on the half-life, however, are not available. This gradual degradation is probably the reason that 1,2-dichloroethane is detected in industrial wastewater as well as in river and tap water. An analysis of drinking water in 80 United States cities revealed concentrations as high as 6 μg/liter in 28 cases, and even 8 mg/liter in 1 case (*180*). In another study, random samples taken from surface water in the vicinity of industrial plants revealed concentrations of 1–2 ppb in 53 of 204 samples (*34*). Measurement values for river and tap water in Europe and Japan were below 1 μg/liter (*42*). It is not entirely impossible that part of this 1,2-dichloroethane concentration was produced secondarily through the chlorination of drinking water. Activated charcoal filters nevertheless trap 90–100% of the 1,2-dichloroethane.

High bioaccumulation in aquatic plants or animals is improbable, even though 1,2-dichloroethane has occasionally been detected in fish and oysters. 1,2-Dichloroethane, however, was also identified in industrial waste dumped in the North Sea (*74*).

A slight risk is associated with the disinfection of food. Several weeks after disinfection, the 1,2-dichloroethane concentration in sacked wheat ranged from 23 to 43 ppm (*208*). Concentrations of 2–23 μg/g were found in several spices that were extracted with 1,2-dichloroethane (*128*). In 1977, registration of pesticides containing 1,2-dichloroethane in the United States numbered 84. Collectively, these pesticides contain more than 2 million pounds of 1,2-dichloroethane (*49*). Little information on this aspect of the environmental hazard is available. High-grade food contamination could not be demonstrated.

2. Workplace Hazard

Workplace hazards arise in the production and further utilization of 1,2-dichloroethane. The number of workers coming in contact with 1,2-dichloroethane is difficult to assess because of its wide range of application. NIOSH speculates that 1.6 million workers in at least 60 branches of industry are exposed to 1,2-dichloroethane and that 18,000 or more are exposed to high concentrations (*121*). The most dangerous exposures are contact with cleaning agents and solvents in the metal and textile industry and disinfectants, since in these uses 1,2-dichloroethane is released into the air. Previously, antirheumatic ointments were a hazard for patients and medical personnel. Antirheumatic ointments containing 80% 1,2-dichloroethane were available on the German and Swiss drug markets until just a few years ago (*150*).

3. Pharmacokinetics

The pharmacokinetics of 1,2-dichloroethane has not been studied. The partition coefficients are blood/air, 19.5 and oil/blood, 23 (*155*). In inhalation and feeding tests with rats, equilibrium was established within 2–3 h, i.e., relatively slowly. After exposure was terminated, the blood concentration rapidly declined in a biphasic elimination curve. The half-lives for the two phases were 6 and 35 min. Most of the 1,2-dichloroethane had been eliminated 18 h after exposure was terminated. Following inhalation or ingestion, labeled 1,2-dichloroethane is distributed primarily in the liver and the kidneys. In the first 48 h after exposure, 30% of ingested, but only 1.8% of inhaled, 1,2-dichloroethane was exhaled unaltered. Most of the 1,2-dichloroethane was metabolized and excreted in the urine; 7.7% was exhaled as CO_2. Within 48 h after inhalation, 96% of the radioactive substances had been eliminated (*146*). The findings after intravenous and intraperitoneal injection were similar. The most important metabolites are chloroacetic acid, *S*-carboxymethylcysteine, and thiodiacetic acid (*212*). Decomposition to ethylene has been demonstrated (*96*). The activity of cytochrome *P*-450 and UDPglucuronyltransferase in rat liver microsomes decreased markedly under the influence of 1,2-dichloroethane (*87*). Rat liver microsomes can transform 1,2-dichloroethane into reactive metabolites that form covalent bonds with macromolecules, particularly DNA, RNA, and proteins. Studies with labeled 1,2-dichloroethane indicated that these covalent bonds were not limited to organs predisposed to cancer; they were also demonstrated in many other kinds of tissues. Pretreatment with phenobarbital intensifies this bonding (*7, 51, 146*).

4. Toxic Effects

a. Animal Tests. For rats, LD_{50} for 30 min exposure was 12,000 ppm; for 7-h exposure, LD_{50} was only 1000 ppm. None of the rats survived 20,000 ppm (*175*). Short anesthesia concentrations are attained with 5000 ppm (*93*). Anesthesia with 1,2-dichloroethane is deeper and more prolonged than with carbon tetrachloride or chloroform. Long-term exposure to concentrations of 500 ppm was lethal for rabbits, guinea pigs, and rats, but not for dogs, cats, and monkeys (*62, 67*).

Many different kinds of damage (i.e., liver and kidney damage, pulmonary edema, adrenal necrosis, intestinal and mesenteric hemorrhages, extremity paralysis, and myocardial necrosis) were observed in individual experimental animals after short- and long-term tests. Liver function impairment was usually less severe and regressed more rapidly than after comparable carbon tetrachloride poisoning (*62, 82, 158, 175*).

Nubeculae could be induced in many experimental animals by inhalation exposure and subcutaneous injection; the nubeculae regressed after 1 week to several months. These alterations were observed in the dog after inhalation exposure as low as 1000 ppm and in the fox after inhalation exposure to 3000 ppm. Since nubeculae also developed when the eyes were covered during inhalation but did not form after direct instillation in the lids, these lesions must be systemic (*61, 82, 177*).

b. Effects in Man. 1,2-Dichloroethane irritates skin and muscosa. Severe necrotic alterations were regularly found in the gastrointestinal tract after accidental or intentional peroral intoxication (*66, 70, 144, 203, 207*). Inflammatory and necrotizing alterations were detected along the respiratory tract after inhalation intoxication. The critical anesthesia limit has not been established, but it is probably around 20,000 ppm (*12, 109*).

Specific organ alterations develop after acute intoxication and chronic exposure, particularly after accidental or intentional peroral intake of liniments containing 1,2-dichloroethane. Degenerative alterations were already detectable in the liver and kidneys of a child who died 21 h after peroral intake (*28*). Liver and kidney damage were also found in most other cases of fatal poisoning. Degenerative alterations have been observed in the brain and myocardium. The frequency of cutaneous bleeding and organ hemorrhages (coagulation consumption coagulopathies) was impressive. Liver damage was frequently observed in survivors of acute intoxication (*3, 40, 90, 100, 144, 161, 175, 203, 213*). In one case, an irreversible cerebral defect with myoclonic syndrome, epileptic seizures, and permanent mental defects developed after peroral intoxication (*25*).

Irreversible organ damage after chronic exposure has not been reported. Nubeculae have not been observed in man.

Standards of permissible exposure are summarized in Table VI.

c. Carcinogenicity. Casuistic reports or epidemiologic studies on the carcinogenicity of 1,2-dichloroethane in man are not available. A few animal experiments, however, have been carried out. After feeding 1,2-dichloroethane to mice, the incidence of benign and malignant tumors was significantly higher in the treated mice than in the control group and was dose dependent. The results in

TABLE VI

Standards of Permissible Exposure for 1,2-Dichloroethane

Index	Permissible exposure	Reference
TLV–TWA	10 ppm (40 mg/m^3)	*185*
TLV–STEL	15 ppm (60 mg/m^3)	*185*
MAK	20 ppm (80 mg/m^3)	*102*
EPA criteria		
Freshwater aquatic life	3900 μg/liter (24 h average)	*170*
	8800 μg/liter (MCC)	*170*
Saltwater aquatic life	880 μg/liter (24 h average)	*170*
	2000 μg/liter (MCC)	*170*
Human health	7 μg/liter	*170*

rats were more differentiated: Male animals developed squamous cell carcinomas of the forestomach and hemangiosarcomas more frequently than the female animals; the incidence of mammary adenocarcinoma and fibroadenoma, however, was higher in the female animals. The effect in mice after intraperitoneal administration was not clear (*115, 184, 200*). A 2-year inhalation study using Swiss mice and Sprague–Dawley rats showed no increase in tumor incidence after exposure to concentrations of 5, 10, 50, and 150 ppm (*99*).

Mutagenicity tests with *S. typhimurium* TA 1530, TA 1535, and TA 100 established that 1,2-dichloroethane without activation is a weak mutagen. The mutagenic effect was markedly intensified by the addition of cytosol and glutathione. Mutagenic effects were demonstrated in *Escherichia coli,* but not in *Streptomyces coelicolor* and *Aspergillus nidulans* (*11, 14, 103, 104, 139, 140*). Increased mutation rates were also observed in barley (*30*). Sex-linked recessive lethals were also found in *D. melanogaster* (*81, 166, 167*). Chloroacetaldehyde, a presumed metabolite of 1,2-dichloroethane, proved to be a potent mutagen in tests with *S. typhimurium* TA 100 (*103*).

1,2-Dichloroethane passes through the placental barrier and accumulates in fetal tissue (*196*). In inhalation tests with rats, the embryonic death rate in the exposed group was 27.9% greater than in the control group (*197*). A further study conducted over a 6-month period revealed reduced fertility and increased perinatal mortality in the first generation, but not in the second generation (*198*). Neither teratogenic nor embryotoxic damage, however, was observed after long-term exposure of pregnant rats and rabbits to 100 or 300 ppm of 1,2-dichloroethane (*141*).

5. *Analysis*

Air samples were trapped against activated charcoal (*142*), desorbed, and analyzed with GC/FID (*59, 122*). Drinking water was analyzed directly with GC/ECD (*124*). Wastewater and food required pretreatment (*27, 69*).

V. 1,1,1-TRICHLOROETHANE (CH_3CCl_3)[5]

A. Technical and Economic Data

1. History

1,1,1-Trichloroethane was first synthesized in 1840. Large-scale production began around 1950 (*143, 190*).

2. Physical Properties

1,1,1-Trichloroethane is a colorless, nonflammable liquid with a sweet, ethereal smell. It is miscible with all common organic solvents and is a good solvent for grease, paraffin, wax, and many other organic substances (*189a, 190, 206*). Its physical properties are given in Table VII.

3. Chemical Properties

1,1,1-Trichloroethane is highly unstable. It splits off hydrogen chloride at room temperature and decomposes into 1,1-dichloroethane and hydrogen chloride at 400°C. In the presence of metallic salts, this process begins at 150°C. Temperature-dependent oxidation to phosgene occurs in air. Hydrogen chloride, phosgene, and dichloroacetylene are produced under UV radiation or increased temperature (*190, 206*).

4. Production and Transport

1,1,1-Trichloroethane can be produced industrially from 1,2-dichloroethane, ethane, or 1,1-dichloroethane. Of the total United States production, 90% is obtained from 1,2-dichloroethane. The most important procedure is induced chlorination of 1,2-dichloroethane in the liquid phase to 1,1,2-trichloroethane, subsequent dehydrochlorination to 1,1-dichloroethylene, and hydrochlorination to 1,1,1-trichloroethane with water-free ferric chloride as catalyst (*190, 191*).

Since 1,1,1-trichloroethane is easily decomposed, it must be stabilized during production. The different requirements for stabilization against metals, light, air, and temperature have contributed to the development and patenting of many stabilizing systems. These systems usually contain 1,4-dioxane, epoxide, alcohols, and nitro compounds (approximately 3–7 vol%). 1,1,1-Trichloroethane must be specially stabilized for the degreasing of metal in the vapor phase (*190*). A recently published study reported the following impurities in 22 samples of technical 1,1,1-trichloroethane: 1,1-dichloroethylene (30–900 μg/ml), 1,1-dichloroethane (11 μg/ml), trichloroethylene (12 μg/ml), 1,1,2-trichloroethane (9 μg/ml). Nitromethane (18 μg/ml), 1,2-epoxybutane (19 μg/ml), butanol (7 μg/ml), and dioxane (10 μg/ml) were identified as stabilizing agents (*60*).

[5]Synonyms: α-trichloroethane, methylchloroform, chloroethene, methyltrichloromethane.

TABLE VII

Physical Properties of 1,1,1-Trichloroethane

Property	Value
Molecular weight	133.42
Density	1371 kg/m^3 (0°C)
Vapor pressure	16 mbar (−20°C)
	49 mbar (0°C)
	133 mbar (20°C)
	320 mbar (40°C)
	627 mbar (60°C)
Melting point	−33°C
Boiling point	74°C (1 bar)
Explosion limit with air	8 vol% (25°C, 1 bar)
Solubility in water	0.44 mass% (20°C)
Solubility of water in 1,1,1-trichloroethane	0.05 mass% (20°C)
Log(octanol/water partition coefficient)	2.17

Stabilized 1,1,1-trichloroethane can be stored in iron containers or, preferably, in high-grade steel tanks to prevent contact with light, air, and moisture. 1,1,1-Trichloroethane should be transported in phosphatized, galvanized, or baked enamel tank cars of high-grade steel (*190*).

5. *Use*

In 1976, 574 million pounds were produced in the United States, of which 64 million pounds were exported (*191*). Slightly more than 200 million pounds were produced in the Federal Republic of Germany and 120 million pounds in Japan (*190, 191*).

1,1,1-Trichloroethane is used primarily as an industrial solvent. In 1974, it was employed in the United States for cold cleaning of metals (37%) and vapor degreasing (34%), as an intermediate product in the production of 1,1-dichloroethylene (23%), and for various other purposes (6%). The percentages for western Europe are approximately the same. The annual consumption of 1,1,1-trichloroethane for vapor degreasing increased by 21% between 1971 and 1974. 1,1,1-Trichloroethane is less corrosive than trichloroethylene or perchloroethylene, for example, and is therefore particularly well suited as a stain remover for delicate textiles.

In 1974, approximately 47% of the total United States production of 1,1-dichloroethylene was synthesized from 1,1,1-trichloroethane.

1,1,1-Trichloroethane is used also in aerosols, in which it functions as low-pressure propellant and, at the same time, as solvent and carrier for active chemicals. It has been estimated that 14 aerosols and 45 other commodities, primarily polishes and stain removers, readily available on the open market in the United States in 1969 contained from 10 to 100% 1,1,1-trichloroethane. Several million pounds are used annually for the production of adhesives. The rest is utilized for products such as lubricants and coolants for steel processing, drain cleaners, shoe polish, ink, and insecticides (*6, 71, 72, 190, 206*).

B. Environmental and Biological Data

1. Environmental Load

1,1,1-Trichloroethane does not occur as a natural product. Its concentrations in the environment are more the fault of the consumer than the producer.

1,1,1-Trichloroethane is relatively unstable and decomposes rapidly in the atmosphere by reacting with atmospheric hydroxyl radicals, ozone, and other substances. The tropospheric half-life is estimated to be 380 days (*43*); the atmospheric lifetime is estimated to be 2 to 5 or 11 years (*18, 98*). Measurements in North America indicated tropospheric concentrations of 145 ± 25 ppt; the values in the lower stratosphere were 50% lower. Concentrations of 65 ppt were measured in the southern hemisphere (*98*). At least 15% of the 1,1,1-trichloroethane in the troposphere enters the atmosphere where it probably decomposes and contributes to the catalytic degradation of the ozone layer. 1,1,1-Trichloroethane is capable of decomposing 20% as much ozone as are Freon 11 and 12, i.e., 1% of the stratospheric ozone is decomposed by 1,1,1-trichloroethane (*43, 106, 132, 174*).

Mean air concentrations for 14 cities and rural areas in the United States ranged between 77 ppt and 0.83 ppb. Similar measurements were obtained in the municipal area of Tokyo, the Republic of Ireland, and South Africa. The 1,1,1-trichloroethane concentration varied markedly during the course of the day and depended on the amount of rain. It declined markedly on sunny days (*22, 50, 95, 126, 151, 169*). A British study of six different towns showed that the air concentration (1–16 ppb) was markedly higher than that in the studies cited above (*156*). According to this study, rainwater concentrations were as high as 90 ng/liter and tap water concentrations (including carbon tetrachloride) as high as 300 ng/liter. The concentrations in seawater (3.3 μg/liter) were even higher; marine sediments contained as much as 5 μg/kg. In a cross-sectional study carried out between 1972 and 1976 in several west European countries, the mean concentrations ranged from 0.1 to 3.0 ppb. The range of distribution for river, lake, and canal water measured at the same time was considerably broader (*21*). Concentrations as high as 16.5 μg/liter were measured in unpurified wastewater (*33*). One group of researchers compared soil from (1) farms and forests far away from any industrial plants, (2) land at some distance from industrial plants, and (3) areas in the immediate vicinity of industrial plants. Mean concentrations of 1, 2.2, and 9.0 $\mu g/m^3$, respectively, were found at the three sites at drilling depths of 15–100 cm (*123*). The same study established dose-dependent leaf damage when the soil was gased under standardized conditions with high doses of a mixture of 1,1,1-trichloroethane, trichloroethylene, and tetrachloroethylene. The doses were 1000 to 3000 times higher than previously detected atmospheric concentrations. 1,1,1-Trichloroethane concentrations ranging from 1 to 10 mg/kg were reported in 12 food samples (*105*). And it was found consistently in many aquatic plants and animals.

2. Workplace Hazard

NIOSH estimates that approximately 2.9 million workers in the United States come in contact with 1,1,1-trichloroethane (*121*). Because of its broad application as solvent, most of these workers are probably exposed to higher doses. High-risk occupations are cleaning and degreasing jobs in the metal and textile industry as well as the production and use of aerosol cans containing 1,1,1-trichloroethane. Decomposition to phosgene in the presence of an open flame, hot metals, and lighted cigarettes is a special hazard. Experimental tests indicated that phosgene was also formed by shortwave radiation during inert gas shielded arc welding (*23*). Stabilizing agents accumulate when 1,1,1-trichloroethane is used for vapor degreasing (*187*).

3. Pharmacokinetics

1,1,1-Trichloroethane can be inhaled, ingested, or absorbed through the skin (*178*). The partition coefficients are blood/air, 3.3 and oil/blood, 108 (*155*). Tests with mice showed that 1,1,1-trichloroethane was distributed in the organs (primarily in the liver, then, in approximately equal amounts, in the brain and kidneys) relatively soon after intake (*65*). Tests with labeled 1,1,1-trichloroethane in rats showed that 98.7% was exhaled unaltered within 25 h after exposure, 0.5% was found as CO_2, and 0.85% appeared in the urine as trichloroethanol (*52*). In contrast to similar compounds, *in vitro* dechlorination in the presence of liver microsomes, NADPH, and oxygen was not demonstrable (*194*). Ikeda and Ohtsuji and other researchers, however, found small amounts of trichloroethanol and trichloroacetic acid in laboratory animals after inhalation and after intraperitoneal injection (*73*). These metabolites are probably produced by hydroxylation.

Similar intake, distribution, and elimination mechanisms can be assumed for man. Percutaneous absorption has been quantitatively demonstrated in man; the amount, however, is insignificant (*148*). After short-term inhalation, labeled 1,1,1-trichloroethane is exhaled rapidly and almost entirely. The relatively rapid exhalation in comparison with other chlorinated hydrocarbons can be explained, for the most part, by the low blood/air partition coefficient of 5 (*113*). The results of an 8 h inhalation test of 72–213 ppm showed that 90% of the inhaled dose had been exhaled within 8 days. In the next 12 days, 1.8% of the retained dose was excreted in the urine as trichloroacetic acid and 4.2% as trichloroethanol (*35*). Similar findings were also obtained in printers who were exposed at their workplace to mean concentrations of 4–53 ppm; the metabolite concentration increased markedly toward the end of the work week (*164*).

The biological half-life of 1,1,1-trichloroethane, therefore, depends primarily on the length and level of exposure and is estimated to be 4–26 h (*45, 110, 112, 164*). The intake of 1,1,1-trichloroethane can be tripled by physical exertion (increased respiratory minute volume) (*111*).

4. Toxic Effects

a. Animal Tests. Compared with other chlorinated hydrocarbons, the irritant effect on the skin and mucosa of rabbits is mild (*29*). After 4 days inhalation of 500 ppm, rats showed no signs of abnormal behavior. After longer inhalation exposure, however, the RNA content in the brain and *P*-450, the microsomal hepatocytochrome, declined somewhat (*156*). Light anesthesia is attained at concentrations as low as 5000 ppm. All animals survived short-term anesthesia concentrations of 18,000 ppm (*1*). Oral LD_{50} for rats and mice is 11 g/kg body weight; intraperitoneal LD_{50} in mice is 5 g/kg body weight (*83,187*).

Organ damage has been reported after acute as well as chronic inhalation. After acute inhalation, however, it was observed only at nearly lethal doses. McNutt *et al.* (*108*) exposed mice to either 250 or 1000 ppm. After 14 days inhalation, they found slight cytoplasmic alterations at concentrations as low as 250 ppm. Hepatocyte necrosis with inflammatory reactions and Kupfer's cell hypertrophy were observed in the group exposed to 1000 ppm. Hepatotoxicity can be somewhat potentiated by enzyme induction with phenobarbital, 3-methylcholanthrene, and other inducers (*17*). No distinct effects could be produced by induction with ethanol (*83*). 1,1,1-Trichloroethane, however, has proved to be a potent enzyme inducer. The metabolism of short-acting barbiturates is increased in rats and mice after inhalation of 2000 ppm and, therefore, the sleep period is shortened. *In vitro* tests have shown that the oxidation of barbiturates and the demethylation of aminopyrene in liver extracts occur more rapidly after inhalation of 3000 ppm 1,1,1-trichloroethane (*91, 92*). The induction effect of 1,1,1-trichloroethane can be inhibited by cyclohexamide and actinomycin, two cytostatically active protein synthesis inhibitors. Lower 1,1,1-trichloroethane concentrations of 500 ppm did not produce enzyme induction after 7 days inhalation (*193*). No intensifying effect was observed with nicotinic hydrochloride (*135*).

In animal experiments, 1,1,1-trichloroethane produced a fall of blood pressure in the left ventricle and the great arteries, as well as cardiac arrhythmias (*182*). Application of low doses (0.25–0.4 cm^3/kg body weight) and subsequent injection of adrenalin in anesthetized dogs resulted in ventricular extrasystoles and ventricular tachycardia (*145, 147*). The thesis that cardiac dysrhythmias are always induced by the excretion of physiologic amounts of catecholamine together with high inhalation doses of 1,1,1-trichloroethane was supported by experiments with mice whose adrenal glands had been removed prior to administration of different adrenalin doses (*64*). Extensive studies on dogs and rabbits have dealt with pressure conditions in the heart and general circulation. With anesthetic concentrations, peripheral vasodilation as well as positively chronotropic and inotropic heart action develop initially; approximately 1 min later, the stroke volume decreases and muscular contraction is reduced. The indirect cardioactive effect in this second phase can be prevented by administration of calcium ions (*63*).

Kidney damage, predominantly swelling of the proximal tubules, develops at an anesthetic concentration equal to the LD_{80}. Necrosis or measurably impaired renal function was not observed (*83a, 133*). 1,1,1-Trichloroethane also appears to influence the immune system. In rats and rabbits exposed for several months to concentrations as low as 18 ppm, the immune response after subcutaneous administration of typhoid vaccine was considerably weaker than that of the control group (*83a*).

b. Effects in Man. 1,1,1-Trichloroethane has a mildly irritant effect on the skin and mucosa, beginning at approximately 500 ppm (*187*). Psychomotor performance was significantly impaired at concentrations exceeding 350 ppm (*44, 153*). Slight disturbances in coordination and pronounced dizziness developed at 900 ppm. Gait disturbances were pronounced at 1900 ppm; test subjects were unable to stand unsupported at 2600 ppm. Anesthetic concentrations range between 10,000 and 26,000 ppm (*148, 187*). Deaths described in the literature (*10, 26, 54, 58, 84, 176, 187*) predominantly involved solvent sniffers or were due to improper use in the household or at the workplace. In most cases, death was due to respiratory paralysis; in a few cases, death was caused by cardiac arrhythmias and ventricular fibrillation. Minimum lethal concentration limits for man are not known.

Because of the low metabolic rate, specific organ damage is extremely unlikely. Acute intoxications are completely reversible. A few cases of reversible liver and kidney damage and one case of chronic liver damage have been described after acute intoxication-induced icteric necrosis (*53, 183*).

Occupational intoxication in chronically exposed workers has not been reported. One epidemiologic study showed no significant differences between the 151 workers examined and the control group (*88*).

Standards of permissible exposure are summarized in Table VIII.

c. Carcinogenicity. The tumor incidence in mice and rats fed various doses of 1,1,1-trichloroethane or in rats after inhalation tests and intratracheal instillation did not differ significantly from that of the control groups. Because of the

TABLE VIII

Standards of Permissible Exposure for 1,1,1-Trichloroethane

Index	Permissible exposure	Reference
TLV–TWA	350 ppm (1900 mg/m^3)	*185*
TLV–STEL	450 ppm (2450 mg/m^3)	*185*
MAK	200 ppm (1080 mg/m^3)	*102*
EPA criteria		
Freshwater aquatic life	5300 μg/liter (24 h average)	*170*
	12,000 μg/liter (MCC)	*170*
Saltwater aquatic life	240 μg/liter (24 h average)	*170*
	540 μg/liter (MCC)	*170*
Human health	15.7 mg/liter	*170*

high death rate of the experimental animals, however, definitive conclusions are not possible at this time. Further studies are in progress (*116, 138*).

The mutagenic effect of 1,1,1-trichloroethane was weak in *S. typhimurium* TA 100, with and without microsomal activation system (*168a*). 1,1,1-Trichloroethane induces *in vitro* transformation in the Fischer rat embryo cell system (*134*).

In addition, the teratogenic effect of 1,1,1-trichloroethane on chick embryos was more potent than the other chlorinated hydrocarbons examined (*32*). This finding, however, was not confirmed by another study in which pregnant rats and mice were exposed to 875 ppm during days 6–15 of gestation for 7 h a day. No damage of any kind was demonstrable in the brood animals or the embryos (*163*).

5. Analysis

Air specimens were trapped against activated charcoal, desorbed, and analyzed with GC/FID, GC/ECD, or GC/MS (*122, 151*). The most reliable method for analyzing water samples is GC/MS (*27*).

VI. 1,1,2-TRICHLOROETHANE ($CHCl_2CH_2Cl$)[6]

A. Technical and Economic Data

1. History

1,1,2-Trichloroethane was first synthesized in 1840 (*143*). Large-scale production began in 1940 (*190*).

[6]Synonyms: β-trichloroethane, ethane trichloride, vinyl trichloride.

2. Physical Properties

1,1,2-Trichloroethane is a colorless, sweet-smelling, nonflammable liquid which is miscible with most organic solvents and forms azeotropes with methanol, tetrachloroethylene, ethanol, and water. In the presence of air, 1,1,2-trichloroethane forms no ignitable mixtures (*56, 189a, 190, 206*). Its physical properties are summarized in Table IX.

3. Chemical Properties

Under exclusion of air and water, 1,1,2-trichloroethane is stable at temperatures below 110°C. It is dehydrochlorinated to almost equal amounts of 1,1-dichloroethylene and 1,2-dichloroethylene at 400°C. The process is accelerated by catalysts. Boiling water produces vigorous hydrolysis (*56, 190*).

4. Production

The most important and economical production process is direct chlorination of ethylene. Various other processes are also employed, but they are not of practical importance (*56, 190*).

5. Use

No information is available on the production of 1,1,2-trichloroethane in the United States and Europe or on the amount imported and exported. Japanese production in 1976 was approximately 72 million pounds.

Most 1,1,2-trichloroethane is formed as an intermediate product in the production of 1,1-dichloroethylene. A very small amount is used as solvent, e.g., for chlorinated rubber and adhesives. Exact data, however, are not available (*56, 72, 190*).

TABLE IX

Physical Properties of 1,1,2-Trichloroethane

Property	Value
Molecular weight	133.41
Density	1141 kg/m^3 (20°C)
Vapor density	4 kg/m^3 (1 bar, boiling point)
Melting point	−37°C
Boiling point	113°C (1 bar)
Vapor pressure	48 mbar (30°C)
	492 mbar (90°C)
	906 mbar (110°C)
Log(octanol/water partition coefficient)	2.17
Solubility in water	0.45% (20°C)
Solubility of water in 1,1,2-trichloroethane	0.05% (20°C)

B. Environmental and Biological Data

1. Environmental Load

1,1,2-Trichloroethane does not occur as a natural product. In light of the low production rates, heavy air contamination is highly unlikely; measurements, however, are not available. Concentrations of 5.4 mg/liter were obtained in industrial wastewater. Concentrations in several random samples of drinking water in the United States and Europe ranged from 0.1 to 8.5 μg/liter (*20, 152*).

2. Workplace Hazard

NIOSH estimates that approximately 112,000 persons have contact with 1,1,2-trichloroethane at their workplace. No measurements are available on mean exposure concentrations. The highest concentrations were found in individuals working around blast furnaces, in steel rolling mills, and in factories manufacturing technical instruments (*121*).

3. Pharmacokinetics

1,1,2-Trichloroethane can be inhaled, ingested, or absorbed through the skin. The partition coefficients are blood/air, 38.6 and oil/blood, 59 (*155*). After intraperitoneal administration of 0.1 to 0.2 g/kg body weight of labeled 1,1,2-trichloroethane in mice, 73 to 87% of the radioactivity was found in the urine and 16 to 22% in the respiratory air (40%, unaltered; 60%, CO_2). The metabolites chloroacetic acid, *S*-carboxymethylcysteine, and thiodiacetic acid were found in the urine. Small amounts of glycolic acid, dichloroethanol, trichloroethanol, oxalic acid, and trichloroacetic acid also were detected, thus indicating that 1,1,2-trichloroethane is metabolized by the formation of chloroacetaldehyde (*211*). In the presence of liver microsomes, NADPH, and oxygen, enzymatic dechlorination could be demonstrated *in vitro*. This process was accelerated by pretreatment with enzyme inducers, phenobarbital, and benzopyrene (*194*). Pharmacokinetic studies on man have not been published.

4. Toxic Effects

a. Animal Tests. Oral LD_{50} in rats is 835 mg/kg body weight; intraperitoneal LD_{50} in mice is 500 mg/kg body weight (*83, 172*). The lethal inhalation concentrations for 3 and 7 h exposures were 18,000 and 14,000 ppm, respectively (*1*). Percutaneous intake was extremely high in guinea pigs (*199*). Dose-dependent liver and kidney damage was demonstrated in different experimental animals after all methods of application. The hepatotoxicity of 1,1,2-trichloroethane was lower than that of chloroform and carbon tetrachloride, but higher than that of 1,1,1-trichloroethane. Pretreatment with enzyme inducers or simultaneous administration of polychlorinated biphenyls and experimental alloxan diabetes intensifies the hepatotoxic effect (*17, 46, 55, 83, 85, 199*).

TABLE X

Standards of Permissible Exposure for 1,1,2-Trichloroethane

Index	Permissible exposure	Reference
TLV–TWA	10 ppm (45 mg/m^3)	*185*
TLV–STEL	20 ppm (90 mg/m^3)	*185*
MAK	10 ppm (45 mg/m^3)	*102*
EPA criteria		
Freshwater aquatic life	310 μg/liter (24 h average)	*170*
	710 μg/liter (MCC)	*170*
Saltwater aquatic life	Not established	*170*
Human health	2.7 μg/liter	*170*

Experiments with fish and aquatic invertebrates indicated that 1,1,2-trichloroethane impaired reproductivity (*149*).

b. Effects in Man. Lethal or anesthetic concentrations for man are not known. Occupational intoxication or epidemiologic studies on exposed groups are not available.

Standards of permissible exposure are given in Table X.

c. Carcinogenicity. After feeding different doses of 1,1,2-trichloroethane to B6C3F$_1$ mice, approximately 50% of the mice in each group were still alive 90 weeks later. The incidence of hepatocellular carcinomas in the treated animals was significantly higher than in the control group. No significant differences, however, were found with a similar experimental model using Osborne–Mendel rats (*117*).

1,1,2-Trichloroethane was not mutagenic in *S. typhimurium* TA 1535, with or without liver microsomes (*140*). No studies on teratogenicity or embryotoxicity are available.

5. *Analysis*

Air samples were trapped against activated charcoal and analyzed with GC/FID (*122*). The most reliable water analyses are obtained with GC/MS (*20*).

VII. 1,1,1,2-TETRACHLOROETHANE (CH_2ClCCl_3)

A. Technical and Economic Data

1. *History*

1,1,1,2-Tetrachloroethane was first synthesized in 1898 (*190*).

2. *Physical Properties*

1,1,1,2-Tetrachloroethane is a colorless, nonflammable, dense liquid (*190*). Its physical properties are given in Table XI.

3. *Chemical Properties*

1,1,1,2-Tetrachloroethane is chemically more stable than 1,1,2,2-tetrachloroethane. It is dehydrochlorinated at 500°C to yield trichloroethane (*190*).

4. *Production*

Pure 1,1,1,2-tetrachloroethane is produced by the addition of chlorine to 1,1-dichloroethylene. It is a by-product of trichloroethylene produced from ethylene (*190*).

5. *Use*

1,1,1,2-Tetrachloroethane is apparently not produced in large amounts or used commercially in the United States or Europe.

B. Environmental and Biological Data

1. *Environmental Load*

Since production is low, 1,1,1,2-tetrachloroethane is not an environmental hazard. Measurement data are not available.

2. *Workplace Hazard*

There are no known jobs at which 1,1,1,2-tetrachloroethane is regularly used.

3. *Pharmacokinetics*

The partition coefficients are blood/air, 30.4, and oil/blood, 142 (*155*). No additional studies on the intake, distribution, and elimination of 1,1,1,2-tetrachloroethane have been published.

4. *Toxic Effects*

a. Animal Tests. Tests with mice, rats, rabbits, and hares indicate that 1,1,1,2-tetrachloroethane has only a slight irritant effect on the skin and mucosa.

TABLE XI

Physical Properties of 1,1,1,2-Tetrachloroethane

Molecular weight	167.68
Melting point	−68°C
Boiling point	129°C (1.013 bar)

TABLE XII

Standards of Permissible Exposure for 1,1,1,2-Tetrachloroethane

Index	Permissible exposure	Reference
TLV–TWA	Not established	*185*
TLV–STEL	Not established	*185*
MAK	Not established	*102*
EPA criteria		
Freshwater aquatic life	420 μg/liter (24 h average)	*170*
	960 μg/liter (MCC)	*170*
Saltwater aquatic life	Not established	*170*
Human health	Not established	*170*

Percutaneous absorption is low. 1,1,1,2-Tetrachloroethane is two to three times less toxic than 1,1,2,2-tetrachloroethane. Histologic examination showed microvacuolization and centrilobular necrosis in the liver. Slight alterations were also found in the myocardium. The offspring of brood animals treated with 1,1,1,2-tetrachloroethane died within 2 days after birth. 1,1,1,2-Tetrachloroethane therefore also passes through the placental barrier. Reproductive disorders, however, were not observed (*136, 188, 189*).

b. Effects in Man. Reports on acute or chronic intoxication in man and epidemiologic studies are not available. Standards of permissible exposure are given in Table XII.

c. Carcinogenicity. Studies on carcinogenicity, mutagenicity, and teratogenicity have not been published.

5. *Analysis*

Special analytical methods for detection of 1,1,1,2-tetrachloroethane are not described in the literature.

VIII. 1,1,2,2-TETRACHLOROETHANE ($CHCl_2CHCl_2$)[7]

A. Technical and Economic Data

1. *History*

1,1,2,2-Tetrachloroethane was first synthesized in 1869. It has been produced commercially since 1908 (*190*).

[7]Synonyms: *sym*-tetrachloroethane, acetylene tetrachloride, 1,1-dichloro-2,2-dichloroethane, ethane tetrachloride.

2. *Physical Properties*

1,1,2,2-Tetrachloroethane is a colorless, dense, nonflammable liquid with a penetrating, sweet, chloroformlike smell. It is miscible with all common organic solvents, and it has the greatest solvent power of all aliphatic chlorinated hydrocarbons. It forms no explosive mixtures with air (*190*). Its physical properties are summarized in Table XIII.

3. *Chemical Properties*

1,1,2,2-Tetrachloroethane is stable in the absence of light, air, and moisture. It decomposes to trichloroethylene and hydrogen chloride only at temperatures above 400°C; this process is demonstrable with suitable catalytic crackers at temperatures as low as 250°C. In the atmosphere, it decomposes to hydrogen chloride and small amounts of phosgene. Dichloroacetyl chloride and phosgene are formed by photooxidation. 1,1,2,2-Tetrachloroethane is resistant to strong acids. Weak alkalis split off trichloroethylene; strong alkalis split off explosive dichloroacetylene. Iron, aluminum, and zinc reduce 1,1,2,2-tetrachloroethane to 1,2-dichloroethylene in the presence of steam (*56, 190*).

4. *Production and Transport*

The most important production process is the blowing of gaseous acetylene and chlorine into liquid tetrachloroethane using ferric chloride as a catalyst. The technical yield is between 90 and 98%. Due to the high cost of acetylene, induced chlorination of ethylene or 1,2-dichloroethane has recently become more important.

The degree of purity of 1,1,2,2-tetrachloroethane is 95–98%. It is stored and transported in iron tanks (*56, 190*).

TABLE XIII

Physical Properties of 1,1,2,2-Tetrachloroethane

Molecular weight	167.86
Density	1597 kg/m^3 (20°C)
Vapor density	5 kg/m^3 (boiling point, 1 bar)
Melting point	−43°C
Boiling point	146°C
Vapor pressure	7 mbar (20°C)
	53 mbar (60°C)
	187 mbar (91°C)
	827 mbar (138°C)
Solubility in water	0.29 mass% (20°C)
Solubility of water in 1,1,2,2-tetrachloroethane	0.03 mass% (20°C)
Log(octanol/water partition coefficient)	2.56

5. Use

Exact production statistics for 1,1,2,2-tetrachloroethane are available for neither the United States nor Europe. It is not produced commercially in Japan. Approximately 440 million pounds of 1,1,2,2-tetrachloroethane was estimated to have been produced in the United States in 1967. Production declined markedly thereafter. The 1974 production was estimated to have been 34 million pounds. By 1976, only one United States company admitted production of 1,1,2,2-tetrachloroethane, but it refused to reveal the amount (*191*).

1,1,2,2-Tetrachloroethane is used almost exclusively as an intermediate product in trichloroethylene production. It is estimated that, in 1967, 85% of all trichloroethylene was produced from 1,1,2,2-tetrachloroethane. As a result of other production processes, however, this percentage fell to 8% in 1974. 1,1,2,2-Tetrachloroethane is used as a mothproofing agent, an insecticide in greenhouses, a grain disinfectant, and a solvent (*56, 190*).

B. Environmental and Biological Data

1. Environmental Load

1,1,2,2-Tetrachloroethane does not occur as a natural product. Most of it enters the environment through production and utilization. In Japan, 1,1,2,2-tetrachloroethane concentrations of 0.01–9.4 ppb were detected in large cities (*127*). In the United States and in Europe, concentrations of 0.01–0.11 μg/liter were obtained in drinking water and 2.2 mg/liter in industrial wastewater (*33, 165*). 1,1,2,2-Tetrachloroethane was also found in industrial vinyl chloride wastes that were dumped in the North Sea (*74*).

2. Workplace Hazard

NIOSH estimates that approximately 11,000 persons have occupational contact with 1,1,2,2-tetrachloroethane (*121*). Production, storage, and transport represent a hazard. No statistics are available on the number of exposed individuals.

3. Pharmacokinetics

1,1,2,2-Tetrachloroethane is primarily inhaled; percutaneous absorption is slight. The partition coefficients are blood/air, 121.4, and oil/blood, 109 (*155*).

Labeled 1,1,2,2-tetrachloroethane was eliminated for 3 days after intraperitoneal injection: 45–61% of the radioactivity was exhaled as CO_2; less than 4% was exhaled unaltered; 23–34% of the radioactivity was excreted in the urine; 16% of the injected dose remained in the body. Dichloroacetic acid, trichloroacetic acid, trichloroethanol, and oxalic acid were detectable in the urine

(*210*). After 8 h inhalation of 200 ppm, trichloroacetic acid and trichloroethanol also were excreted in the urine.

Comparative volunteer tests with other labeled halogenated hydrocarbons showed that, after short-term inhalation, 1,1,2,2-tetrachloroethane has the lowest elimination rate of all compounds; only approximately 3% had been exhaled within 1 h (*113*).

4. *Toxic Effects*

a. Animal Tests. After 4 h inhalation, LC_{50} for male rats was 8.6 mg/liter (*210*). Oral LD_{50} for rats was 250 mg/kg body weight; intraperitoneal LD_{50} for mice was 820 mg/kg body weight (*48, 188*).

1,1,2,2-Tetrachloroethane is a strong irritant of skin and mucosa. Maximum anesthetic concentrations for experimental animals are unknown.

1,1,2,2-Tetrachloroethane has a potent hepatotoxic effect. After just 3 h inhalation of 800 ppm, triglycerides and phospholipids increased significantly in mouse livers; the maximum concentration was attained after 25 h. The alterations, however, were not as severe as those developing after carbon tetrachloride exposure. Benzopyrene hydroxylase and *p*-nitroanisole-*O*-demethylase concentrations declined 50% 24 h after feeding one single dose of 437 mg/kg body weight to rats (*192*). After long-term exposure, inflammatory alterations were detectable in the liver. This effect was intensified by raising the ambient temperature. With high doses, the mortality rate in mice also depends on the time of day. Fatty degeneration of the renal tubules as well as icteric necrosis were found after acute intoxication (*160, 209*).

b. Effects in Man. 1,1,2,2-Tetrachloroethane is a strong irritant of skin and mucosa and a potent anesthetic. Minimum anesthetic concentrations for man have not been established.

Many severe or fatal intoxications resulting from accidental inhalation or ingestion at the workplace were reported prior to 1930. Pronounced to extremely severe toxic alterations of the liver and toxic fatty degeneration of the renal tubules were always present in such cases. Severe cerebral damage and peripheral neuroparalysis have also been reported. Alterations of the nervous system were observed after long-term exposure (*47, 97, 214*). Recent epidemiologic studies are not available.

Standards of permissible exposure are given in Table XIV.

c. Carcinogenicity. The results of ingestion tests with mice showed a significant dose-dependent incidence of hepatocellular carcinoma. Ingestion tests with rats, however, showed no significant difference between treated and control groups in the incidence of such carcinoma (*118*). The difference between mice

TABLE XIV

Standards of Permissible Exposure for 1,1,2,2-Tetrachloroethane

Index	Permissible exposure	Reference
TLV–TWA	5 ppm (35 mg/m³)	*185*
TLV–STEL	10 ppm (70 mg/m³)	*185*
MAK	1 ppm (7 mg/m³)	*102*
EPA criteria		
Freshwater aquatic life	170 μg/ml (24 h average)	*170*
	380 μg/ml (MCC)	*170*
Saltwater aquatic life	70 μg/ml (24 h average)	*170*
	160 μg/ml (MCC)	*170*
Human health	1.8 μg/ml	*170*

receiving an intraperitoneal injection and the control group was also not significant. Since the death rate of the animals was high, these findings should be regarded with reservation (*184*). 1,1,2,2-Tetrachloroethane had a mutagenic effect on *S. typhimurium* strains TA 1530 and TA 1535, but not on TA 1538. Mutagenic and cytotoxic effects of 1,1,2,2-tetrachloroethane were also demonstrable in other bacterial systems (*14, 16*).

Deformities were observed in embryos and young animals after treatment of pregnant mice with daily 1,1,2,2-tetrachloroethane doses of 300 mg/kg body weight (*159*).

5. *Analysis*

Air samples were trapped against activated charcoal, desorbed, and analyzed with GC/FID (*122*). The most reliable water analyses are provided by GC/MS (*124*).

IX. PENTACHLOROETHANE ($CHCl_2CCl_3$)[8]

A. Technical and Economic Data

1. *History*

Pentachloroethane was first synthesized in 1840. Its industrial importance is not known (*143, 190*).

[8]Synonym: ethane pentachloride.

TABLE XV

Physical Properties of Pentachloroethane

Molecular weight	202.31
Density	1680 kg/m^3 (20°C)
Vapor density	6 kg/m^3 (1 bar)
Melting point	−29°C
Boiling point	161°C (1 bar)
Vapor pressure	5 mbar (20°C)
	35 mbar (60°C)
	79 mbar (80°C)
	173 mbar (100°C)
Solubility in water	0.05 mass% (20°C)
Solubility of water in pentachloroethane	0.24 mass% (20°C)

2. *Physical Properties*

Pentachloroethane is a colorless, dense, nonflammable liquid with a sweet–sour smell. It does not form explosive mixtures with air and is miscible with all solvents (*190*). Its physical properties are summarized in Table XV.

3. *Chemical Properties*

In the absence of air and moisture, pentachloroethane is stable. In the presence of water, it hydrolyzes at normal temperatures. It is split by weak alkalis or heat to tetrachloroethylene and hydrogen chloride, and to small amounts of trichloroethylene and chlorine (*190, 206*).

4. *Production and Transport*

Pentachloroethane can be produced by the addition of chlorine to trichloroethylene using ferric chloride as a catalyst and by induced chlorination of 1,2-dichloroethane with ethylene. Pentachloroethane must be stabilized. After stabilization, it can be stored in iron tanks for extended periods of time (*190, 206*).

5. *Use*

Pentachloroethane is used almost exclusively as an intermediate product in tetrachloroethylene production. Production statistics are not available. Since tetrachloroethylene is now produced via considerably more economical processes, the annual production of pentachloroethane in the United States and Europe is probably low. Statistics on its use in other fields are not available.

B. Environmental and Biological Data

1. *Environmental Load*

Studies on the environmental load are not available.

2. *Workplace Hazard*

The production of pentachloroethane can be hazardous.

3. *Pharmacokinetics*

Studies on the intake, distribution, and elimination of pentachloroethane are not available.

4. *Toxic Effects*

a. Animal Tests. Pentachloroethane is a strong irritant to skin and mucosa. It is a more potent anesthetic than chloroform. The LD_{50} for dogs after intravenous injection is 100 mg/kg body weight and in mice after inhalation, LD_{50} is 35 mg/liter (*9, 93*). The anesthetic dose ranges from 13 to 37 mg/liter, depending on the length of anesthesia. Inhalation of 120 ppm for 8 h a day over a period of 23 days was tolerated by cats without any signs of CNS impairment (*94*). Pronounced toxic alterations, however, were found in the liver, kidneys, and lungs. Fatty degeneration of the liver and toxic alterations of the kidneys and lungs were observed in dogs after 3 weeks exposure (*75*).

b. Effects in Man. Toxic damage in man after acute or chronic exposure has not been reported. Epidemiologic studies are not available.

Standards of permissible exposure are given in Table XVI.

c. Carcinogenicity. Studies on carcinogenicity, mutagenicity, and teratogenicity have not been published.

5. *Analysis*

Special analytical methods for detection of pentachloroethane are not cited in the literature.

TABLE XVI

Standards of Permissible Exposure for Pentachloroethane

Index	Permissible exposure	Reference
TLV–TWA	Not established	*185*
TLV–STEL	Not established	*185*
MAK	5 ppm (40 mg/m³)	*102*
EPA criteria		
Freshwater aquatic life	440 μg/liter (24 h average)	*170*
	1000 μg/liter (MCC)	*170*
Saltwater aquatic life	38 μg/liter (24 h average)	*170*
	87 μg/liter (MCC)	*170*
Human health	Not established	*170*

X. HEXACHLOROETHANE (CCl_3CCl_3)[9]

A. Technical and Economic Data

1. History

It is not known exactly when hexachloroethane was first synthesized. Commercial production began around 1920 (*190*).

2. Physical Properties

Hexachloroethane is a colorless crystal with a camphorlike smell. It occurs in three forms: rhombic (below 46°C), triclinic (46–71°C), and cubic (above 71°C). It forms azeotropes with numerous organic compounds (*190*). Its physical properties are summarized in Table XVII.

3. Chemical Properties

When hexachloroethane is slowly warmed, it sublimes without melting or decomposing. Larger amounts of chlorine are split off at temperatures above 250°C. Hexachloroethane is acid and alkali resistant at higher temperatures; in the presence of water, however, it corrodes iron. In the presence of metals, it dechlorinates to tetrachloroethylene and metal chlorides. It disproportionates to tetrachloroethylene and carbon tetrachloride at temperatures above 400°C (*190*).

4. Production

Hexachloroethane is produced by chlorination of tetrachloroethylene with ferric chloride at temperatures of 100 to 140°C (*190*). No information is available on impurities, stabilizing agents, or special requirements for storage and transport.

5. Use

Significant quantities of hexachloroethane are apparently not produced in the United States. In 1967, undisclosed amounts were produced by only one company. Production has since been discontinued and the requirements are being filled by imports (*71*). In 1976, the United States imported 1.46 million pounds; 760,000 from Great Britain and 700,000 from France. The annual Japanese production for 1975 was 600,000–1,000,000 pounds.

Hexachloroethane is used primarily for the production of smoke candles and grenades by intensively mixing it with zinc, magnesium, and aluminum dust plus an oxygen compound. After ignition, this mixture forms fuming metallic chloride. Small quantities are used also for degassing in aluminum and magnesium

[9]Synonyms: 1,1,1,2,2,2-hexachloroethane, hexachloroethylene, carbon hexachloride, ethane hexachloride, perchloroethane.

TABLE XVII

Physical Properties of Hexachloroethane

Molecular weight	236.76	
Density	2091 kg/m^3 (20°C)	
Vapor density	6.3 kg/m^3 (sublimation point, 1 bar)	
Melting point	188°C	
Boiling point	185°C	
Vapor pressure	0.29 mbar	(20°C)
	1.33 mbar	(40°C)
	24 mbar	(80°C)
	116 mbar	(120°C)
Solubility in water	50 mass-ppm (20°C)	
Log(octanol/water partition coefficient)	3.34	

smelting, as ignition suppressant in flammable liquids, as a component of high-pressure lubricants, as plasticizer in cellulose esters, as additive in fire extinguishing agents, as mothproofing agent, and as a helminthicide in veterinary medicine (*56, 190,* 206).

B. Environmental and Biological Data

1. Environmental Load

Hexachloroethane does not occur as a natural product. No reports on its detection in the atmosphere have been published. It, however, was identified in wastewaters from paper mills (concentration < 1 μg/liter) and in samples of tap water from 4 of 13 cities (concentration of 0.03–4.3 μg/liter). The concentrations in samples of river water and industrial wastewaters ranged from 4.4 to 8.4 μg/liter. It, however, could only be detected in 1 of 204 surface water samples from highly industrialized areas (*34, 77, 78, 165*).

2. Workplace Hazard

NIOSH estimates that approximately 1500 workers come in contact with hexachloroethane. Paperboard mills are particularly hazardous (*121*).

3. Pharmacokinetics

Ingested hexachloroethane is rapidly distributed in the body; the highest concentration is found in fatty tissue and the lowest in muscle tissue (*38*). The partition coefficients have not been established. Labeled hexachloroethane was metabolized slowly by rats: 14–24% of the radioactivity was exhaled as carbon dioxide, hexachloroethane, tetrachloroethylene, and 1,1,2,2-tetrachloroethane. Only 5% was excreted in the urine 3 days after ingestion. The following metabolites were identified: trichloroethanol (1.3%), dichloroethanol (0.4%), trichloroacetic acid (1.3%), dichloroacetic acid (0.8%), monochloroacetic acid (0.7%), and oxalic acid (0.1%); the rest remained in the body (*76*). Dechlorination of hexachloroethane was observed in rabbit liver homogenates (*13*). Formation of pentachloroethane and tetrachloroethylene was also detected in sheep (*38*).

Nothing is known about the pharmacokinetics of hexachloroethane in man.

4. *Toxic Effects*

a. Animal Tests. Hexachloroethane is a mild irritant to the skin and mucosa; it is a more potent anesthetic than chloroform. Severe inhalation intoxication was observed in rats at 5900 ppm. LD_{50} for rodents is between 1000 and 7000 mg/kg (*201*). Microsomal monooxygenase activity in rat liver declined 50% after administration of oral doses as low as 2.5 g/kg body weight (*192*). A substantial induction of hepatic enzymes, however, cannot be demonstrated (*201*). Liver damage was observed with higher doses (*38, 201*).

b. Effects in Man. Acute and chronic effects in man have not been described. Epidemiologic studies are not available. Standards of permissible exposure are given in Table XVIII.

c. Carcinogenicity. After 78 weeks, the incidence of dose-dependent hepatocellular carcinoma was significantly higher in $B6C3F_1$ mice fed various doses of hexachloroethane than in the control groups. This effect could not be demonstrated under similar test conditions in Osborne–Mendel rats, in which tumor incidence in the test animals was not significantly higher than in the control group. Due to the high death rate of the experimental animals, however, no conclusions can be drawn from these tests (*119*).

TABLE XVIII

Standards of Permissible Exposure for Hexachloroethane

Index	Permissible exposure	Reference
TLV–TWA	1 ppm (10 mg/m^3)	*185*
TLV–STEL	3 ppm (30 mg/m^3)	*185*
MAK	1 ppm (10 mg/m^3)	*102*
EPA criteria		
Freshwater aquatic life	62 μg/liter (24 h average)	*170*
	140 μg/liter (MCC)	*170*
Saltwater aquatic life	7 μg/liter (24 h average)	*170*
	16 μg/liter (MCC)	*170*
Human health	5.9 μg/liter	*170*

Mutagenicity tests on five strains of *S. typhimurium* and on *Saccharomyces cerevisiae* D4 were negative.

Teratogenic damage was not demonstrable in another study using Sprague–Dawley rats (*201*).

5. Analysis

Air samples were trapped against activated charcoal, desorbed, and analyzed with GC/FID (*122*). The most reliable water analyses were obtained with GC/MS (*34*).

XI. GENERAL CONCLUSIONS

Hazards by chlorinated ethanes can be classified into three categories: (1) global environmental hazards, (2) local environmental hazards, and (3) hazards to limited groups of persons.

A global environmental hazard is thinkable only for 1,1,1-trichloroethane because it is theoretically able to degrade the ozone layer in the atmosphere. Further investigations, therefore, are necessary to clarify this hypothesis.

Local environmental hazards are most likely with 1,1,1-trichloroethane and 1,2-dichloroethane. Both compounds are produced in large quantities and produce a measurable load for several ecological systems in environments near and far from their production site. The concentrations detected to now are in a range that excludes any hazards to vegetation, animals, or humans, but our range of knowledge is based on only a few individual investigations in some countries. They should be enlarged by systematic measurements in air, water, and soil on an international basis. No results of measurements are available about the environmental load of monochloroethane. This fact is surprising as monochloroethane production is the second largest of all chlorinated ethanes. Measurements of the atmospheric load seem particularly necessary, whereas the contamination of soil and water is probably very low. All other chlorinated ethanes are produced only in low quantities. It is unlikely that they are of any ecological importance.

Theoretically, hazards to limited groups of persons may exist in the workplace for all chlorinated ethanes used. NIOSH estimates that about 5 million workers in the United States are exposed. Of great importance are 1,1,1-trichloroethane, 1,2-dichloroethane, monochloroethane, and 1,1,2-trichloroethane. The other chlorinated ethanes are used in the workplace only in small quantities, but, in individual cases, they may cause serious intoxications. It is difficult to assess the general risk because of the large range of variation in the exposure conditions, but the concentrations in the workplace are usually in the parts per million range, whereas the loads for the general population are in the parts per trillion or parts per billion range. A human risk seems to exist only in the workplace.

References

1. Adams, E. M., Spencer, H. C., Rowe, V. K., and Irish, D. D. (1950). *Arch. Industr. Hyg. Occup. Med.* **1,** 225–236.
2. Adrian, J. (1967). "The Pharmacology of Anesthetic Drugs." Thomas, Springfield, Illinois.
3. Akimov, G. A., and Kolesnichenko, J. P. (1978). *Arkh. Patol.* **40,** 49–53.
4. Ames, B. N., Durston, W. E., Yamasaki, E., and Lee, F. D. (1973). *Proc. Natl. Acad. Sci. U.S.A.* **70,** 2281–2285.
5. Astrand, J. (1975). *Scand. J. Work, Environ. Health* **1,** 199–218.
6. Aviado, D. M., Zakhari, S., Simaan, J. A., and Ulsamer, A. G. (1976). "Methyl Chloroform and Trichloroethylene in the Environment." Cleveland, Ohio.
7. Banerjee, S., Van Duuren, B. L., and Oruambo, J. (1980). *Cancer Res.* **40,** 2170–2173.
8. Barnes, G. (1979). *Int. J. Addict.* **14,** 1–26.
9. Barsoum, G. S., and Saad, K. (1934). *Q. J. Pharm. Pharmacol.* **7,** 205.
10. Bass, M. (1970). *J. Am. Med. Assoc.* **212,** 2075–2079.
11. Bignami, M., Cardamone, G., Carere, A., Couba, P., Doglioti, E., Morpurgo, G., and Ortali, V. A. (1977). *Mutat. Res.* **46,** 243–244.
12. Brass, K. (1949). *Dtsch. Med. Wochenschr.* **74,** 553–554.
13. Bray, H. G., Thorpe, W. V., and Vallance, D. K. (1952). *Biochem. J.* **51,** 183–201.
14. Brem, H., Stein, A. B., and Rosenkranz, H. S. (1974). *Cancer Res.* **34,** 2576–2579.
15. Büch, H. P., and Büch, U. (1980). *In* "Allgemeine und Spezielle Pharmakologie und Toxikologie" (W. Forth, D. Henschler, and W. Rummel, eds.), p. 384. Bibliograph. Inst., Mannheim/Wien/Zürich.
16. Callen, D. F., Wolf, C. R., and Philpot, R. M. (1980). *Mutat. Res.* **77,** 55–64.
17. Carlson, G. P. (1973). *Life Sci.* **13,** 67–73.
18. Chang, J. S., and Penner, J. E. (1978). *Atmos. Environ.* **12,** 1867–1874.
19. "Chloroethanes: Review of Toxicity" (1978). NIOSH Curr. Intell. Bull. No. 27. U.S. Dept. Health, Ed., and Welfare, Washington, D.C.
20. Coleman, W. E., Lingg, R. D., Melton, R. G., and Kopfler, F. C. (1976). *In* "Identification and Analysis of Organic Pollutants in Water" (L. H. Keith, ed.). Ann Arbor Sci. Publ., Ann Arbor, Michigan.
21. Correia, Y., Martens, G. J., van Mensch, F. H., and Whim, B. P. (1977). *Atmos. Environ.* **11,** 1113–1116.
22. Cox, R. A., Derwent, R. G., Eggleton, A. E. J., and Lovelock, J. E. (1976). *Atmos. Environ.* **10,** 305–308.
23. Dahlberg, A., Christiansen, V. O., and Eriksson, E. A. (1973). *Ann. Occup. Hyg.* **16,** 41–46.
24. Davidson, B. M. (1926). *J. Pharmacol. Exp. Ther.* **26,** 37.

24a. Dilling, W. L., Tefertiller, N. B., and Kallos, G. J. (1975). *Environ. Sci. Technol.* **9,** 833–838.

25. Dorndorf, W., Kresse, M., Christian, W., and Katritzki, G. (1975). *Arch. Psychiatr. Nervenkr.* **220,** 373–379.
26. Dornette, W. H. L., and Jones, J. P. (1960). *Anesth. Analg.* **39,** 249.
27. Dowty, B. J., Carlish, D. R., and Laseter, J. L. (1975). *Environ. Sci. Technol.* **9,** 762–765.
28. Duerwald, W. (1954). *Arch. Toxicol.* **15,** 144–150.
29. Duprat, P., Delsaut, L., and Grediski, D. (1976). *Eur. J. Toxicol. Environ. Hyg.* **9,** 171–177.

30. Ehrenberg, L., Osterman-Golkar, S., Singh, D., and Lundqvist, U. (1974). *Radiat. Bot.* **15,** 185–194.
31. Elfskind, L. (1928). *Bruns' Beitr. Klin. Chir.* **167,** 251.
32. Elovaara, E., Hemminki, K., and Vainio, H. (1978). *Toxicology* **12,** 111–120.
34. Ewing, B. B., Chian, E. S. K., Cook, J. C., Evans, C. A., Hopke, P. K., and Perkins, E. G. (1977). EPA-560/6-77-015. U.S. Environ. Prot. Agency, Washington, D.C.
35. Fernandez, J. G., and Humbert, B. E. (1977). *Arch. Mal. Prof. Med. Trav. Secur. Soc.* **38,** 415–425.
36. Fishbein, L. (1980). *Banbury Rep.* **5,** 227–238.
37. Fishbein, L. (1979). "Potential Industrial Carcinogens and Mutagens." Elsevier, Amsterdam.
38. Fowler, J. S. K. (1969). *Br. J. Pharmacol.* **35,** 530–542.
39. Fox, J. L. (1977). *Chem. Eng. News* **72,** 34–46.
40. Freundt, K. J., Eberhardt, H., and Walz, V. M. (1963). *Arch. Gewerbepathol. Gewerbehyg.* **20,** 41–48.
41. Frey, E. (1912). *Biochem. J.* **40,** 29.
42. Fujii, T. (1977). *J. Chromatogr.* **139,** 297–302.
43. Gäb, S. (1981). *WaBoLu—Ber.* **3,** 55–61.
44. Gamberale, F., and Hultengren, M. (1973). *Work, Environ. Health* **10,** 82–92.
45. Gazzaniga, G., Binaschi, S., Sportelli, A., and Riva, M. (1969). *Boll.—Soc. Ital. Biol. Sper.* **45,** 97–99.
46. Gehring, P. J. (1968). *Toxicol. Appl. Pharmacol.* **13,** 287–298.
47. Gobbato, F., and Slavich, G. (1968). *Med. Lav.* **59,** 667.
48. Gohlke, R., Schmidt, P., and Bahmann, H. (1977). *Z. Gesamte Hyg. Ihre Grenzgeb.* **23,** 278–282.
49. Gold, L. S. (1980). *Banbury Rep.* **5,** 209–225.
50. Grimsrud, E. P., and Rasmussen, R. A. (1975). *Atmos. Environ.* **9,** 1014–1017.
51. Guengerich, F. P., Crawford, W. M., Domoradzki, J. Y., Macdonald, T. L., and Watanabe, P. G. (1980). *Toxicol. Appl. Pharmacol.* **55,** 303–317.
52. Hake, C. L., Waggoner, T. B., Robertson, D. N., and Rowe, V. K. (1960). *Arch. Environ. Health* **1,** 101–105.
53. Halevy, J., Pitlik, S., Rosenfeld, J., and Eitan, B. D. (1980). *Clin. Toxicol.* **16,** 467–472.
54. Hall, F. B., and Hine, C. H. (1966). *J. Forensic Sci.* **11,** 404–413.
55. Hanasono, G. K., Witschi, H., and Plaa, G. L. (1975). *Proc. Soc. Exp. Biol. Med.* **149,** 903–907.
55a. Hansch, C., Vittoria, A., Silipo, C., and Jow, P.Y. C. (1975). *J. Med. Chem.* **18,** 546–548.
56. Hardie, D. W. F. (1964). *In* "Encyclopedia of Chemical Technology" (R. E. Kirk and D. F. Othmer, eds.), pp. 149–170. Wiley, New York.
57. Hasanen, E., Soininen, V., Pyysalo, H., and Leppamaki, E. (1979). *Atmos. Environ.* **13,** 1217–1220.
58. Hatfield, R., and Maykowski, R. T. (1970). *Arch. Environ. Health* **20,** 279–281.
59. Henschler, D., ed. (1980). "Analytische Methoden zur Prüfung gesundheitsschädlicher Arbeitsstoffe. Vol. 1, Luftanalysen." Verlag Chemie, Weinheim.
60. Henschler, D., Reichert, D., and Metzler, M. (1980). *Int. Arch. Occup. Environ. Health* **47,** 263–268.
61. Heppel, L. A., Neal, P. A., Endicott, K. M., and Porterfield, V. T. (1944). *Arch. Ophthamol.* **32,** 391–394.
62. Heppel, L. A., Neal, P. A., Perrin, T. L., Endicott, K. M., and Porterfield, V. T. (1946). *J. Ind. Hyg.* **28,** 113.
63. Herd, P. A., Lipski, M., and Martin, H. F. (1974). *Arch. Environ. Health* **28,** 227–233.

64. Hermansen, K. (1970). *Acta Pharmacol. Toxicol.* **28,** 17–27.
65. Hes, J. P., Cohn, D. F., and Streifler, M. (1979). *Isr. Ann. Psychiatry Relat. Discip.* **17,** 122–125.
66. Hinkel, G. K. (1965). *Dtsch. Gesundheitswes.* **20,** 1327–1331.
67. Hofmann, H. Th., Birnstiel, H., and Jobst, P. (1970). *Arch. Exp. Pathol. Pharmakol.* **266,** 360.
68. Holmberg, B., Jakobson, J., and Sigvardsson, K. (1977). *Scand. J. Work, Environ. Health* **3,** 43–52.
69. Horwitz, W., ed. (1975). ''Official Methods of Analysis of the Association of Official Analytical Chemists,'' 12th ed. Assoc. Off. Anal. Chem., Washington, D.C.
70. Hueper, W. C., and Smith, C. (1935). *Am. J. Med. Sci.* **189,** 778–784.
71. IARC (1979). *IARC Monogr. Eval. Carcinog. Risk Chem. Humans* **19.**
72. IARC (1980). *IARC Monogr. Eval. Carcinog. Risk Chem. Humans* **20.**
73. Ikeda, M., and Ohtsuji, H. (1972). *Br. J. Industr. Med.* **29,** 99–104.
74. Jensen, S., Lange, R., Berge, G., Palmork, K. H., and Renberg, K. (1975). *Proc. R. Soc. London, Ser. B.* **189,** 333–346.
75. Joachimoglu, G. (1921). *Berl. Klin. Wochenschr.* **58,** 147.
76. Jondorf, W. R., Parke, D. V., and Williams, R. T. (1957). *Biochem. J.* **65,** 14.
77. Keith, L. H. (1976). *Environ. Sci. Technol.* **10,** 555–564.
78. Keith, L. H., Garrison, A. W., Allen, F. R., Carter, M. H., Floyd, T. L., Pope, J. D., and Thruston, A. D. (1976). *In* ''Identification and Analysis of Organic Pollutants in Water'' (L. H. Keith, ed.). Ann Arbor Sci. Publ., Ann Arbor, Michigan.
79. Kellam, R. G., and Dusetzina, M. (1980). *Banbury Rep.* **5,** 265–274.
80. Killian, H., and Weese, H. (1954). ''Die Narkose.'' Thieme, Stuttgart.
81. King, M. T., Beikirch, A., Eckhardt, K., Gocke, E., and Wild, D. (1979). *Mutat. Res.* **66,** 33–44.
82. Kistler, G. H., and Lückhardt, A. B. (1929). *Curr. Res. Anesth. Analg.* **8,** 65.
83. Klaasen, C. D., and Plaa, G. L. (1966). *Toxicol. Appl. Pharmacol.* **9,** 139–151.
83a. Klaasen, C. D., and Plaa, G. L. (1967). *Toxicol. Appl. Pharmacol.* **10,** 119–131.
84. Kleinfeld, M., and Feiner, B. (1966). *J. Occup. Med.* **8,** 358–364.
85. Kluwe, W. M., Herrmann, C. L., and Hook, J. B. (1979). *J. Toxicol. Environ. Health* **5,** 605–615.
86. König, R. (1933). *Langenbecks Arch. Chir.* **99,** 147.
87. Kokarovtseva, M. G. (1978). *Ukr. Biokhim. Zh.* **51,** 10–13.
88. Kramer, C. G., Ott, M. G., and Fulkerson, I. E. (1978). *Arch. Environ. Health* **33,** 331–342.
89. Kretzschmar, J. G., Peperstraete, H., and Rymen, T. (1976). *Extern* **5,** 147–178.
90. Kuoni, J. (1980). *Praxis* **69,** 1225–1231.
91. Lal, H., and Shah, H. (1967). *Toxicol. Appl. Pharmacol.* **10,** 389.
92. Lal, H., Olshan, A., Puri, S., Shah, H., and Fuller, G. C. (1969). *Toxicol. Appl. Pharmacol.* **14,** 625.
93. Lazarew, N. W. (1929). *Arch. Exp. Pathol. Pharmakol.* **141,** 19.
94. Lehmann, K. B., and Schmidt-Kehl, L. (1936). *Arch. Hyg. Bakteriol.* **72,** 327.
95. Lillian, D., Singh, H. B., Appleby, A., Lobban, L., Arnts, R., Gumpert, R., Hague, R., Toomey, J., Kozazis, J., Antell, M., Hansen, D., and Scott, B. (1975). *Environ. Sci. Technol.* **9,** 1042–1048.
96. Liversey, J. C., and Anders, M. W. (1979). *Drug Metab. Dispos.* **7,** 199–203.
97. Lobo-Mendonça, R. (1963). *Br. J. Ind. Med.* **20,** 50–56.
98. Lovelock, J. E. (1977). *Ecotoxicol. Environ. Saf.* **1,** 399.
99. Maltoni, C., Valgimigli, L., and Scarnatto, C. (1980). *Banbury Rep.* **5,** 3–29.

100. Martin, G., Knorpp, K., Huth, K., Heinrich, F., and Mittermeyer, C. (1968). *Dtsch. Med. Wochenschr.* **93,** 2002–2010.
101. Maugh, T. H. (1974). *Science* **183,** 94.
102. "Maximale Arbeitsplatzkonzentrationen" (1981). Boldt Publ., Boppard, Fed. Rep. of Germany.
103. McCann, J., Simmon, V., Streitweiser, D., and Ames, B. N. (1975). *Proc. Natl. Acad. Sci. U.S.A.* **72,** 3190–3193.
104. McCann, J., Spingarn, N. E., Lobori, J., and Ames, B. N. (1975). *Proc. Natl. Acad. Sci. U.S.A.* **72,** 979–983.
105. McConnell, G., Ferguson, D. M., and Pearson, C. R. (1975). *Endeavour* **34,** 13–18.
106. McConnell, J. C., and Schiff, H. J. (1978). *Science* **199,** 174.
107. McDonald, J. R., Gendolfi, A. J., and Sipes, J. G. (1980). *Fed. Proc. Fed. Am. Soc. Exp. Biol.* **39,** 38–48.
108. McNutt, N. S., Amster, R. L., McConnell, E. E., and Morris, F. (1975). *Lab. Invest.* **32,** 642–654.
109. Menschick, H. (1957). *Arch. Gewerbepathol. Gewerbehyg.* **15,** 241–252.
110. Monster, A. C. (1979). *Int. Arch. Occup. Environ. Health* **42,** 311–317.
111. Monster, A. C., Boersma, G., and Steenway, H. (1979). *Int. Arch. Occup. Environ. Health* **42,** 293–311.
112. Monster, A. C., and Houtkooper, J. M. (1979). *Int. Arch. Occup. Environ. Health* **42,** 319–323.
113. Morgan, A., Black, A., and Belcher, D. R. (1970). *Ann. Occup. Hyg.* **13,** 219–233.
114. Müller, J. (1925). *Arch. Exp. Pathol. Pharmakol.* **109,** 276.
115. National Cancer Institute (1978). "Bioassay of 1,2-Dichloroethane for Possible Carcinogenicity," DHEW Publ. No. (NIH)78-1361. U.S. Dept. Health, Ed., and Welfare, Washington, D.C.
116. National Cancer Institute (1977). "Bioassay of 1,1,1-Trichloroethane for Possible Carcinogenicity," DHEW Publ. No. (NIH)77-803. U.S. Dept. Health, Ed., and Welfare, Washington, D.C.
117. National Cancer Institute (1978). "Bioassay of 1,1,2-Trichloroethane for Possible Carcinogenicity," DHEW Publ. No. (NIH)78-1324. U.S. Dept Health, Ed., and Welfare, Washington, D.C.
118. National Cancer Institute (1978). "Bioassay of 1,1,2,2-Tetrachloroethane for Possible Carcinogenicity," DHEW Publ. No. (NIH)78-827. U.S. Dept. Health, Ed., and Welfare, Washington, D.C.
119. National Cancer Institute (1978). "Bioassay of Hexachloroethane for Possible Carcinogenicity," DHEW Publ. No. (NIH)78-1318. U.S. Dept. Health, Ed., and Welfare, Washington, D.C.
120. National Cancer Institute (1978). "Chemicals Being Tested for Carcinogenicity by the Carcinogenesis Testing Program." DHEW, National Institutes of Health, Washington, D.C.
121. National Institute for Occupational Safety and Health (1974). "National Occupational Hazard Survey," Vol. 1, DHEW Publ. No. (NIOSH)74-127. U.S. Dept. Health, Ed., and Welfare, Washington, D.C.
122. National Institute for Occupational Safety and Health (1977). "NIOSH Manual of Analytical Methods," DHEW (NIOSH) Publ. No. 77-157B. U.S. Dept. Health, Ed., and Welfare, Washington, D.C.
122a. Neely, W. B., Branson, D. R., and Blau, G. E. (1974). *Environ. Sci. Technol.* **8,** 1113–1115.
123. Neumayer, V. (1981). *WaBoLu Ber.* **3,** 24–31.

124. Nicholson, A. A., Meresz, O., and Lemyk, B. (1977). *Anal. Chem.* **49,** 814–819.
125. Nuckolls, A. H. (1955). Underwriter's Laboratory Rep. No. 2375 (Original work published 1933). *Cited in* "The Halogenated Hydrocarbons, Toxicity and Potential Dangers" (W. F. van Oettingen, ed.), Publ. Health Serv. Publ. No. 414. U.S. Publ. Health Serv., Washington, D.C.
126. Ohta, T., Morita, M., and Mizoguchi, J. (1976). *Atmos. Environ.* **10,** 557–560.
127. Okuno, T., Tsuij, M., Shintani, Y., and Watanabe, H. (1974). *Chem. Abstr.* **87,** 72564f.
128. Page, B. D., and Kennedy, B. P. (1975). *J. Assoc. Off. Anal. Chem.* **58,** 1062.
129. Pearson, C., and McConnel, G. (1975). *Proc. R. Soc. London Ser. B.* **189,** 305–332.
130. PED Co. Environmental, Inc. (1979). "Monitoring of Ambient Levels of EDC Near Production and User Facilities," EPA Rep. No. 600/4-79-029. Environ. Prot. Agency (Off. Res. Develop.), Research Triangle Park, North Carolina.
133. Plaa, G. L., and Larson, R. F. (1965). *Toxicol. Appl. Pharmacol.* **7,** 37–44.
134. Price, P. J., Hassett, C. M., and Mansfield, J. I. (1978). *In Vitro* **13,** 290–300.
135. Priestly, B. G., and Plaa, G. L. (1976). *Arch. Int. Pharmacodyn. Ther.* **223,** 132–141.
136. Prost, G., Rigaud, M., and Pelletier, N. (1977). *Arch. Mal. Prof. Med. Trav. Secur. Soc.* **38,** 205–225.
137. Purchase, I. F. M., Longstaff, E., Ashby, J., Styles, J. A., Anderson, D., Lefevre, P. A., and Westwood, F. R. (1976). *Nature (London)* **264,** 624–627.
137a. Radding, S. B., Liu, D. H., Johnson, H. L., and Mill, T. (1977). "Review of the Environmental Fate of Selected Chemicals," EPA-560/5-77-003. U.S. Environ. Prot. Agency (Office of Toxic Substances), Washington, D.C.
138. Rampy, L. W., Quast, I. F., Leong, B. K. I., and Gehring, P. I. (1978). *Proc. Int. Congr. Toxicol., 1st, 1977,* p. 27.
139. Rannug, U., and Beije, B. (1979). *Chem.—Biol. Interact.* **24,** 265.
140. Rannug, U., Sundvall, A., and Ramel, C. (1978). *Chem.—Biol. Interact.* **20,** 1–16.
141. Rao, K. S., Murray, I. S., Deacon, M. M., Calhoun, L. L., and Young, J. T. (1980). *Banbury Rep.* **5,** 149–161.
142. Reckner, L. R., and Sachdev, I. (1975). DHEW (NIOSH) Publ. No. 75-184. U.S. Dept. Health, Ed., and Welfare, Washington, D.C.
143. Regnault, V. (1840). *Ann Chem. Pharm.* **33,** 310–334.
144. Reinfried, H. (1958). *Dtsch. Gesundheitswes.* **13,** 778–779.
145. Reinhardt, C. F., Mullin, L. S., and Mayfield, M. E. (1972). *Toxicol. Appl. Pharmacol.* **22,** 305.
146. Reitz, R. H., Fox, T. R., Domoradzki, J. Y., Quast, J. F., Langvardt, P., and Watanabe, P. G. (1980). *Banbury Rep.* **5,** 135–144.
147. Rennick, B. R., Malton, S. D., Moe, G. K., and Seevers, M. H. (1949). *Fed. Proc. Fed. Am. Soc. Exp. Biol.* **8,** 327.
148. Riihimaki, V., and Pfäffli, P. (1978). *Scand. J. Work, Environ. Health* **4,** 73–85.
149. Rosenberg, R., Grahn, O., and Johansson, L. (1975). *Water Res.* **9,** 607–612.
150. "Rote Liste" (1971). Editio Cantor, Aulendorf, Fed. Rep. of Germany.
151. Russel, J. W., and Shadoff, L. A. (1977). *J. Chromatogr.* **134,** 375–384.
152. Safe Drinking Water Committee (1977). "Drinking Water and Health." Nat. Acad. Sci., Washington, D.C.
153. Salvini, M., Binaschi, S., and Riva, M. (1971). *Br. J. Ind. Med.* **28,** 286–292.
154. San, R. H. C., and Stich, H. F. (1975). *Int. J. Cancer* **16,** 284–291.
155. Sato, A., and Nakajima, T. (1979). *Arch. Environ. Health* **34,** 69–75.
156. Savolainen, H., Pfäffli, P., Tengen, M., and Vainio, H. (1977). *Arch. Toxicol.* **38,** 229–237.
157. Sayers, R. R., Yant, W. P., Thomas, B. H., and Bürger, L. B. (1929). *U.S. Public Health Bull.* **185.**

158. Sayers, R. R., Yant, W. P., White, C. P., and Patty, F. A. (1930). *Public Health Rep.* **45,** 225.
159. Schmidt, R. (1976). *Biol. Rundsch.* **14,** 220–223.
160. Schmidt, P., Burck, D., and Buerger, A. (1980). *Z. Gesamte Hyg. Ihre Grenzgeb.* **26,** 167–172.
161. Schönborn, H., Prellwitz, W., and Baum, P. (1970). *Klin. Wochenschr.* **48,** 822–824.
162. Schwetz, B. A., Leong, B., and Gehring, P. J. (1974). *Toxicol. Appl. Pharmacol.* **28,** 452–464.
163. Schwetz, B. A., Leong, B. K., and Gehring, P. J. (1975). *Toxicol. Appl. Pharmacol.* **32,** 84–96.
164. Seki, Y., Urashima, Y., Aikawa, H., Matsumura, H., Ichikawa, Y., Hiratsuka, F., Yoshioka, Y., Shimbo, S., and Ikeda, M. (1975). *Int. Arch. Arbeitsmed.* **34,** 39–49.
165. Shackelford, W. M., and Keith, L. H. (1976). "Frequency of Organic Compounds Identified in Water," EPA-600/4-76-062. U.S. Environ. Prot. Agency, Athens, Georgia.
166. Shakarnis, V. F. (1969). *Genetika* **5,** 89–95.
167. Shakarnis, V. F. (1970). *Vestn. Leningr. Univ. Ser. Biol.* **25,** 153–156.
168. Shmuter, L. M. (1973). *Zh. Mikrobiol., Epidemiol. Immunobiol.* **50,** 104–109.
168a. Simmon, V. F., Kaukanen, K., and Tardiff, R. G. (1977). *In* "Progress in Genetic Toxicology" (D. Scott, B. A. Bridges, and F. H. Sobels, eds.), pp. 249–258. Elsevier, Amsterdam.
169. Singh, H. B., Salas, L. I., and Cavanagh, L. A. (1977). *J. Air Pollut. Control Assoc.* **27,** 332–336.
170. Sittig, M. (1980). "Priority Toxic Pollutants. Health Impacts and Allowable Limits." Noyes Data Corp., New Jersey.
171. Smith, H. F. (1956). *Am. Ind. Hyg. Assoc., Q.* **17,** 129.
172. Smyth, H. F., Carpenter, C. P., Werl, C. S., Pozzani, U.S., Striegel, J. A., and Nycum, J. S. (1969). *Am. Ind. Hyg. Assoc. J.* **30,** 470–476.
173. Spector, W. S., ed. (1956). "Handbook of Toxicology," Vol. 1. Saunders, Philadelphia.
174. Spence, J. W., and Houst, P. L. (1978). *J. Air Pollut. Control Assoc.* **28,** 250–253.
175. Spencer, H. C., Rowe, V. K., Adams, E. M., McCollister, D. D., and Irish, D. D. (1951). *Arch. Ind. Hyg. Occup. Med.* **4,** 482.
176. Stahl, C. J., Fatteh, A. V., Dominguez, A. M. (1969). *J. Forensic Sci.* **14,** 393–397.
177. Steindorff, K. (1922). *Arch. Ophthalmol.* **109,** 253–264.
178. Stewart, R. D. (1968). *Ann. Occup. Hyg.* **11,** 71–79.
179. Suta, B. (1969). "Assessment of Human Exposures to Atmospheric Ethylene Dichloride." SRI International, Menlo Park, California.
180. Symons, J. M., Bellar, T. A., Carswell, J. K., DeMarco, J., Kropp, K. L., Robeck, G. G., Seeger, D. R., Slocum, C. J., Smith, B. L., and Stevens, A. A. (1975). *J. Am. Water Works Assoc.* **67,** 634.
181. Takeuchi, Y. (1966). *Jpn. J. Ind. Health (Sangyo Igaku)* **8,** 371–374.
182. Taylor, G. I., Drew, R. T., Lores, E. M., and Clemmer, T. A. (1976). *Toxicol. Appl. Pharmacol.* **38,** 379–387.
183. Texter, E. C., Grunow, W. A., and Zimmerman, H. J. (1979). *Clin. Res.* **27,** 684A.
184. Theiss, J. C., Stoner, G. D., Shimkin, M. B., and Weisburger, E. K. (1977). *Cancer Res.* **37,** 2717–2720.
185. "Threshold Limit Values (TLVs) for Chemical Substances in the Workroom Air" (1981). ACGIH, Cincinnati, Ohio.
186. Tomokuni, K. (1970). *Acta Med. Okayama* **24,** 315–322.
187. Torkelson, T. R., Oyen, F., McCollister, D. D., and Rowe, V. K. (1958). *Am. Ind. Hyg. Assoc. J.* **19,** 353–362.

188. Truhaut, R., Nguyen, P. L., and Dutertre, H. (1974). *Arch. Mal. Prof. Med. Trav. Secur. Soc.* **35,** 593–608.
189. Truhaut, R., Thevenin, M., and Warnet, M. (1975). *Eur. J. Toxicol. Environ. Hyg.* **8,** 175–179.
189a. Tute, M. S. (1971). *Adv. Drug. Res.* **6,** 1–77.
190. "Ullmanns Enzyklopädie der technischen Chemie," 4th ed. (1975). Vol. 9, pp. 420–442. Verlag Chemie, Weinheim.
191. United States International Trade Commission (1977). "Synthetic Organic Chemicals, U.S. Production and Sales, 1976," USITC Publ. No. 833. U.S. Govt. Printing Office, Washington, D.C.
192. Vainio, H., Parkki, M. G., and Marniemi, J. (1976). *Xenobiotica* **6,** 599–604.
193. Van Dyke, R. A., and Chenoweth, M. B. (1965). *Anesthesiology* **26,** 348–357.
194. Van Dyke, R. A., and Wineman, C. G. (1971). *Biochem. Pharmacol.* **20,** 463–470.
195. Vogel, E., and Sobels, F. H. (1976). *In* "Chemical Mutagens: Principles and Methods for Their Detection" (A. Hollaender, ed.), Vol. 4, pp. 93–142. Plenum, New York.
196. Vozovaya, M. A., and Malyarova, L. K. (1975). *Gig. Sanit.* **6,** 94.
197. Vozovaya, M. A. (1976). *Gig. Sanit.* **6,** 100–102.
198. Vozovaya, M. A. (1974). *Gig. Sanit.* **7,** 25.
199. Wahlberg, J. E. (1976). *Ann. Occup. Hyg.* **19,** 115–119.
200. Ward, J. M. (1980). *Banbury Rep.* **5,** 35–49.
201. Weeks, M. H., Angerhofer, R. A., Bishop, R., Thomasino, J., and Pope, C. R. (1979). *Am. Ind. Hyg. Assoc. J.* **3,** 187–199.
202. Weiss, F. (1958). *Dtsch. Gesundheitswes.* **13,** 185–188.
203. Weiss, F. (1957). *Arch. Gewerbepathol. Gewerbehyg.* **15,** 253–264.
204. Williams, R. T. (1959). "Detoxication Mechanisms." Chapman & Hall, London.
205. Wilson, J. G., and Fraser, F. C. (1977). "Handbook of Teratology." Plenum, New York.
206. Winnacker, K., and Küchler, L. (1972). *Chem. Technol., 3. Neubearb. Aufl.* **4,** 45–51.
207. Wirtschafter, Z. I., and Schwartz, E. D. (1939). *J. Ind. Hyg.* **21,** 126–131.
208. Wit, S. L., Besemer, A., Das, H., Goedkoop, W., Loosjes, F. E., and Meppelink, E. R. (1969). Rep. No. 36/69. Natl. Inst. Public Health, Bilthoven, Netherlands.
209. Wolff, D. L., and Siegmund, R. (1978). *Biol. Zentralbl.* **97,** 345–352.
210. Yllner, S. (1971). *Acta Pharmacol. Toxicol.* **29,** 499–512.
211. Yllner, S. (1971). *Acta Pharmacol. Toxicol.* **30,** 248–256.
212. Yllner, S. (1971). *Acta Pharmacol. Toxicol.* **30,** 257–265.
213. Yokaiden, R. E., and Babcock, J. R. (1973). *Arch. Environ. Health* **26,** 281–284.
214. Zollinger, F. (1931). *Arch. Gewerbepathol. Gewerbehyg.* **2,** 298–325.

Chemical Substance Index

A

ABS, *see* Alkylbenzene sulfonates
Acetaminophen, 353, 390
Acetyl chloride, 403
Acetylene, 414, 432
Acrolein, 169
Actinomycin, 424
Adrenalin, 424, 425
Aflatoxins, 182
  aflatoxin B_1, 196, 215
Aldicarb, 128–131, 135
Alkylamines, 414
Alkylbenzene sulfonates, 237
Alkyllead, 15, 42, 51, 408, *see also* specific compound
Alkylmercury, 14, 42, 46, *see also* specific compound
Alkyltin, 47, *see also* specific compound
Aluminum, 18, 25, 26, 30, 31, 34, 35, 51
Aluminum oxide, 17, 157, 160, 163, 166, 361
Aluminum phosphate, 17
Amines, 403, *see also* specific compound
  aromatic, 143, 185, 219
  vasoactive, 192
Aminopyrine, 424
2-Amino-4,6,9-trinitroperimidine, 13
Ammonia, 237, 241, 243, 364, 373, 380
Ammonium acetate, 19
Ammonium sulfate, 6, 10, 12, 13
Ammonium thiocyanate, 169
Anhydrite, *see* Calcium sulfate
Aniline, 383
Anthracene, 220
Antimony, 11, 35, 43–46, 53
Antimony(III), 44
Antimony(V), 44
Arochlor, 222
Arochlor 1254, 168
Arsenic, 34, 35, 38, 43–48, 50, 53, 116, 158, 216
Arsenic(III), 38, 44, 47, 48
Arsenic(V), 44, 47, 48
Arsenic trioxide, 157, 158
Asbestos, 6, 125, 143–148, 150–152, 157, 173, 216, 385
  amosite, 148, 156
  chrysotile, 146, 148, 152, 156–157, 166–167
  crocidolite, 148, 152, 161
Ascorbic acid, 362, 366
Asphalt, 212
Aspirin, 390
Azo dyes, 384

B

B[*a*]P, *see* Benzo[*a*]pyrene
Barbiturates, 424
BCME, *see* Bis(chloromethyl) ether
Benz[*a*]anthracene, 163, 202, 212, 220
Benzene, 122, 385
Benzo[*a*]fluoranthrene, 163
Benzo[*k*]fluoranthene, 212, 220
Benzo[*g,h,i*]perylene, 151, 220
Benzo[*a*]pyrene, 147–164, 184, 196, 202, 212, 220, 428
Beryllium, 34
Bis(chloromethyl) ether, 343, 352, 371
N-N'-Bis(hydroxymethyl)urea, 350
Bismuth, 34, 35, 43, 44, 46, 53
Bisulfite, 170
Bromine, 18, 265, 299
Butanediol, 344
Butanol, 420

C

Cadmium, 7, 8, 17, 18, 20–22, 25, 28–30, 33–38, 42, 43, 48–50, 53

Cadmium compounds, 8, 25
 $Cd(OH)_2$, 8
Cadmium ions, 26
Cadmium oxide, 8, 172
Caffeine, 383
Calcite, 5, 9, 10, 22
Calcium, 18, 30, 35
Calcium carbonate, 17, *see also* Calcite
Calcium sulfate, 10, 14
Carbon, elemental, 14, 167, 170, 171
 particles, 145, 150, 151, 157, 158, 160, 161, 166, 170, 171
Carbon dioxide, 255, 267–269, 273, 274, 289, 300–306, 308–310, 313, 314, 316, 317, 319, 320, 414, 439
Carbon disulfide, 113, 131–134
Carbon hexachloride, *see* Hexachloroethane
Carbonic acid, 237
Carbon monoxide, 143, 171, 265, 289, 291, 300, 301, 414
Carbon oxides, 403
Carbon tetrachloride, 261, 262, 272, 289, 299, 403, 418, 422, 428, 434, 438
S-Carboxymethylcysteine, 417, 428
Catecholamine, 424
Cellulose ester, 6, 439, *see also* specific compound
Cerium, 34,
Cesium, 34
CFC, *see* Chlorofluorocarbons
CFC-11, *see* Trichlorofluromethane
CFC-12, *see* Dichlorodifluoromethane
CFC-113, *see* Trichlorotrifluoroethane
CFC-114, *see* Dichlorotetrafluoroethane
CFC-115, *see* Chlorotrifluoroethylene
Chitosan, 35, 37
Chlorinated acetic acids, 403
Chlorinated camphenes, 210
Chlorinated ethanes, 401–448
Chlorinated hydrocarbons, 168, 184, 206, 215, 401–448, *see also* specific compound
 aromatic, 405
Chlorinated hydrochloric ether, *see* 1,1-Dichloroethane
Chlorinated paraffins, 168
Chlorine, 18, 237, 241, 254, 257, 262, 265, 294–296, 302, 414, 430, 432, 436, 438
 gas, 125
 radicals, 261–265, 271, 281, 284, 286
Chlorine monoxide, 254, 258, 261–264, 271, 281, 284–286, 296, 302, 307
Chlorine nitrate, 258, 261–263, 286, 287, 290, 307, 309
Chlorine oxides, 253, 257–264, 271, 306, 313
Chloroacetaldehyde, 419, 428
Chloroacetic acid, 417, 428, 440
Chlorocarbons, 261, 308, 316, *see also* specific compound
Chlorodifluoromethane, 299, 307
Chloroethane, *see* Monochloroethane
Chloroethanes, *see* Chlorinated ethanes
Chloroethene, *see* 1,1,1-Trichloroethane
Chloroethyl, *see* Monochloroethane
Chlorofluorocarbons, 254, 255, 261, 262, 264, 267, 273, 275, 279–280, 289, 290, 294–299, 302, 304–310, 313, 314, 316–320, 326, 330
Chloroform, 414, 418, 428, 437, 440
Chloromethane, 261, 262, 289, 352, 353
Chlorotrifluoroethylene, 299, 307
Chromic acid, 163
Chromium, 30, 34, 35, 43, 46, 243, 384
Chromium(III), 5, 37, 43, 47
Chromium(VI), 5, 37, 43
Chromium carbonyl, 152
Chrysene, 212, 220
Coal dust, 146, 170
Coal tar, 371, 372
Cobalt, 30, 34, 35, 38
Cobalt oxide, 146, 166
Copper, 17, 18, 20–22, 28–30, 32–41, 43, 48–51, 53, 158, 215–217, 236, 239, 241, 243
Copper clays, 26
Copper compounds, 14
Corticosteroids, 353, 390
Croton oil, 187
Cupric ions, 40
Cuprous ions, 40
Cyclohexamide, 424
Cysteine, 35

D

DBC, *see* Dibenzo[*c,g*]carbazole
DDE, 168
DDT, 112, 168, 215, 243, 244
DEN, *see* Diethylnitrosamine
Detergents, 241
Dialkylmercury, 14
Diamine, 35
Dibenz[*a,h*]anthracene, 151, 163, 212, 220

Dibenzo[*c,g*]carbazole, 151, 163, 164
Dibenz[*a,i*]pyrene, 163
Dibromoethane, 352
Dichloroacetic acid, 433, 440
Dichloroacetyl chloride, 403, 432
Dichloroacetylene, 420, 432
Dichlorodifluoromethane, 261, 262, 279, 280, 294, 296–299, 307, 308
α-Dichloroethane, *see* 1,1-Dichloroethane
α,β-Dichloroethane, *see* 1,2-Dichloroethane
sym-Dichloroethane, *see* 1,2-Dichloroethane
1,1-Dichloroethane, 402, 411–413, 420
1,2-Dichloroethane, 402, 405, 414–420, 432, 436, 441
Dichloroethylene, *see* 1,2-Dichloroethane
1,1-Dichloroethylene, 420, 421, 427, 430
1,2-Dichloroethylene, 427, 432
Dichlorotetrafluoroethane, 299, 307
Diesel oil, 210, 212
O,O-Diethyl-*O*-(*p*-methylsulfinyl)phenyl phosphorothioate, *see* Fensulfothion
Diethylnitrosamine, 165–166, 173, 371, 372, 374
Dihalomethanes, 352
Dimethyl arsonate, 47
7,12-Dimethylbenz[*a*]anthracene, 152, 154, 158, 163, 204, 224
Dimethyldiethyllead, 15
Dimethylmercury, 14
Dimethylnitrosamine, 165, 196, 352
Dimethylsulfoxide, 352
1,4-Dioxane, 420
Dioxin, 113, 134–136
Diquat, 236
Dithiocarbamate, 35
DMBA, *see* 7,12-Dimethylbenz[*a*]anthracene
DMN, *see* Dimethylnitrosamine
DMSO, *see* Dimethylsulfoxide
Dopamine, 192
Dyes, 387

E

Epichlorohydrin, 383
Epinephrine, 192
1,2-Epoxybutane, 420
Ethane hexachloride, *see* Hexachloroethane
Ethane pentachloride, *see* Pentachloroethane
Ethane trichloride, *see* 1,1,2-Trichloroethane
Ethanol, 367, 381, 383, 427
Ethyl cellulose, 408
Ethyl chloride, *see* Monochloroethane
Ethylene, 414, 417, 427, 430, 432, 436
Ethyleneamine, 415
Ethylene chloride, *see* 1,2-Dichloroethane
Ethylene dichloride, *see* 1,1-Dichloroethane *and* 1,2-Dichloroethane
Ethylene glycol, 169
Ethylhydroxyethyl cellulose, 408
Ethylidene dichloride, *see* 1,1-Dichloroethane

F

Feldspars, 17
Fenitrothion, 243
Fensulfothion, 167, 168
Ferric chloride, 414, 420, 432, 436, 438
Flue dust, 158
Fluoranthene, 212
Fluorine, 265
Fluorocarbons, 265, 269, 307
Fly ash, 10, 13, 170, 172–174
Formaldehyde, 122, 143, 169, 337–392
Formalin, 338, 364, 373, 377, 381, 382
Formate, 348, 350, 351
Formic acid, 351, 373
N-Formylcysteine, 350
Freons, 403, 404, 422
Fuel oil, 199, 212
Fulvates, 30, 49
Fulvic acid, 30, 31
Furanase, 224, 225

G

Germanium, 43, 44, 46, 53
Glass fiber, 6, 12, 27, 29, 383
Glycerin, 169, 381
Glycol dichloride, *see* 1,2-Dichloroethane
Goethite, 10
Gold, 35

H

Halomethanes, 185
HCFC-22, *see* Chlorodifluoromethane
Hematite, 150, 152, 153, 155–160, 162–166, 170, 216
Herbicides, 168, 241
Hexachloroethane, 402, 438–441

1,1,1,2,2,2-Hexachloroethane, *see* Hexachloroethane
Hexachloroethylene, *see* Hexachloroethane
Hexamethylenetetramine, 344, 364–366, 373, 380
Hexamethylenetetramine–resorcinol resin, 361, 379, 380
Hexamethylphosphoramide, 352
HMPA, *see* Hexamethylphosphoramide
HMT, *see* Hexamethylenetetramine
HR, *see* Hexamethylenetetramine–resorcinol
Humic acids, 28, 30
Hydrocarbons, 259, 343, *see also* specific compound
  chlorinated, *see* Chlorinated hydrocarbons
  halogenated, 434
  oxygenated, 385
  petroleum, 241, 243
Hydrochloric acid, 12, 20, 169, 171–172, 374
Hydrochloric ether, *see* Monochloroethane
Hydrogen, 265, 273
Hydrogen chloride, 261, 263, 281, 286, 370, 371, 402, 403, 408, 411, 414, 420, 432, 436
Hydrogen cyanide, 380
Hydrogen fluoride, 265
Hydrogen oxides, 253, 254, 257–264, 267, 271, 279, 281, 283–284, 291, 300, 306, 317
Hydrogen oxychloride, 260–263, 286, 287, 290
Hydrogen peroxide, 20, 257, 260, 263
Hydrogen peroxide radical, 254, 257, 259–263, 281, 283, 284
Hydroxylamine hydrochloride, 20, 21
Hydroxyl radicals, 260–265, 267, 279, 281, 283–284, 296, 301, 328, 403, 422
6-Hydroxymethyl-2[2-(5-nitro-2-furyl)vinyl]pyridine, *see* Furanase
N-(Hydroxymethyl)urea, 350

I

India ink, powder, 154, 158, 163
Indium, 34
Insecticides, 244
Iodine, 190, 192, 265
Iodo-2′-deoxyuridine, 222
Iron, 21, 25, 26, 29, 30, 33–38, 49–52, 438
Iron compounds, 14, 52
  Fe_3O_4, 17
  α-Fe_2O_3, 17
  $FeSO_4$, 17
  iron(III) hydroxide, 33, 35, 36
IUdR, *see* Iodo-2′-deoxyuridine

K

Kerosene, 236

L

Lanthanum, 35
Lead, 3, 7, 9, 15–18, 20–23, 25, 28–30, 33–38, 42–44, 46, 48–53, 116, 121, 408, 415
  automotive, 18
  elemental, 16, 17
Lead carbonates, 41, 51
Lead chloride, 3
Lead compounds, 6–7, 15–17, 22–23, 25, *see also* specific compound
  labeled, 3
  leaded gasoline additives, 6, 15
  lead oxides, 53, 157, 172
  $PbBrCl$, 7
  $PbBr_xCl_y$, 10
  $PbBrCl\cdot(NH_4)_2BrCl$, 7
  $PbBrCl\cdot 2NH_4Cl$, 7
  α-$2PbBrCl\cdot NH_4Cl$, 7
  $PbCl_2$, 10
  $PbCO_3$, 26
  $2PbCO_3\cdot Pb(OH)_2$, 17
  $PbFe(OH)_3$, 26
  PbO, 8
  PbO_2, 23
  Pb_3O_4, 17
  $Pb(OH)Br$, 10
  $(PbO)_2PbBrCl$, 10
  $PbO\cdot PbSO_4$, 17, 23
  PbS, 8, 23
  $PbSO_4$, 7, 8, 16, 17, 22, 23
  $PbSO_4\cdot(NH_4)_2SO_4$, 7, 16, 17
Leaded gasoline, 408
Lead fulvates, 43
Lead humates, 26, 41, 43
Lead sulfide, 3
Lindane, 168

M

Magnesium, 35
Magnesium chloride, 169

Magnesium oxide, 157, 166
Manganese, 21, 25, 26, 29, 30, 33–35, 38, 46, 50, 52
Manganese dioxide, 170, 172
Manganese ions, 26
Manganese oxyhydroxides, 33
3-MC, *see* 3-Methylcholanthrene
Melamine–formaldehyde, 360, 377
2-Mercaptoethanol, 352
Mercuric chloride vapor, 14
Mercuric ions, 41
Mercury, 14–15, 29, 34, 35, 38, 41–44, 46, 47, 53
  vapor, 14
Mercury(I), 47
Mercury(II), 47
Mercury compounds, 14, 15, 26, *see also* specific compound
Metals, 1–61, 142, 172–173, 206, 237, 438, *see also* specific element
Metal bisulfate, bivalent, 12
Metal chlorides, 438
Metal oxides, 145, 167, 172–174
Metal sulfates, bivalent, 12
Methane, 255, 259, 265, 266, 269, 272, 273, 275, 277–280, 289–291, 301, 302, 304–306, 308–310, 313–317, 319, 320, 326
Methanediol, *see* Methylene glycol
Methanol, 342, 352, 353, 367, 381, 427
Methoxychlor, 168
2-Methylanthracene, 212
Methylchloroform, *see* 1,1,1-Trichloroethane
3-Methylcholanthrene, 151, 152, 163, 424
Methylene glycol, 341
N-Methyl-*N*′-nitro-*N*-nitrosoguanidine, 196, 224, 368, 374
N-Methyl-*N*-nitrosourethane, 166–167
Methyltrichloromethane, *see* 1,1,1-Trichloroethane
Micas/clay minerals, 17
Minerals, natural, 9, *see also* specific substance
Mineral oil, 169
Mirex, 243
MNNG, *see* *N*-Methyl-*N*′-nitro-*N*-nitrosoguanidine
MNU, *see* *N*-Methyl-*N*-nitrosourethane
Molybdenum, 34, 35
Monochloroacetic acid, *see* Chloroacetic acid
Monochloroacetyl chloride, 403
Monochloroethane, 402, 408–410, 441
Monomethyl arsonate, 47
Monomethylmercury, 14, 42, 47
Morpholine, 240
Muriatic ether, *see* Monochloroethane

N

Naphthalene, 196
Nickel, 20–21, 34, 35, 38, 46, 171–172
Nickel compounds, 26
Nickel oxides, 146, 163, 166, 172
Nickel sulfide, 157
Nicotinic hydrochloride, 424
Niobium, 35
Nitrate radicals, 258, 271, 287
Nitric acid, 12, 143, 169, 258, 260, 264, 281, 285–286, 328
Nitric oxide, 258–264, 271, 272, 281, 282, 285, 291, 302
Nitrilotriacetic acid, 41, 42
Nitrogen, molecular, 289
Nitrogen dioxide, 169, 171, 254, 258–263, 268, 270, 271, 281–283, 289, 291, 302, 306, 414
Nitrogen oxides, 253, 254, 257–264, 268, 270, 271, 281–283, 285–287, 291–294, 302, 306, 313, 316, 317, 326
Nitrogen pentoxide, 258, 263, 271, 283, 286, 287, 290
Nitromethane, 420
Nitrosamines, 143, 216
N-Nitroso compounds, 142, 164–166
Nitrous acid, 143
Nitrous oxide, 254, 255, 258, 264, 266, 269, 273, 275, 277–280, 289, 290, 292–295, 302, 304–306, 308–310, 313, 314, 316, 317, 319, 320, 326
Norepinephrine, 192
NTA, *see* Nitrilotriacetic acid

O

Octanol/water, 88, 94, 184
Oil, crude, 244
Olive oil, 224
Organochromium, 47
Organometallic complexes, 30, 33, 36, 37, 42, 45, 52, *see also* specific compound
Organotin halides, 45
Oxalic acid, 433

Oxine, 35
Oxygen
  dissolved, 239, 241
  species, in the atmosphere, 253, 255–262, 268, 272, 282, 289, 291, 300
Ozone, 251–336, 403, 422, 441

P

PAH, *see* Polynuclear aromatic hydrocarbons
Paint, 126, 192
Paraformaldehyde, 341
Particles, 10, 25, 141–174
  mercury, 14
  sulfur-containing, 10, 52
PCBs, *see* Polychlorinated biphenyls
Pentachloroethane, 402, 435–437, 440
Pentaerythritol, 344
Perchloroethane, *see* Hexachloroethane
Perchloroethylene, 421
2-Perimidinylammonium bromide, 12
2-Perimidinylammonium sulfate, 13
Pernitric acid, 258, 260, 263, 271, 283, 287, 302
Perylene, 212, 220
Pesticides, 65, 81, 142, 164, 167–169, 173, 210, 241, 404, 416, *see also* specific substance
  acidic, 167
PF, *see* Phenol–formaldehyde resin
Phenanthrene, 220
Phenobarbital, 424, 428
Phenol, 237, 379, 381–383
Phenol–formaldehyde resin, 338, 339, 344, 360, 379, 380, 382
Phenolics, 241, 403
Phosgene, 403, 404, 408, 414, 420, 423, 432
Pigments
  inorganic, 385
  organic, 385
Plastic dust, 383
Platinum, 35
Polyamine–polyurea, 35
Polycarbonate, 6
Polychlorinated biphenyls, 93, 168, 184, 206, 213, 215, 428
Polyethylene glycols, 29
Poly(maleic anhydride), 35, 37, 51
Polynuclear aromatic hydrocarbons, 142–144, 147–164, 173, 174, 183, 184, 187, 196, 201, 202, 204–206, 212, 219
  metabolites of, 205
    polar, 206
Polypropylene, 196
Polystyrene, 6
Polytetrafluoroethylene, 6
Polyvinyl chloride, 171–172
Potassium, 25
Potassium chloride, 169
Potassium sulfate, 10
PTFE, *see* Polytetrafluoroethylene
Pulp mill effluent, 239, 241
Pyrene, 163

Q

Quartz, 5, 9

R

Radon, 145
Radon daughters, 145, 146
Resorcinol, 380

S

Salicylic acid, 35
Scandium, 34
Selenium, 34, 35, 43, 44, 46, 53, 216
Selenium(IV), 44
Serotonin, 192
Silica, 122, 145, 150, 157, 168, 216
Silicon, 18
Silicon dioxide, 17, 163
Silver, 34, 35
Silver nitrate, 367
Silyl xanthate, 35
Simazine, 236
Sodium, 30, 35, 265
Sodium chloride, 10, 169–171, 265, 357
Sodium formate, 348
Sodium hydroxide, 265
Sodium hypochlorite, 387
Sodium sulfate, 10
Sodium vanadate, 169
Stirofos, 168
Styrene, 383
Sulfates, 7–8, 10, 12–13, 52, 53, 171, *see also* specific compound

atmospheric, 12–13
NH_4HSO_4, 12
$(NH_4)SO_4$, 12
$(NH_4)_2SO_4$, 12–13
Sulfide, 13
Sulfite salts, 12, 13
Sulfur, 13, 171
Sulfur dioxide, 3, 13, 169–171, 322
Sulfuric acid, 3, 12–13, 170, 171, 264

T

Talc, 157, 160
Tars, 416
Tellurium, 44
Tetraalkyllead, 15–16, 45, 46, *see also* specific compound
1,1,1,2-Tetrachloroethane, 402, 429–431
1,1,2,2-Tetrachloroethane, 402, 405, 430–435, 439
Tetrachloroethylene, 415, 427, 436, 438–440
Tetradecanoylphorbol acetate, 187–188, 368, 374
Tetraethyllead, 15, 16, 47, 408
Tetramethyllead, 15, 47
Thallium, 34, 38
Thiazolidine-4-carboxylic acid, 350
Thiodiacetic acid, 417, 428
Thiols, 35, 37
Thorium, 34
Tin, 34, 35, 38, 43–46, 53
methylated, 46
Tin(IV), 45, 46
Titanium, 9, 18, 25, 26, 35
Titanium dioxide, 10, 157
Tobacco smoke, 143–148, 164, 173
condensate, 143, 154
Toluene, 382
Toxaphene, 210, 212
TPA, *see* Tetradecanoylphorbol acetate
S-Triazines, 168
Trichloroacetic acid, 423, 428, 433, 434, 440
Trichloroacetyl chloride, 403
α-Trichloroethane, *see* 1,1,1-Trichloroethane
β-Trichloroethane, *see* 1,1,2-Trichloroethane
1,1,1-Trichloroethane, 261, 262, 267, 289, 290, 299, 402, 412, 415, 416, 420–426, 428, 441
1,1,2-Trichloroethane, 402, 405, 414, 420, 426–429, 441
Trichloroethanol, 423, 428, 433, 434, 440
Trichloroethylene, 415, 420, 421, 423, 430, 432, 433, 436
Trichlorofluoromethane, 261, 262, 279, 294, 296–299, 307, 308
Trichlorotrifluoroethane, 299, 307
Triethylene glycol, 169
Triethylmethyllead, 15
Trimethylethyllead, 15
Trimethylpropane, 344
Trioxane, 341
Trioxymethylene, *see* Trioxane
Tungsten, 34, 35

U

UF, *see* Urea–formaldehyde resin
Uranium, 34, 35, 145, 146
Urea–formaldehyde resin, 338–340, 342, 344–348, 360, 361, 377, 378, 382, 389

V

Vanadium, 9, 34, 35
Vinyl chloride, 411, 414–416, 433
Vinyl trichloride, *see* 1,1,2-Trichloroethane

W

Water vapor, emissions, 320

Y

Yttrium, 34, 35

Z

Zinc, 8, 9, 17, 18, 20–22, 28, 30, 33–38, 42, 49, 50, 53, 237, 241, 242
Zinc compounds, 8, 14
$(NH_4)_2Zn(SO_4)_2$, 14
zinc fulvates, 26
zinc oxide, 8, 14, 171
zinc sulfate, 14
zinc sulfide, 8
Zirconium, 35

Subject Index

A

Active radicals in atmosphere, modeling for, 280–286
Air, scoring systems for hazard assessment in, 98
Air pollutants, particulate, determination of, 13–14
Airborne particles, physicochemical speciation of, 3–14
Aquatic animals
  industrial chemical effects on behavior of, 233–250
  neoplasia of as carcinogen indicator, 181–232
    chemical agent exposure and, 195–196
    diagnostic aspects, 189–195
    from environmental pollution, 183–184
    epizootiologic studies, 196–217
    from viruses, 186
Asbestos, effect on smoking, 144–145
Atmosphere
  ambient, modeling of, 269–289
    applications, 273–275
    long-lived chemical species, 275–289
    potential-change calculations, 289–306
  transport of chemical species in, 265–267
Atmospheric sulfates, inorganic speciation of, 12–13

B

Behavior of fish
  industrial chemical effects on, 233–250
  stress behavior, 241–243
  variability in, 235–237
Benzo[*a*]pyrene, particulate matter effects on exposure to, 155–163
Black River bullheads, neoplasms in as carcinogen indicators, 202–206
Bullheads
  neoplasms in as carcinogen indicators, 202–206
    experimental induction, 219–222

C

Cancer, as environmental disease, 184–185
Carbon dioxide, increase of, in atmosphere, 300
Carbon disulfide, heart disease and, medical records of, 131–134
Carbon monoxide, increase of, in atmosphere, 300–301
Carcinogens
  aquatic animal neoplasia as indicator of, 181–232
  chlorinated ethanes as, 406–407, 418–419, 425–426, 429, 434–435, 440–441
  evaluation scoring systems for hazard assessment of, 103–104
  formaldehyde as, 369–375
Carcinogenesis, environmental, 184–189
  of aquatic animals, 181–232
  initiation, promotion, and metabolism in, 187–189
  oncogenes and viruses in, 185–187
Chemical species, long-lived, in ambient atmosphere, 275–289
Chemicals, mediation of toxicological properties of, by particulate matter, 141–180
Chloride, in atmosphere, modeling for, 284–285
Chlorinated ethanes, 401–448
  analytical methods for, 407
  carcinogenicity of, 406–407
  1,1-dichloroethane, 411–413
  1,2-dichloroethane, 414–419
  environmental and biological data on, 403–408

environmental load of, 403–404
as environmental hazards, 441
hexachloroethane, 438–441
history of, 402
impurities in, 403
monochloroethane, 408–411
pentachloroethane, 435–437
pharmacokinetics of, 404–405
production of, 402
properties of, 402
standards of permissible exposure to, 407
1,1,1,2-tetrachloroethane, 429–430
1,1,2,2-tetrachloroethane, 431–435
toxic effects of, 405–407
1,1,1-trichloroethane, 420–426
1,1,2-trichloroethane, 426–429
uses of, 403
Chlorofluorocarbons (CFCs)
in atmosphere, modeling for, 279–280
perturbations of, in stratosphere, 294–299
Chloroxy radical, in atmosphere, modeling for, 284–285
Copper mining waters, fish neoplasms in from lake polluted with, 215–217
Croaker, experimental induction of pigmented lesions in, 222–225

D

1,1-Dichloroethane
environmental load of, 412
pharmacokinetics of, 412
physical properties of, 412
standards of permissible exposure to, 413
technical and economic data on, 411
toxic effects of, 413
workplace hazard of, 412
1,2-Dichloroethane
analysis of, 419
carcinogenicity of, 418–419
environmental load of, 415–416
pharmacokinetics of, 417
physical properties of, 415
standards of permissible exposure to, 419
technical and economic data for, 414–415
toxic effects of, 417–419
uses of, 415
as workplace hazard, 417
Dioxin, exposure to, medical record use in assessment of, 134–136
Dumpsites, scoring systems for hazard assessment of, 99–100

E

Electron spectroscopy (ESCA), in inorganic physicochemical speciation, 4
Environmental carcinogenesis, 184–189
Environmental effects, scoring systems for hazard assessment of, 102–103
Environmental pollution
aquatic animal neoplasia and, 183–184
direct evidence, 218–225
indirect evidence, 218
Exposure, scoring systems for hazard assessment of, 100–102

F

Fish
industrial chemical effect on behavior of, 233–250
neoplasms in as carcinogen indicators, 196–199
case histories, 202–216
Flatfishes, hepatomas in as carcinogen indicators, 206–208
Food and food additives, scoring systems for hazard assessment of, 100
Formaldehyde
carcinogenicity of, 369–375
chemical forms of, 341
chemical properties of, 341–343
chemical reactivity of, 342–343
dermal studies on, 364
epidemiology of, 375–389
case-control and cohort studies, 383–389
case reports, 376–378
cross-sectional studies, 378–383
human exposure studies, 375–376
genetic effects of, 367–369
health effects of, 337–400
human exposure to, 346–347
hypersensitization to, 359
in vitro studies on, 367
ingestion studies on, 364–366
inhalation studies on, 361–364
injection studies on, 366–367

metabolism of, 348–353
interconversions, 349
product uses of, 345
regulatory activities on, 340
reproductive effects of, 361–367
sources and exposure of, 343–348
teratogenic effects of, 361–367
toxicology of, 353–359
acute toxicity, 353

G

Gases and vapors, particulate matter effects on exposure to, 169–172
Genetic effects of formaldehyde, 367–369
General Foods Corporation computerized medical record, 118
Greenhouse effect, description of, 302–307

H

Halogen perturbations, in stratosphere, 294–299
Hazard assessment, scoring systems for, 63–109
Hazardous chemical exposures
medical record role in evaluation of, 111–139
feasibility of use, 123
speed of detection by, 121–123
Health and environmental effects literature, computerized abstracts of, 84–86
Hexachloroethane
analysis of, 441
environmental load of, 439
pharmacokinetics of, 439–440
physical properties of, 439
standards of permissible exposure to, 440
technical and economic data for, 438–441
toxic effects of, 440–441
as workplace hazard, 439
Hudson River tomcod, hepatomas in as carcinogen indicators, 212–215
Hydrochloric acid in atmosphere, modeling for, 286
Hydrogen peroxide radical in atmosphere, modeling for, 283–284
Hydroxy radical in atmosphere, modeling for, 283–284
Hyperplasia, neoplasia and, 190–193

I

Industrial chemicals, effects on fish behavior, 233–250
preference or avoidance, 237–241
stress behavior, 241–243
Inorganic compounds, physicochemical speciation of, 1–61

L

Lead, inorganic speciation of, in air, 15–16
Lead compounds, physicochemical speciation of, 6–7

M

Medical record, role in evaluation of hazardous chemical exposures, 111–139
case studies, 128–136
carbon disulfide and heart disease, 131–134
dioxin exposure, 134–136
pesticide contamination of drinking water, 129–131
characteristics of, as a data source, 115
methodological considerations, 124–129
cross-sectional vs. cohort design, 124–125
data validation, 127
objective assessment, 125–127
potential confounding factors, 127–128
potential uses of, 113–115
recommendations for, 136–137
Mercury, inorganic speciation of, in air, 14–15
Metal compounds, particulate matter effects on exposure to, 172–173
Metal vapors and alkyls, determination of, in air, 14–15
Methane in atmosphere
increase in, 301–302
modeling for, 277–279
Minerals, physicochemical speciation of, 8, 9
Mollusks, neoplasms in as carcinogen indicator, 199–202
Monochloroethane
environmental load of, 409
pharmacokinetics of, 409–410
standards of permissible exposure to, 411

technical and economic data on, 408–409
toxic effects of, 410
workplace hazard from, 409

N

Neoplasia
hyperplasia and, 190–193
mimicry of, in fish, 190–193
Nitric oxide, in atmosphere, modeling for, 281–282, 285
Nitrogen dioxide, in atmosphere, modeling for, 281–282
N-Nitroso compounds, particulate matter effects on exposure to, 164–167

O

Odd chlorine, chemistry of, 261–263
Odd hydrogen, chemistry of, 259–261
Odd nitrogen
chemistry of, 258–259
in stratosphere, perturbations in, 291–294
Oncogenes, in environmental carcinogenesis, 185–187
Ozone
chemistry and physics of, 255–269
production and loss of, 255–257
in stratosphere, modification by man, 251–336
atmosphere modeling, 269–289
greenhouse effect, 302–306
measurements, 320–328
radiation effects, 267–269

P

Paraformaldehyde, properties of, 341
Particulate matter
mediation of toxicological properties of chemicals by, 141–180
gases and vapors, 169–172
metal compounds, 172–173
N-nitroso compounds, 164–167
pesticides, 167–169
polynuclear aromatic hydrocarbons, 147–164
tobacco smoke, 143–147
Pentachloroethane
physical properties of, 436
standards of permissible exposure to, 437
technical and economic data for, 435–436
toxic effects of, 437
as workplace hazard, 437
Perturbations in stratosphere, 289–290
Pesticides, particulate matter effects on exposure to, 167–169
Physicochemical speciation of inorganic compounds, 1–61
airborne particles, 3–14
atmospheric sulfates, 12–13
metal vapors and alkyls, 14–16
natural waters, 24–51
sediments and soils, 18–24
single-particle techniques, 8–12
street dusts, 16–18
by X-ray powder diffraction, 4–8
Polynuclear aromatic hydrocarbons (PAH)
animal studies on, 154–164
benzo[*a*]pyrene, 155–163
environmental occurrence of, 147–149
in vitro methods for, 152–154
intratracheal instillation technique for tests on, 154–155
particulate-enhanced uptake of, into membranes, 150–152
in sediments, 220
Puget Sound flatfishes, hepatomas in as carcinogen indicator, 206–208

R

Radiation effects on stratospheric ozone, 267–269
Reproduction, formaldehyde effects on, 361–367

S

Salamanders, hepatomas in as carcinogen indicators, 208–212
Scanning electron microscopy (SEM)
in inorganic physicochemical speciation, 4, 8–9
of street dusts, 18
Scoring systems for hazard assessment, 63–109

applications of, 106–107
bias in, 80
comparison of system capabilities, 97–106
comprehensive systems, 104–106
systems focused on a particular medium, 98–100
criteria for, 91–97
for assessment of biological effects, 95–97
for assessment of environmental effects, 97
for assessment of exposure, 92–94
data availability, 79
expert judgment role, 79–80
factors to consider, 64
needs of the program, 79
steps involved in, 80–91
compilation of candidate list, 80
data collection, 82–83, 87
scoring, 87–89
weighting and combining scores, 89–91
survey of, 64
table, 65–78
Sediments, inorganic speciation of, 18–24
Sink gases in atmosphere, modeling for, 280–286
Sink species, temporary, in atmosphere, 286–287
Soils, inorganic speciation of, 18–24
Smoking
asbestos effect on, 144–145
mineral particulate effect on, 145–146
Speciation, importance of, 2
Stratospheric ozone modification by man, 251–336
future research on, 328–330
model calculations for potential changes, 289–306
multiple perturbations and, 306–320
ozone measurements, 320–328
Street dusts, inorganic speciation of, 16–17
Stress behavior of aquatic animals, biological monitoring of, 241–243
Sulfates
atmospheric
inorganic speciation of, 12–13
physiocochemical speciation of, 7, 8
Sulfuric acid, atmospheric, determination of, 12–13

T

Teratogen, formaldehyde as, 361–367
1,1,1,2-Tetrachloroethane
pharmacokinetics of, 430
physical properties of, 430
standards of permissible exposure to, 431
technical and economic data for, 429–430
toxic effects of, 430–431
1,1,2,2-Tetrachloroethane
analysis of, 435
environmental load of, 433
pharmacokinetics of, 433–434
standards of permissible exposure for, 435
technical and economic data for, 431–433
toxic effects of, 434–435
workplace hazard of, 433
Tiger salamanders, hepatomas in as carcinogen indicators, 208–212
Tobacco smoke
composition of, 143
mineral particulates and, 145–147
Tomcod, hepatomas in as carcinogen indicators, 212–215
Toxicity, scoring systems for hazard assessment of, 103
Toxicology data, secondary sources of, 83
Trace metals, in natural waters, 32–34
1,1,1-Trichloroethane
analysis of, 426
as environmental hazard, 441
environmental load of, 422–423
pharmacokinetics of, 423–424
physical properties of, 421
standards of permissible exposure to, 426
technical and economic data on, 420–421
toxic effects of, 424–425
uses of, 421–421
workplace hazard of, 423
1,1,2-Trichloroethane
analysis of, 429
environmental load of, 428
pharmacokinetics of, 428
physical properties of, 427
standards of permissible exposure to, 429
technical and economic data for, 426–427
toxic effects of, 428–429
workplace hazard of, 428
Trioxane, properties of, 341

U

Urea, formaldehyde reaction with, 342
Urea–formaldehyde foam insulation (UFFI)
  formaldehyde from, 345, 348
  regulations for, 340

V

Viruses, environmental carcinogenesis and, 185–187

W

Water
  centrifugation of, 25
  comprehensive speciation schemes, 48–51
  dialysis of, 31
  electrodeposition studies on, 42–43
  filtration of, 25–29
  gas/liquid chromatography of, 43–46
  gel filtration chromatography of, 31–32
  generic speciation of, 25–32
  high-performance liquid chromatography of, 46–47
  inorganic speciation of, 18–24
  ion-exchange chromatography of, 34–37, 47–48
  ion-selective electrode studies on, 43
  pesticide contamination of, medical record studies of, 128–131
  polarography and voltametry of, 37–42
  scoring systems for hazard assessment in, 98–99
  trace metals in, 32–34
  ultrafiltration of, 29–30
Water vapor, effect on stratosphere, 300–302

X

X-ray powder diffraction (XRD)
  of airborne particles, 4–8
  of street dusts, 16–17